PC UW
LLYFRGELL
LIBRARY
ABERYSTWYTH

Holocene Land–Ocean Interaction and Environmental Change around the North Sea

Geological Society Special Publications

Series Editors

A. J. HARTLEY

R. E. HOLDSWORTH

A. C. MORTON

M. S. STOKER

It is recommended that reference to all or part of this book should be made in one of the following ways:

SHENNAN, I. & ANDREWS, J. (eds) 2000. *Holocene Land–Ocean Interaction and Environmental Change around the North Sea*. Geological Society, London, Special Publications, **166**.

HORTON, B .P., EDWARDS, R. J. & LLOYD, J. M. 2000. Implications of a microfossil-based transfer function in Holocene sea-level studies. *In*: SHENNAN, I. & ANDREWS, J. (eds) *Holocene Land–Ocean Interaction and Environmental Change around the North Sea*. Geological Society, London, Special Publications, **166**, 41–54.

GEOLOGICAL SOCIETY SPECIAL PUBLICATION NO. 166

Holocene Land–Ocean Interaction and Environmental Change around the North Sea

EDITED BY

IAN SHENNAN
University of Durham, UK

AND

JULIAN ANDREWS
University of East Anglia, UK

2000

Published by

The Geological Society

London

THE GEOLOGICAL SOCIETY

The Geological Society of London was founded in 1807 and is the oldest geological society in the world. It received its Royal Charter in 1825 for the purpose of 'investigating the mineral structure of the Earth' and is now Britain's national society for geology.

Both a learned society and a professional body, the Geological Society is recognized by the Department of Trade and Industry (DTI) as the chartering authority for geoscience, able to award Chartered Geologist status upon appropriately qualified Fellows. The Society has a membership of 8600, of whom about 1500 live outside the UK.

Fellowship of the Society is open to those holding a recognized honours degree in geology or cognate subject and who have at least two years' relevant postgraduate experience, or who have not less than six years' relevant experience in geology or a cognate subject. A Fellow with a minimum of five years' relevant postgraduate experience in the practice of geology may apply for chartered status. Successful applicants are entitled to use the designatory postnominal CGeol (Chartered Geologist). Fellows of the Society may use the letters FGS. Other grades of membership are available to members not yet qualifying for Fellowship.

The Society has its own publishing house based in Bath, UK. It produces the Society's international journals, books and maps, and is the European distributor for publications of the American Association of Petroleum Geologists, (AAPG), the Society for Sedimentary Geology (SEPM) and the Geological Society of America (GSA). Members of the Society can buy books at considerable discounts. The publishing House has an online bookshop (*http://bookshop.geolsoc.org.uk*).

Further information on Society membership may be obtained from the Membership Services Manager, The Geological Society, Burlington House, Piccadilly, London W1V 0JU, UK. (Email: *enquiries@geolsoc.org.uk*: tel: +44 (0)171 434 9944).

The Society's Web Site can be found at *http://www.geolsoc.org.uk/*. The Society is a Registered Charity, number 210161.

Published by The Geological Society from:
The Geological Society Publishing House
Unit 7, Brassmill Enterprise Centre
Brassmill Lane
Bath BA1 3JN, UK

(*Orders*: Tel. +44 (0)1225 445046
Fax +44 (0)1225 442836)
Online bookshop: *http://bookshop.geolsoc.org.uk*

First published 2000

British Library Cataloguing in Publication Data
A catalogue record for this book is available from the British Library.

ISBN 1-86239-054-1
ISSN 0305-8719

Typeset by Aarontype Ltd, Bristol, UK

Printed by Whitstable Litho, UK

Distributors

USA
AAPG Bookstore
PO Box 979
Tulsa
OK 74101-0979
USA
Orders: Tel. +1 918 584-2555
Fax +1 918 560-2652
Email *bookstore@aapg.org*

Australia
Australian Mineral Foundation Bookshop
63 Conyngham Street
Glenside
South Australia 5065
Australia
Orders: Tel. +61 88 379-0444
Fax +61 88 379-4634
Email *bookshop@amf.com.au*

India
Affiliated East-West Press PVT Ltd
G-1/16 Ansari Road, Daryaganj,
New Delhi 110 002
India
Orders: Tel. +91 11 327-9113
Fax +91 11 326-0538

Japan
Kanda Book Trading Co.
Cityhouse Tama 204
Tsurumaki 1-3-10
Tama-shi
Tokyo 206-0034
Japan
Orders: Tel. +81 (0)423 57-7650
Fax +81 (0)423 57-7651

Contents

An introduction to Holocene land–ocean interaction and environmental change around the western North Sea

I. SHENNAN[1] & J. ANDREWS[2]

[1] *Environmental Research Centre, Department of Geography, University of Durham, Durham DH1 3LE, UK*
[2] *School of Environmental Sciences, University of East Anglia, Norwich NR4 7TJ, UK*

The majority of the research presented in this Special Publication arises from the Land–Ocean Evolution Perspective Study (LOEPS), one component of the Land–Ocean Interaction Study (LOIS), Phase 1 of which ended in 1998. It is therefore appropriate to introduce this research in the context of LOIS as a whole before summarizing the main conclusions relating to LOEPS.

An overview of the Land–Ocean Interaction Study

The Land–Ocean Interaction Study (LOIS), was a seven-year Natural Environment Research Council (NERC) funded Community Research Project (CRP). The detailed planning for LOIS began in 1990 and continued into 1991 as a series of workshops, which laid the foundations for the LOIS Science Plan (1992). The LOIS CRP was conceived as a collaborative multidisciplinary study to be undertaken by UK scientists from NERC institutions and the higher education institutes (HEI). The coastal zone was to be studied in an integrated way, to provide a holistic view of the way coastal systems work and to demonstrate how they might respond to future changes resulting from human activities. It was also anticipated that LOIS research would interface with other contemporary NERC CRPs, including the North Sea Project (see e.g. Charnock *et al.* 1994), the Biogeochemical Ocean Flux Study (see, e.g. introduction in Savidge *et al.* 1992), and the Terrestrial Initiative in Global Environmental Research (see e.g. Oliver *et al.* 1999).

The objectives of LOIS were set out in the science plan (LOIS Science Plan 1992) and then in revised form in the implementation plan (LOIS Implementation Plan 1994).

(1) To estimate the contemporary fluxes of momentum and materials (sediments, nutrients, contaminants) into and out of the coastal zone, including transfers via rivers coasts, ground-water, the atmosphere and the shelf–ocean boundary.
(2) To characterize the key physical and biogeochemical processes that govern coastal morphodynamics and the functioning of coastal ecosystems, with particular reference to the effects of variations in sediment supply and inputs of pollutants.
(3) To describe the evolution of coastal systems from Holocene to recent (*sic*, we assume this was meant to mean present) in response to changes in relative sea-level and the impact of human activities.
(4) To develop coupled land–ocean models to simulate the transport, transformation and fate of materials in the coastal zone, and provide the basis for predicting hydrological, geomorphological and ecological conditions under different environmental scenarios for the next 50–100 years.

It was hoped that these objectives would be achieved by the implementation of four interlinked, and in some cases overlapping component studies as described in Fig. 1. Much of the LOIS research was centred on the River–Atmosphere–Coast Study (RACS) site (Fig. 2), which included the east coast of England between Berwick-upon-Tweed and Great Yarmouth, to include the various river catchments and the adjoining area of the North Sea.

The Land–Ocean Evolution Perspective Study

The Land–Ocean Evolution Perspective Study (LOEPS) was charged with meeting objective 3 of LOIS (see above). It was clear that understanding the history of material fluxes over long time scales was a prerequisite for making

From: SHENNAN, I. & ANDREWS, J. (eds) *Holocene Land–Ocean Interaction and Environmental Change around the North Sea*. Geological Society, London, Special Publications, **166**, 1–7. 1-86239-054-1/00/$15.00

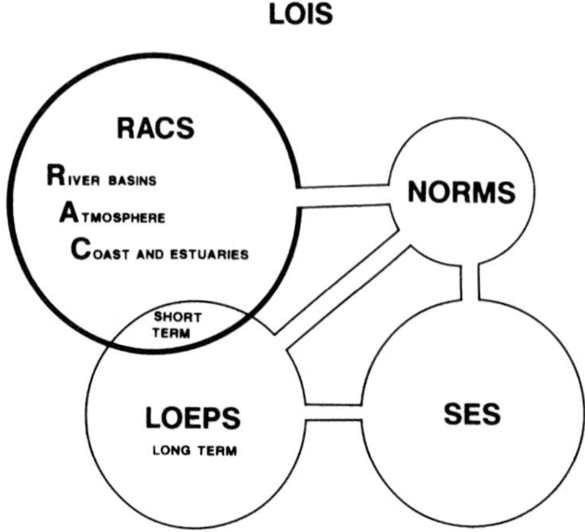

Fig. 1. Schematic representation of the interrelation between the four component studies of LOIS. RACS, River–Atmosphere–Coast Study; NORMS, North Sea Modelling Study; SES, Shelf Edge Study; LOEPS, Land–Ocean Evolution Perspective Study. The overlap between RACS and LOEPS reflected overlap on decadal to centennial time-scales.

sensible predictions about the way the coastal zone might respond to future environmental changes. It was envisaged that five LOEPS objectives would be addressed (LOIS Implementation Plan 1994).

(1) To determine, through study of the Holocene sedimentary record and changing coastal disposition, how sediment fluxes between the land and ocean have been influenced by changes in sea level, climate, geomorphology and land-use.
(2) To determine the regional history of sediment fluxes, sources and sinks at the RACS site, with particular reference to the relative importance of fluvial, coastal and sea-bed sediment sources.
(3) To determine the historical components of relative sea-level change along the coast, enabling refined predictions for the next 50–100 years.
(4) To improve absolute dating of Holocene sedimentary sequences.
(5) To model Holocene tidal and storm circulation affecting sedimentation in collaboration with the North Sea Modelling Study (NORMS) (see Fig. 1).

The overall aim of LOEPS was thus to describe the evolution of coastal systems over the last 10 000 years in response to changes in natural climatic conditions, changes in relative sea-level and the changes wrought by human activities. This was to be achieved through special topic thematic studies, typically at HEIs, and through a Core Programme of central and co-ordinating functions performed by the British Geological Survey (BGS). The Core Programme was to: (a) compile and maintain the LOIS geoscience database; (b) administer and manage sediment sampling and curation, including a drilling campaign of new cored boreholes both offshore and onshore; (c) co-ordinate a radiocarbon dating programme through the NERC East Kilbride Laboratory; (d) compile and refine the regional Holocene stratigraphy of the RACS area in both a national and European context; (e) to assemble data on the regional history of sediment flux in the RACS area, its sources and sinks through the Holocene to the present day.

Implementation of LOEPS research

Following on from the LOIS Science Plan (1992), LOIS research began in 1993 and the LOIS Implementation Plan (1994) was published. By this time the details of the LOEPS had crystallized into 15 special topics (Table 1), co-ordinated by the Core Programme as envisaged in the science plan. Specific objectives were focused to reflect the research teams assembled and centred on five key areas.

Objectives 1 and 2 were essentially unchanged from the planning stage (see above) and have involved the Core Programme at BGS, special topics concentrating largely on new core material from the Tees Estuary (**Plater *et al.***), the Humber estuary (**Andrews *et al.***; **Metcalfe *et al.***; **Rees *et al.***; **Ridgway *et al.***), the Lincolnshire Marshes, the Fenland (**Brew *et al.***), North Norfolk (**Andrews *et al.***), the river catchments draining into the Humber and Tees (**Macklin *et al.***), lake systems within the Humber catchment, and the integration of the data from all these areas. Sediment source, sink and flux information was aided by a geochemical study as part of the Core Programme (see e.g. Plater *et al.*; Rees *et al.*, Ridgway *et al.*).

The other objectives were defined later, at the stage of finalizing the LOIS Implementation Plan in 1994. Objective 3 to determine the historical components of relative sea-level change along the coast, enabling refined predictions for the next 50–100 years, required the synthesis of data collected by many of the research teams (**Shennan *et al.***). There have been four major elements to address objective 4: a large, co-ordinated accelerator mass spectrometry (AMS) radiocarbon dating project (**Shenan *et al.***); development of

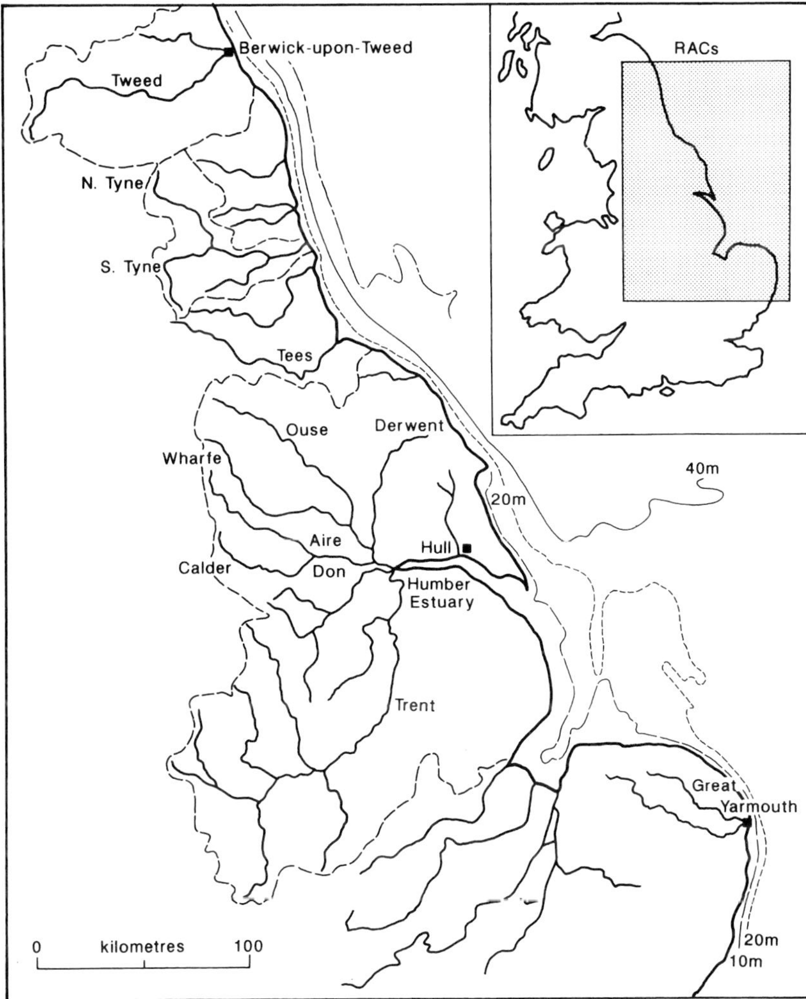

Fig. 2. Map of the RACS study site (shaded area on inset) showing principal catchments studied (pecked line).

new techniques in luminescence dating (**Bailiff & Tooley**; **Clarke & Rendell**) application of palaeomagnetic techniques to coastal clastic sediments (**Ridgway *et al.***); geochemical and isotope techniques for the last 150 years (**Andrews *et al.***; **Plater *et al.***). Objective 5 involved three main elements. **Horton *et al.*** report new approaches to identify and date different tide levels from Holocene sequences and **Shennan *et al.*** model Holocene tidal circulation affecting in the western North Sea. The final element, to model both Holocene tidal and storm circulation and how in combination they affect sedimentation, forms part of Phase 2 of LOIS, which continues to AD 2000.

A key element to the success of LOEPS has been the integration and interaction between all elements of the special topics and the Core Programme. The papers presented at the symposium and the contributions to this volume aim to reflect this integration rather than summarize the results of individual research projects or the achievement of a single LOEPS objective. The contributions are arranged into four thematic sections:

- techniques;
- Humber catchment;
- other areas within the RACS study site;
- regional scale analyses.

Table 1. *The 15 thematic special topics*

LOIS project No.	Principal investigators	Special topic title
12	M. G. Macklin & J. Ridgway	Holocene and historic environmental change in the Yorkshire Ouse, Tees and Tweed basins and its influence on sediment and chemical fluxes to east coast estuaries and the coastal zone
31a	J. E. Andrews, T. D. Jickells, B. A. Maher, A. Grant, P. F. Dennis & R. M. Middleton	Organic carbon, nutrient and metal contents and storage in saltmarsh and estuarine sediments of the Humber
32	G. M. Harwood†, J. E. Andrews, P. N Chroston, B. M. Funnell, A. C. Kendall, B. A. Maher, P. Balson, I. K. Bailiff, C. Bristow M. J. Tooley & G. B. Shimmield	Sedimentary evolution of the North Norfolk barrier island coastline in the context of Holocene sea-level change
33	A. J. Plater	Relating post-glacial sediment fluxes in the Tees Estuary to changes in sea-level, coastal morphology and catchment land-use
41	H. Rendell, P. Townsend & R. Parish	Development of a methodology for luminescence dating of Holocene sediments
65	J. R. L. Allen, A. Parker & K. Pye	The Wash–Fenland embayment: sediment sources and supply in the Holocene
75	J. Orford, P. Wilson & A. Wintle	Recent environmental history of coastal dune fields in north Norfolk and northeast Northumberland in relation to land–sea interactions
78	F. Oldfield, R. W. Battarbee, R. Thompson & G.A. Wolff	A lake-sediment-based study of the Holocene history, flux and characterization of fine, particulate, terrestrially derived sediments in the Humber region
240	I. K. Bailiff & M. J. Tooley	Development of a methodology for luminescence dating of Holocene sediments.
272	J. Sheail	Documentary evidence of changes in the fluxes of the riverine and coastal ecosystems
283	A. Wintle	Development of a methodology for luminescence dating of Holocene sediments
313	I. Shennan	Differential crustal movements within the RACS study site (Berwick-Upon-Tweed to north Norfolk)
316	I. Shennan & R. T. R. Wingfield†	Modelling Holocene depositional regimes in the western North Sea at 1 ka time intervals
346	P. S. Balson & D. S. Brew	Sediment provenance and palaeogeographical evolution of the Wash embayment
348	S. Metcalfe, S. Ellis, J. Pethick, I. Shennan & M. J. Tooley	Holocene evolution of the Humber Estuary

† Deceased.

Techniques

Because some of the methodologies and techniques were common to a number of the special topics, the paper by **Ridgway *et al.*** was conceived in part to communicate the details of various methods, such that the information is not repeated in all of the individual papers. However, **Ridgway *et al.*** are also able to demonstrate how the ensemble of techniques are applied at a regional level. A number of cores from the Holocene of the Humber were chosen to illustrate how the multi-technique approach yields data sets that are reinforcing, leading to confident and powerful environmental reconstructions.

Horton *et al.* present quantitative methods, transfer functions for fossil foraminifera assemblages calibrated from contemporary analogues, that together with AMS radiocarbon dating of calcareous foraminifera enable a greater range of Holocene sediments than previously available to be used as indicators of past tide levels. Wider application of these techniques offers new directions for research in sea-level reconstruction at scales ranging from individual estuaries to regional phenomena such as changes in tidal parameters during the Holocene and differential effects of glacio- and hydro-isostasy. These data are used in other contributions to this volume.

A major area of concern was to try and improve the methodologies for dating minerogenic sediments within the Holocene coastal sediments, especially those that were water-laid. To this end, considerable effort was directed towards improving the methodologies for luminescence dating. The achievements in this area are reflected in two papers that specifically address methodology (**Bailiff & Tooley**; **Clarke & Rendell**), while other aspects of method development and application are implicit in the contributions from **Orford *et al.***, **Andrews *et al.***, (Norfolk) and **Plater *et al.*** **Clarke & Rendell** focused on the use of alkali feldspars as ideal 'dosimeters', utilizing a better understanding of the characteristics of the feldspars to optimize the luminescence signal for dating marine coastal zone sediments. **Bailiff & Tooley's** contribution centres on a Fenland core where radiocarbon-dated organic intercalations occur at various levels within an otherwise minerogenic core. Infra-red-stimulated luminescence (IRSL) of the silt fraction of these water-laid sediments gives ages that are largely consistent with the radiocarbon dates. Chronological resolution of 1 ka or better is probably achievable by the (IRSL) method employed.

Humber Estuary

The Holocene sediments of the Humber catchment and estuary and their evolution were very poorly understood before LOEPS. This led to a number of allied studies focused on the Humber system. Using the Ouse system as an example from the Humber catchment, **Macklin *et al.*** investigated the geomorphological, geochemical and chronological elements of fluvial sedimentary sequences from the upland to the estuarine lowland. They reveal a complicated relationship between river response and environmental forcing parameters such as land-use and climate change. They suggest that for much of the Holocene, sediment delivery from the Ouse catchment to the Humber Estuary was relatively low. **Rees *et al.*** used the new sediment cores from the LOEPS Core Programme to establish a lithostratigraphic and chemostratigraphic framework for the Holocene sequences of the infilled estuary. Eight characteristic sediment suites were identified, which show the progressive influence of marine sediments as sea level rose. In addition, widespread erosive episodes have left distinct geomorphic surfaces, while partially removing or redistributing older sediments. The preserved volumes of the sediment suites were calculated, while **Metcalfe *et al.*** studied the environmental facies evolution of the Holocene sediment prism as a whole. Sixteen environmental facies were identified mainly by diatom and pollen data, and using the radiocarbon chronologies to constrain sea-level history, maps of the changing environments and geography were constructed. In closely related work, **Andrews *et al.*** used various geochemical data from the Humber cores to reconstruct the storage history of organic carbon, nutrient elements and sulphur. This data, while contributing to the environmental facies identification, was used principally to construct one of the first well-constrained Holocene organic carbon budgets for a temperate estuary. A major result of this allied Humber research work has been to identify the clear effect of human activity on material and chemical flux in the late Holocene. The modern managed estuary has almost no space to store sediments or attendant chemicals, whereas the pre-reclamation system was a large sediment and material sink.

Other areas within the RACS study site

The coastal morphology and Holocene evolution of the Tees Estuary, studied by **Plater *et al.***, like other coastal regions studied under LOEPS, has been influenced by sea-level change and human activity. The Tees area has been rebounding since the removal of glacial ice, such that the sea-level rise was decelerating between 8 and 3 ka BP. The sedimentary sequence is thus more strongly influenced by riverine-derived sediments, in contrast to sequences further south. Human activity and climatic changes have probably also influenced sediment flux, and the record of human activity is archived as metal pollution. The Holocene evolution of the north Norfolk barrier coast in the south of the study area contrasts strongly with the more estuarine sites. **Andrews *et al.*** demonstrate that the structure of the pre-Holocene surface is not a simple shelf, but contains a buried trough feature, probably an old river valley or glacial outwash feature. Details of the sandy barrier facies in this area were recorded for the first time, and the overall control on sedimentation is proposed to be autocyclic, superimposed on a facies evolution governed broadly by sea-level rise. **Orford *et al.*** show that the initiation and survival of coastal dune sequences in Northumberland and north Norfolk relate to macroscale relative sea-level changes over the last 4 ka. Because of differential isostatic effects (see **Shennan *et al.***) the Northumberland dunes formed earlier and have responded to relative sea-level fall, while those in north Norfolk are much younger, forming and surviving under dominant relative sea-level rise. Both dune systems appear to respond to shorter-term disturbances, such as the Little Ice Age, and may also indicate small-scale variations in relative sea-level. The largest inland Holocene sediment sink on the east coast, the Fenland embayment, is analysed by **Brew *et al.***; their geochemical analyses suggest a general consistency of sediment provenance. They identify three macroscale episodes related to varying responses between sedimentation and relative sea-level change: initial and rapid transgression; sediment infilling of the embayment; deposition of intertidal clastic sediments alternating with peat accumulation.

Regional scale analyses

The final two contributions analyse sea-level data from the whole study area. In the first, **Shennan *et al.*** quantify the isostatic effect of the glacial rebound process, including both the ice (glacio-isostatic) and water (hydro-isostatic) load contributions, showing a *c.* 20 m range at 8 cal. ka BP from north to south in the RACS area. By 4 cal. ka BP relative sea-level in Northumberland was above present, whereas in areas to the south relative sea-level was below present throughout the Holocene. Estimates for pre-industrial relative sea-level change range from $1.04 \pm 0.12\,\mathrm{mm\,a^{-1}}$ in the Fenland to $-1.30 \pm 0.68\,\mathrm{mm\,a^{-1}}$ (i.e. sea-level fall) in north Northumberland although this may overestimate the current rate of sea-level fall. Local-scale processes identified include possible differential isostatic effects within the Humber Estuary and the Fenland, tide-range changes during the Holocene, and the effects of sediment consolidation. These processes help explain the variation in altitude between sea-level reconstructions derived from index points taken from basal peats and those from peats intercalated within thick sequences of Holocene sediments. In the second paper, **Shennan *et al.*** use data from the RACS site and cores from the floor of the North Sea taken as part of the LOEPS Core Programme. The full data set enables the development and testing of models of the palaeogeographies of coastlines in the western North Sea and models of tidal range changes through the Holocene epoch. Key stages include a western embayment off northeast England as early as 10 ka BP; the evolution of a large tidal embayment between eastern England and the Dogger Bank before 9 ka BP with connection to the English Channel prior to 8 ka BP; and Dogger Bank as an island at high tide by 7.5 ka BP and totally submerged by 6 ka BP. After 6 ka BP the major changes in palaeogeography occurred inland of the present coast of eastern England. The models predict tidal ranges smaller than present in the early Holocene, with only minor changes since 6 ka BP.

This volume is just one output from LOEPS. As with any large science programme, much of the detailed research will be published in forthcoming scientific journal articles, many of which are referenced in the individual contributions here. Data arising from LOEPS are compiled and maintained at the LOEPS Data Centre at BGS, Keyworth, UK and will be published on a CD-ROM in 1999. This extensive LOEPS knowledge base can now be taken forward and factored into modelling studies and management strategies intended to promote improved environmental conditions in our rivers, estuaries and low lying coasts.

The research presented in this Special Publication stems largely from a symposium held at the Geological Society on 7 September 1998, jointly supported by the Geological Society and the Quaternary Research Association.

In addition to the commitment of the research teams, in particular the various post-doctoral and post-graduate research assistants, the success of LOEPS also stems from the effort and vision of the LOEPS Steering Committee and N. McCave, LOEPS Scientific Chairman up to 1996. We hope that our colleagues on the Steering Committee, R. Arthurton, I. Bailiff (from 1996), P. Balson, C. Evans, M. Macklin (from 1996), F. Oldfield (to 1996), J. Pethick and M. Tooley, will consider the book a fitting reflection of the imagination and hard work they put in to LOEPS. We are grateful to Lisa Tempest for her contribution to the organization of the symposium and the production of this Special Publication.

Finally, it is fitting to mention that two members of the LOEPS research community, Gill Harwood and Robin Wingfield, died before they were able to see the final results of their work. Gill was seriously ill throughout the planning and early years of LOIS research, and died in 1996. Robin completed his commitment to both special topic and Core Programme research and was present at the Geological Society Symposium in September 1998, but sadly died in Spring 1999 before seeing his results in print. We salute their efforts and hope that this volume is a fitting memorial to their memory.

References

CHARNOCK, H., DYER, K. R., HUTHNACE, J. M., LISS, P. S., SIMPSON, J. H. & TETT, P. B. 1994. *Understanding the North Sea System*. The Royal Society, Chapman & Hall, London.

LAND–OCEAN INTERACTION STUDY 1992. *Science Plan for a Community Research Project*. Natural Environment Research Council, Swindon.

——1994. *Implementation Plan for a Community Research Project*. Natural Environment Research Council, Swindon.

OLIVER, H. R., CHALONER, W. G., & ROSE, J. 1999. TIGGER: NERC-stimulated research into the global perspective of terrestrial global environmental research. *Journal of the Geological Society of London*, **156**, 341–344.

SAVIDGE, G., TURNER, D. R., BURKILL, P. H., WATSON, A. J., ANGEL, M. V., PINGREE, R. D., LEACH, H. & RICHARDS, K. J. (1992). The BOFS 1990 Spring bloom experiment: temporal evolution and spatial variability of the hydrographic field. *Progress in Oceanography*, **29**, 235–281.

Analysis and interpretation of Holocene sedimentary sequences in the Humber Estuary

J. RIDGWAY,[1] J. E. ANDREWS,[2] S. ELLIS,[3] B. P. HORTON,[4] J. B. INNES,[4] R. W. O'B. KNOX,[1] J. J. McARTHUR,[4,5] B. A. MAHER,[2] S. E. METCALFE,[6] A. MITLEHNER,[6] A. PARKES,[3,7] J. G. REES,[1] G. M. SAMWAYS,[2,8] & I. SHENNAN[4]

[1] *British Geological Survey, Keyworth, Nottingham NG12 5GG, UK (e-mail: j.ridgway@bgs.ac.uk)*

[2] *University of East Anglia, School of Environmental Sciences, Norwich NR4 7TJ, UK*

[3] *University of Hull, School of Geography and Earth Resources, Cottingham Road, Hull HU6 7RX, UK*

[4] *University of Durham, Department of Geography, South Road, Durham DH1 3LE, UK*

[5] *Present address: Meteorological Office, Beaufort Park, East Hampstead, Wokingham RG40 3DN, UK*

[6] *University of Edinburgh, Department of Geography, Drummond Street, Edinburgh EH8 9XP, UK*

[7] *Present address: Northsea Software Systems, 18 Newlands House, Newlands Science Park, Inglemire Lane, Hull HU6 7TQ, UK*

[8] *Present address: Badley Ashton and Associates, Winceby House, Winceby, Horncastle, Lincolnshire LN9 6PB, UK*

Abstract: The interpretation of the Holocene evolution of the Humber Estuary has been made possible only through integrated multidisciplinary studies involving *inter alia*: drilling, to obtain sedimentary records of the Holocene Estuary fill; multi-element, carbon–nitrogen–sulphur and stable carbon isotope geochemistry; heavy and clay mineralogy; palaeomagnetism; radio-carbon dating; and pollen, diatom and foraminiferal studies. Eight chemostratigraphic suites and 14 palaeo-environments have been recognized. Sediment types, environments of deposition and provenance change in response to rising sea-level, showing a range from freshwater fluvial deposition of locally derived terrestrial sediment to intertidal and subtidal deposition of sediments from marine sources. The methods used are illustrated with reference to sediment cores from inner and outer estuary locations. The results show that Holocene environmental characterization is most secure when a number of different, but complementary, techniques are used. The integration of radiocarbon dates with palaeomagnetic and geochemical data improves the understanding of the presence and significance of time breaks, which is crucial to constraining sedimentation rates and material budgets.

The Land–Ocean Evolution Perspective Study (LOEPS) component of the Land–Ocean Interaction Study (LOIS) project, perhaps more than any other part of LOIS had to adopt an integrated, multidisciplinary approach in order to interpret the historical record preserved in the Holocene sediments of the east coast of England. The work on the Humber Estuary fill provides a good example of the advantages of such multidisciplinary research. Holocene sediments are rarely exposed and the Holocene history of the Humber region has had to be reconstructed through detailed analysis of cores from a series of boreholes drilled through the Holocene sedimentary prism. A stratigraphy has been developed, based largely on geochemistry, clay mineralogy and heavy mineralogy; diatom, foraminiferal and pollen analysis, coupled

From: SHENNAN, I. & ANDREWS, J. (eds) *Holocene Land–Ocean Interaction and Environmental Change around the North Sea*. Geological Society, London, Special Publications, **166**, 9–39. 1-86239-054-1/00/$15.00

Table 1. *Summary of methods used in the interpretation of the Humber cores*

Technique	Purpose	Sampling	Methodology	Output
Drilling	Provide complete sedimentary sections through Holocene sequences and material for magnetic measurements	Locations determined from inspection of existing maps and drilling logs	Shell and auger percussion rig with a modified vibrocore barrel Combined Stitz percussion/piston corer for orientated cores	9 cm core in opaque or transparent plastic liner
Levelling	Provide precise Ordnance Datum (OD) heights of borehole tops	All borehole sites	Nikon Totalstation levelling system used to relate borehole tops to benchmarks	Altitudes of borehole tops in relation to OD
Logging	Provide a basic sedimentological description of each core	Split cores at British Geological Society (BGS) core store	Visual inspection and measurement of core recovery, thickness of lithologies, depth of boundaries, etc.	Interpreted core logs
Particle size analysis (PSA)	Detailed sedimentological characterization	Sub-samples from cores	Wet and dry sieving and weighing, Micromeritics Sedigraph analysis	Weights at 1 Φ intervals plotted on histograms
Palaeomagnetism (palaeosecular variation, PSV) and magnetic susceptiblity	To help constrain the age of the sediment and to identify sedimentary packages and hiatuses (i.e. time gaps)	Continuous sampling of orientated core sections using 2.2 cm polystyrene cubes	GM400 cryogenic magnetometer, Molyneux af demagnetiser, Bartington susceptibility bridge	NRMs (natural remnant magnetizations: declination, inclination or dip and intensity of magnetization ($mA\,m^{-1}$). Magnetic susceptibility: ($m^3\,kg^{-1}$)
Geochemistry	To aid stratigraphic correlation, provenance studies and assessment of anthropogenic effects	Continuous and/or composite samples of cores based on logged sedimentary units	Pilot study with portable X-ray fluorescence (XRF). Main study using XRF and ICP-MS	Multi-element analyses
Clay mineralogy	To aid stratigraphic correlation, provenance studies	Sub-samples from cores	Separation of clay fraction by pipette and centrifuge. Quantitative XRD using Phillips PW1130/00 X-ray diffractometer (2–20° 2θ using CuK_α radiation) with glycolation and heating (to 350°C) stages	Quantitative estimates of clay mineral species proportions

Interpretation	Benefits	Constraints	References
Drilling stopped at base of Holocene deduced from presence of bedrock, till or glacio-fluvial gravels	Penetrates clays to gravels. Usually good core recovery	Compaction of near-surface fine-grained sediments and peats. Some disturbance, particularly of sands and gavels. Penetration and recovery affected by water-table in sands and gravels. Relatively imprecise depth measurements	Ridgway *et al.* (1998)
	Accurate altitudes to relate other measurements to	Availability of Ordnance Survey bench marks. Other types (e.g. Environmental Agency) may be less reliable	Ridgway *et al.* (1998)
Lithologies based on modified Folk scheme. Colours based on Munsell chart scheme	Minimum of equipment required. Non-destructive	Subjective and may vary with logger	Ridgway *et al.* (1998)
Folk classification of samples	Quantitative classification from clay to gravel	Destructive of cores, time consuming, quality control difficult	Ridgway *et al.* (1998)
Down-core profiles of declination and inclination cross compared with UK PSV master curve. Magnetic susceptibility values indicate the concentration of magnetizable minerals present within each samples interval	Potentially rapid, non-destructive technique using a whole core scanner	Potential for compaction and rotation during coring. Destructive if whole core scanner not available. Difficult to resolve PSV 'slave' and 'master' records if slave record was not formed by continuous sedimentation	Turner & Thompson (1981, 1982)
Comparison of geochemical signatures using spidergrams. Element and element ratio scatterplots. Summnary statistics, correlation coefficients, cluster analysis, principal component analysis	Quantitative, precise and accurate analyses. Can be compared with other geochemical data sets. Applicable to wide range of sediment types	Subjective sampling interval may cross genuine stratigraphic boundaries. Depending on range of elements more than one technique may be needed. Grain size influences concentrations	Ridgway *et al.* (1988)
Estuarine sediment clay mineral composition was compared with potential source material and between cores and sedimentary units	Direct link to potential sources	Chiefly applicable to fine grained sediments	Ridgway *et al.* (1998)

(*continued*)

Table 1. (*continued*)

Technique	Purpose	Sampling	Methodology	Output
Heavy mineralogy	To aid stratigraphic correlation, provenance studies and assessment of anthropogenic effects	Continuous and/or composite samples of cores based on logged sedimentary units	Ultrasonic cleaning, wet sieving, separation (63–125 μm fraction) in bromoform (sp. gr. 2.90) and mounting in Canada Balsam for study by optical microscope	Relative abundances of heavy minerals
Diatom analysis	To help define palaeoenvironments and validate sea-level index points	Sub-samples from cores	Separation and mounting in Naphrax followed by identification and counting by optical microscope	Counts of identified species
Foraminiferal analysis	To help define palaeoenvironments and validate sea-level index points	Sub-samples from cores	Wet sieving followed by identification and counting by optical microscope	Counts of identified species
Pollen analysis	To help define palaeoenvironments, validate sea-level index points and provide relative dating	Sub-samples from cores	Chemical separation followed by identification and counting by optical microscope	Counts of identified taxa
Carbon–nitrogen–sulphur chemistry	To establish C–N–S inventory of Holocene sediments for flux calculations and to aid environmental characterization	Representative samples of main lithofacies types from cores	Elemental analyser	Weight % total C, N, S and organic C
Stable carbon isotopes	To help identify source of organic matter and aid environmental characterization	Representative samples of main lithofacies types from cores	Mass spectrometer	δ^{13}C relative to Vienna Pee Dee belemnite (VPDB) scale
Radiocarbon dating	To provide an independent chronology	Sub-samples from cores: peat, wood, shells, etc.	Accelerator mass spectrometry of small organic samples, conventional ^{14}C dating of large samples	Age determination with 1 and 2σ error limits

Interpretation	Benefits	Constraints	References
Classification into stable and unstable species to distinguish between till and local sandstone origin	Direct link to potential sources, relatively simple and cheap technique	Non-quantitative. Not suitable for very fine-grained sediments. Less discriminating than geochemistry	Ridgway *et al.* (1998)
Classification into Holobian environment groups	Sensitive indicator of palaeosalinity, enabling reconstruction of environmental onditions	Preservation may be poor, particularly in coarse-grained clastic sediments	Hustedt (1953, 1957), van der Werff & Huls (1958–1974), Hendey (1964), Beyens & Denys (1982), Batterbee (1986), Hartley (1986), Krammer & Lange–Bertalot (1986, 1988, 1990), Round *et al.* (1990), Denys (1991), Vos & de Wolf (1993), Sims (1996)
Classification into intertidal and subtidal environments using cluster analysis and detrended correspondence analysis	Semi-quantitative palaeoenvironmental reconstructions	Not suitable for freshwater sediments. Subject to post-depositional changes	Scott & Medioli (1980, 1986), de Rijk (1995), Horton (1997)
Classification into pollen and spore groups	Reconstruction of ecological plant communities and vegetation change, and relative dating	Not very suitable in non-organic sediments	Moore *et al.* (1991)
Organic C and C/N ratio used to distinguish marine, saltmarsh and freshwater sources	Quantitative, precise and accurate analyses. Can be compared with other geochemical data sets. Applicable to wide range of sediment types	Least successful in sandy sediments. Data could be inconclusive in very oxidized sediments	Verado *et al.* (1990)
Interpreted with reference to globally significant data sets for marine and terrestrial organic matter types	Quantitative, precise and accurate analyses. Can be compared with other geochemical data sets. Applicable to wide range of sediment types	Expensive and time consuming analyses requiring specialist laboratory. Least successful in sandy sediments with low organic content	Fichez *et al.* (1993)
Calculation of calibrated age from radiocarbon date	Facilitates correlation of events and processes between cores	Only suitable for organic samples. Contamination possible. Shell material may need correction for marine reservoir effect. Wide age range may mean imprecise date	Stuiver & Reimer (1993), Hughen *et al.* (1998)

with carbon–nitrogen–sulphur geochemistry and carbon stable isotope work, have been used to interpret palaeoenvironments; radio-carbon dating and palaeomagnetism have helped to establish a time frame; and the whole has been linked to provide evidence of sea-level at various times through the Holocene and to allow the reconstruction of palaeogeographies.

This paper briefly describes the methods used and then provides examples of their application to representative cores from inner and outer estuary environments. The paper is intended also to provide a summary of methodologies to which other papers in this volume will refer.

Techniques

The techniques employed are summarized in Table 1 and described in more detail below.

Drilling

Borehole sites for the LOEPS drilling programme were selected on the basis of existing borehole logs and previous research on the Holocene succession by the authors. The location of the holes (prefixed HMB) drilled in the Humber region are shown in Fig. 1.

The drilling technique was based upon a typical shell and auger rig, but utilizing a marine vibrocore barrel. Core recovery was very good, in most cases being virtually 100%. Most holes were cased with steel tubing after removal of the core barrel and the hole then cleaned to the depth of penetration ready for the next 1 m core barrel (or in some cases 1.5 m) to be hammered in (a core run).

Levelling

The tops of all boreholes were levelled to Ordnance Datum (OD) to provide standard elevations against which stratigraphic horizons and sea-level indicators could be referenced. The elevation of the ground level at borehole sites was determined using a Nikon Totalstation DTM-A10LG levelling system. The difference in altitude between the instrument and a pole-mounted prism is automatically calculated and logged on the basis of their separation (measured by microwaves) and angular difference from the vertical. The Totalstation and prism were set up at the borehole site and at an Ordnance Survey

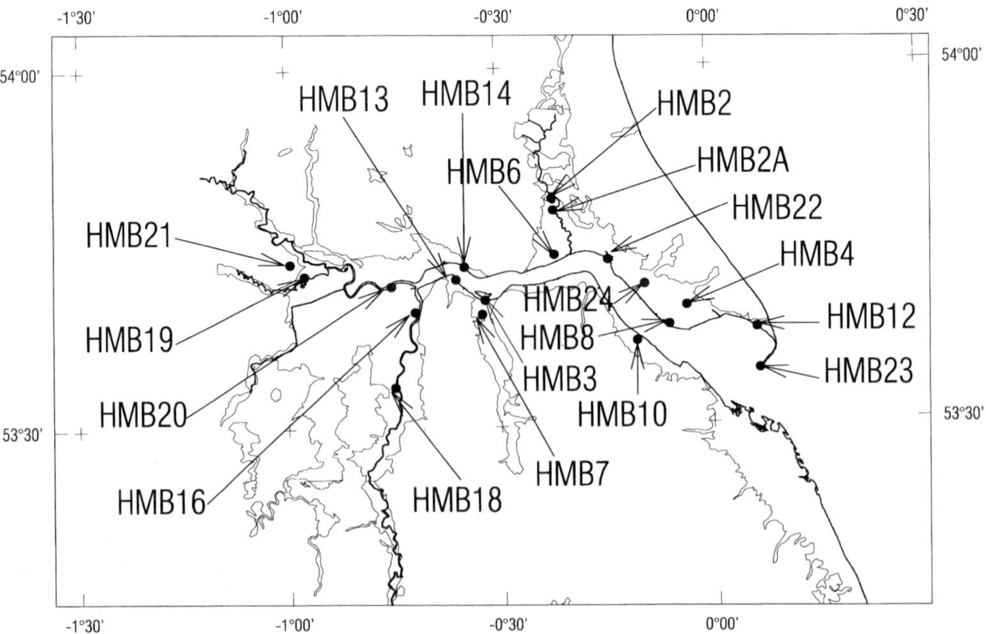

Fig. 1. Locations of the boreholes drilled in the Humber region.

(OS) benchmark respectively (or vice versa); if it was not possible to accurately sight one from the other, the difference in altitude between these and intermediate stations was determined. The difference in altitude between borehole site and benchmark was calculated on both outward and inward legs. If the difference between these values was less than 2 mm, the mean was taken as being within acceptable error, and the altitude of the borehole site determined relative to OD.

Logging

Cores were logged according to a protocol (Ridgway *et al.* 1998) to ensure a degree of consistency between logs recorded by different workers. The protocol was developed largely to ensure a uniformity of approach to the effects of volume change, core loss and other features, which are a consequence of the drilling technique and have to be accounted for in interpreting the core and driller's logs. Two logs were made: a sample log and an interpreted log, serving different functions.

The sample log is a record of what material actually exists. In any run it shows the thickness of material preserved as core, as well as the thickness of additional material recovered, but not captured in the polycarbonate liner (e.g. sediment in the core barrel below the core catcher), normally stored in a separate labelled bag. Apart from thicknesses of material preserved, the log shows lithological boundaries, sedimentary structures and biogenic features.

The interpreted log is a reconstruction of the sequence penetrated by the borehole, using all available information. The aim of the log is to establish the true depth of any horizon within a core, with a reasonable degree of accuracy, allowing a good estimate of the elevation of the horizon to be calculated (e.g. −1.12 m OD). A typical log is shown in Fig. 2.

On completion, core logs were distributed to the LOEPS community and the cores made available for sampling.

Particle size analysis (PSA)

Particle size analysis facilitated the sedimentological characterization of the Holocene sediments and provided an important input to the stratigraphic interpretation. PSA was performed by wet-sieving of sub-samples to separate the <63 and >63 μm fractions. The latter was then dried, dry sieved and weighed at 1.0 Φ intervals. The <63 μm fraction was dried, weighed, dispersed in a solution of sodium hexametaphosphate and the proportions of silt and clay determined using the Micromeritics Sedigraph 5100^{TM} particle size analysis system.

Palaeomagnetism

Palaeomagnetic data were collected for two purposes.

(1) Most cores lacked a reliable radiocarbon chronology in their upper parts, typically because peat horizons or *in situ* shells were absent. This prompted exploration of the possibility of using palaeomagnetic secular variation (PSV) records (Turner & Thompson 1981, 1982) to constrain the chronology. The aim was to obtain magnetic declination and inclination records from orientated sediment cores, for comparison with the UK 'master curve' of palaeosecular variation (Turner & Thompson 1982), thereby obtaining a chronology for the Humber sediments. Where possible, the PSV-derived chronology was verified with radiocarbon dates from carefully selected *in situ* rootlets at sedimentological surfaces, autochthonous organic matter in lake and freshwater marsh facies and freshwater gastropod shell carbonate.
(2) Major excursions in the palaeomagnetic data were also used, and are probably most informative in indicating discrete sedimentary packages and hiatuses.

Orientated cores were obtained using a combined Stitz percussion/piston corer. Natural remanent magnetizations (NRMs) of individual samples were analysed using a GM400 cryogenic magnetometer at the University of East Anglia (UEA), with a sensitivity level of 0.01×10^{-8} A m^{-2}. One in ten samples from each core section was subjected to alternating frequency demagnetization at steps of 5, 10, 20 and 40 mT, to identify the number of components contributing to the NRM. The magnetic susceptibility of each sample was measured using a Bartington bridge. Selected samples were also subjected to additional rock magnetic measurements, including incremental acquisition of isothermal remanence and anhysteretic magnetizations.

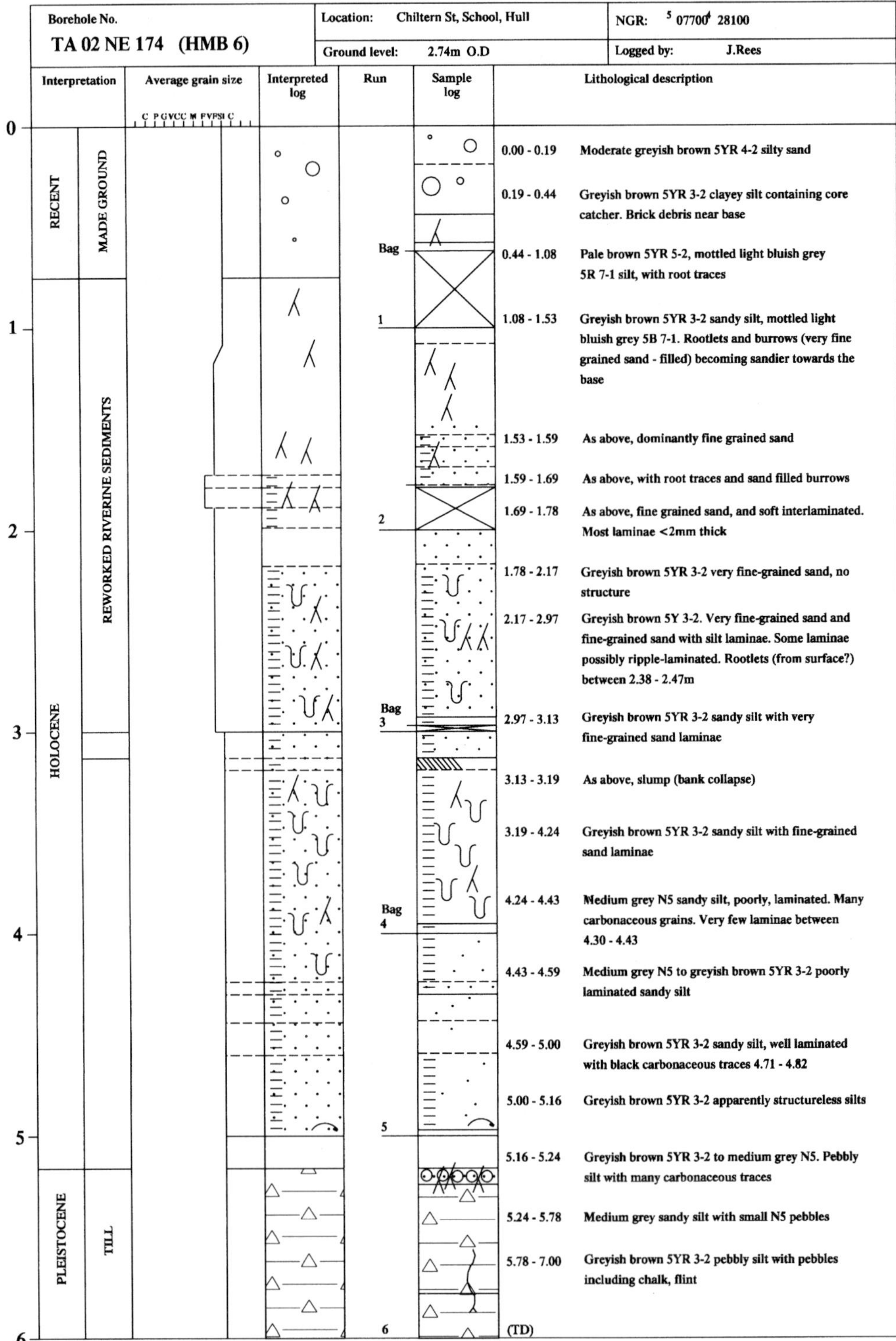

Fig. 2. A typical core log showing sample and interpreted logs and the kind of detail recorded.

Multi-element geochemistry

The most important factors determining the distribution of elements in sediments and sedimentary rocks are: (a) the mineralogical and chemical composition of the source materials; (b) the partitioning of the elements between sediment and surface or ground-water during deposition or diagenesis. Thus in some cases the geochemistry largely reflects provenance, while in others it is determined predominantly by conditions of deposition or diagenesis (see Haslam & Plant 1990). Within the Holocene of the UK the bulk of the sediments are of clastic origin with a chemistry controlled largely by source and depositional processes.

The geochemical characteristics of the Holocene sediments assist in developing a stratigraphy and in determining provenance. The sampling interval is of great importance in a programme intended to geochemically characterize sedimentary units. Because the range of geochemical variation within the Holocene sediments under consideration was unknown, a pilot study was undertaken, employing a portable X-ray fluorescence (XRF) analyser. This technique allowed the range of geochemical variation within and between recognizable sedimentary units to be assessed by performing analyses on the surfaces of the split cores without disturbing the sediments.

A range of cores from the four main LOIS study areas (Teeside, Humber, Fenland and north Norfolk) was examined. Depending on the thickness of the sedimentary unit, analyses were performed near the top, in the middle and near the base. The equipment, a portable Spectrace SP9000 energy dispersive XRF spectrometer, sampled an area of approximately 1 cm diameter and care was taken to avoid inhomogeneities when placing the instrument on the core surface. The technique is described in more detail by Ridgway *et al.* (1998).

The elements determined were K, Ca, Ti, Mn, Fe, Zn, Sr, Zr, Rb and Ba, and the degree of chemical variation in the sedimentary units was examined by comparison of spidergrams, in which element values, normalized to a suitable datum (in this case upper crustal average from Wedepohl 1995), were plotted on a logarithmic scale on the y-axis against individual element position on the x-axis. Examination of spidergrams for 27 cores showed that the chemistry of composite samples from distinct lithological units should provide a basis for the development of a chemostratigraphy for the Holocene sediments of the Humber area (Fig. 3).

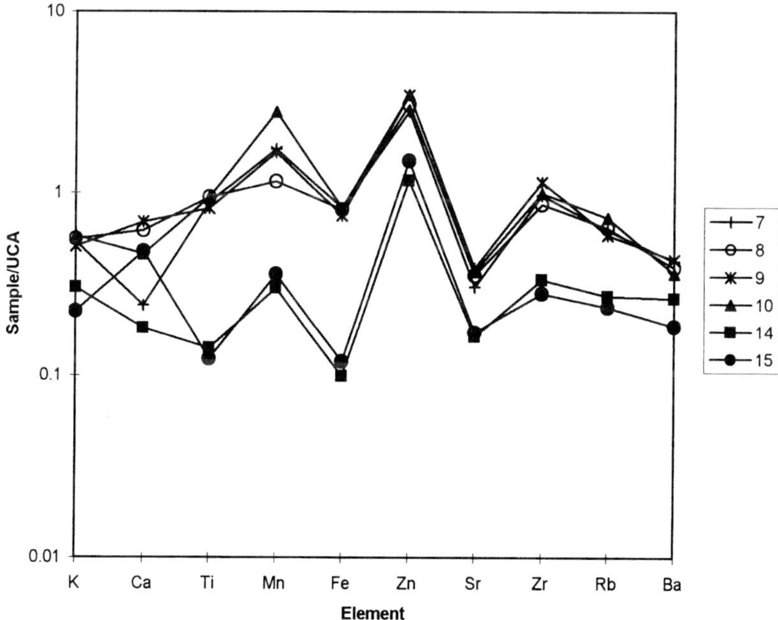

Fig. 3. Spidergram showing distinctive grouping of geochemical signatures from portable XRF data on HMB13.

Following the successful pilot study, major lithological units were sampled for a laboratory-based XRF study at the British Geological Surveyused by (BGS).

The general principles of the system used, which enables elements in the range F to U to be determined with a higher degree of accuracy and precision than is possible with a portable instrument, are described by Ingham & Vrebos (1994). Analyses, for MgO, Al_2O_3, P_2O_5, K_2O, CaO, TiO_2, MnO, Fe_2O_3 (total), Cr, Co, Ni, Cu, Zn, As, Rb, Sr, Y, Zr, Ba, Pb, La and Ce, were carried out using three sequential, fully automatic wavelength-dispersive XRF spectrometers. Further details of the method, including quality control procedures, are given in Ridgway *et al.* (1998).

Comparison of spidergrams (see above), based on the multi-element XRF data, allowed the chemical signatures from individual samples to be grouped into distinctive chemostratigraphic groups. In the Humber region these groups or suites were given informal stratigraphic names and their validity further tested by the examination of scatterplots of element or element ratio pairs (Rees *et al.* this volume).

Clay mineralogy

Clay mineralogy was used to investigate the derivation of the sediment clay fraction. Samples were taken from HMB10 and in the outer Humber and from HMB16 and 20 in the inner Humber. With respect to provenance, samples from potential source materials were taken from BGS reference archives, including North Sea sediments offshore of the Humber and the Skipsea Till from Holderness, and also from floodplain sediments of the catchment areas of three of the main tributaries (Ouse, Don and Trent) of the inner estuary.

Air-dried samples were gently crushed in an agate mortar. The disaggregated samples were then suspended in distilled water; a few drops of Dispex were added to prevent clay flocculation. The suspensions were stirred and allowed to stand for 4 h, after which time the top 5 cm was extracted by pipette, concentrated by centrifuging and the supernatant fluid poured off. The resulting slurries were mounted on slides and allowed to dry overnight at room temperature. The air-dried samples then were scanned on a Phillips PW1130/00 X-ray diffractometer from 2–20° 2θ using CuK_a radiation, then glycolated and re-run, and heated to 350°C and re-run again. Clay contents were quantified by the measurement of peak areas and the results normalized to 100%.

Heavy mineralogy

Heavy mineral suites in sediments reflect the heavy mineral content of the source area and, like geochemistry, can be used to characterize sedimentary units and to provide information on provenance. Anthropogenic influences can be recognized through the identification of industrial related components, such as slag.

Samples for heavy mineral analysis were taken over the same sampling intervals as used for geochemical analysis. The samples were ultrasonically cleaned and wet sieved. Heavy minerals were separated from the 63–125 μm grain size fraction by gravity-settling in bromoform (sp. gr. 2.90). Heavy mineral fractions were mounted in Canada Balsam and analyses carried out by conventional optical microscopy; this involved a count of 200 non-opaque detrital grains per sample (where grain recovery permitted) to determine the overall composition of the suite.

The relative abundance of unstable to stable minerals was determined as a means of assessing the relative contributions to the assemblages of northerly derived tills and local pre-Quaternary sandstones. In order to achieve this, an unstable : stable mineral index (USi) has been devised

$$\mathrm{USi} = 100 \times \frac{\mathrm{Ca + Px + Ep}}{\mathrm{Ca + Px + Ep + Ap + Ru + To + Zr}}$$

where Ca = calcic amphibole; Px = pyroxene; Ep = epidote; Ap = apatite; Ru = rutile; To = tourmaline; and Zr = zircon. Other minerals, such as garnet, were omitted from the calculation since they were known to occur in both the sandstones and the tills.

Diatom analysis

Diatom analysis was used to determine palaeoenvironmental conditions.

Organic matter was oxidized from samples using hydrogen peroxide (Battarbee 1986). Cleaned solutions from each sub-sample were evaporated on two coverslips at different concentrations and were mounted in Naphrax, a high-refraction mounting medium. Diatom assessment and counting was done by optical microscopy and, where possible, a minimum of 200 diatom valves was counted. Species identifications wereas made using van der Werff & Huls (1958–1974), Hendey (1964), Hartley (1986), Krammer & Lange-Bertalot (1986, 1988, 1990), Round *et al.* (1990) and Sims (1996). Results of these counts were expressed as a percentage of total diatom valves (%TDV). Environmental assessments were based on the criteria of Beyens & Denys (1982), Denys (1991) and Vos & de Wolf (1993), with additional environmental information based on lithological evidence.

Raw counts were converted into percentages and taxa grouped according to salinity categories using TILIA 2.4.

The salinity preferences of the diatom species were classified according to the halobian groups of Hustedt (1953, 1957) and are summarized below:

(a) polyhalobian (marine), i.e. >30‰;
(b) mesohalobian (marine–brackish), i.e. 0.2–30‰;
(c) oligohalobian–halophilous (brackish), i.e. optimum in slightly brackish water;
(d) oligohalobian–indifferent (fresh–brackish), i.e. optimum in freshwater, but tolerant of slightly brackish water;
(e) halophobous, i.e. exclusively freshwater.

Foraminiferal analysis

Pollen and diatom assemblages are the most commonly used sea-level indicators (SLI) and are frequently used to infer the palaeoenvironments (Hartley 1986; Denys 1991; Waller 1998). Although these indicators are employed in both North American and European studies, the use of foraminifera has become increasingly important because the well-defined foraminiferal zones that subdivide the intertidal zone enhance the identification of fossil marsh deposits, providing accurate indicators of former sea level during the Holocene. Although calcareous foraminifera are often present within the contemporary marsh assemblages, they are largely ignored in the production of SLIs due to problems of preservation of the calcareous tests in fossil saltmarsh deposits (Scott & Medioli 1980; Scott & Leckie 1990; Jennings & Nelson 1992). Calcareous tests are rapidly destroyed after death through dissolution. Dissolution occurs near the sediment–water interface where a lowered pH associated with organic matter decomposition and sulphide–ammonia oxidation drives the reaction (Green *et al.* 1993). Scott & Medioli (1978, 1986) stated that assemblages of agglutinated saltmarsh foraminifera are the most accurate sea-level indicators on temperate coastlines and that such assemblages exhibit a strong correlation with elevation above mean tide level (MTL). Furthermore, the assemblages are well preserved, easily detectable in fossil deposits and occur in high numbers (100–200 cm^{-3}), thereby providing a good statistical base for palaeoenvironmental interpretations.

In terms of foraminiferal distributions, the intertidal zone can be divided into two parts: firstly, an agglutinated assemblage that is restricted to the vegetated marsh; secondly, a calcareous assemblage that dominates the mudflats and sandflats of the intertidal zone (Horton *et al.* 1999*a*, *b*). The agglutinated assemblage is commonly employed as an SLI. Saltmarsh foraminiferal zonation is a significantly more accurate indicator of sea level than undifferentiated marsh deposits, because well-defined zones subdividing the marsh increase the vertical resolution of the deposits (Scott & Medioli 1978).

In this study sample prep-aration and taxonomy follows Scott & Medioli (1980), de Rijk (1995) and Horton (1997).

Pollen analysis

The analysis of fossil pollen and spore assemblages is a sensitive technique for the reconstruction of past plant communities (Moore *et al.* 1991), sea-level history and coastal zone palaeogeography (e.g. Rodwell 1995; Long *et al.* 1998; Waller 1998).

The primary use of pollen analysis within this study has been to assign the organic units within sediment cores to depositional wetland environments in order to reconstruct the palaeogeography of the coastal zone. Where the pollen data can demonstrate a gradual transition from an organic saltmarsh deposit to a clastic estuarine sediment, the contact is validated as a sea-level index point that formed at or near mean high water of spring tides (MHWST) and which can be assigned to a secure reference tidal level. Where the pollen data show a peat of freshwater origin with no saltmarsh indicators but overlain by estuarine sediment, a depositional hiatus or eroded contact is indicated, and the degree of erosion can be estimated by the successional status of the freshwater assemblage relative to MHWST. Thus reedswamp implies less erosion than would fen carr, under conditions of a positive sea-level tendency. Even where no tendency can be observed, a truncated freshwater peat is still useful in establishing a limiting value, below which sea level must have lain at that time.

Pollen analysis was also used to evaluate the accuracy of radiocarbon dates. Where several key sites have been dated using radiocarbon (e.g. Beckett 1981), features such as the *Alnus* rise or the *Ulmus* decline can be correlated between pollen diagrams and, although not synchronous, provide a good framework for chronological comparison.

Pollen preparation and identification in this study followed the standard procedures outlined in Moore *et al.* (1991).

Carbon, nitrogen and sulphur concentrations

Carbon, nitrogen and sulphur (CNS) concentrations in Holocene Humber sediments were measured principally so that a Holocene storage inventory could be established (Andrews *et al.* this volume). In addition, variation in C/N and C/S ratios allowed identification of sources of organic matter and their variation over time. These data were required to demonstrate the role of temperate estuarine sediments in the global coastal-zone flux of C N and S. The CNS values when combined with carbon isotope values also allowed for more precise environmental facies determinations.

Total carbon, nitrogen and sulphur concentrations were measured with a Carlo Erba EA 1108 elemental analyser with combustion at 1020°C, using sulphanilamide as the standard. Organic carbon (C_{org}) abundances of selected samples were determined by treating samples with sulphurous acid, to remove carbonate, prior to elemental analysis (Verado *et al.* 1990). Replicate analysis of laboratory standards with compositions close to the samples (3.50 wt% total C, 2.0 wt% C_{org}, 0.15 wt% total N and 0.50 wt% total S) gave 1σ precisions of ±0.04 wt% total C, ±0.12 wt% C_{org}, 0.006 wt% total N and ±0.06 wt% total S. The total S values should reflect total S in the sample, even though reduced reactive phases such as iron monosulphides will have oxidized, probably to elemental sulphur (see Canfield *et al.* 1986).

Stable carbon isotopes

Organic carbon $\delta^{13}C$ values for bulk organic matter in the Holocene Humber sediments were measured to identify the source of organic matter (terrestrial versus marine), and to show how that source changed with time. Combined CNS and $\delta^{13}C$ values also allowed for more precise environmental facies determinations.

Organic carbon $\delta^{13}C$ was measured on selected samples that had been leached with 10% HCl to remove carbonate. CO_2 was evolved from organic matter by sealed tube combustion at 900°C using cuprous oxide as the oxidant (method in Fichez *et al.* 1993). Isotope ratios were measured with a VG Sira Series II mass spectrometer. The machine was calibrated using the National Bureau of Standards NBS19 and NBS18 standards and results are expressed relative to the Vienna Pee Dee belemnite (VPDB) scale. Replicate analyses of an L-alanine laboratory standard gave a 2σ precision of ±0.1‰ for organic carbon $\delta^{13}C$. CO_2 yield (volume) was also measured during isotopic analysis as an alternative method to determine C_{org} content. There is generally good agreement between C_{org} abundances determined by elemental analysis and CO_2 yield during isotope analysis, except for intertidal mud (ITM) samples (see below), probably due to the pres-ence of small amounts of ferroan dolomite that were not completely removed by the sulphurous acid pretreatment prior to elemental analysis.

Radiocarbon dating

Radiocarbon dating has been used to provide an independent chronological framework for the study of past environmental changes under LOIS. Accelerator mass spectrometry (AMS) dating of well over 400 small-volume samples was carried out at the University of Arizona NSF AMS facility via the radiocarbon laboratory at East Kilbride. Most samples comprised bulk organic material and were selected to date key stratigraphic changes in sediment columns, following the validation of the sample's importance by biostratigraphic analyses and the establishment of a guide age by relative pollen dating. Organic–clastic contacts validated as sea-level index points were a priority, as was the date of basal peat formation as an indicator of sea-level tendency or as a limiting date and as part of palaeogeographic mapping. Organic silts and clays, or plant material within them, which were identified as probable saltmarsh surfaces, were also selected for dating. These also assisted with calculating sediment accumulation rates within long clastic sequences. Shells, both freshwater and marine, were also dated for this purpose, with reservoir effect corrections applied where appropriate. A much smaller number of conventional radiocarbon assays were also obtained from East Kilbride, primarily on wood or peat samples from alluvial sediments. Calibration of radiocarbon dates followed Stuiver & Reimer (1993) and, when required, Hughen *et al.* (1998).

Sea-level methodology

A consistent methodology has been followed in the collection and analysis of data for the reconstruction of sea-level history, which employs litho- and biostratigraphic data at the site scale to establish the relationship between sedimentation, water table height and sea-level changes (Shennan 1982, 1986). These data allow an indicative meaning to be assigned to a sample, which is the relationship between the environment of deposition of the sample and a reference water-level within the tidal cycle. This reference tide level varies with the type of individual sample, although many samples selected are from peat–clastic contacts, which represent the change in depositional regime that occurs around MHWST. Others have a less secure reference tide level and an error term, the indicative range, which is defined for each type of sample used and is, therefore, also applied. Each index point must also have assigned a tendency of sea-level movement. A positive tendency represents an increase in marine influence and a negative tendency a decrease. When a known age, altitude, indicative meaning and tendency have been established for a sample, it is considered validated as a sea-level index point, which may be used in the reconstruction of sea-level history. These site-scale tendencies derived from index point data are used for between-site correlation and for recognition of any wider significance within the sea-level data for a particular time or area. Samples that cannot be quantitatively related to a reference water level and indicative range, such as peats forming in totally freshwater environments or in a raised bog environment above the limit of all tides, are also useful since they provide maximum elevation of sea level at the time, and are referred to as limiting data.

Application to the Humber cores

The application of the above techniques and methodologies to the study of the Holocene sediments of eastern England can be illustrated with reference to three cores from the Humber Estuary region: HMB10, HMB12 and HMB16 (Fig. 1). Both HMB10 (core length 10.96 m) and HMB12 (8.11 m) are from the eastern or outer Estuary near Immingham ([^{5}20900^{4}15200] south bank) and Lockham ([^{5}39000^{4}17200], north bank), respectively. HMB16 (17.96 m) comes from near Garthorpe in the lower Trent Valley on the south side of the western or inner estuary [^{4}8593^{4}1918]. The core logs for the three boreholes are shown in Figs 4–6, with an indication of the average grain size, the interpreted chemostratigraphic suites and palaeoenvironments, and the distribution of radiocarbon dates expressed in thousands of calibrated years before present (BP). Palaeomagnetic records are shown for HMB12 and HMB16, and CNS records for HMB10 and HMB16.

In general terms, chemostratigraphic suites, defined purely on the basis of chemistry, are found to have specific palaeoenvironmental characteristics identified on the basis of diatom, foraminiferal and pollen evidence. Changes in chemistry and palaeoenvironment can be recognized even when no significant changes in grain size are seen and are confirmed by CNS and isotope data. Radiocarbon and palaeomagnetic studies provide a temporal framework for the sequences and demonstrate the time significance of stratigraphic breaks.

Chemostratigraphy and palaeoenvironments

Eight chemostratigraphic suites have been recognized in the Holocene sediments of the Humber Estuary (Rees *et al.* this volume). Of these, four, i.e. Basal, Butterwick, Garthorpe and Saltend, are intersected by the three cores under consideration. PSA results are not mentioned specifically in the following sections, but information from these studies is implicit in much of the discussion.

Inner estuary: HMB16. Borehole HMB16 was terminated at the junction between sands, containing shell fragments, and the Triassic Mercia Mudstone Group. Approximately 3 m of medium–coarse sand, with shell fragments, granules and pebbles of rock fragments and mudstone rip-up clasts from the underlying Triassic sediments, overly the Mercia Mudstones before giving way to a basal peat. This sand is not shown in Fig. 4, but, together with approximately 1.8 m of coarse silt above the peat, it forms part of the Basal Suite. This suite is found in nine of the estuary boreholes and is comprised of sand and silt bodies, with very variable geochemical signatures (Fig. 7), which are interpreted to have been deposited in a fluvial environment and derived from local sources, probably Vale of York drift. Diatom and pollen data confirm a freshawater origin for the basal peat and overlying silt, but no data are available for the basal sand. The peat contains two distinctive pollen assemblages, which record a transition from an oak–hazel fenwood to an alder carr, indicating a rising water-table. Above the peat, the sandy silt contains abundant and diverse diatom assemblages. These are dominated by freshwater oligohalobian forms, with high frequencies of: *Cyclotella meneghiniana* (planktonic); *Ellerbeckia arenaria*, *Gyrosigma attenuatum*, (benthic epipelic); *Amphora ovalis*, *Cocconeis placentula*, *Cymbella aspera*, *C. cistula*, *Rhopalodia gibba* and *Synedra ulna* (all epiphytic). There is a decline in these species towards the top of the unit and a concurrent increase in the brackish taxa: *Campylodiscus echeneis*, *Cyclotella striata* and *Surirella striatula*, pointing to a rising water table, with increasing marine influence. Other fossils, including ostracods and gastropods, were recovered from this silt, and analyses of these and of the sediment matrix indicate a body of standing water within a fresh–brackish marsh, with some fen carr woodland and other macrophytic vegetation.

The Basal Suite in HMB16 is overlain by the geochemically distinct Butterwick Suite (Fig. 7). Although there is no significant change in grain size (Fig. 4), there is a corresponding transition from marsh to river channel conditions shown by a decline in epiphytic diatom taxa, with a concurrent increase in brackish planktonic and sediment-inhabiting species, suggesting that more open water became established towards the top of the Basal Suite. In the laminated silts with root traces of the Butterwick Suite, diatom recovery declines markedly, with assemblages dominated by the brackish planktonic species *Cyclotella striata*, with smaller frequencies of freshwater species such as *Cymbella cistula* and *Pinnularia maior*; indicating the presence of a river channel, subject to limited salinity fluctuations.

The junction between the Butterwick and overlying Garthorpe Suites is marked by a change in geochemistry (Fig. 8) and an increasing marine influence shown by a marked decline in the fresh–brackish species *Cyclotella striata* and the presence of mudflat diatom species *Tryblionella navicularis*, *T. punctata* and *Campylodiscus echeneis* and, higher in the sequence, the aerophilous saltmarsh species *Diploneis interrupta*. Detrital wood fragments are common and the environment is interpreted as a carr with saltmarsh influence. The thin peat at about −2.5 m OD, still within the Garthorpe Suite (Fig. 4), contains a range of mesohalobian diatoms, primarily *Diploneis interrupta, Nitzschia navicularis* and *Navicula peregrina*, with some polyhalobian forms like *Diploneis smithii* and *Paralia sulcata*. Saltmarsh pollen taxa (Chenopodiaceae, *Artemisia* and *Plantago maritima*) are found near the base of the peat, but the saltmarsh influence declines in mid peat to be replaced by Cyperaceae pollen, indicative of a coastal reedswamp. The appearance of *Navicula peregrina* shows the presence of nutrient-rich standing water (Krammer & Lange-Bertalot 1990). Above the reedswamp peat the diatom assemblages point to alternating mudflat and saltmarsh conditions, and the upper part of the Garthorpe Suite is considered to have been deposited in a low saltmarsh environment.

The topmost 40 cm of the Garthorpe Suite show a change, albeit slight, in chemistry, and this is preferred as the boundary with the overlying Saltend Suite rather than the erosional base of the sand unit at approximately OD (Fig. 4). This is supported by palaeomagnetic evidence discussed below. The differences in geochemical signatures between the Saltend and Garthorpe Suites are shown in Fig. 8. The presence of burrows in the upper part of the Garthorpe Suite suggests that any chemical similarity to the Saltend Suite here can be ascribed to bioturbation. Within the Saltend Suite a developing marine influence is shown by increased frequencies of marine and marine–brackish planktonic diatoms, which suggest the presence of an intertidal channel, probably a palaeochannel of the River Trent.

Fig. 4. Interpreted log of HMB16, plotted relative to ordnance datum. (**a**) Magnetic data. Environmental interpretations (RCM, OHF, etc. see below for definitions) are based on integrated data, but particularly lithological and biofacies information. (**b**) CNS geochemical data. Stable carbon isotope data are given on the weight per cent C_{org} plot (−24.6, etc.) and are in parts per thousand relative to VPDB. (**c**) Key to ornament and symbols used. Radiocarbon dates are given in thousands of calibrated years before present (BP). Dates of 8.4, 8.0 and 3.7 ka BP were made on *in situ* peats; 7.4 and 7.8 ka BP are best estimates of the age of the FWM unit based on gastropod shell carbonate and autochthonous organic matter; 6.7 and 6.4 ka BP are maximum ages based on well-preserved wood samples, ?4.5 ka BP is a best estimate based on *in situ Phragmites* material, and 2.1 ka BP on *in situ* rootlets. Chemostratigraphic suites: BA, Basal Suite; BU, Butterwick Suite; GA, Garthorpe Suite; SA, Saltend Suite. Note that Holocene sediment (river channel sands of the BA suite), not shown here, continue below the basal peat to −14.7 m OD. OHF, oak, hazel fenwood; AC, alder carr; FWM, freshwater marsh; LSM, lower saltmarsh; ITCS, intertidal channel sands; ITMF, intertidal mudflat; RCM, river channel muds. For key to ornaments and symbols see Fig. 4c.

(a)

GRAIN SIZE
INTERPRETED LOG
Fs Sl
INCLINATION (5pt running mean) degrees
50 60 70 80
Magnetic susceptibility (Dimensionless)
0 200 400
Magnetic intensity (mA/m)
0 200 400 600
+3.0
+2.0
+1.0
0.0
-1.0
-2.0
-3.0
-4.0
-5.0
-6.0
-7.0
-8.0
-9.0
-10.0
-11.0
ITMF
ITCS
LSM
P
AC
HOLOCENE
RCM
FWM
OHF/AC
2.1
3.7
?4.5
6.4
6.7
7.4
7.8
8.0
8.4
SA
GA
BU
BA
Ka
Cal years
BP

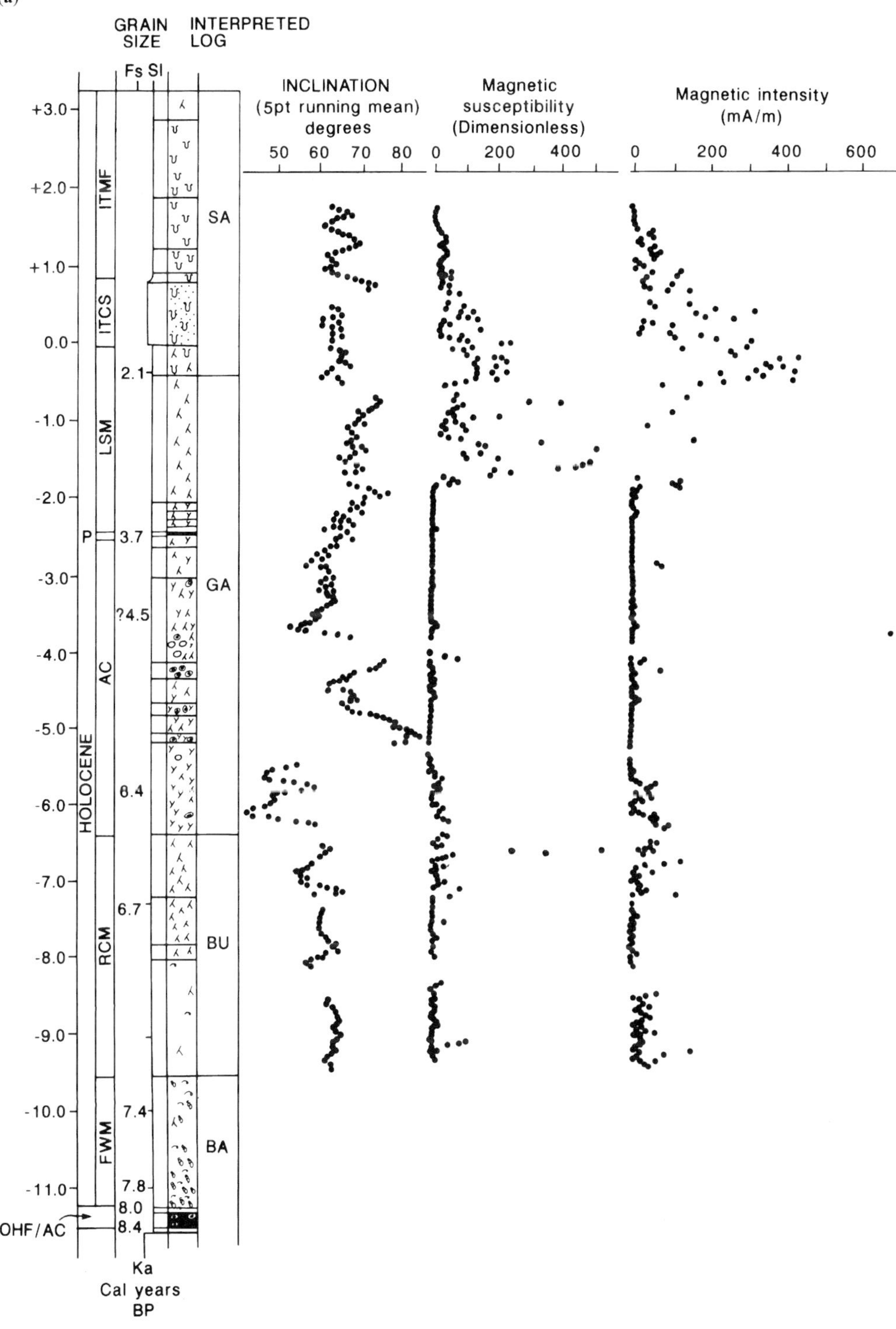

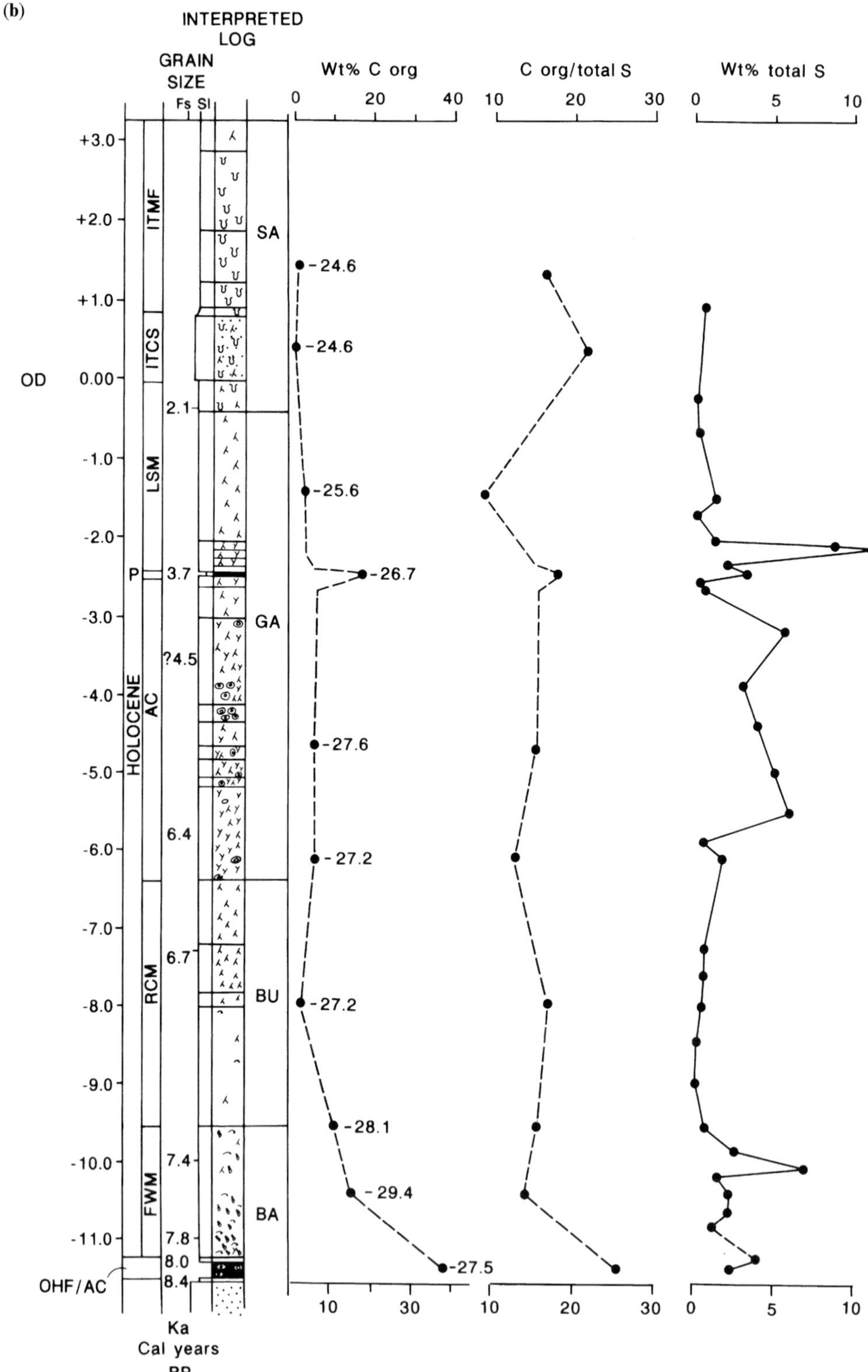
(b)
INTERPRETED LOG
GRAIN SIZE
Fs Sl
Wt% C org
C org/total S
Wt% total S
OD
HOLOCENE
ITMF
ITCS
LSM
P
AC
RCM
FWM
OHF/AC
SA
GA
BU
BA
2.1
3.7
?4.5
6.4
6.7
7.4
7.8
8.0
8.4
-24.6
-24.6
-25.6
-26.7
-27.6
-27.2
-27.2
-28.1
-29.4
-27.5
Ka Cal years BP

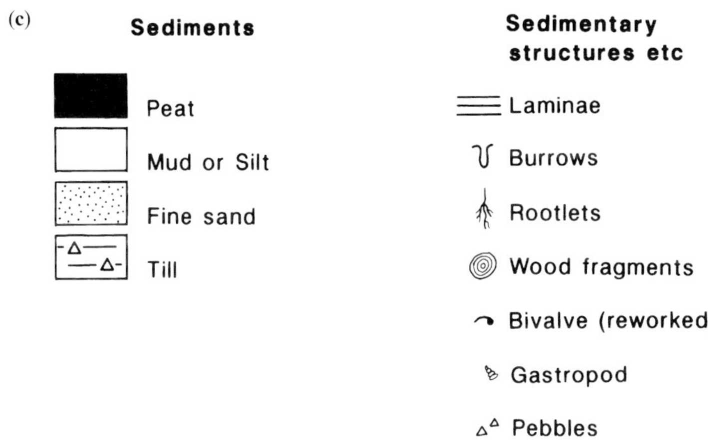

The environment near the base of the Saltend Suite is, therefore, interpreted as intertidal channel sands (Fig. 4). These are overlain by strongly laminated silts with burrows, which contain only a few poorly preserved robust diatoms, including the channel and mudflat taxa *Paralia sulcata*, *Cyclotella striata* and *Tryblionella navicularis*. The topmost part of HMB16 is thus considered to have been laid down in an intertidal mudflat.

Outer estuary: south bank, HMB10. At the base of borehole HMB10, 1.75 m of sediments assigned to the Garthorpe Suite overly a thin (5 cm) horizon of weathered till, above till proper. A 9-cm-thick peat horizon occurs some 50 cm above the weathered till (Fig. 5). Diatom preservation in this sequence is generally poor and environmental interpretation correspondingly difficult. Pollen and diatoms from the peat indicate a saltmarsh, with high numbers of the marine–brackish epipelic diatom species *Campylodiscus echeneis* and *Navicula gregaria* suggesting a standing body of brackish water. Pollen and foraminifera were not preserved in the lower part of the peat, but the upper part is shown to be of intertidal origin by high frequencies of *Jadammina macrescens* and lesser *Miliammina fusca*. Foraminiferal test linings were also abundant in these samples and support for a saltmarsh origin is found in records of probable saltmarsh pollen taxa, including Chenopodiaceae, *Aster*- and *Taraxacum*-types. Near the upper contact the peat pollen assemblage contains high *Tilia* (lime), *Ulmus* (elm), *Quercus* (oak) and *Alnus* (alder). High Gramineae (grass) pollen and freshwater aquatic taxa like *Typha latifolia*, *T. angustifolia* and *Menyanthes* suggest coastal reedswamp and fen conditions nearby (Fig. 9). The silt overlying the peat contains mesohalobian diatom forms such as *Tryblionella navicularis* and *T. punctata* suggesting proximity to and intertidal environments and the sequence as a whole is interpreted to have accumulated in a high saltmarsh environment grading into low saltmarsh near the junction with the overlying Saltend Suite (Fig. 5).

The erosional base of the Saltend Suite is characterized by a change in both geochemistry (Fig. 8) and environmental conditions, marked by a highly mixed assemblage of marine, brackish and freshwater diatom taxa. The high numbers of freshwater taxa (including *Achnanthes clevei*, *Navicula pupula* and *Nitzschia archibaldii*) are difficult to interpret, as they occur within a unit of fine-grained, laminated sandy silt. This factor, along with other dominant diatoms, including *Rhaphoneis amphiceros*, *Delphineis surirella*, *Thalassiosira eccentrica* and *Paralia sulcata*, is suggestive of an intertidal channel or flat environment. It is possible that the freshwater species were introduced into this environment via freshwater artesian wells or ‘blow wells’; chalk-water springs occurring where high ground-water pressure has forced a flow path upward through the confining till and Holocene deposits, thereby providing the necessary conditions for freshwater diatoms. A similar situation has been documented from other areas that are underlain by chalk and till, e.g. Holderness on the east coast north of the Humber and northeast Norfolk (Berridge & Pattison 1994). Above −3.0 m OD, a laminated silt contains a predominance of marine planktonic and tychoplanktonic diatoms, in particular *Paralia sulcata*, *Actinoptychus senarius*,

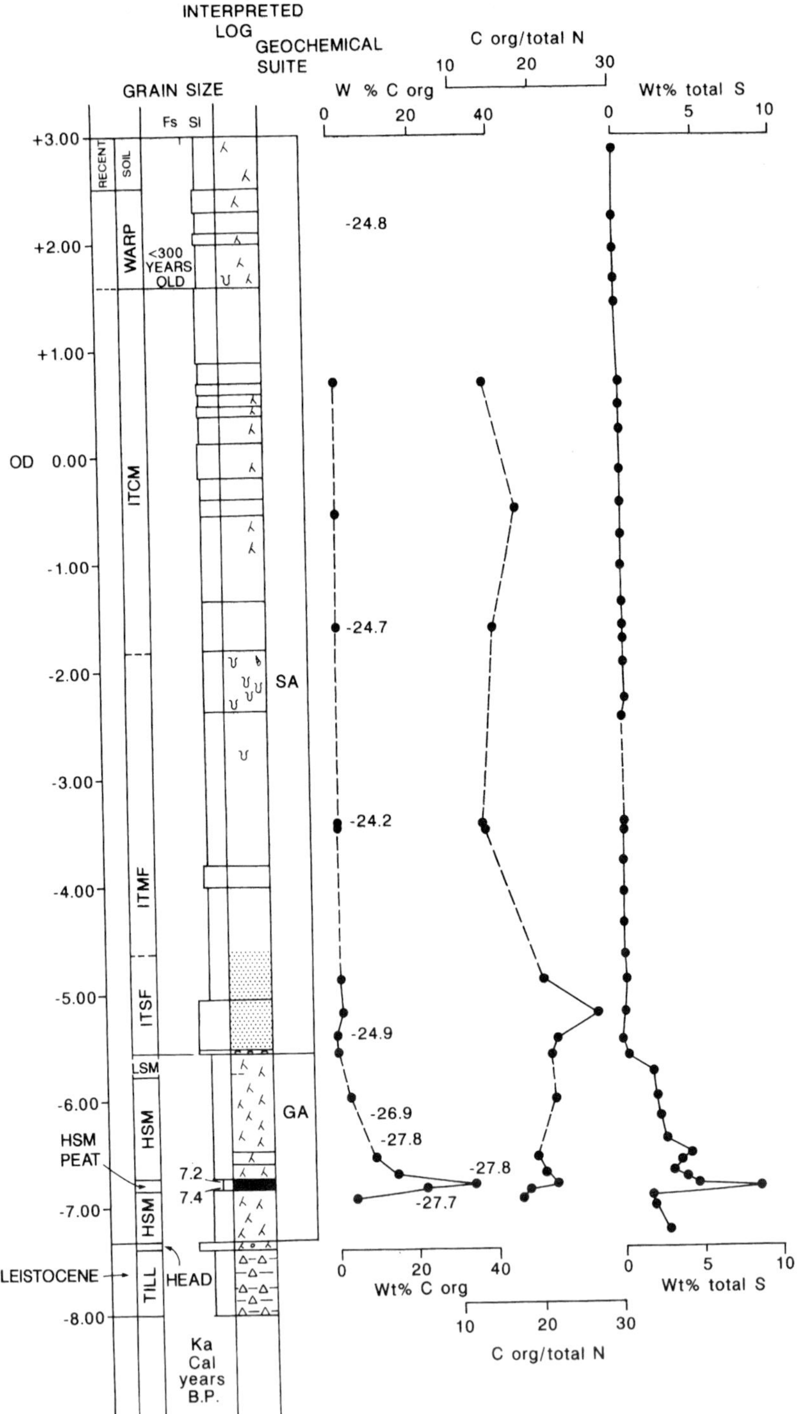

Fig. 5. Interpreted log of HMB10, plotted relative to ordnance datum and showing CNS geochemical data. Environmental interpretations (HSM, ITCM, etc.) are based on integrated data, but particularly lithological and biofacies information. Radiocarbon dates on the peat are given in thousands of calibrated years before present (BP). WARP, Sediment introduced by artificial flooding behind man-made embankments. Chemostratigraphic suites: GA, Garthorpe Suite; SA, Saltend Suite. Stable carbon isotope data are given on the weight per cent C_{org} plot (−24.9, etc.) and are in parts per thousand relative to VPDB. HSM, high saltmarsh; ITSF, intertidal sand flat; ITCM, intertidal channel muds; other abbreviations as for Fig. 4.

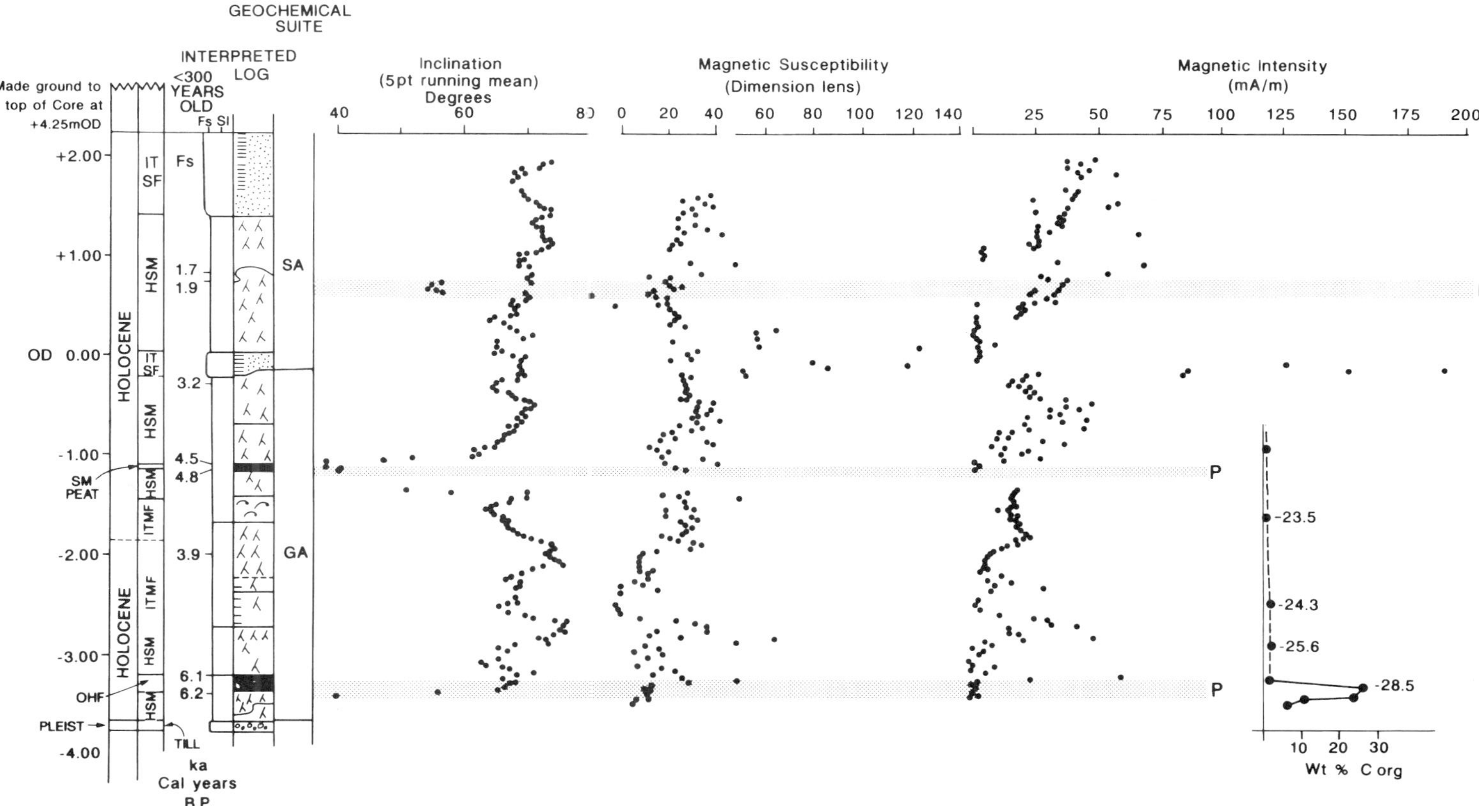

Fig. 6. Interpreted log of HMB12, plotted relative to ordnance datum, showing magnetic data. Environmental interpretations (HSM, OHF, etc.) are based on integrated data, but particularly lithological and biofacies information. Radiocarbon dates are given in thousands of calibrated years before (BP). Dates of 6.4, 6.1, 4.8 and 4.5 ka BP were made on *in situ* peats; 1.7–1.9 ka BP on slumped/reworked peat; 3.2 ka BP on *in situ* rootlets, and 3.9 ka BP on *in situ Scrobicularia* shell. Note that this core continues to +4.25 m OD as made ground (not shown). Chemostratigraphic suites: GA, Garthorpe suite; SA, Saltend suite. Other abbreviations as for Figs 4 and 5.

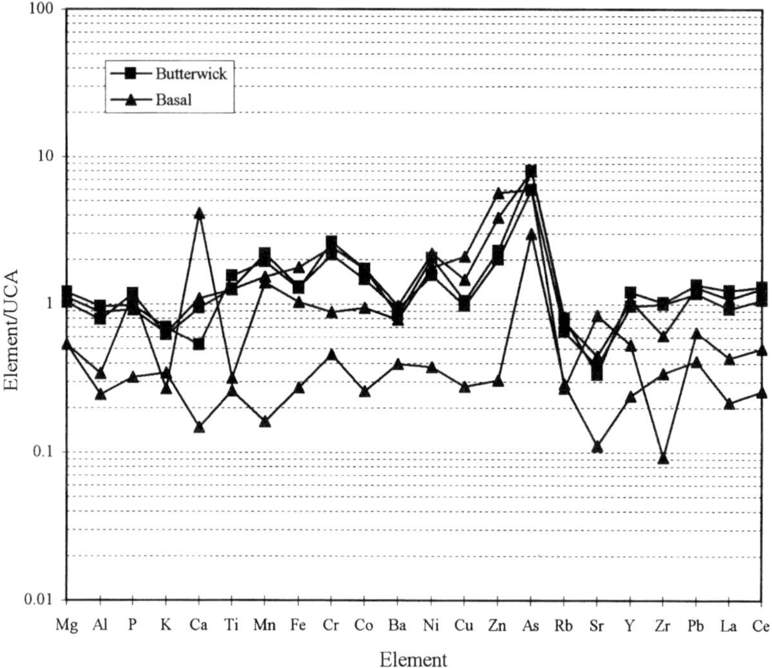

Fig. 7. Spidergram showing differences between the Basal and Butterwick Suites in HMB16.

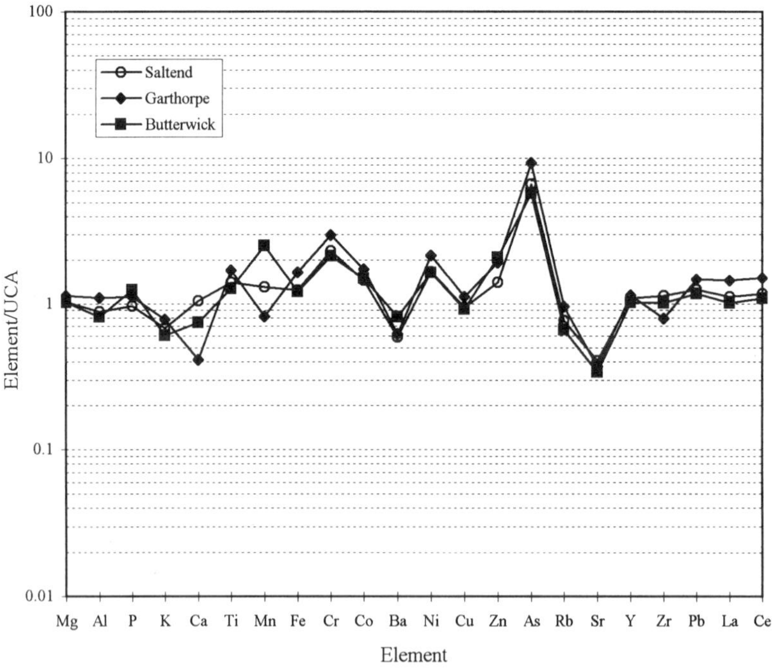

Fig. 8. Spidergram showing differences between the Butterwick, Garthorpe and Saltend Suites in HMB16.

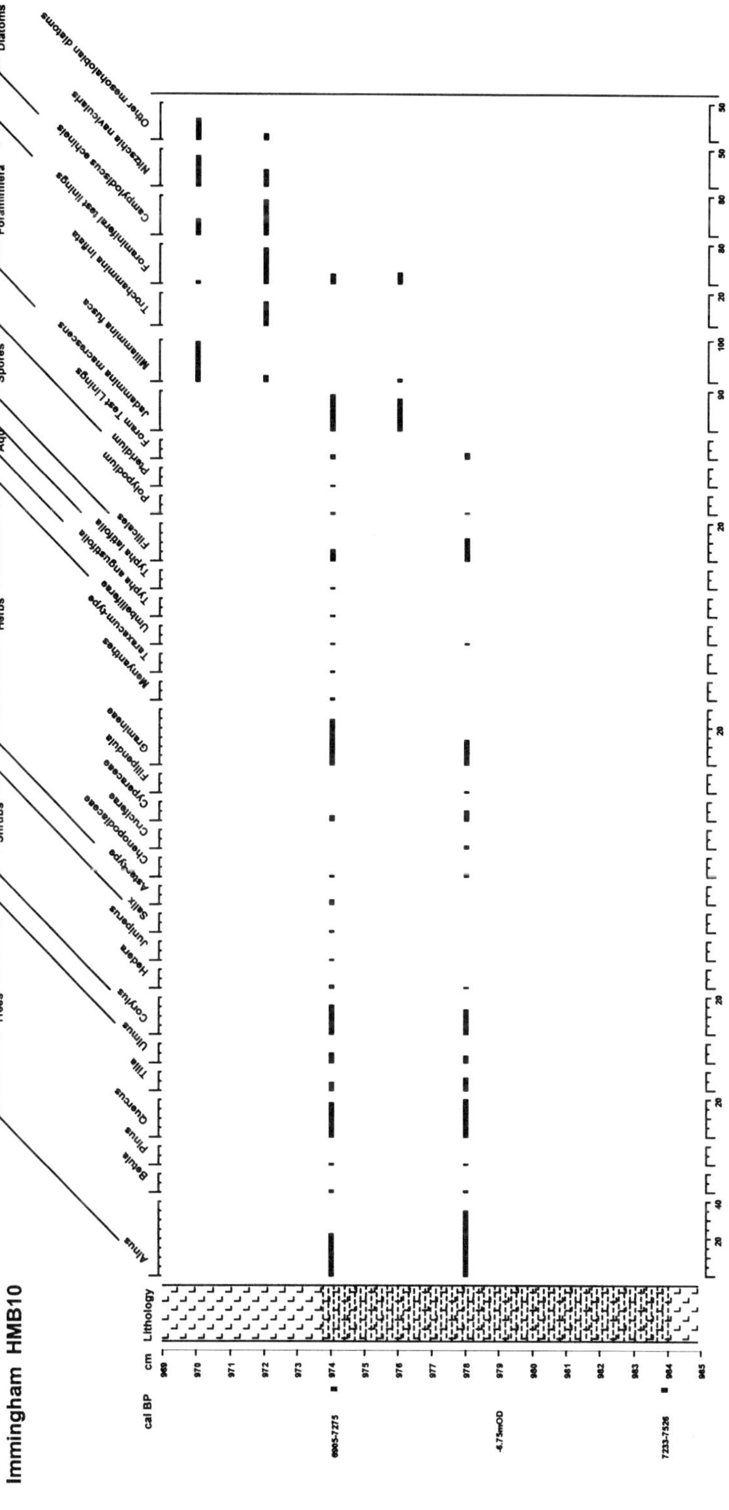

Fig. 9. Biota assemblage diagram for part of HMB10 (Immingham).

Biddulphia alternans and *Delphineis surirella*; these and the lithology, which changes to more laminated silt and sand, suggest the presence of an intertidal channel, with increasingly marine conditions becoming established. The uppermost part of the core, above approximately 1.5 m OD, although similar in geochemistry to the Saltend Suite sediments, may represent land reclaimed by the process of 'warping' (Gaunt 1994).

Outer estuary: north bank, HMB12. In HMB12, sediments considered to belong to the Garthorpe Suite extend upwards from a weathered till to an erosive junction with the Saltend Suite at just below OD (Fig. 6). The lowermost sediments are barren, but diatom assemblages recovered from the base of the peaty silt with wood fragments, some 25 cm above the till, are dominated by freshwater and brackish taxa including *Navicula peregrina* (brackish epipelic), *Pinnularia nobilis* (freshwater epipelic) and *Cymbella aspera* (freshwater epiphytic). All of these prefer organic-rich eutrophic environments and a body of standing water with abundant macrophyte vegetation is indicated. Pollen extracted from the same levels confirm this interpretation, with high frequencies of oak, and a fresh–brackish oak fenwood is suggested as the likely palaeoenvironment. The top of the peaty silt yields a slightly different diatom assemblage, with more brackish species including *Navicula elegans* and *N. digitoradiata* in addition to *N. peregrina*, indicating slightly higher salinities suggestive of a rising water-table. This is confirmed by pollen from this interval, which shows high frequencies of Gramineae (grasses). A reedswamp is thus suggested. Above the peat, there is a major increase in both numbers and preservation of diatoms. Some components of the preceding assemblage continue to dominate (*Navicula elegans, N. peregrina*), whilst the following marine and marine–brackish taxa appear: *Cocconeis scutellum, Tryblionella granulata, T. navicularis, Nitzschia scalaris* and *Diploneis interrupta*. These show a continuation of a nutrient-rich environment. However, the lithology changes to a grey clayey silt with rootlets and this, together with the appearance of the marine diatoms, suggests the presence of a saltmarsh with a strong marine influence, which probably encroached across the underlying reedswamp during a period of sea-level rise. This unit extends up through the core to −1.75 m OD, with diatoms declining in both numbers and state of preservation. It is probable that the saltmarsh environment became progressively nutrient-poor as sedimentation proceeded. From −1.75 to −1.48 m OD diatoms exhibit a sharp decline in both numbers and state of preservation, with only very few robust species, such as *Diploneis smithii*, present. The lithology, a laminated sandy silt with shell bands, suggests the presence of a mudflat. A rare example of a regressive contact occurs at −1.48 m OD, with a return to saltmarsh conditions shown by the reappearance of clayey silt with root traces. Diatoms increase markedly in both numbers and state of preservation, the assemblage being similar to that in the lower saltmarsh unit, but with far lower abundances of *Navicula peregrina* and *Tryblionella navicularis*. *Nitzschia elegans* is absent. There are also significantly greater frequencies of: *Caloneis westii, Diploneis interrupta, D. smithii* and *Paralia sulcata*. These point to greater marine influence, but with periods of exposure to subaerial conditions as indicated by the high abundance of the aerophilous species *Diploneis interrupta*. A high saltmarsh environment is envisaged for this stage of sedimentation. Diatom preservation and numbers decline markedly above −1.14 m OD, with the assemblages dominated by the marine tychoplanktonic species *Paralia sulcata* and the epipelic taxa *Caloneis westii, Diploneis didyma* and *Scolioneis tumida*, perhaps indicating deepening water conditions.

A fine sand unit of the Saltend Suite overlies the Garthorpe saltmarsh silts with an erosional contact at −0.25 m OD (Fig. 6). The sand, which extends upwards almost to OD, contains only a few poorly preserved diatoms, dominated by marine planktonic (*Triceratium favus*) and tychoplanktonic taxa (*Delphineis surirella, Paralia sulcata, Rhaphoneis amphiceros*). These indicate a subtidal channel, or possibly a sandflat. The absence of channel structures would suggest the latter environment. Above OD, silt with rootlets reappears, extending upwards to 1.4 m OD. Diatoms are very poor in both numbers and preservation, with *Paralia sulcata* and the epipelic taxa *Caloneis westii* and *Tryblionella navicularis* predominating. A dissolution assemblage is indicated. Geochemistry indicates a high C:N ratio and this, together with the presence of rootlets, suggests a saltmarsh environment. No diatoms were recovered above 1.5 m OD where the lithology changes to a fine laminated sand, interpreted as a sandflat. This is succeeded at 2.15 m OD by made-ground.

Heavy mineral and clay mineralogy

The mean USi (see earlier methods section) for each of the chemostratigraphic suites can be used to provide an indication of provenance. The Basal Suite in HMB16 has an index of only ten and is characterized by relatively stable heavy minerals comparable with those found in the Triassic Sherwood Sandstone Group, a common rock type of the Humber catchment. In contrast, the Saltend Suite has a USi of 81 and a high proportion of unstable heavy minerals, including metamorphic types, which suggest derivation from the Scottish Highlands. These unstable minerals are most likely of first cycle origin and may have arrived in the Humber Estuary by long-shore drift or in glacial till. If the former, the Saltend Suite has a strong marine component (see Rees *et al.* this volume).

Clay mineralogy provides an additional means of characterizing sediment suites; the Saltend Suite, for instance, is distinguished from the Garthorpe Suite by containing a greater proportion of expansible clays (Rees *et al.* 1999).

Time frame: the radiocarbon and palaeomagnetic record

The distributions of radiocarbon dates in the three cores are shown in Figs 4–6, where they can be compared with magnetic susceptibility, intensity and inclination records for HMB12 and HMB16. In this paper, calibrated radiocarbon ages are given as the central value of the 2σ range and the full data for dates are given in Table 2.

Due to sample recovery problems (probably attributable primarily to core rotation during the coring process), the absolute records of magnetic declination appear to have been corrupted, rendering comparison with the UK master record unreliable. Inclination records obtained from below *c.* 6 m drill depth also appear incompatible with the master curve; they may have been affected physically by drilling and/or compaction, or chemically, by post-depositional diagenesis. However, the inclination data obtained down to 6–7 m (drill depth) appear more robust and correlation with the master curve has been attempted.

Within approximately the upper 6 m of the cores, abrupt changes in the inclination values, magnetic intensity and magnetic susceptibility probably identify subtle sedimentological breaks, and some are accompanied by shifts in the C/N geochemistry of the sediments. These breaks may therefore represent time gaps corresponding to periods of non-deposition, or resulting from sediment erosion. It is not ideal to attempt palaeomagnetic correlation between a discontinuous sedimentary sequence, and the continuous Holocene UK master curve and the Humber palaeomagnetic data may be most informative in indicating discrete sedimentary packages and hiatuses. However, preliminary attempts to use the data for comparative dating purposes are discussed below.

In the outer Humber area PSV work was done on duplicates of cores HMB14 and HMB12 (discussed here). In HMB12 (Fig. 6) the magnetic record begins below made ground at +2.25 m OD. It is clear that peaty horizons in the core at +0.75, −1.20 and −3.40 m OD have low magnetic susceptibility and intensity, whilst minerogenic horizons have higher values. The peats are also associated with marked swings to lower values in the inclination data. These are probably artefacts arising from the very weak magnetic properties of the organic-rich sediment. Sandy horizons are characterized by the highest susceptibility and intensity records, reflecting increased concentrations of magnetic minerals. The highest values, around −0.20 m OD, correlate with the change from the Garthorpe to Saltend geochemical suites and may identify a major time break.

Between +2.25 and −0.40 m OD, the inclination values vary only slightly around a value of 70°. Comparison with the inclination record of the UK master curve suggests that deposition of the sediment with this inclination could have occurred at 0.2, or 1.2, or between 2.0 and 3.9 cal. ka BP. A radiocarbon date at +0.75 m OD (on a reworked peat) suggests an age around 1728–1941 cal. a BP. Roots at −0.30 m OD give a date of 3261 cal. a BP, consistent with the older PSV age range. The swing in inclination values to 40° at the thin peat at −1.20 m OD (dated at about 4.7 cal. ka BP) is probably an artefact (see above) and the inclination data below this peat are not compatible with the master curve, negating sensible age predictions.

In the inner Humber area, palaeomagnetic measurements were done on duplicates of cores HMB19 and HMB16 (discussed here). HMB16, located in the lower Trent Valley, has quite an extensive set of radiocarbon dates sampled to corroborate the PSV record. A feature of the HMB16 magnetic record is the very low susceptibility and intensity record below −2.00 m OD. This correlates with the switch from minerogenic low saltmarsh (LSM) sediments to older organic-rich carr sediments with low contents of magnetic minerals. This change to weakly magnetic sediments suggests that the inclination data are unlikely to be reliable below −2.00 m OD.

From +1.77 to −0.51 m OD the inclination data, when compared with the master curve,

Table 2. *Radiocarbon ages and calibration data for specific dates quoted in the text*

Site	Material	Laboratory code	^{14}C age $\pm 1\sigma$ (a BP)	Calibrated age (a BP)†			Altitude (m OD)	Notes
				Max.	Mean	Min.		
Lockham, HMB12	Decomposed peat (top)	AA25561	1830 ± 45	1868	1728	1619	0.75	Slumped but close to *in situ*
Lockham, HMB12	Decomposed peat (base)	AA25560	2005 ± 45	2047	1941	1836	0.75	Slumped but close to *in situ*
Lockham, HMB12	Roots (*in situ*)	AA25559	3060 ± 50	3370	3261	3082	−0.29	
Lockham, HMB12	Amorphous peat (*in situ*)	AA23890	4040 ± 65	4816	4479	4354	−1.16	
Lockham, HMB12	Amorphous peat (*in situ*)	AA23891	4235 ± 60	4871	4829	4569	−1.27	
Lockham, HMB12*	*Scrobicularia* sp. Shell‡	AA29908	3950 ± 70	4120	3910	3718	−2.01	
Lockham, HMB12	Wood fen peat	AA22672	5325 ± 50	6271	6138	5944	−3.3	
Lockham, HMB12	Wood fen peat	AA23434	5425 ± 70	6391	6238	5997	−3.45	
Garthorpe, HMB16	Roots (in situ)	AA25564	2160 ± 60	2327	2138	1985	−0.56	
Garthorpe, HMB16	Saltmarsh peat	AA23437	3425 ± 65	3836	3664	3475	−2.56	
Garthorpe, HMB16*	*Phragmites* rhizome	AA29902	4025 ± 55	4805	4480	4356	−3.37	Questionable due to ? sea-water influence
Garthorpe, HMB16*	Wood fragment	AA29903	5600 ± 60	6492	6385	6289	−5.88	
Garthorpe, HMB16*	Wood fragment	AA29904	5935 ± 60	6890	6750	6653	−7.43	
Garthorpe, HMB16	Autochthonous organic matter	AA25565	6540 ± 60	7524	7391	7281	−9.76	
Garthorpe, HMB16	Peat (undifferentiated)	AA25585	7265 ± 60	8135	8034	7922	−11.43	
Garthorpe, HMB16	Peat (undifferentiated)	AA25586	7745 ± 60	8580	8464	8374	−11.65	
Immingham, HMB10	Peat (undifferentiated)	AA23433	6520 ± 75	7526	7387	7233	−6.77	
Immingham, HMB10	Peat (undifferentiated)	AA23432	6245 ± 80	7275	7170	6905	−6.66	

* Sampled on duplicate core taken for palaeomagnetic work.
† The calibrated ages shown are the age ranges that contain 95.4% of the area under the probability curve. Ages in this table were calibrated with CALIB 3.0 using the bidecadal atmospheric curve (Stuiver & Reimer 1993; and references therein) except for sample marked ‡, which was calibrated with the marine calibration data set (Stuiver & Braziunas 1993).

suggest an age of <0.5 cal. ka BP (Fig. 4). At −0.51 m OD, a 5° offset in inclination data from 65 to 70° probably identifies an erosion surface and possible time break. This break comes in the muddy unit, below the marked increase in grain size at approximately OD (Fig. 4), where an erosion surface might be expected, but corresponding with a change in chemistry between the Garthorpe and Saltend Suites. While the break can be seen in core, it is the magnetic data that suggest the time significance of the surface. Just below −0.51 m OD, the inclination data indicate possible ages of *c.* 1.1, 2.0 or 4.3 cal. ka BP. A radiocarbon date of 2138 cal. a BP taken on roots at −0.56 m OD, just below the erosion surface, suggests a date of around 2.0 cal. ka BP is correct. At −2.59 m OD, a marked swing in the inclination data suggests a PSV date of *c.* 4.7 cal. ka BP. However, a 1-ka mismatch with a radiocarbon date of 3664 cal. a BP on peat at this level suggests that the PSV record is not reliable, perhaps due to the low magnetic intensity. The peat radiocarbon date plots in a sensible position on an age–depth plot for the core, suggesting that the date is reasonable.

Below −2.59 m OD, the inclination data can only be interpreted to give much older dates than the associated radiocarbon dates (described below) allow, suggesting that the inclination data are unreliable. However, radiocarbon dates on various targets below −5.00 m OD give a reasonable chronology. Two radiocarbon dates on well-preserved wood fragments between −5.80 and −7.50 m OD give maximum ages of 6400 and 6750 cal. a BP, and the basal oak fen and alder carr peat between −11.44 and −11.66 m OD is dated between 8034 and 8464 cal. a BP. Various dates on shell carbonate and autochthonous organic matter from a gastropod-bearing freshwater marsh sequence above the peat (between −9.76 and −11.66 m OD) suggest an age between *c.* 8.0 and 7.0 cal. ka BP.

CNS and isotope data: interpretation

Inner estuary: HMB16. In the inner Humber, core HMB16 contains a varied depositional sequence and a concerted effort was made to constrain the chronology and time breaks using radiocarbon and palaeomagnetic techniques. Below −2.56 to −11.70 m OD (the base of the basal peat), the core is dominated by C_{org}-rich (3–40 wt%) fluvial deposits including alder carr (AC), river channel muds (RCM), freshwater marsh (FWM) and the very organic-rich basal peat. The basal peats have CNS values similar to other Humber peats (see HMB10 and HMB12 below). However, the FWM, RCM and AC sediments have low C_{org}/total N ratios (values around 13–17), suggesting a lower content of woody material. These values, interpreted alone, might suggest incorporation of marine organic matter; however, the associated $\delta^{13}C_{org}$ values (typically −27 to −30‰) clearly identify the C_3 terrestrial organic matter source. The total S values are also diagnostic, being low (<0.6 wt%) in the RCM sediments, typically between 1 and 2 wt% in the FWM sediments, but higher in the basal peat and organic-rich AC sediments, again suggesting a large component of organically bound S (Fig. 4).

Above −2.56 m OD (probably an erosion surface based on palaeomagnetic data) the sequence is clearly transgressive with LSM sediments (probably around 3.0–2.0 cal. ka BP) replaced at an erosion surface by young (probably <0.5 ka based on PSV) intertidal sediments at −0.51 m OD to the top of the core. These more marine saltmarsh and mudflat sediments have 'marine' CNS and isotopic values, similar to those described in HMB10 and HMB12.

Outer estuary: south bank, HMB10. The up core change of measured parameters in HMB10 (Fig. 5) is probably reasonably representative of the outer Humber. Here the basal high saltmarsh (HSM) peats (−6.8 to −6.75 m OD; mean age of 7170–7387 cal. a BP) contain up to 34.4 wt% C_{org}, and in the overlying HSM sediments (to −6.58 m OD) this decreases to about 9.5 wt% C_{org}. All of these HSM sediments have C_{org}/total N ratios between 17.5 and 22 suggesting a strong component of woody terrestrial saltmarsh vegetation, and the $\delta^{13}C_{org}$ values between −27 and −27.8‰ confirm the presence of C_3 terrestrial plants (see e.g. Deines 1980). This basal Holocene sequence, younging from the dated peat horizon at −6.75 m OD, and probably representing progressive marine transgression of HSM, is cut by a sharp sedimentological boundary at −5.50 m OD. Here, the older Holocene sediments of the Garthorpe Suite are overlain by intertidal sands and then muds of the Saltend Suite. The Saltend Suite has an erosional base (Rees *et al.* this volume) and may be much younger than the underlying Garthorpe Suite. These intertidal flat and channel sediments are reasonably homogeneous, both geochemically and sedimentologically, giving way at +1.84 m OD to reclaimed sediment at the top of the core. The C_{org} content of the Saltend Suite sediments is typically low, around 1.2–1.5 wt% and lower in the basal sandy part. However, the C_{org}/total N ratios are high in the lowest part of the unit (−5.56 to −4.76 m OD) suggesting proximity to a source of C_3 saltmarsh organic detritus. Above −4.76 m OD, the muds have lower C_{org}/total N ratios, between 17 and 13, values indicative of a stronger component of more labile marine-derived organic matter (typical C_{org}/total N ratios are around 6–9, see discussion in Andrews *et al.* this volume). The $\delta^{13}C_{org}$ values above −5.50 m OD are consistent around −24 to 25‰, confirming the increased marine influence. (Marine organic matter probably has a $\delta^{13}C_{org}$ value around −19‰ (Andrews *et al.* this volume.)

Total S in HMB10 is highest in the organic-rich basal peat and is likely to be largely organically bound S (see Andrews *et al.* this volume). High S values above and below the peat may be due to S migration (see Andrews *et al.* this volume), although HSM and LSM facies typically have values around 2 wt% in any case. Intertidal facies have low sulphur values, typically <0.5 wt%.

These data show quite clearly: (a) how the amount of organic matter and S storage has varied in the core; and (b) how the changing source of organic matter (terrestrial versus more marine influence) can be identified. A disadvantage of the data in HMB10 is that, excepting the basal peat, there is no chronology for the core, making it difficult to integrate changes in storage over time. An attempt to address this was made by improving the chronology and identifying time breaks in HMB12 and HMB16 with both palaeomagnetic data and more complete radiocarbon dating.

Outer estuary: north bank, HMB12. It appears from the radiocarbon data on autochthonous organic matter (backed in part by the PSV data), that sedimentation in HMB12 began with basal oak fenwood peat some 6.0 cal. ka BP, and that punctuated sedimentation was occurring at times between 4800 and 1728 cal. a BP. Calculation of *net* storage of constituents between dated horizons is therefore straightforward. Above the basal peat, sedimentation switched between intertidal mud–sand flat and high saltmarsh environments. There is no clear record of continued marine transgression, suggesting that this area was not in the axial part of the estuary, but perhaps in a marginal channel on the estuary flank.

Interpretation of the CNS geochemical data (not all shown on Fig. 6) is similar to that in HMB10; the oak hazel fenwood (OHF) peats and HSM sediments have relatively high C_{org}/total N ratios (values in the twenties) and $\delta^{13}C_{org}$ values around −26 to −28‰ (Fig. 6), whereas the intertidal sediments have lower C_{org}/total N ratios (values around 12–14) and less negative $\delta^{13}C_{org}$, reflecting the increased component of marine organic matter. The C_{org} content is high in the basal OHF peats (up to 26 wt%; Fig. 6), but typically between 1 and 2 wt% above the basal OHF peat, where it is higher in intertidal mudflat (ITMF) and HSM sediments and lower in the sandy facies. The total S data show similar relations to HMB10, being high in the organic-rich sediments and low in the intertidal sediments.

Sea-level index points

In accordance with the sea-level methodology described above, the biostratigraphic data from the three Humber cores may be used to evaluate the status of the radiocarbon dated contacts as index points for the reconstruction of sea-level history. Each core provides particular types of sea-level data point (Metcalfe *et al.* this volume). At core HMB10 in the outer estuary, the upper contact of the basal peat, dated 6245 ± 80 a BP (6905–7275 cal. a BP), is a transgressive sea-level index point formed around MHWST during a positive tendency of sea-level movement, as saltmarsh indicators near the top of the peat and above show its deposition within intertidal environments. The lower peat contact, dated 6520 ± 75 a BP (7233–7526 cal. a BP), is also a sea-level index point, with freshwater peat forming over non-marine sediments due to rising water tables under a positive sea-level tendency, then passing upwards into saltmarsh peat (Figs 5 and 9).

Core HMB12, also in the outer estuary, records the continued rise of sea level. The basal peat, shown by the biostratigraphy to be entirely saltmarsh in origin, also records two sea-level index points formed under a positive tendency. The upper contact, dated 5325 ± 50 a BP (6271–5944 cal. a BP), is a transgressive index point formed around MHWST. The lower contact is dated 5425 ± 70 a BP (6391–5997 cal. a BP) and records saltmarsh peat formation at MHWST over a fluvial sequence. The upper, intercalated peat at HMB12, also described by Metcalfe *et al.* (this volume), provides two gradual contacts which are sea-level index points formed around MHWST, the upper transgressive contact dated 4040 ± 65 a BP (4816–4354 cal. a BP) and the lower regressive contact dated 4235 ± 60 a BP (4871–4569 cal. a BP) (Fig. 6). As the

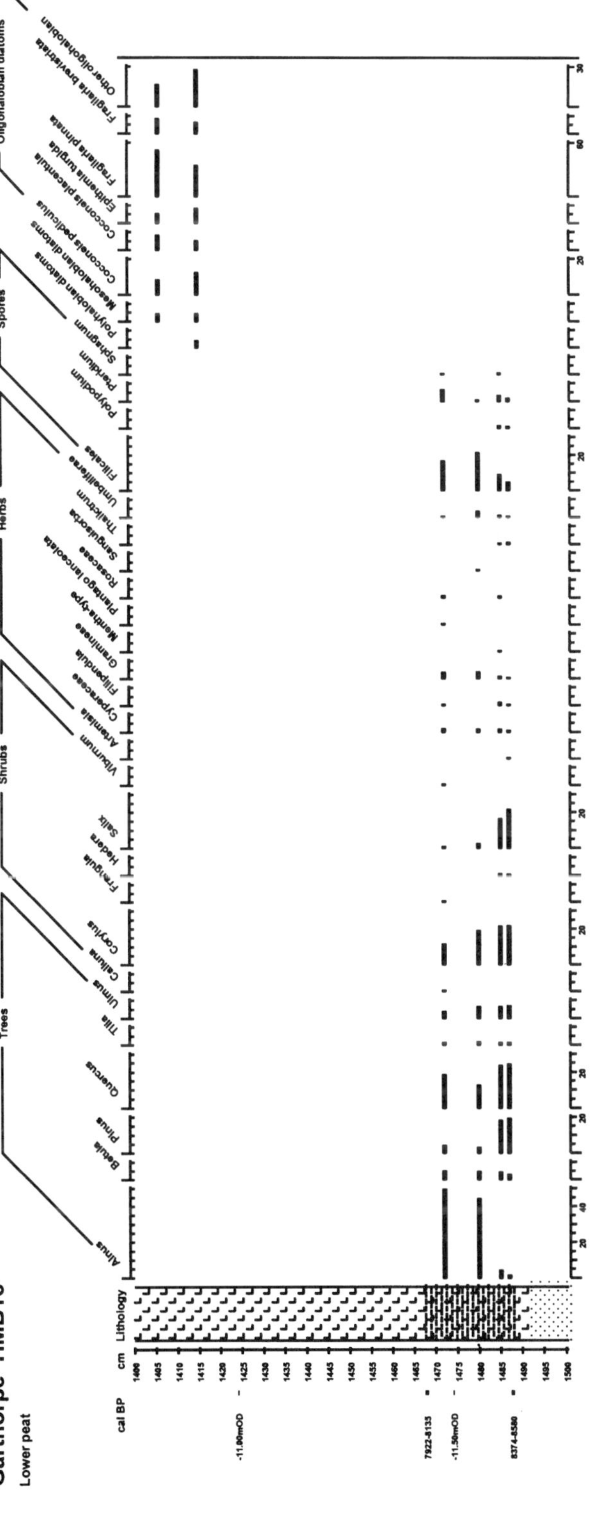

Fig. 10. Biota assemblage diagram for the lower peat in HMB16 (Garthorpe).

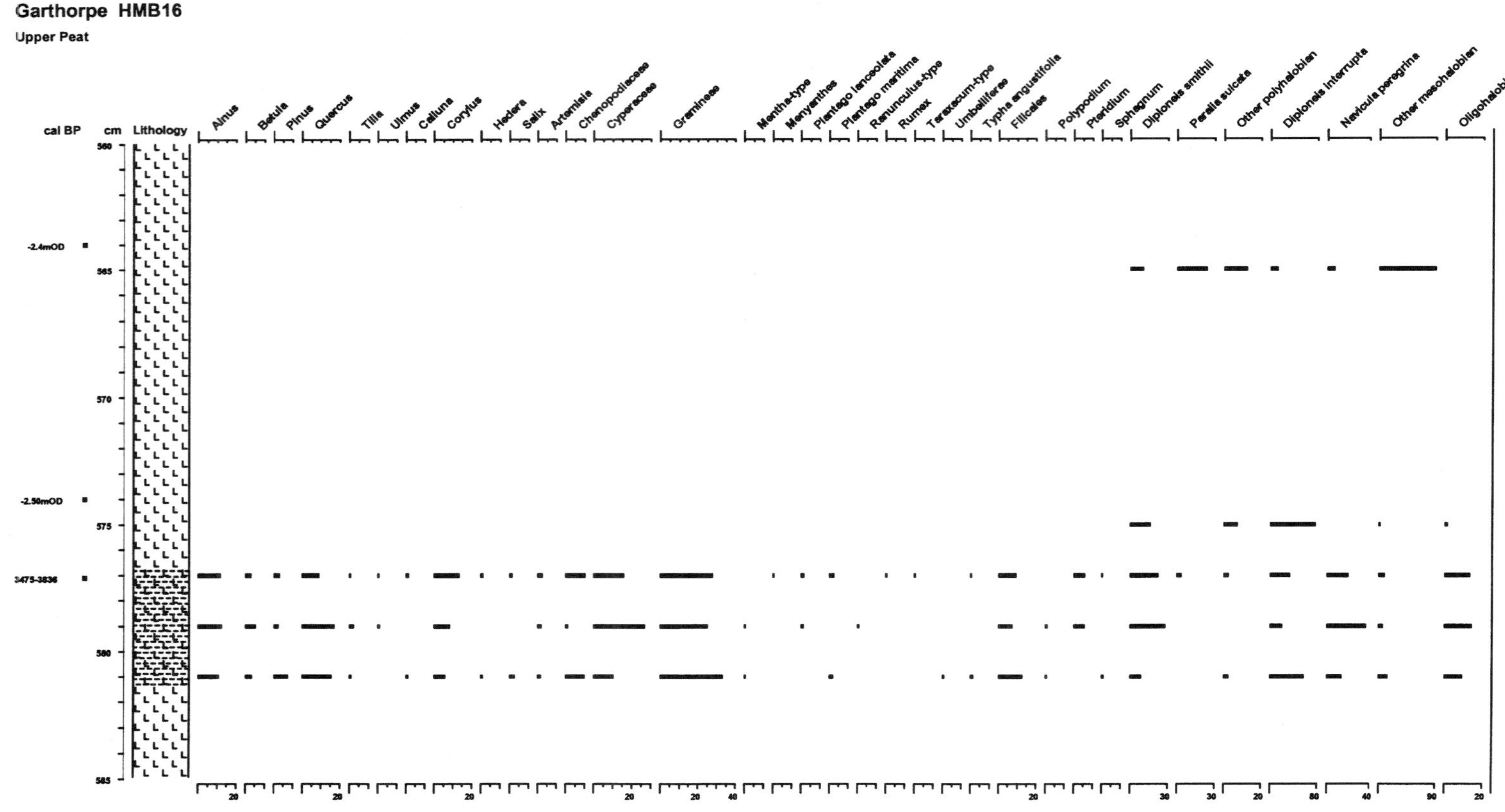

Fig. 11. Biota assemblage diagram for the upper peat in HMB16 (Garthorpe).

data from HMB10 and HMB12 are similar, biostratigraphic diagrams from HMB12 are not shown.

In contrast to these outer estuary cores, the basal peat at HMB16 in the inner estuary contains no evidence at all of marine influence, and formed entirely under freshwater conditions (Fig. 10). In this core both the upper (7265 ± 60 a BP; 7922–8135 cal. a BP) and lower (7745 ± 60 a BP; 8374–8580 cal. a BP) dates form only limiting values upon basal peat within a fluvial sequence and cannot be related to a former tide level. The upper, intercalated peat at HMB16 has been shown by the biostratigraphic evidence to be of saltmarsh origin and is stratified within estuarine, silty clays. The upper contact, dated 3425 ± 65 a BP (3475–3836 cal. a BP), is a transgressive sea-level index point formed around MHWST. The lower, regressive contact was not dated (Fig. 11). Sea-level index points such as the above are combined to form a sea-level curve which defines the history of sea-level change within an estuary or region such as the Humber (Long *et al.* 1998; Metcalfe *et al.* this volume).

Summary

This paper highlights the multidisciplinary approach that the LOEPS researchers have adopted in their study of the Holocene sediments of the Humber. It is also representative of the type of integration that was achieved throughout the LOEPS study area.

The results emphasize the following points.

(1) That Holocene environmental characterization is most secure when a number of different, but complementary, techniques are used. In this case, the integration of sedimentological, pollen, diatom, foraminiferal and varied geochemical data converge to give confident palaeoenvironmental characterization.

(2) That chronology in cored Holocene sequences can be integrated with palaeomagnetic and geochemical data to give a better understanding of the presence and significance of time breaks. In many cases these breaks are not obvious in cored sequences, especially in clay and silt-dominated sections that might otherwise be taken to represent continuous sedimentation intervals. In some instances these breaks may correspond to regionally significant surfaces, and those in this study are defined by changes in geochemical character. The recognition of time breaks is crucial to constraining sedimentation rates and material budgets.

H. Glaves, M. Slater, R. Newsham (all of BGS), J. Owen and O. Forster assisted with the levelling. P. King and T. Hardman (UEA) assisted with CNS elemental analyses and grain size analysis, and P. Dennis was involved in pro-ducing the isotope data. Technical support from UEA technicians is very much appreciated, especially P. Judge, who drafted some of the diagrams. This research was supported by NERC LOIS Special Topic allocations GST/02/736 and 766 and the paper is LOIS publication number 593. J. Ridgway, J. G. Rees and R. W. O'B. Knox publish by permission of the Director, British Geological Survey.

References

ANDREWS, J. E., SAMWAYS, G., DENNIS, P. F. & MAHER, B. A. 2000. Origin, abundance and storage of organic carbon and sulphur in the Holocene Humber Estuary – emphasising human impact on storage changes. *This volume.*

BATTARBEE, R. W. 1986. Diatom analysis. *In:* BERGLUND, B. E. (ed.) *Handbook of Holocene Palaeoecology and Palaeohydrology*. J. Wiley, Chichester, 527–570.

BECKETT, S. C. 1981. Pollen diagrams from Holderness, north Humberside. *Journal of Biogeography*, **8**, 177–198.

BERRIDGE, N. G. & PATTISON, J. 1994. *Geology of the Country around Grimsby and Patrington*. Memoirs of the British Geological Survey, Sheets 90, 91, 81 and 82 (England and Wales), Stationary Office, London.

BEYENS, L. & DENYS, L. 1982. Problems in diatom analysis of deposits: allochthonous valves and fragmentation. *Geolologie en Mijnbouw*, **61**, 159–162.

CANFIELD, D. E., RAISWELL, R., WESTRICH, J. T., REAVES, C. M. & BERNER, R. A. 1986. The use of chromium reduction in the analysis of reduced inorganic sulphur in sediments and shales. *Chemical Geology*, **54**, 149–155.

DEINES, P. 1980. The isotopic composition of reduced organic carbon. *In:* FRITZ, P. & FONTES, J. C. (eds) *Handbook of Environmental Isotope Geochemistry*. **1**, Elsevier, Amsterdam, 329–406.

DENYS, L. 1991. A check-list of the diatoms in the Holocene deposits of the Western Belgian coastal plain with a survey of their apparent ecological requirements. I. Introduction, ecological code and complete list. *Belgische Geologische Dienst, Professional Paper 1991/2*, **246**, 1–41.

DE RIJK, S. 1995. *Agglutinated Foraminifera as Indicators of Salt Marsh Development in Relation to Late Holocene Sea-level Rise*. PhD thesis, Free University, Amsterdam.

FICHEZ, R., DENNIS, P., FONTAINE, M. F. & JICKELLS, T. D. 1993. Isotopic and biochemical composition of particulate organic material in a shallow water estuary (Great Ouse, North Sea, England). *Marine Chemistry*, **43**, 263–276.

GAUNT, G. D. 1994. *Geology of the Country around Goole, Doncaster and the Isle of Axholme*. Memoirs of the British Geological Survey, Sheets 79 and 88 (England and Wales), Stationery Office, London.

GREEN, M. A., ALLER, R. C. & ALLER, J. Y. 1993. Carbonate dissolution and temporal abundances of foraminifera in Long Island Sound sediments. *Limnology and Oceanography*, **38**, 331–345.

HARTLEY, B. 1986. A check-list of the freshwater, brackish and marine diatoms of the British Isles and adjoining coastal waters. *Journal of the Marine Biologists Association, UK*, **66**, 531–610.

HASLAM, H. W. & PLANT, J. A. 1990. *Rock Geochemistry: Guidelines for the Acquisition and Interpretation of Lithogeochemical Data. Part 1: General. Part 2: Sedimentary Rocks*. British Geological Survey, Technical Report, WP/90/1.

HENDEY, N. I. 1964. *An Introductory Account of the Smaller Algae of British Coastal Waters, Part v Bacillariophyceae*. Stationery Office, London.

HORTON, B. P. 1997. *Quantification of the Indicative Meaning of a Range of Holocene Sea-level Index Points from the Western North Sea*. PhD thesis, University of Durham.

——, EDWARDS, R. J. & LLOYD, J. M. 1999*a*. Reconstruction of former sea levels using a foraminiferal-based transfer function. *Journal of Foraminiferal Research*, **29**, 117–129.

——, —— & ——1999*b*. UK intertidal foraminiferal distributions: implications for sea-level studies. *Marine Micropaleonotology*, **36**, 205–223.

HUGHEN, K. A., OVERPECK, J. T., LEHMAN, S. J., KASHGARIAN, M., SOUTHON, J. R. & PETERSON, L. C. 1998. A new ^{14}C calibration data set for the last deglaciation based on marine varves. *Radiocarbon*, **40**, 483–494.

HUSTEDT, F. 1953. Die Systematik der Diatomeen in ihren Bezeihungen zur Geologie und Okologie nebst einer Revision des Halobien-systems. *Svensk Botanisk Tidskrift*, **47**, 509–519.

——1957. Die Diatomeenflora des Fluss-systems der Weser im Gebeit der Hansestadt Bremen. *Abhandlungen Naturwissenschaftlichen Verein, Bremen*, **34**, 181–440.

INGHAM, M. N. & VREBOS, B. A. R. 1994. High productivity geochemical XRF analysis. *Advances in X-ray Analysis*, **37**, 717–724.

JENNINGS, A. E. & NELSON, A. R. 1992. Foraminiferal assemblage zones in Oregon tidal marshes – relation to marsh floral zones and sea level. *Journal of Foraminiferal Research*, **22**, 13–29.

KRAMMER, K. & LANGE-BERTALOT, H. 1986. *Bacillariophyceae. Süsswasserflora von Mitteleuropa, Band 2 Teil 1: Naviculaceae*. Gustav Fischer Verlag, Jena.

—— & ——1988. *Bacillariophyceae. Süsswasserflora von Mitteleuropa, Band 2 Teil 2: Bacillariaceae, Epithemiaceae, Surirellaceae*. Gustav Fischer Verlag, Jena.

—— & ——1990. *Bacillariophyceae. Süsswasserflora von Mitteleuropa, Band 2 Teil 3: Centrales, Fragilariaceae, Eunotiaceae*. Gustav Fischer Verlag, Jena.

LONG, A. J., INNES, J. B., KIRBY, J. R., LLOYD, J. M., RUTHERFORD, M. M., SHENNAN, I. & TOOLEY, M. J. 1998. Holocene sea-level change and coastal evolution in the Humber estuary, eastern England: an assessment of rapid coastal change. *The Holocene*, **8**, 229–247.

METCALFE, S. E., ELLIS, S., HORTON, B., INNES, J. B., MCARTHUR, J. J., MITLEHNER, A., PARKES, A., PETHICK, J. S., REES, J. G., RIDGWAY, J., RUTHERFORD, M. M., SHENNAN, I. & TOOLEY, M. J. 2000. The Holocene evolution of the Humber Estuary: reconstructing change in a dynamic environment. *This volume*.

MOORE, P. D., WEBB, J. A. & COLLINSON, M. E. 1991. *Pollen Analysis*. Blackwell, Oxford.

REES, J. G., RIDGWAY, J., KNOX, R. W. O'B., ELLIS, S., NEWSHAM, R. & PARKES, A. 1999. Holocene sediment storage in the Humber Estuary. *This volume*.

RIDGWAY, J., REES, J. G., GOWING, C. J. B., INGHAM, M. N., COOK, J. M., KNOX, R. W. O'B., BELL, P. D., ALLEN, M. A. & MOLINEAUX, P. J. 1998. *Land-Ocean Evolution Perspective Study (Loeps) Core Programme, Geochemical Studies, 1: Methodology*. Britsh Geological Survey, Technical Report, WB/98/55.

RODWELL, J. S. 1995. *British Plant Communities*, **4**, *Aquatic Communities, Swamps and Tall-herb Fens*. Cambridge University Press, Cambridge.

ROUND, F. E., MANN, D. G. & CRAWFORD, R. M. 1990. *The Diatoms: Biology and Morphology of the Genera*. Cambridge University Press, Cambridge.

SCOTT, D. B. & MEDIOLI, F. S. 1978. Vertical zonation of marsh foraminifera as accurate indicators of former sea levels. *Nature*, **272**, 528–531.

—— & ——1980. Quantitative studies of marsh foraminifera distribution in Nova Scotia: implications for sea-level studies. *Journal of Foraminiferal Research, Special Publication*, **17**, 1–58.

—— & ——1986. Foraminifera as sea-level indicators. *In*: VAN DE PLASSCHE, O. (ed.) *Sea-level Research: a Manual for the Collection and Evaluation of Data*. Geobooks, Norwich, 435–456.

SCOTT, D. K. & LECKIE, R. M. 1990. Foraminiferal zonation of Great Sippwissett Salt Marsh (Falmouth, Massachusetts). *Journal of Foraminiferal Research*, **20**, 248–266.

SHENNAN, I. 1982. Interpretation of Flandrian sea-level data from the Fenland, England. *Proceedings of the Geologists' Association*, **93**, 53–63.

——1986. Flandrian sea-level changes in the Fenland I and II. *Journal of Quaternary Science*, **1**, 119–154, 155–179.

SIMS, P. A. (ed.) 1996. *An Atlas of British Diatoms*. Koeltz Scientific Books, Königstein.

STUIVER, M. & BRAZIUNAS, T. F. 1993 Modeling atmospheric ^{14}C influences and ^{14}C ages of marine samples back to 10 000 BC. *Radiocarbon*, **35**, 137–189.

—— & REIMER, P. J. 1993. Extended ^{14}C data base and revised CALIB 3.0 ^{14}C age calibration program. *Radiocarbon*, **35**, 215–230.

TURNER, G. M & THOMPSON, R. 1981, Lake sediment record of the geomagnetic secular variation in Britain during Holocene times. *Geophysical Journal of the Royal Astronomical Society*, **65**, 703–725.

—— & ——1982. Detransformation of the British geomagnetic secular variation record for Holocene times. *Geophysical Journal of the Royal Astronomical Society*, **70**, 789–792.

VAN DER WERFF, A. & HULS, H. 1958–1974. *Diatomeen flora van Nederland*, **1–10**. De Hoef, the Netherlands.

VERADO, D. J., FROELICH, P. N., & MCINTYRE, A. 1990. Determination of organic carbon and nitrogen in marine sediments using the Carlo Erba NA-1500 analyzer. *Deep-Sea Research*, **37**, 157–165.

VOS, P. C. & DE WOLF, H. 1993. Diatoms as a tool for reconstructing sedimentary environments in coastal wetlands; methodological aspects. *Hydrobiologia*, **269/270**, 285–296.

WALLER, M. P. 1998. An investigation into the palynological properties of fen peat through multiple pollen profiles from south-eastern ngland. *Journal of Archaeological Science*, **25**, 631–642.

WEDEPOHL, K. H. 1995. The composition of the continental crust. *Geochimica et Cosmochimica Acta*, **59**, 1217–1232.

Implications of a microfossil-based transfer function in Holocene sea-level studies

B. P. HORTON,[1] R. J. EDWARDS[2] & J. M. LLOYD[1]

[1] *Environmental Research Centre, Department of Geography, University of Durham, South Road, Durham DH1 3LE, UK (e-mail: B.P.Horton@durham.ac.uk)*
[2] *Faculteit der Aardwetenschappen, Vrije Universiteit, 1081 HV Amsterdam, Netherlands*

Abstract: Fifty-two sea-level index points are described from samples collected within the Land–Ocean Interaction Study area. The vertical relationship between relative sea-level and a reference water level for each index point was estimated using two contrasting methods: a lithological-based approach, which is routinely employed in sea-level studies, and a foraminiferal-based transfer function. Comparison of the two methods reveals that the range of the former is 0.14 ± 0.09 m smaller than the latter because the foraminiferal-based transfer function takes into account differences in tidal ranges between study sites. Furthermore, the reference water-level estimates of transgressive index points using the foraminiferal-based transfer function are on average 0.19 ± 0.12 m higher than those of the lithological-based approach. This may be due to the rapid response time of foraminiferal assemblages relative to lithological indicators or the uneven spatial sampling within the contemporary foraminiferal data set. Whilst these inter-method differences are small in magnitude, they are comparable in size to the scale of changes under investigation by recent high-resolution sea-level studies. In contrast, the reference water levels of both methods are comparable for regressive and basis index points. Index points from clastic sediments were also produced using the foraminiferal-based transfer function. Calcareous foraminifera from intertidal environments can be used to produce indicative meanings and supply material for accelerator mass spectrometry radiocarbon dating. This method expands the range of stratigraphic sequences that can be employed in sea-level reconstruction by redressing the over-reliance on transgressive and regressive contacts.

Many studies have sought to reconstruct variations in relative sea-level (RSL) using microfossil data (e.g. diatoms, foraminifera and pollen) contained in a range of Holocene sedimentary deposits (Jelgersma 1961; van de Plassche 1982; Shennan 1982, 1986; Shennan *et al.* 1983; Tooley 1986; Long *et al.* 1998; Zong & Horton 1999). These microfossil data can provide information on a diverse range of processes such as changes in ice sheet extent, crustal movements, coastal evolution and sedimentary processes, which are vital for engineers and decision makers alike. Regardless of their application however, these microfossil data and their associated RSL reconstructions are all subject to fundamental errors associated with the precise determination of age and altitude.

One major source of altitudinal error is introduced when attempting to quantify the indicative meaning. The indicative meaning of a sea-level indicator describes the vertical relationship between the local environment in which it accumulated and a contemporaneous reference water level (Shennan 1982, 1986; van de Plassche 1986). It is defined in terms of the modern vertical range occupied by the sea-level indicator (the indicative range) measured relative to a given tide level (the reference water-level) such as mean high water spring tide (MHWST). Since sea-level trends are seldom inferred from a single type of dated material, differences in their indicative meanings must be considered when compiling data to reconstruct vertical movements of RSL.

Whilst the last decade has seen a shift towards increasingly high-resolution studies of RSL (Thomas & Varekamp, 1991; Varekamp *et al.* 1992; Nydick *et al.* 1995; Edwards, 1998), the accuracy and precision attainable is dictated in part by the quality of contemporary investigations of the relationship between RSL, environmental conditions and the succession and seasonal variations of microfossil assemblages. This study applies a foraminiferal-based transfer function (hereafter referred to as FBTF) developed by Horton (1997) and Horton *et al.* (1999*a*)

From: SHENNAN, I. & ANDREWS, J. (eds) *Holocene Land–Ocean Interaction and Environmental Change around the North Sea*. Geological Society, London, Special Publications, **166**, 41–54. 1-86239-054-1/00/$15.00

to produce a range of sea-level index points (SLIs) from material collected as part of the Land–Ocean Interaction Study (LOIS). In this study we compare the indicative meanings generated by this method with those produced by a traditional lithological-based approach (hereafter referred to as LBA) developed during the International Geological Correlation Programme Projects 61 and 200 (Preuss 1979; van de Plassche 1982, 1986; Shennan 1982, 1986; Tooley 1982; Shennan *et al.* 1983). The relative performance of both methods is evaluated and potential implications of FBTF to Holocene sea-level studies are presented. Of greatest significance is the ability to produce SLIs from minerogenic sequences in addition to the traditional transgressive and regressive contacts of intercalated and basal peats. The FBTF enables quantifiable records of RSL change to be produced based on a firm understanding of the relationship between contemporary foraminiferal assemblages and altitude relative to the tidal frame.

Background

Fossil pollen and diatom assemblages are routinely employed as sea-level indicators to reconstruct the indicative meaning of SLIs (Nelson & Kashima 1993). More recently, the use of fossil saltmarsh foraminiferal assemblages has become increasingly common because of their potential to quantify the indicative meaning of a range of Holocene sea-level index points more precisely than is possible through the use of vertically zoned floral indicators. Scott & Medioli (1978, 1986) stated that assemblages of agglutinated saltmarsh foraminifera are the most accurate sea-level indicators on temperate coastlines and that such assemblages exhibit a strong correlation with elevation above mean tide level (MTL), although the width of foraminiferal zones does not vary directly with tidal range. For example, in Nova Scotia, Scott & Medioli (1980) suggested that the former position of highest astronomical tide (HAT) can potentially be estimated to a precision of ±5 cm by identifying the upper boundary of foraminiferal Zone IA (monospecific assemblages of *Jadammina macrescens*).

Foraminifera have been employed to 'validate' SLIs by analysing the changes in composition of assemblages with different salinity preferences (Shennan *et al.* 1996). In the last few years, these techniques have been used to evaluate earthquake-induced coseismic subsidence (Guilbault 1995, 1996; Nelson *et al.* 1996), the indicative meaning of SLIs (Long *et al.* 1998), and small-scale changes in RSL (Thomas & Varekamp 1991; Varekamp *et al.* 1992; Nydick *et al.* 1995; Edwards 1998). The quality of these studies depends upon the accuracy and precision of RSL reconstruction. Despite this, there remains a lack of statistical-based analyses of foraminiferal assemblages and their associated indicative meanings in the UK (Horton 1997).

Methods

Shennan (1982, 1986) compiled data from a range of sources to permit quantification of the

Table 1. *Indicative range and reference water-level for commonly dated materials (Shennan, 1982)*

Dated material	Indicative range* (cm)	Reference water-level†
Phragmites or monocot peat		
Directly above saltmarsh deposit	±20	[(MHWST ± HAT)/2] – 20 cm
Directly below saltmarsh deposit	±20	MHWST – 20 cm
Directly above fen wood deposit	±20	MHWST – 10 cm
Directly below fen wood deposit	±20	[(MHWST + HAT)/2] – 10 cm
Directly above and below saltmarsh deposit	±40	MHWST
Middle of layer	±70	Infer from stratigraphy
Fen wood peat		
Directly above *Phragmites* or saltmarsh deposit	±20	(MHWST + HAT)/2
Directly below *Phragmites* or saltmarsh deposit	±20	MHWST
Basis peat‡		
Directly below *Phragmites* or saltmarsh deposit	±20	MHWST
Directly below fen wood deposit	±80	MTL–MHWST

* The indicative range (given as a maximum) is the most probable vertical range in which the sample occurs.
† The reference water-level is given as a mathematical expression of tidal parameters ± an indicative difference. The indicative difference is the distance from the mid-point of the indicative range to the reference water-level.
‡ A basis peat is a basal peat formed as the result of rising relative sea-level.

indicative meanings for lithostratigraphical and biostratigraphical contexts commonly used as sources of material for radiocarbon dating (Table 1). The LBA assumes, where no hiatus is present, that intercalated sedimentary facies formed in environments that existed side by side in space. In this way, by quantifying the range of elevations across which the transition between one sub-environment and another occurs, it is possible to assign an indicative range to a lithostratigraphical or biostratigraphical contact. Thus, whilst the indicative range of *Phragmites* is stated to be 0.70 m (Preuss 1979), Shennan (1982) refined this value to 0.20 m where a *Phragmites* peat rests conformably above or below a saltmarsh deposit.

Recent work by Horton (1997) and Horton *et al.* (1999*a*) has developed statistical-based methods to reconstruct indicative meanings based upon the relationship between foraminiferal assemblages and elevation with respect to the tidal frame. Foraminiferal and altitudinal data were collected from ten contemporary UK intertidal environments (Fig. 1). The altitudinal

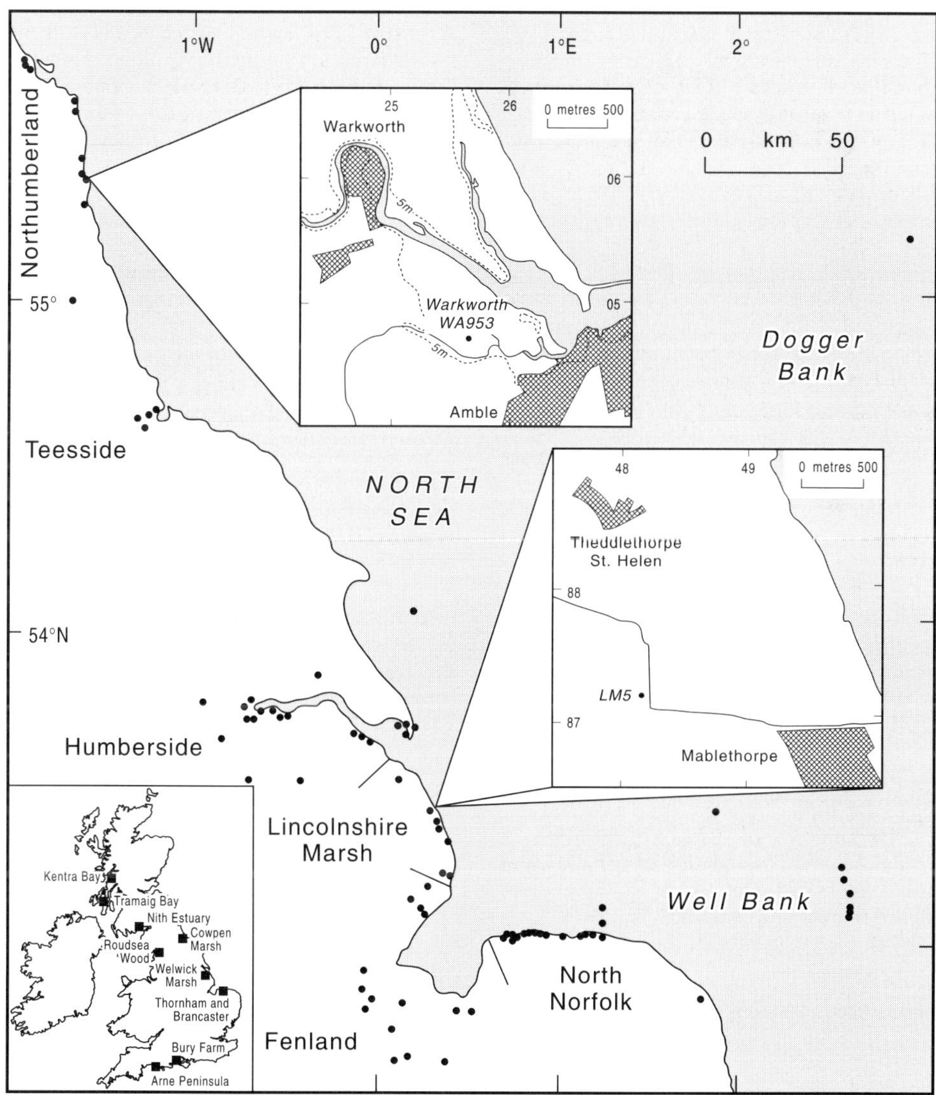

Fig. 1. Map of the UK showing the location of boreholes in the LOIS study area with inserts for contemporary intertidal study environments and fossil sites (Warkworth and Theddlethorpe).

data from different marshes were expressed as a standardized water-level index (SWLI), which takes account of differences in tidal range between sites and defines elevation with respect to MTL (Horton 1997; Horton *et al.* 1999*a*; Zong & Horton 1999). MTL is used instead of mean sea-level (MSL), because MSL does not leave a long term geological record in sediments or landforms. Therefore, sea-level curves using MSL are actually based upon MTL (Jardine 1986).

The FBTF was then developed to reconstruct indicative meanings using weighted averaging (WA) regression and calibration with inverse and classical deshrinking (Birks 1995; Jones & Juggins 1995; Gasse *et al.* 1997; Juggins & ter Braak 1997). Statistical measures suggest that precise reconstructions of indicative meanings are possible (inverse $r = 0.88$; classical $r = 0.89$) together with sample-specific standard errors of prediction for individual fossil samples (bootstrapped SE_{pred}). The SWLIs are subsequently back-transformed relative to Ordnance datum (UK national levelling datum) and expressed in metres to produce reference water levels and associated indicative ranges. The back-transformation is dependent on the local tidal range. Furthermore, the establishment of sea-level index points is no longer stratigraphically constrained but can be applied to complete fossil sequences permitting reconstruction of changing water depths throughout entire cores.

In this study, LBA and FBTF were evaluated by applying them to fossil material collected as part of the LOIS project. Sampling locations were grouped into the following six geographical areas: Northumberland, Teesside, Humber Estuary, Lincolnshire Marshes, Fenland and north Norfolk (Fig. 1).

Samples for foraminiferal analysis were prepared using standard procedures (Horton *et al.* 1999*b*). The foraminiferal taxonomy follows Murray (1971, 1979) and de Rijk (1995) (Table 2). Foraminiferal assemblages are expressed as a percentage of dead foraminiferal tests (Horton *et al.* 1999*b*). The preservation of tests was generally very good in contemporary and fossil deposits. As a result, a minimum count of 300 tests was possible for most samples (following Patterson & Fishbein 1989). Radiocarbon age estimates were converted into calendar years *via* the program CALIB 3.0 (Stuiver & Reimer 1993) using the bidecadal dataset, intercept method (A) and an error multiplier of one.

Results

A total of 52 SLIs are reported in this study using results collected by the Environmental Research Centre (ERC), University of Durham and other LOIS partners. Further information regarding their indicative meanings is referenced in Table 3. The sea-level data are primarily derived from the transgressive and regressive contacts of intercalated–basal peats, reflecting the traditional requirements of LBA. To facilitate comparison between the indicative meanings, LBA and FBTF are applied to an intercalated basal peat sequence. In addition, the indicative meanings of clastic sequences are estimated.

Intercalated peat

Two SLIs (WA953/526 and WA953/539) were produced from an intercalated peat taken from Warkworth, Northumberland, UK. The field site is immediately downstream of Warkworth where the River Coquet emerges from an incised meander into a small area of lowland, in the lee of an area of drift-covered bedrock and coastal dunes (Fig. 1). The piston core WA953 was used for lithostratigraphical and biostratigraphical analyses [NV 2519 0516].

Table 2. *Foraminiferal taxonomy*

Species	Citation
Ammonia beccarii var *limnetes* (Todd & Bronniman)	*Ammonia beccarii* var *limnetes*: Murray (1971)
Elphidium williamsoni (Williamson)	*Elphidium williamsoni*: Murray (1979)
Haynesina germanica (Ehrenberg)	*Protelphidium germanicum*: Murray (1979)
Jadammina macrescens (Brady)	*Jadammina macrescens*: Murray (1971) Murray (1979) de Rijk (1995)
Miliammina fusca (Brady)	*Miliammina fusca*: Murray (1971) Murray (1979) de Rijk (1995)
Trochammina inflata (Montagu)	*Trochammina inflata*: Murray (1971) Murray (1979) de Rijk (1995)

Table 3. *The indicative meaning of SLIs collected by ERC and other LOIS partners based on a lithological-based approach (LBA) and a foraminiferal-based transfer function (FBTF)*

No.*	Site	Core	Description	LBA		FBTF	
				RWL† (m OD)	IR‡ (m)	RWL† (m OD)	IR‡ (m)
Northumberland							
1	Alnmouth	AL951/180	Transgressive contact	2.15	0.20	2.39	0.25
2	Alnmouth	AL951/244	Basis peat	2.35	0.20	2.40	0.25
3	Bridge Mill	BM957/132	Transgressive contact	2.20	0.20	2.39	0.27
4	Cresswell Pond	CP958/43	Basis peat	2.45	0.20	2.49	0.27
5	Broomhouse Farm	BR968/72	Regressive contact	2.49	0.20	2.43	0.27
6	Broomhouse Farm	BR968/86	Transgressive contact	2.20	0.20	2.43	0.27
7	Warkworth	WA942/78	Transgressive contact	2.25	0.20	2.56	0.27
8	Warkworth	WA953/526	Transgressive contact	2.25	0.20	2.29	0.26
9	Warkworth	WA953/539	Regressive contact	2.55	0.20	2.59	0.27
Teesside							
10	Holme Fleet	HMBB5/527	Regressive contact	2.86	0.20	2.80	0.30
11	Portrack Marsh	PMC5/454	Regressive contact	2.86	0.20	2.88	0.30
12	Teesside industrial	T2/1365	Transgressive contact	2.55	0.20	2.79	0.29
Humber Estuary							
13	Dirtness Levels	DL961/332	Transgressive contact	4.00	0.20	4.40	0.40
14	Dunswell	HMB2/384	Transgressive contact	3.40	0.20	3.81	0.37
15	South Marsh	HMB5/864	Transgressive contact	3.20	0.20	3.45	0.40
16	Immingham	HMB10/974	Transgressive contact	3.20	0.20	3.59	0.40
Lincolnshire Marshes							
17	Marshchapel	LM2/993	Transgressive contact	2.80	0.20	3.09	0.36
18	Marshchapel	LM2/996	Basis peat	3.00	0.20	3.05	0.36
19	Sand-le-mere	SM954/250	Transgressive contact	2.65	0.20	2.97	0.36
20	Sand-le-mere	SM954/268	Transgressive contact	2.65	0.20	3.05	0.36
21	Sand-le-mere	SM955/17	Regressive contact	3.16	0.20	3.03	0.33
22	Theddlethorpe	LM5/601	Transgressive contact	3.05	0.20	3.12	0.35
23	Theddlethorpe	LM5/632	Regressive contact	3.32	0.20	3.19	0.35
	Theddlethorpe	LM5/1215	Calcareous foraminifera	–	–	1.12	0.40
	Theddlethorpe	LM5/1265	Calcareous foraminifera	–	–	0.93	0.40
	Theddlethorpe	LM5/1281	Calcareous foraminifera	–	–	1.28	0.40
24	Theddlethorpe	LM5/1287	Transgressive contact	3.05	0.20	2.88	0.35
	Theddlethorpe	LM5/1295	Basis peat	3.25	0.20	–	–
Fenland							
25	Adventurers' Land	F21/446	Transgressive contact	3.57	0.20	3.82	0.40
26	Adventurers' Land	F21/477	Regressive contact	3.95	0.20	3.84	0.40
27	Cowbit	F16/738	Transgressive contact	3.60	0.20	3.79	0.43
28	Gedney Fen	F8/297	Transgressive contact	3.60	0.20	3.66	0.43
29	Gosberton	F17/565	Transgressive contact	3.60	0.20	3.67	0.43
30	Gosberton	F17/571	Regressive contact	4.01	0.20	3.97	0.43
31	South Lynn	F15/496	Transgressive contact	3.57	0.20	3.55	0.37
32	South Lynn	F15/528	Regressive contact	3.95	0.20	3.84	0.40
33	South Lynn	F15/1069	Transgressive contact	3.57	0.20	3.54	0.37
34	South Lynn	F15/1081	Regressive contact	3.77	0.20	3.96	0.40
35	South Lynn	F19/1058	Transgressive contact	3.60	0.20	3.79	0.43
36	Wrangle Bank	F4/344	Transgressive contact	3.28	0.20	3.33	0.38
37	Wrangle Bank	F4/352	Regressive contact	3.48	0.20	3.62	0.40
38	Wrangle Lowgate	F5/1350	Interval (whole of layer)	3.15	0.40	3.21	0.38
39	Wrangle Lowgate	F5/1385	Transgressive contact	2.95	0.20	3.16	0.38
North Norfolk							
40	Blakeney	NNC2/815	Transgressive contact	2.35	0.20	2.47	0.27
41	Blakeney	NNC2/823	Regressive contact	2.66	0.20	2.46	0.27
42	Brancaster	NNC29/875	Transgressive contact	2.85	0.20	3.01	0.32
43	Salthouse	NNC40/554	Transgressive contact	2.35	0.20	2.59	0.34
44	Salthouse	NNC40/575	Regressive contact	2.66	0.20	2.52	0.29
45	Scolt Head	NNC28/1103	Transgressive contact	3.25	0.20	3.32	0.38
46	Thornham	NNC35/386	Transgressive contact	2.85	0.20	3.11	0.35
47	Thornham	NNC35/592	Transgressive contact	2.85	0.20	2.96	0.32
Offshore							
48	Dogger Bank	55/+02/213VE/406	Transgressive contact	2.20	0.20	2.37	0.27

* No., Number of SLI used in Fig. 4.
† RWL, present altitude of reference water-level (m OD).
‡ IR, indicative range (m).

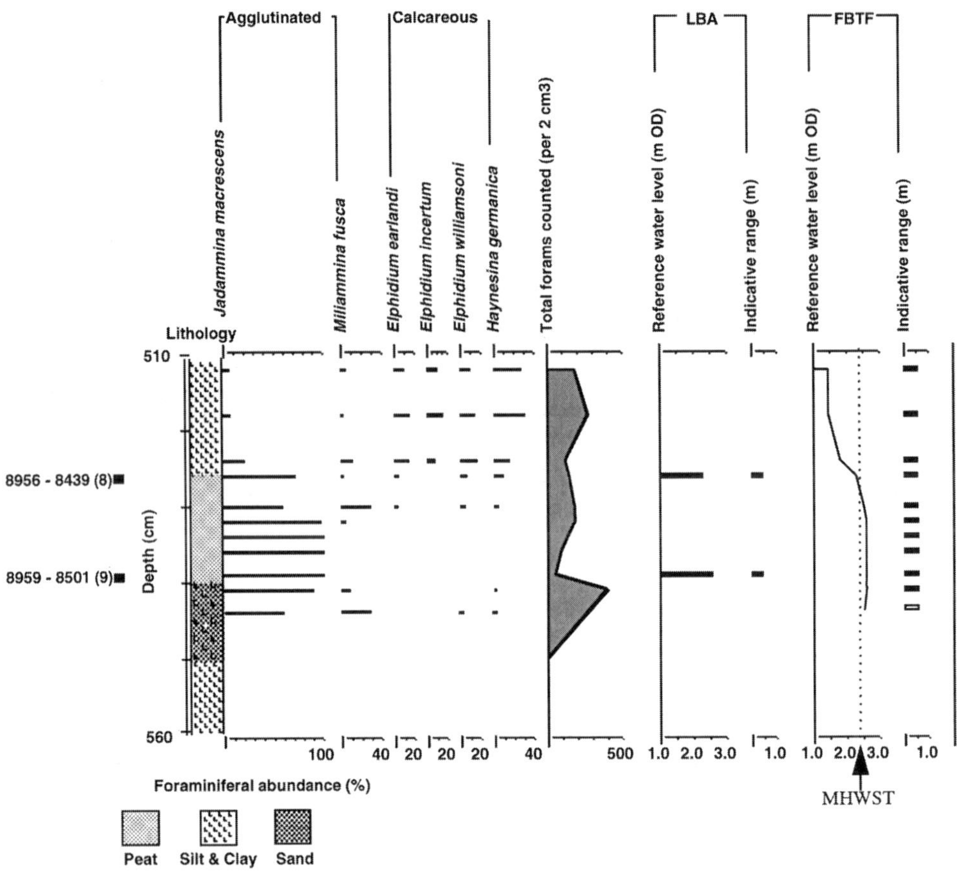

Fig. 2. Warkworth WA953 foraminiferal diagram. Foraminiferal abundance is calculated as a percentage of dead foraminiferal tests (only species greater than 10% are shown). Reference water-levels and error ranges are estimated using a lithological-based approach (LBA) and a foraminiferal-based transfer function (FBTF). The radiocarbon dates (expressed in calibrated years BP), sea-level index point numbers (used in Fig. 4) and MHWST are shown. The stratigraphy is drawn according to Troels-Smith (1955).

The lithostratigraphy (Fig. 2) comprises a diamicton consisting of a stiff pink clay at the base, which is overlain by a sandy clay with organic patches. This unit grades into an amorphous peat 539 cm below ground surface and is subsequently overlain by a clayey peat at a depth of 532 cm. Overlying this unit is a blue-grey clayey silt containing the remains of *Phragmites*.

The biostratigraphy indicates that foraminiferal tests are absent from the diamicton at the base of the core (Fig. 2). The overlying sandy clay is dominated by *Jadammina macrescens* with low frequencies of *Miliammina fusca*, *Haynesina germanica* and *Ammonia beccarii* var. *limnetes*. The peat, which is further dominated by *J. macrescens*, replaces this unit. These assemblages suggest a higher–high saltmarsh environment. Above the peat unit, in the clayey peat, the percentage of *J. macrescens* falls as the abundance of *M. fusca* and numerous calcareous species increases. *Haynesina germanica* and *Elphidium* species, which suggest a tidal flat environment, dominate the clayey silt.

Sample WA953/539 is taken from the regressive contact between the sandy clay and overlying peat at a depth of 539 cm. Sample WA953/526 is taken from the transgressive contact at 526 cm depth, between a peat and the overlying clayey silt. Lithostratigraphical and biostratigraphical evidence indicates both samples are *Phragmites* or monocot peats resting directly above (WA953/539) and below (WA953/526) clastic saltmarsh deposits. From these data, LBA estimates the indicative meaning of WA953/539 to be 2.55 ± 0.20 m OD, whilst the WA953/526 is predicted to be 2.25 ± 0.20 m OD.

The FBTF described by Horton (1997) and Horton *et al.* (1999*a*) were used to produce predicted SWLIs from the fossil assemblages present within the core. In this case, analyses were not restricted to single levels at the transgressive or regressive contacts, but rather consider the sediment column as a whole or at least those parts containing sufficient foraminifera (Fig. 2). These data indicate that the maximum reference water-level occurs within the peat between 539 and 534 cm depth, corresponding to an estimated indicative meaning of 2.59 ± 0.27 m OD (local MHWST is at 2.45 m OD). The estimates of indicative meaning decline within the upper clayey peat and clayey silt to reach a minimum of 1.44 ± 0.27 m OD at 512 cm depth. However, the SWLIs of samples from the lower clastic unit are exceptionally high: indeed the sample at 541 cm depth is above MHWST. This is caused by the low abundance of calcareous foraminifera within the clastic unit and hence, is dominated by agglutinated foraminifera (*J. macrescens*) with optima high in the tidal range. As a result, the change in SWLI across the regressive contact is negligible compared to the transgressive contact. In contrast to the reference water-level, the indicative range only varies through the core from ±0.26 to ±0.27 m because all fossil samples possess good modern analogues.

The FBTF assigns sample WA953/539 an indicative meaning of 2.59 ± 0.27 m OD, whilst sample WA953/526 is given an estimate of 2.29 ± 0.26 m OD. A comparison with the estimates produced by LBA indicates that the reference water-levels produced by FBTF are both 4 cm higher, whilst the indicative ranges are 6 and 7 cm larger for WA953/526 and WA593/539, respectively.

Basal peat

The second set of index points (LM5/1295 and LM5/1287) were produced from a basal peat collected at Theddlethorpe, Lincolnshire Marshes, UK. The field site [TF 4823 8716] is situated 1 km southeast of Theddlethorpe and approximately 1.5 km landward of the Lincolnshire coast (Fig. 1). A piston core LM5 was extruded for lithostratigraphical and biostratigraphical analyses.

The lithostratigraphy (Fig. 3) comprises a glacial diamicton at the base of the core, overlain by a thin, well-humified basal peat found between 1295 and 1287 cm below the ground surface. The peat is overlain by an olive-grey silty clay with some organic material and numerous mollusc fragments.

The biostratigraphical data (Fig. 3) demonstrate that foraminiferal tests are absent from the diamicton and the base of the peat. The agglutinated species *J. macrescens*, *M. fusca* and *T. inflata* dominate the remaining peat samples, suggesting a saltmarsh environment. In the silty clay, the agglutinated species are rapidly replaced by calcareous species. The abundance of calcareous foraminifera continues to increase within the silty clay and is dominated by *A. beccarii*, *H. germanica* and *Elphidium* species, which suggest a tidal flat environment.

Sample LM5/1295 is taken at a depth of 1295 cm from a basis peat directly below a *Phragmites* or clastic saltmarsh deposit. Sample LM5/1287 is taken at a depth of 1287 cm from a *Phragmites* or monocot peat directly below a clastic saltmarsh deposit. From these data, LBA estimates the indicative meaning of LM5/1295 as 3.25 ± 0.20 m OD and assigns LM5/1287 a value of 3.05 ± 0.20 m OD.

Once again, the fossil assemblages present within the core were used to estimate the indicative meanings *via* FBTF for a range of samples (Fig. 3). These data indicate that the maximum tide level occurs at a depth of 1291 cm within the basal peat and is assigned an indicative meaning of 3.18 ± 0.35 m OD, which is above local MHWST (3.15 m OD). Values of SWLI decline markedly above the transgressive contact and continue to fall within the silty clay to a minimum of 0.93 ± 0.40 m OD at a depth of 1265 cm.

The absence of foraminifera in sample LM5/1295 precludes an estimation of indicative meaning *via* FBTF, but a value of 3.10 ± 0.35 m OD is given for LM5/1287. In this instance, the reference water level and indicative range of FBTF are, respectively, 5 cm higher and 15 cm larger than those produced by LBA.

The estimation of indicative meaning from clastic sequences

In the past, collection of material for SLIs has concentrated on the transition between organic and minerogenic sedimentation (see Table 1) since the indicative meaning of these stratigraphic contexts can be defined using LBA. The LBA did not attempt to define the indicative meaning of minerogenic sediments because a suitable dating technique was not then available. Consequently, current understanding of sea-level change is based upon data from a very restricted range of depositional conditions.

However, the development of the accelerator mass spectrometry (AMS) technique for radiocarbon dating has greatly increased the range of

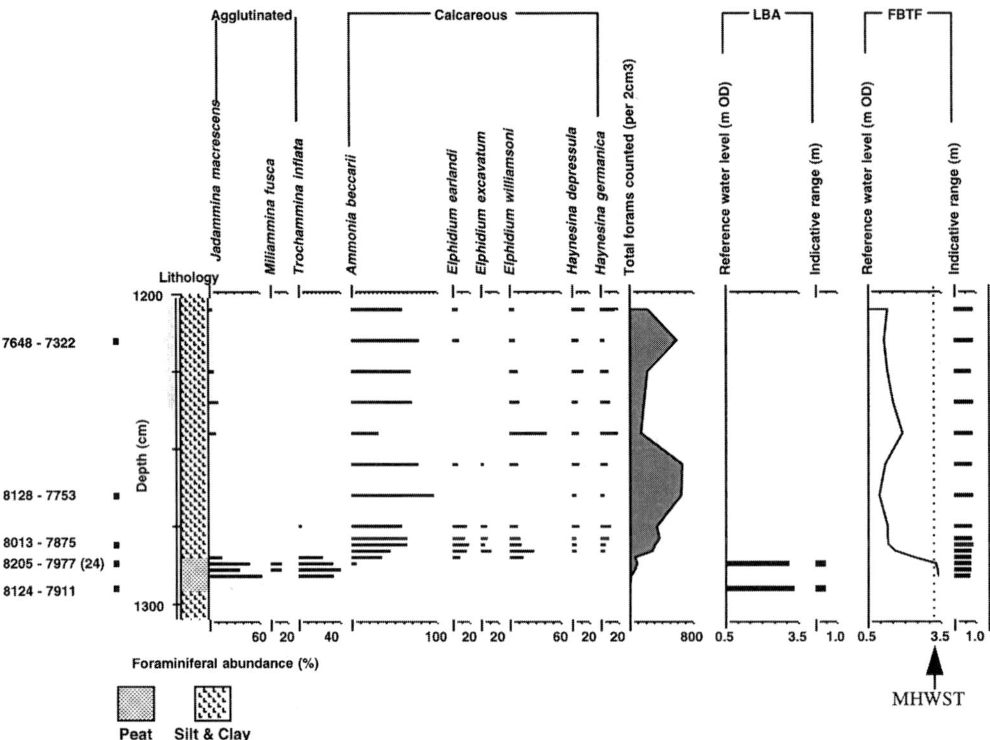

Fig. 3. Theddlethorpe LM5 foraminiferal diagram. Foraminiferal abundance is calculated as a percentage of dead foraminiferal tests (only species greater than 10% are shown). Reference water-levels and error ranges are estimated using a lithological-based approach (LBA) and a foraminiferal-based transfer function (FBFT). The radiocarbon dates (expressed in calibrated years BP), sea-level index point number (used in Fig. 4) and MHWST are shown. The stratigraphy is drawn according to Troels-Smith (1955).

datable sedimentary deposits (Hajdas *et al.* 1995; Jiang *et al.* 1997). Previous research has employed the calcium carbonate of mollusc and foraminifera as sources of AMS dates from inner-shelf, outer-shelf and deep sea sediments (e.g. Austin *et al.*, 1995; Heier-Nielsen 1995; Kristensen *et al.* 1995). In a similar way, calcareous foraminiferal assemblages can be used to produce both age estimates and indicative meanings from intertidal clastic sequences *via* FBTF. To compare the calcareous marine ^{14}C ages with the continental chronologies from 'peat dates', all ^{14}C dates obtained on foraminifera were corrected for the 'marine reservoir age effect'. It was assumed that the 400 year. reservoir effect (R) estimated by Bard *et al.* (1990) for North Sea water was applicable to the river–atmosphere–coast study (RACS) area. However, this estimate was from the open-ocean samples and, therefore, may be very different in coastal regions. Nevertheless, this value was applied to all calcareous AMS dates with the simplifying assumption of a deviation $\Delta R = 0$ (Heier-Nielsen 1995; Kristensen, *et al.* 1995; Jiang *et al.* 1997).

Three minerogenic samples were selected from core LM5 described above, and foraminifera were recovered for AMS dating (Table 4). The dated material from LM5/1215 and LM5/1281 consisted of monospecific assemblages of *A. beccarii*, whilst the sample from LM5/1265 contained a mixed foraminiferal assemblage including *A. beccarii*, and *Haynesina* and *Elphidium* species. Each foraminiferal sample was paired with a sample of the gastropod *Hydrobia ulvae* in order to distinguish between possible causes of observed age anomalies. It is assumed that *H. ulvae* are *in situ* (Heier-Nielsen 1995; Kristensen *et al.* 1995).

The age differences of the foraminiferal and *H. ulvae* dates of core LM5 are shown in Table 4. The mean foraminiferal dates of depths 1215 and 1265 cm are 181 and 220 cal. a BP younger than the equivalent *H. ulvae* dates, respectively, whereas the mean foraminiferal date of 1281 cm is 124 cal. a BP older. However, the age ranges

Table 4. *Radiocarbon dates from Theddlethorpe, Lincolnshire Marshes (the dates are corrected a marine reservoir of 400 years)*

Sample	Laboratory code	Age (^{14}C a BP ± 1σ)	Calibrated age (cal. a BP ± 2σ)	δ^{13}C (±0.1‰)	RWL‡ (m OD)
LM5b/1215*	AA23938	6650 ± 110	7648–7322	−3.8	1.12 ± 0.40
LM5b/1215b†	AA23937	6860 ± 70	7793–7539	−2.7	1.12 ± 0.40
LM5b/1265*	AA23940	7170 ± 95	8128–7753	−4.9	0.93 ± 0.40
LM5b/1265b†	AA23941	7260 ± 100	8287–8033	−5.2	0.93 ± 0.40
LM5b/1281a*	AA23943	7155 ± 55	8013–7875	−6.2	1.28 ± 0.40
LM5b/1281b†	AA23942	7060 ± 65	7945–7694	−4.4	1.28 ± 0.40

* a, foraminiferal dates.
† b, hydrobia ulvae dates.
‡ RWL, present altitude of reference water-level and indicative range.

overlap at ±2 standard deviations. Furthermore, the age anomalies from LM5 are less than many previous studies (Heier-Nielsen 1995; Kristensen *et al.* 1995; Knudsen *et al.* 1996), which suggests that the AMS dating of calcareous foraminiferal assemblages from intertidal sediment is a viable addition to the traditional use of transgressive and regressive contacts in sea-level chronologies.

Discussion

A comparison of indicative meanings estimated by LBA and FBTF

The case studies above are illustrations of intercalated and basal peat sequences. The LBA and FBTF are now applied to 52 SLIs collected as part of the LOIS project. The results are divided into three groups on the basis of their stratigraphic context: transgressive; regressive, and basis. Comparison of the two methods is summarized in Table 5 and reveals that the indicative ranges of FBTF are greater than LBA in all contexts with an average difference of 0.14 ± 0.09 m. The magnitude of the difference between methods is most pronounced in macrotidal areas (e.g. Cowbit F16, where the tidal range is 8.56 m and the difference amounts to 0.23 m).

The cause of these observed differences in estimates of indicative range becomes apparent when the details of each method are examined. The data used to develop LBA were originally developed with the objective of reconstructing Holocene sea-level changes in the Fenland (Shennan 1982, 1986). The indicative range is selected from a variety of constants depending upon the stratigraphic context of a sample (Table 1). The use of constant values is equivocal when applied to areas outside the Fenland since, intuitively, the vertical range of an indicator will be greater in areas with larger tidal ranges. For example, Zong (1993) calculated the indicative range of a *Phragmites* or monocot peat directly below a clastic saltmarsh deposit in Morecambe Bay to be 60 cm (40 cm larger

Table 5. *Summary statistics for calculating the indicative meaning for the total combined dataset and transgressive, regressive and basis subsets using a lithological-based approach (LBA) and a foraminiferal-based transfer function (FBTF)**

Classification	RWL† (m)	IR‡ (m)		
	D§ (II-I)	LBA	FBTF	D§ (II-I)
Total	0.11 ± 0.16	0.20	0.35 ± 0.08	0.14 ± 0.09
Transgressive	0.19 ± 0.12	0.20	0.35 ± 0.10	0.16 ± 0.10
Regressive	−0.05 ± 0.10	0.20	0.34 ± 0.06	0.14 ± 0.06
Basis	0.04 ± 0.01	0.20	0.30 ± 0.04	0.10 ± 0.04

* The mean and standard deviation are shown.
† RWL, present altitude of reference water-level.
‡ IR, indicative range.
§ D, Difference between methods.

than LBA). This difference is best attributed to the relatively large tidal amplitude in Morecambe Bay (*c.* 10.5 m) when compared with the Fenland (*c.* 8.5 m).

In contrast, the data used to compile FBTF are derived from ten sites that vary from microtidal to macrotidal in nature (Horton 1997; Horton *et al.* 1999*a*, *b*). Furthermore, estimates of indicative range are expressed in terms of a SWLI interval that contains within it an inherent consideration of tidal characteristics. For this reason, the vertical interval of the indicative range estimated by FBTF is greater.

A second important feature is the altitude of the reference water levels. Shennan (1982, 1986) stressed that the calculation of the indicative meaning is dependent on the type of stratigraphic contact under consideration. Furthermore, the reference water-level should be expressed as a mathematical function of tidal parameters (e.g. the mid-point between MHWST and MTL). This may account for the fact that, in general, differences between methods for estimates of reference water level are small (Table 5). Nevertheless, the reference water levels estimated by FBTF are slightly lower in regressive contexts and significantly higher in transgressive ones. The basis dates are near-identical.

The altitudes of transgressive (Fig. 4a) dates for FBTF are 0.19 ± 0.12 m higher than LBA. The LBA estimates the tide level for transgressive dates as MHWST minus 20 cm (Table 1). This tide level approximates the *in situ* drowning of a saltmarsh or freshwater marsh and the formation of a tidal flat. The tide levels for transgressive dates using FBTF occur at higher altitudes because of properties of the fossil data and problems associated with the contemporary data.

Firstly, it is generally acknowledged within fossil deposits that there is a lag between a movement of RSL and the response of a saltmarsh to this change. Lag effects have been identified in modelling studies of saltmarsh response to fluctuations in RSL (Allen 1990; 1995; French 1993; French *et al.* 1995). Furthermore, it is acknowledged that biostratigraphic indicators such as foraminifera may respond more rapidly to these changes (Long 1992; Allen 1995; Reed 1995). The results of this study appear to demonstrate the more rapid response time of foraminiferal assemblages relative to lithostratigraphic indicators because FBTF estimates the tide level for a foraminiferal assemblage at one point in time and space and not for a change in assemblage from one environment to another. In this way during a transgressive event the foraminifera are capable of recording its early stages and, therefore, yield higher reference water level estimates than those from the lithostratigraphic response to a change in depositional environment.

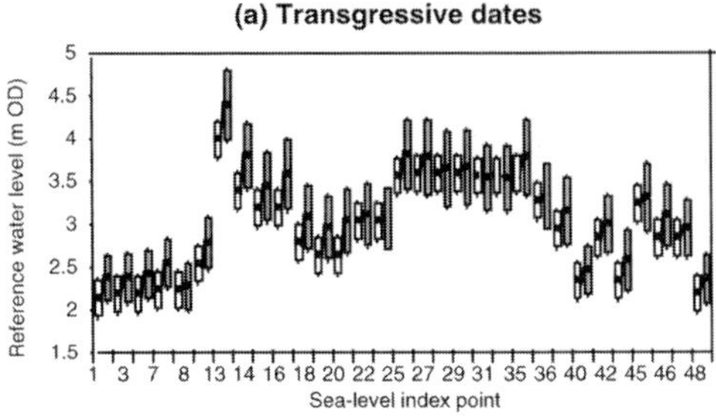

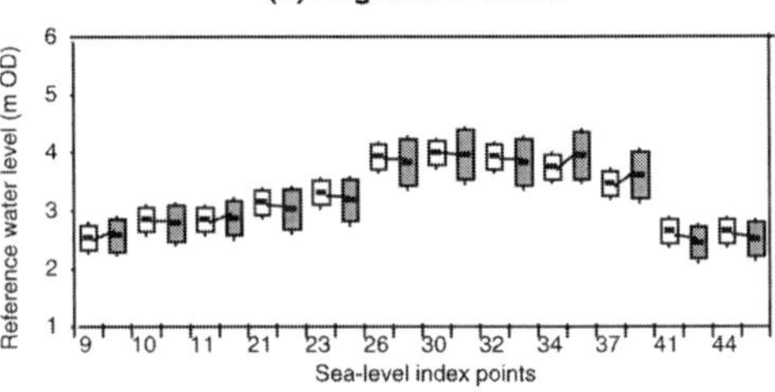

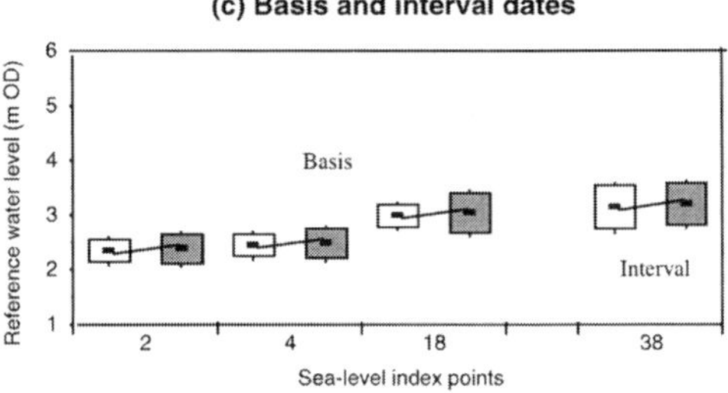

Fig. 4. The indicative meanings of (**a**) transgressive, (**b**) regressive and (**c**) basis dates using lithological-based approach (clear) and foraminiferal-based transfer function (shaded). Their reference water-levels and indicative ranges are shown. See Table 3 for SLI numbers.

Secondly, the contemporary data set consists of two sites with a modern analogue of a transgressive coastline. Arne Peninsula and Bury Farm are located on the south coast which is experiencing regional subsidence of between 0.2 and 1.4 mm a^{-1} (Shennan 1989). These sites contribute only 16% of the contemporary data. Thus, indicative meanings calculated by FBTF may not have sufficient data to differentiate between transgressive and regressive dates (Horton *et al.* 1999*a*).

The reference water-levels of FBTF are comparable to LBA for regressive dates (Fig. 4b).

The LBA estimates the tide level of regressive contacts to be the mid-point between MHWST and HAT minus 20 cm. This tide level corresponds to *in situ* saltmarsh growth. The tide levels for regressive dates using FBTF occur at similar altitudes because the majority of marshes within the contemporary data set are from regressive shorelines (for example, Nith Estuary and Kentra Bay). However, it should be noted that, whilst on average FBTF estimates are lower than those of LBA by only 0.05 ± 0.10 m the difference can be up to 0.20 m. The lower altitudes of FBTF may also be attributed to a lag between RSL and saltmarsh response. FBTF will record lower estimates of reference water-levels in regressive scenarios because of the reduced lag.

Implications for Holocene sea-level studies

Allen (1995) noted that model simulations of saltmarsh accretion indicate not only the existence of a lag between a change in RSL and its recognition in a sedimentary sequence, but also that the magnitude of this lag is variable. Furthermore, the altitude at which lithostratigraphic contacts form is sensitive to forcing factors such as the turbidity of tidal waters and consequently these facies have no fixed indicative meaning and poorly constrain the timing and magnitude of changing RSL. Allen (1995) suggested that future research should seek to obtain sea-level information from a greater variety of stratigraphic settings, and that the development of microfossil-based transfer functions to reconstruct a tide level were a way of achieving this goal. The FBTF employed here is an example of this advocated approach since it directly fixes samples relative to the tidal frame. However, the discrepancies of indicative meanings between LBA and FBTF are in the order of centimetres to decimetres and as such may not be a significant source of error in classic sea-level studies concerned with change throughout the Holocene as a whole. In these instances it is not uncommon to be dealing with vertical changes in sea level of many metres and so the errors introduced between reconstruction methods will be proportionally very small. However, decimetre-scale variations are important when identifying oscillations in RSL rise during the Holocene (Nydick *et al.* 1995). Furthermore, a new generation of high-resolution sea-level studies is seeking to quantify low magnitude changes in RSL that have occurred during the last 2–3 ka where decimetre-scale variations are of interest (Varekamp *et al.* 1992; Allen 1997; Edwards 1998). These studies commonly employ saltmarsh foraminifera as sea-level indicators and are, therefore, well suited to the adoption of FBTF-type approach.

Aside from the increases in precision outlined above, another important contribution of FBTF to sea-level studies is the establishment of radiocarbon-dated SLIs from minerogenic sediments. Calcareous foraminifera from intertidal environments can be used to produce indicative meanings and supply material for AMS radiocarbon dating. Therefore, FBTF expands the range of stratigraphic sequences that can be employed in sea-level reconstruction by redressing the over-reliance on transgressive and regressive contacts.

Whilst FBTFs offer a range of advantages over traditional methods of estimating indicative meaning, there are some problems associated with such an approach. Limitations regarding the contemporary data set and the quantitative techniques were discussed by Horton (1997) and Horton *et al.* (1999*a*). In particular, there is at present an uneven vertical distribution of samples within the contemporary data. The majority of samples (85%) are taken from above a SWLI of 160 (approximately mean high water neap tide). Furthermore, there are possible problems associated with infaunal lifestyle and bioturbation of foraminifera (Collison 1980; Saffert & Thomas 1998; Horton *et al.* 1999*b*).

There are also problems associated with the fossil data. Firstly, the majority of SLIs collected by the ERC and other LOIS partners are not based on foraminifera. Either they have been sampled for other biostratigraphical techniques (e.g. diatoms), or if they have been sampled, the foraminiferal tests are in low number or absent. The FBTF is only applicable to sample counts greater than 40 individuals. Secondly, some foraminiferal-predicted changes in SWLI between peat and clastic units are surprisingly low, contradicting the other lithostratigraphical and biostratigraphical evidence (e.g. regressive contact of WA953). An absence or low abundance of calcareous foraminifera within clastic units can produce an insensitivity to changes in SWLI, since fossil assemblages are dominated by agglutinated specimens with optima high in the tidal range. Such agglutinated dominance can arise in clastic units originally characterized by calcareous taxa where dissolution of tests has occurred (Scott & Medioli 1980; Scott & Leckie 1990; Jennings & Nelson 1992; Green *et al.* 1993). In such instances, the foraminiferal assemblages violate one of the basic assumptions of quantitative palaeoenvironmental reconstructions, which states that the taxa in the contemporary data set and their ecological responses to the environmental variable(s) of interest have not changed significantly over the time span represented by the fossil assemblage (Imbrie & Kipp 1971; Birks *et al.* 1990; Birks 1995). Whilst this problem may affect certain portions of the stratigraphic record, it is unlikely to be a significant factor in the highest marsh environments where calcareous foraminifera are absent, or in the low intertidal–subtidal environments where preservation of calcareous tests is usually high (Horton 1997; Horton *et al.* 1999*b*). In fact, SLIs that are established in minerogenic settings require large numbers of calcareous foraminifera for dating and will by necessity only come from areas where dissolution is not significant.

Foraminiferal-based statistical methods of sea-level reconstruction, such as FBTF, have advantages in terms of precision, speed of response and applicability over traditional methods currently employed in sea-level research. With an increasing demand for high-resolution studies of sea-level change, particularly those seeking to quantify the link between variations in ocean level and climate, it is imperative that the new generation of techniques employed are of the highest possible precision and accuracy. A vital step towards the realization of this aim is the collection of more surface data to improve the range of modern analogues covered by contemporary data sets.

Conclusion

(1) Fifty-two SLIs were collected from samples within the LOIS study area. The indicative meaning of each index point has been estimated where possible by two contrasting methods, a traditional LBA routinely employed in sea-level studies, and a FBTF developed by Horton (1997) and Horton *et al.* (1999*a*). A comparison of the two methods reveals that the indicative range of FBTF is larger than LBA and that reference water level estimates of FBTF are on average 0.19 ± 0.12 m higher than those of LBA for samples from transgressive contacts. Whilst these inter-method differences are small in magnitude, they are comparable in size to the scale of changes under investigation by recent high-resolution sea-level studies. In contrast, the reference water-levels of both methods are comparable for regressive and basis index points.

(2) An advantage of FBTF over traditional approaches is its ability to reconstruct water-level changes from minerogenic sediments. Calcareous foraminifera from intertidal environments can be used to produce indicative meanings and supply material for AMS radiocarbon dating. In this way, FBTF expands the range of stratigraphic sequences that can be employed in sea-level reconstruction by redressing the over-reliance on transgressive and regressive contacts.

This research is supported by the Land–Ocean Interaction Study (LOIS) Community Research Programme (LOIS publication no. 541), carried out under a special topic award from the National Environment Research Council (Contract number GST/02/0761). Special acknowledgements are given to: S. Grayson, F. Green and J. R. Kirby for their skills in the field; I. Shennan, A. J. Long, Y. Zong, H. J. B. Birks, D. S. Brew and S. Juggins for their suggestions; and the two reviewers, J. Murray and E. Thomas, for their comments. The authors also thank the Environmental Research Centre, Quaternary Laboratory and Cartographic section of the Department of Geography, University of Durham.

References

ALLEN, J. R. L. 1990. Salt-marsh growth and stratification: a numerical model with special reference to the Severn Estuary, southwest Britain. *Marine Geology*, **95**, 77–96.

——1995. Salt-marsh growth and fluctuating sea level: implications of a simulation model for Flandrian coastal stratigraphy and peat based sea-level curves. *Sedimentary Geology*, **100**, 21–45.

——1997. On the minimum amplitude of regional sea-level fluctuations during the Flandrian. *Journal of Quaternary Science*, **12**, 501–505.

AUSTIN, W. E. N., BARD, E., HUNT, J., KROON, D. & PEACOCK, D. 1995. The ^{14}C age of the Icelandic Vedde Ash: implications for Younger Dryas marine reservoir age corrections. *Radiocarbon*, **37**, 53–62.

BARD, E., LABEYRIE, L. D., PICHON, J. J., LABRACHERIE, M., ARNOLD, M., MOYES, J. & DUPLESSY, J. C. 1990. The last deglaciation in the southern and northern hemispheres: a comparison based on oxygen isotopes, sea surface temperature estimates and accelerator ^{14}C dating from deep-sea sediments. *In*: BLEIL, U. & THIEDE, J. (eds) *Geological History of the Polar Oceans: Arctic versus Antarctic*. Kluwer Academic Press, Dordrecht, 405–415.

BIRKS, H. J. B. 1995. Quantitative palaeoenvironmental reconstructions. *In*: MADDY, D. & BREW, J. S. (eds) *Statistical Modelling of Quaternary Science Data*. Technical Guide No. 5 Quaternary Research Association, Cambridge, 161–236.

——, LINE, J. M., JUGGINS, S., STEVENSON, A. C. & TER BRAAK, C. J. F. 1990. Diatom and pH reconstruction. *Philosophical Transactions of the Royal Society of London*, **327**, 263–278.

COLLISON, P. 1980. Vertical distribution of foraminifera off the coast of Northumberland, England. *Journal of Foraminiferal Research*, **10**, 75–78.

DE RIJK, S. 1995. *Agglutinated Foraminifera as Indicators of Salt Marsh Development in Relation to Late Holocene Sea-Level Rise*. PhD thesis, Free University, Amsterdam.

EDWARDS, R. J. 1998. *Late Holocene Relative Sea-level Change and Climate in Southern Britain*. PhD thesis, University of Durham.

FRENCH, J. R. 1993. Numerical simulations of vertical marsh growth and adjustments to accelerated sea-level rise, north Norfolk, UK. *Earth Surface Processes and Landforms*, **18**, 63–83.

——, SPENCER, T., MURRAY, A. & MOELLER, I. 1995. *Aspects of the Geomorphology and Ecology of the North Norfolk Coast*. Spring field meeting.

GASSE, F., BARKER, P., GELL, P. A., FRITZ, S. C. & CHALIE, F. 1997. Diatom-inferred salinity in palaeolakes: An indirect tracer of climate change. *Quaternary Science Reviews*, **16**, 547–563.

GREEN, M. A., ALLER, R. C. & ALLER, J. Y. 1993. Carbonate dissolution and temporal abundances of foraminifera in Long Island Sound sediments. *Limnology and Oceanography*, **38**, 331–345.

GUILBAULT, J., CLAGUE, J. J. & LAPOINTE, M. 1995. Amount of subsidence during a Late Holocene earthquake – evidence from fossil tidal marsh foraminifera at Vancouver Island, west coast of Canada. *Palaeogeography, Palaeoclimatology, Palaeoecology*, **118**, 49–71.

——, ——, ——1996. Foraminiferal evidence for the amount of coseismic subsidence during a Late Holocene earthquake on Vancouver Island, west coast of Canada. *Quaternary Science Reviews*, **15**, 913–937.

HAJDAS, I., ZOLITSCHAKA, B., IVY-OCH, S. D., BEER, J., LEROY, S. A. G., NEGENDANK, J. W., RAMRATH, M. & SUTER, M. 1995. AMS radiocarbon dating of annually laminated sediments from lake Holzmaar, Germany. *Quaternary Science Reviews*, **14**, 137–143.

HEIER-NIELSEN, S. 1995. *The Improvement of Marine Sediment Chronology by Comparative AMS ^{14}C Dating*. PhD thesis, University of Aarhus.

HORTON, B. P. 1997. *Quantification of the Indicative Meaning of a Range of Holocene Sea-level Index Points from the Western North Sea*. PhD thesis, University of Durham.

——, EDWARDS, R. J. & LLOYD, J. M. 1999*a*. Reconstruction of former sea-levels using a foraminiferal-based transfer function. *Journal of Foraminiferal Research*, **29**, 117–129.

——, ——, ——1999*b*. UK intertidal foraminiferal distributions: implications for sea-level studies. *Marine Micropaleontology*, **36**, 205–223.

IMBRIE, J. & KIPP, N. G. 1971. A new micropalaeontological method for quantitative paleoclimatology: application to a late Pleistocene Caribbean core. *In*: TUREKIAN, K. K. (ed.) *The Late Cenozoic Glacial Ages*. Yale University Press, New Haven and London, 71–181.

JARDINE, W. G. 1986. Determination of altitude. *In*: van de PLASSCHE, O. (ed.) *Sea-level Research: a Manual for the Collection and Evaluation of Data*. Geobooks, Norwich, 569–590.

JELGERSMA, S. 1961. Holocene sea-level changes in the Holocene. *Mededelingen van die Geologische Stichting*, **CVI**, 1–100.

JENNINGS, A. E. & NELSON, A. R. 1992. Foraminiferal assemblage zones in Oregon tidal marshes – relation to marsh floral zones and sea level. *Journal of Foraminiferal Research*, **22**, 13–29.

JIANG, H., BJORCK, S. & KNUDSEN, K. L. 1997. A palaeoclimatic and palaeoceanographic record of the last 11 000 ^{14}C years from the Skagerrak–Kattegat, northeastern Atlantic margin. *The Holocene*, **7**, 301–311.

JONES, V. J. & JUGGINS, S. 1995. The construction of diatom-based chlorophyll transfer function and its application at three lakes on Signy Island (maritime Antarctic) subject to differing degrees of nutrient enrichment. *Freshwater Biology*, **34**, 433–445.

JUGGINS, S. & TER BRAAK, C. J. F. 1997. *CALIBRATE*, Department of Geography, University of Newcastle.

KNUDSEN, K. L., CONRADSEN, K., HEIER-NIELSEN, S. & SEIDENKRANTZ, M.-S. 1996. Paleoenvironments in the Skagerrak-Kattegat basin in the eastern North Sea during the last deglaciation. *Boreas*, **25**, 65–77.

KRISTENSEN, P. S., HEIER-NIELSEN, S. & HYLLEBERG, J. 1995. Late-Holocene salinity fluctuations in Bjornsholm Bay, Limfjorden, Denmark, as deduced from micro- and macrofossil analysis. *The Holocene*, **5**, 313–322.

LONG, A. J. 1992. Coastal responses to changes in sea level in the East Kent Fens and southeast England, UK over the last 7500 years. *Proceedings of the Geologists' Association*, **103**, 187–199.

——, INNES, J. B., KIRBY, J. R., LLOYD, J. M., RUTHERFORD, M. M., SHENNAN, I. & TOOLEY, M. J. 1998. Holocene sea-level change and coastal evolution in the Humber Estuary, eastern England: an assessment of rapid coastal change. *The Holocene*, **8**, 229–247.

MURRAY, J. W. 1971. Living foraminiferids of tidal marshes: a review. *Journal of Foraminiferal Research*, **1**, 153–161.

——1979. *British Nearshore Foraminiferids*. Academic Press, London, 68pp.

—— & KASHIMA, K. 1993. Diatom zonation in southern Oregon tidal marshes relative to vascular plants, foraminifera and sea level. *Journal of Coastal Research*, **9**, 673–697.

NELSON, A. R., JENNINGS, A. E. & KASHIMA, K. 1996. An earthquake history derived from stratigraphic and microfossil evidence of relative sea-level change of Coos Bay, Southern Coastal Oregon. *Geological Society of American Bulletin*, **108**, 141–154.

NYDICK, K. R., BIDWELL, A. B., THOMAS, E. & VAREKAMP, J. C. 1995. A sea-level rise curve from Guildford Connecticut, USA. *Marine Geology*, **124**, 137–159.

PATTERSON, R. T. & FISHBEIN, E. 1989. Re-examination of the statistical methods used to determine the number of point counts needed for micropaleontological quantitative research. *Journal of Paleontology*, **63**, 245–248.

PREUSS, H. 1979. Progress in computer evaluation of sea level data within the IGCP Project No. 61. *In*: *Proceedings of the 1978 International Symposium of Coastal Evolution in the Quaternary*. Sao Paulo, Brazil, 104–134.

REED, D. J. 1995. The response of coastal marshes to sea-level rise. *Earth Surface Processes and Landforms*, **20**, 39–48.

SAFFERT, H. L. & THOMAS, E. 1998. Living foraminifera in salt marsh peat cores: Kelsey Marsh (Clinton, CT) and the Great Marshes (Barnstable, MA). *Marine Micropaleontology*, **33**, 175–202.

SCOTT, D. B. & MEDIOLI, F. S. 1978. Vertical zonation of marsh foraminifera as accurate indicators of former sea levels. *Nature*, **272**, 528–531.

—— & ——1980. Quantitative studies of marsh foraminifera distribution in Nova Scotia: implications for sea-level studies. *Journal of Foraminiferal Research, Special Publication*, **17**, 1–58.

—— & ——1986. Foraminifera as sea-level indicators. *In*: VAN DE PLASSCHE, O. (ed.) *Sea-level Research: a Manual for the Collection and Evaluation of Data*. Geobooks, Norwich, 435–456.

SCOTT, D. K. & LECKIE, R. M. 1990. Foraminiferal zonation of Great Sippwissett Salt Marsh (Falmouth, Massachusetts). *Journal of Foraminiferal Research*, **20**, 248–266.

SHENNAN, I. 1982. Interpretation of the Flandrian sea-level data from the Fenland, England. *Proceedings of the Geologists' Association*, **93**, 53–63.

——1986. Flandrian sea-level changes in the Fenland, II: Tendencies of sea-level movement, altitudinal changes and local and regional factors. *Journal of Quaternary Science*, **1**, 155–179.

——1989. Holocene crustal movements and sea-level changes in Great Britain. *Journal of Quaternary Science*, **4**, 77–89.

——, LONG, A. J., RUTHERFORD, M. M., GREEN, F. M., INNES, J. B., LLOYD, J. M., ZONG, Y. & WALKER, K. J. 1996. Tidal marsh stratigraphy, sea-level change and large earthquakes: a 5000 year record in Washington, USA. *Quaternary Science Reviews*, **15**, 1023–1059.

——, TOOLEY, M. J., DAVIS, M. J. & HAGGART, B. A. 1983. Analysis and interpretation of Holocene sea-level data. *Nature*, **302**, 404–406.

STUIVER, M. & REIMER, P. J. 1993. Extended ^{14}C database and revised CALIB 3.0 ^{14}C calibration program. *Radiocarbon*, **35**, 215–230.

THOMAS, E. & VAREKAMP, J. C. 1991. Palaeoenvironmental analysis of marsh sequences (Clifton, Connecticut): Evidence for punctuated rise in relative sea-level during the Holocene. *Journal of Coastal Research Special Issue*, **11**, 125–158.

TOOLEY, M. J. 1982. Sea-level changes in northern England. *Proceedings of the Geologists' Association*, **93**, 43–51.

——1986. Sea levels. *Progress in Physical Geography*, **10**, 120–129.

TROELS-SMITH, J. 1955. Characterisation of unconsolidated sediments. *Danmarks Geologiske Undersogelse Series, IV*, **3**, 1–73.

VAN DE PLASSCHE, O. 1982. *Sea-level Change and Water Movements in the Netherlands During the Holocene*. PhD thesis, Free University, Amsterdam.

—— (ed.) 1986. *Sea-level Research: a Manual for the Collection and Evaluation of Data*. Geobooks, Norwich. 617pp.

VAREKAMP, J. C., THOMAS, E. & VAN DE PLASSCHE, O. 1992. Relative sea-level rise and climate change over the last 1500 years. *Terra Nova*, **4**, 293–304.

ZONG, Y. 1993. *Flandrian sea-level changes and impacts of projected sea-level rises on the coastal lowlands of Morecombe Bay and the Thames Estuary, UK*. PhD Dissertation, University of Durham.

—— & HORTON, B. P. 1999. Diatom-based tidal-level transfer functions as an aid in reconstructing Quaternary history of sea-level movements in Britain. *Journal of Quaternary Science*, **14**, 153–167.

Luminescence dating of fine-grain Holocene sediments from a coastal setting

I. K. BAILIFF[1] & M. J. TOOLEY[2]

[1] *Luminescence Dating Laboratory, Environmental Research Centre, University of Durham, Dawson Building, South Road Durham DH1 3LE, UK (e-mail: Ian.Bailiff@durham.ac.uk)*
[2] *School of Geography, Kingston University, Kingston upon Thames KT1 2EE, UK*

Abstract: A detailed study of a core of Holocene age from a site (Adventurers' Land, F21) in the Fenlands, UK, based on the measurement of infra-red-stimulated luminescence (IRSL) with silt-sized fractions of water-laid deposits has yielded a sequence of 24 luminescence ages. An additive dose procedure was applied to evaluate the palaeodose, and the annual dose was assessed on the basis of radioactivity and elemental analyses of samples taken from the core. The proportions by weight of the mineral fraction of water and organic material within the sediments ranged from *c.* 20 to 170% for the former and between *c.* 10 and 25% for the latter. Overall, the sequence of IRSL age evaluations is consistent with the calibrated radiocarbon ages obtained for three intercalated organic horizons within the core. Although the luminescence ages for F21A do not possess a resolution that would enable rates of sedimentation between the organic sediments to be determined with confidence, the general trend of the luminescence ages with depth suggests a chronological resolution of the order of 1 ka or better.

Establishing a chronology for the emplacement of marine clastic sediments in coastal lowlands and on the continental shelf has hitherto relied upon the dating of *in situ* organic deposits from palaeolagoons and saltmarshes, and shells from palaeotidal flats (Tooley 1993). By obtaining radiocarbon age determinations for suitable organic material from the organic deposits above and below such clastic sediments, a rate of sedimentation can be obtained providing it is assumed that the rate was constant. However, from previous work (Tooley 1978*a*, *b*) it was clear from the changing physical properties, structures and micro- and macro-sub-fossil content of the clastic sediments within such sequences that there were significant changes in the rates of delivery of the sediments throughout the period. Consequently, to establish the chronology and rates of these facies changes new approaches to dating the clastic sediments directly are required, complementing the well-established and widely applied method of radiocarbon dating for the organic deposits. With such an objective in mind, this paper describes an experimental study designed to explore the potential of luminescence dating to provide absolute dates for the emplacement of clastic sediments from a coastal setting.

Luminescence dating techniques have been applied successfully to wind-blown deposits above the water-table and some work has been done on water-laid deposits of fluvial and lacustrine origin (for a recent review see Prescott & Robertson 1997). Work by Lian *et al.* (1995) with organic-rich coastal sediments from the western coastal region of North America established an important methodological foundation for the luminescence dating of coastal sediments of Quaternary age based on the use of thermoluminescence (TL) and optically stimulated luminescence (OSL) techniques. In the latter, infra-red stimulated luminescence (IRSL) was measured. Successful applications of IRSL techniques applied to the dating of different types of sediment from terrestrial archaeological deposits of Holocene age are reported by Lang & Wagner (1996) and Rees-Jones & Tite (1997). However, it was considered that further methodological research was required to evaluate the efficacy of dating coastal sediments of Holocene age for use in the Land–Ocean Interaction Study (LOIS). A consortium of three laboratories was established under the Land–Ocean Evolution Perspective Study (LOEPS) component of LOIS at the Universities of Aberystwyth, Durham and Sussex to address these problems.

From: SHENNAN, I. & ANDREWS, J. (eds) *Holocene Land–Ocean Interaction and Environmental Change around* the North Sea. Geological Society, London, Special Publications, **166**, 55–67. 1-86239-054-1/00/$15.00

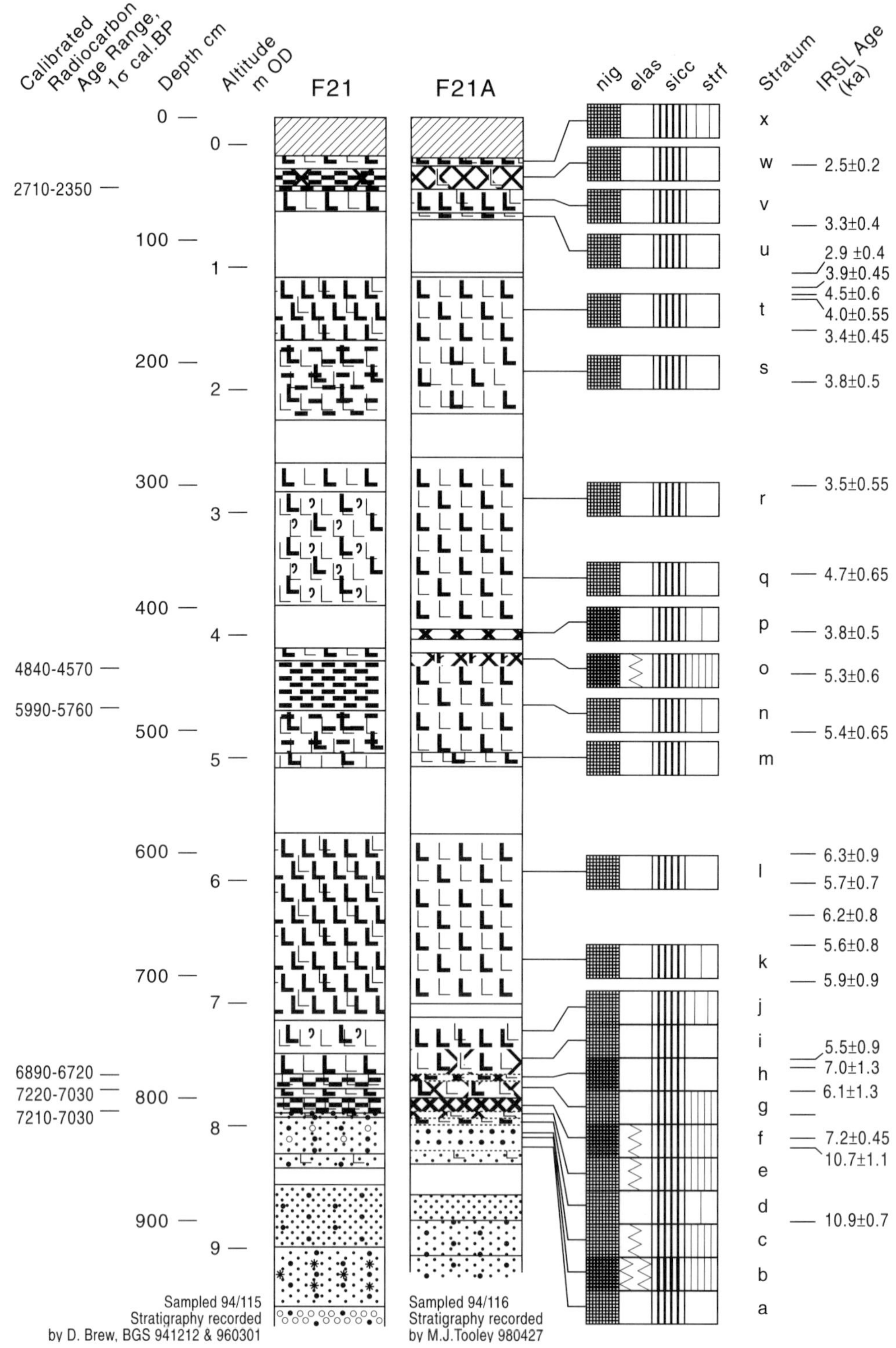

ADVENTURERS' LAND, CAMBRIDGESHIRE - TF35770185
Calibrated Radiocarbon Age Range, 1σ cal.BP
Depth cm
Altitude m OD
F21
F21A
nig
elas
sicc
strf
Stratum
IRSL Age (ka)
2710-2350
4840-4570
5990-5760
6890-6720
7220-7030
7210-7030
0
100
200
300
400
500
600
700
800
900
0
1
2
3
4
5
6
7
8
9
x
w
v
u
t
s
r
q
p
o
n
m
l
k
j
i
h
g
f
e
d
c
b
a
2.5±0.2
3.3±0.4
2.9 ±0.4
3.9±0.45
4.5±0.6
4.0±0.55
3.4±0.45
3.8±0.5
3.5±0.55
4.7±0.65
3.8±0.5
5.3±0.6
5.4±0.65
6.3±0.9
5.7±0.7
6.2±0.8
5.6±0.8
5.9±0.9
5.5±0.9
7.0±1.3
6.1±1.3
7.2±0.45
10.7±1.1
10.9±0.7
Sampled 94/115
Stratigraphy recorded
by D. Brew, BGS 941212 & 960301
Sampled 94/116
Stratigraphy recorded
by M.J.Tooley 980427

Four of the objectives within the research proposed by the consortium are given below.

(1) The development of a methodology for the dating of marine clastic sediments of Holocene age (last 10 ka) using luminescence techniques, especially optically stimulated luminescence.
(2) Identification of suitable minerals for dating by characterization of the luminescence properties of quartz and feldspar in alluvial, coastal marine and offshore sedimentary deposits in the river–atmosphere–coast study (RACS) area in eastern England (Berwick-upon-Tweed to Great Yarmouth).
(3) Testing of the methodology with sediments from a range of different facies and sites in the RACS area.
(4) The Aberystwyth (Orford *et al.* this volume) and Durham (Andrews *et al.* 1999) laboratories were to participate in dating programmes as part of the collaborative projects in the LOEPS parts of the LOIS research programme.

In this paper, the results of an investigation of the application of luminescence techniques to the dating of the silt-size fraction of sediments from a site in the Fenlands, with dating control provided by calibrated radiocarbon ages, are presented and evaluated. This site proved to be the only one amongst a large number sampled within the RACS area to provide the minimum of three *in situ* organic layers for dating control.

Stratigraphy and lithology

The area on Adventurers' Land [TF3577 0185] from where the cores were taken has long been known for its variety of organic and inorganic sediments. Skertchly (1877) recorded from the River Nene Valley unconsolidated sediments up to 6 m thick, 'in which the marine and freshwater beds have alternated in rapid succession' (p. 141). Godwin & Clifford (1938) incorporated Skertchly's results into their sections across the southern Fenland, based on a series of borings made by the River Nene Catchment Board from Dog-in-a-Doublet sluice to Pear Tree Hill (Godwin & Clifford 1938; Fig. 27, p. 370). They noted that the lower peat divided to form three peat beds separated by fen clay attaining maximum thicknesses overall of about 12 m down to about −10 m Ordnance Datum (OD). They concluded that, 'the area must therefore have been a centre of equilibrium about which the conditions of fresh and brackish water fluctuated repeatedly during the period of land sea-level change that culminated in the phase of the most widespread sheet of fen clay' (p. 372). It was these descriptions that led one of the authors (MJT) to the area in 1975, when some 725 cm of unconsolidated sediments were proved to −7.8 m OD at [TF3567 0182] and the existence of Godwin & Clifford's three lower peat beds confirmed (Tooley 1978*a*, p. 162). Subsequent sampling failed to recover the two intermediate organic beds, but a U4 core from the basal deposits yielded material for three radiocarbon assays undertaken by M. A. Geyh at Hanover. At an adjacent site Shennan (1980, 1982) recovered a core, using a modified Livingstone piston corer, from which a limited number of pollen and diatom analyses were obtained together with five radiocarbon age determinations.

Core F21 (British Geological Survey (BGS) registered borehole TF30SE/51) and an adjacent core F21A (BGS registered borehole TF30SE/50) were taken by the BGS: the latter was reserved for luminescence sampling and was consequently obtained in opaque liners. After extrusion, the cores were double-wrapped in opaque polythene liners, sealed and stored at 4°C. All subsequent sampling of the core was performed under dim red-lighting conditions and transported to the laboratory in opaque and water-tight containers. Since recording the stratigraphy under subdued red light was difficult, the stratigraphy of F21A was recorded (see Fig. 1) at the BGS in daylight, using the scheme proposed by Troels-Smith (1955).

Luminescence dating methodology

The nature of the Holocene sediments in core F21A imposes certain initial constraints concerning the type of luminescence technique that

Silt Clay Sand Gravel  Peat

Fig. 1. Stratigraphy and lithology of Cores F21 and F21A, Adventurers' Land, Cambridgeshire, as recorded using the scheme of Troels-Smith (1955). The representation of the main lithological units is given in the Key. The assessment of the physical properties of the sediment for core F21A is indicated to the right of the diagram where: *nig(ror)*, *elas(ictas)*, *sicc(itas)* and *strf(stratificatio)* indicate degrees of darkness, elasticity, dryness and stratification respectively and where degrees are shown by the density of the pattern on a five point scale.

may be applied. One of the main divisions in luminescence techniques is between the grain-size range selected for dating measurements; fine grains (*c.* 4–11 μm diameter) and coarse grains (>90 μm diameter). The nature of the silts in the intertidal deposits in core F21A dictates that a fine-grain approach is employed. Both the luminescent minerals, feldspar and quartz, are usually present in such sediments. It is possible to measure selectively luminescence from feldspars without mineral separation by use of near infra-red stimulation (800–900 nm), which results in the production of optically stimulated luminescence (OSL). Where infra-red stimulation is employed the luminescence is referred to as infra-red stimulated luminescence (IRSL), and measurement of IRSL under continuous stimulation as a function of time yields an IRSL decay curve (Fig. 2). The form of the decay curve reflects the liberation of charges trapped at defect sites (referred to as traps) in the crystal structure and their capture at other defect sites (referred to as luminescence centres) and the concomitant emission of luminescence.

The application of IRSL techniques to fine-grain samples has been favoured because of the generally much higher intensity of emissions from feldspars compared with quartz for grains in this size range and the avoidance of the need to use complex mineral separation techniques (Aitken 1998). The dating of water-lain sediments by OSL provides a significantly higher likelihood that the sediments were effectively zeroed before burial than could be obtained if thermally stimulated luminescence (TL) procedures were employed (see e.g. Bailiff 1992; Wintle 1997), although it should be noted that the verification of effective zeroing for fine-grain sediments by means of laboratory testing alone is equivocal.

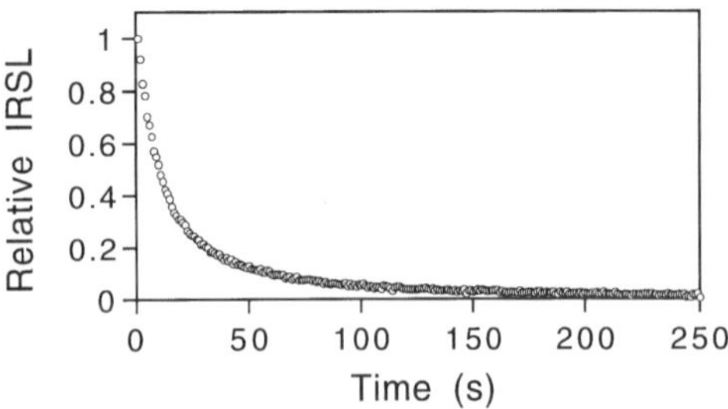

Fig. 2. Example of an IRSL decay curve recorded using a fine-grain disc prepared using sediment from core F21A. The luminescence is recorded under constant infra-red stimulation for a period of 250 s, during which the sample is held at a temperature of 75°C.

Principles of age evaluation

The luminescence age equation, expressed here in its simplest form

$$\text{Luminescence age} = \frac{\text{Palaeodose}}{\text{Annual dose}} = \frac{P}{\dot{D}_{\text{tot}}} \quad (1)$$

is evaluated by the experimental determination of the two principal physical quantities, the palaeodose, P, and the effective annual dose, $\dot{D}_{\text{tot}}$. In terms of age evaluation, the palaeodose is the radiation dose absorbed by mineral grains since burial (assuming no relict trapped charge) and is evaluated by the application of luminescence techniques to selected minerals extracted from the sediment.

The annual dose is the sum of the radiation dose arising from naturally occurring radionuclides within the sediment environment (the lithogenic component) and that arising from cosmic rays (the cosmogenic component). As the depth below the ground surface of the sampled location increases with age, the cosmic-ray dose-rate also progressively decreases (Prescott & Hutton 1994) with age. It is therefore not appropriate to assume an average value of cosmic dose-rate in Equation 1. Following Lian *et al.* (1995) the cumulative dose, D_{cos}, due to cosmic radiation can be estimated provided assumptions are made concerning the development of overburden with time. Equation 1 may be re-written as

$$\text{Luminescence age} = \frac{P - D_{\text{cos}}}{\dot{D}_{\alpha\beta\gamma}} \quad (2)$$

where $\dot{D}_{\alpha\beta\gamma}$ is the effective annual dose due to U, Th and K.

There are several factors related to the nature of the deposition: the effectiveness of zeroing, burial history and the recovery of sediments, which potentially affect the evaluation of the age equation. By testing sediments with age control, such as those provided in core F21A, we aimed to test assumptions incorporated within currently formed experimental procedures. In the following sections are discussed: (a) the relevant laboratory procedures employed to determine the component quantities needed to evaluate the age equation; (b) the factors affecting the calculated luminescence age; (c) a comparison of the suite of luminescence ages with the chronological markers provided by the calibrated radiocarbon age ranges for the intercalated peats.

Experimental

For all the samples discussed in this paper, a variant of the fine-grain luminescence technique (see e.g. Wintle 1997) was employed, being the appropriate approach where silt-sized sediments are recovered.

Samples

Sub-samples (typically *c.* 30 g) taken for dating tests were cut from selected depth ranges (see Table 1) in the core and divided (vertically) into two sub-samples intended for measurement of: (a) moisture content and natural radioactivity; and (b) luminescence. In the majority of cases the core thickness of samples was 1 cm. An additional quantity of sediment was also taken from each sampled level for mineral composition analyses by powder X-ray diffraction (XRD). XRD analyses for samples taken from the main lithological units in the core indicated: (a) a dominant presence of quartz throughout the core; (b) the presence of calcite, dolomite, orthoclase and kaolinite from the sub-surface to a depth of *c.* 700 cm; (c) negligible feldspar at depths below 800 cm. Holt (pers. comm.) also measured SiO_2 values, and confirmed its dominance; in the pre-Holocene sediments at depths greater than 810 cm values in excess of 90% were revealed, whereas in the Holocene they were 54–63%. A check for evidence of chemical weathering (Parish 1994) and the general consistency of grain-size separation was also performed with a selection of samples from various depths using scanning electron microscopy (SEM). The sample condition was found to be satisfactory, with little evidence of chemical weathering.

Procedures

Preparation of samples. Samples intended for luminescence measurements were dried in air at 50°C, followed by gentle crushing and sieving to ensure the particles were less than 90 μm diameter. Standard H_2O_2 and HCl washing treatments were employed to remove organic matter and carbonates, respectively, if detected. The 4–11 μm fraction of the treated sediment was extracted by settling in acetone (no flocculation was encountered) and deposited onto aluminium alloy discs following the standard fine-grain procedure (Aitken 1998). Samples for annual dose assessment were weighed, dried at 50°C for at least 14 days and periodically weighed until a stable weight was obtained to determine the as-cored moisture content. The same samples were then prepared for measurements related to the determination of the annual

Table 1. *Luminescence ages, core F21A*

Sample No.	Depth (cm)	Luminescence age (ka)	Overall error (ka)	Random error (ka)
F21A-				
1	39–40	2.5	0.2	0.15
74	74–75	3.3	0.4	0.25
130	130–131	2.9	0.4	0.25
2AU	144–145	3.9	0.45	0.3
2A	145–146	4.5	0.6	0.4
2AB	146–147	4.0	0.55	0.4
175	165–166	3.4	0.45	0.3
2B	213–214	3.8	0.5	0.2
300	300–301	3.5	0.55	0.2
350	350–351	4.7	0.65	0.4
3B	412–417	3.8	0.5	0.2
4	468–474	5.3	0.6	0.3
500	500–501	5.4	0.65	0.35
600	603–604	6.3	0.9	0.65
625	625–626	5.7	0.7	0.35
650	650–651	6.2	0.8	0.3
675	675–676	5.6	0.8	0.55
700	705–706	5.9	0.9	0.55
6A	769–774	5.5	0.9	0.3
775	775–776	7.0	1.3	0.6
6B1	790–795	6.1	1.3	0.55
6B3	829–834	7.2	0.45	0.3
840	840–841	10.7	1.1	1.0
900	900–901	10.9	0.7	0.5

dose by crushing (and subsequently ball-milling for 10–15 min). Following radioactivity measurements, the weight loss due to the presence of organic material and structural water was determined after each of two stages of ashing comprising 24 h at 500°C and 2 h at 950°C.

Luminescence measurements. The IRSL measurements were performed with a commercially available semi-automated reader (Risø National Laboratory type DA-12) using a near infra-red emitting diode (IRED) stimulation source (880Δ40 nm). The luminescence was detected after passing through a broad-band filter (6 mm thick Schott BG-39). Known doses of beta and alpha ionizing radiation were delivered using calibrated $^{90}Sr/^{90}Y$ and ^{241}Am sources, respectively; the former having been calibrated on several occasions against a Secondary Standard Dosimetry Laboratory gamma photon source (discussed most recently in Göksu *et al.* 1995) and the latter by comparison with a calibrated ^{241}Am source located at the Research Laboratory for Archaeology and the History of Art, at the University of Oxford.

The fine-grain (fg) discs were first subjected to a short-duration stimulation and measurement of IRSL (typ. 1s) to provide a means of correcting for differences in the quantity of luminescing minerals between aliquots (referred to as normalization). A correction factor for the depletion in trapped charge caused by the use of the normalization procedure was obtained by repeated short-duration IRSL measurements with a separate set of fg discs, which had not been subjected to a pre-annealing treatment (see below).

The palaeodose was evaluated using a multiple aliquot additive dose procedure from which additive beta and alpha dose-response curves were obtained. This procedure is based on characterizing the increase in luminescence to additional known radiation doses (delivered by either alpha or beta radiation) and referred to as a dose-response curve (Fig. 3). The selection of the range of additional (known) dose to be applied was based on establishing, at least, a threefold change in IRSL.

Following the completion of the administration of laboratory dose, all fg discs were subjected to a thermal pre-annealing treatment comprising storage at 150°C for 2 h followed by three days at 100°C and finally at least 24 h at ambient temperatures. The samples were placed in containers in a fixed configuration within the annealing oven, and the temperature history logged throughout. After the pre-annealing treatment the IRSL was measured at a sample temperature of 75°C for 250 s. The background (due to scattered light from the stimulation source and photomultiplier dark count) subtracted from the recorded IRSL signal was established following prolonged infra-red stimulation; typically, an exposure of 1 ks was sufficient to reduce the signal to a level close to instrumental background.

The dose-response curves were obtained using a procedure of data analysis based on the measurement of the integrated IRSL (0–100 s).

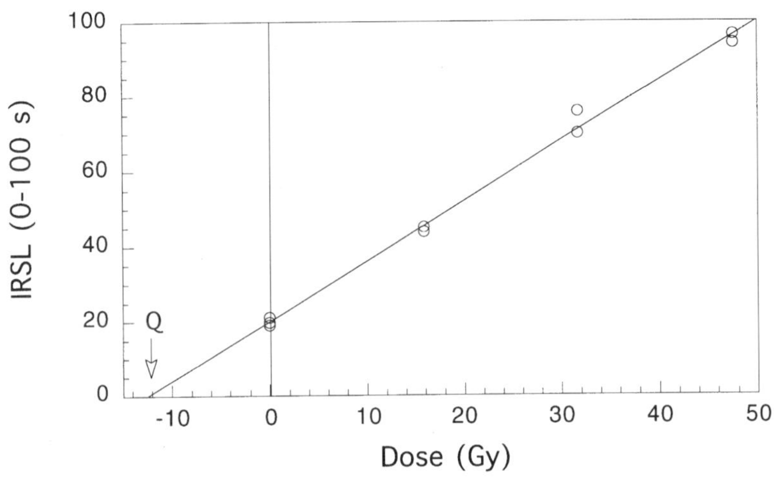

Fig. 3. An example of a dose-response curve that has been fitted to data obtained following the additive dose procedure. The interception of the extrapolated (linear) curve fitted to the data with the dose axis yields the value of Q as indicated.

Table 2. *Annual dose, cosmic dose and palaeodose values for samples from core F21A*

Sample No.	Annual dose, $\dot{D}_{\alpha\beta\gamma}$ (Gy ka^{-1})	α* (%)	β* (%)	γ* (%)	D_{cos} (Gy)	F_w† (%)	F_o† (%)	Q_β (Gy)	dQ (Gy)
F21A-									
1	3.43 ± 0.27	19	53	29	0.5	51	11	9.2	0.5
74	3.11 ± 0.29	18	51	31	0.65	64	9	11.0	1.4
130	2.70 ± 0.28	22	46	32	0.7	74	18	8.6	1.3
2AU	2.54 ± 0.26	19	46	34	0.8	73	9	10.7	0.7
2A	2.38 ± 0.24	21	50	29	0.8	71	7	11.4	0.6
2AB	2.58 ± 0.27	18	49	32	0.8	76	10	11.2	1.2
175	2.49 ± 0.28	17	51	32	0.8	84	14	9.4	0.8
2B	2.37 ± 0.30	17	56	26	0.8	94	9	9.8	0.3
300	2.63 ± 0.40	19	52	29	0.8	116	13	9.9	0.4
350	2.30 ± 0.26	18	51	31	0.85	82	9	11.8	0.8
3B	2.38 ± 0.31	18	52	30	0.85	98	15	9.9	0.4
4	2.87 ± 0.28	16	55	29	0.95	70	9	16.2	1.0
500	2.76 ± 0.32	17	53	30	0.95	84	13	15.8	1.4
600	2.32 ± 0.25	18	51	31	1.05	77	10	15.6	1.2
625	2.24 ± 0.25	18	51	31	1.05	81	10	13.8	0.5
650	2.14 ± 0.26	20	51	29	1.05	89	10	14.3	0.4
675	2.18 ± 0.24	18	50	31	1.05	79	2	13.2	1.2
700	2.17 ± 0.29	21	48	31	1.05	100	12	13.9	1.1
775	1.90 ± 0.33	24	48	28	1.1	132	22	14.4	0.9
6A	2.00 ± 0.35	19	53	28	1.1	132	25	12.0	0.9
6B1	1.01 ± 0.22	16	57	28	1.1	171	3	7.2	0.8
6B3	2.93 ± 0.14	26	43	32	1.1	16	5	22.3	1.0
840	2.98 ± 0.14	31	38	31	1.5	9	8	33.5	2.8
900	2.36 ± 0.11	23	47	30	1.5	14	5	27.3	1.0

* Percentage contributions to the annual dose due to the three main radiation types.
† The ratio of water (F_w) and organic material (F_o) to ashed sediment (950°C) by weight, respectively.

The values of the dose-axis intercepts of the additive beta dose and additive alpha dose-response curves are denoted here as Q_β and Q_α, respectively; the values of Q_β are listed in Table 2 for each sample. For all the samples listed in Table 2 the additive dose growth characteristics obtained using beta and alpha radiation were linear within experimental error and the quantities Q_β and Q_β were obtained by extrapolation of the two types of dose-response curve to the dose axis (Fig. 3). The extent of low dose supralinearity was checked in several cases and found not to be significant with the limits of experimental error; thus the value of Q_β was taken to correspond to the palaeodose. The a value (Aitken 1983) was calculated using the value of Q_α and the unsealed alpha count-rate; it is a measure of the efficiency of alpha relative to beta radiation in generating latent luminescence and is required to evaluate the component of annual dose due to alpha radiation (Aitken 1985). The a values (average value of 0.082 ± 0.009; standard deviation, s.d. 24) are consistent with previously measured values obtained for IRSL measurements with fine-grain sediment samples (e.g. Rees-Jones & Tite 1997); their values reflect the substantially reduced efficiency of alpha radiation (<10%) compared with beta or gamma radiation for the same absorbed dose.

Tests for the loss of latent signal due to anomalous fading (Wintle 1977), performed with a separate set of unirradiated and beta-irradiated fg discs, indicated no significant loss of latent IRSL signal. The IRSL was measured following storage for three to six months and the application of pre-annealing at the end of the storage period.

Luminescence ages. The luminescence ages calculated using Equation 2 are listed in Table 1 along with two error terms: the overall error column includes random sources of error and random error includes systematic errors based on the scheme of error assessment described by Aitken (1985). The former term is used when examining differences between luminescence ages obtained by the same laboratory and the latter when comparing luminescence and independent dating evidence, such as the calibrated radiocarbon age ranges for F21A.

Annual and cosmic dose. Measurements related to the assessment of the annual dose were performed using dried samples, as discussed above. A combination of experimental techniques was applied to determine the various components of the annual dose, including: thick source alpha counting (TSAC; see e.g. Aitken 1985), beta-TLD (thermoluminescence dosimetry; Bailiff & Aitken 1980) and X-ray fluorescence (XRF). The annual dose, $\dot{D}_{\alpha\beta\gamma}$, for each sample is given in Table 2. Conversion of TSAC activities and potassium concentration to dose-rate were performed using coefficients obtained by Adamiec & Aitken (1998). Corrections for the presence of water and organic material within the sediment samples were applied in the assessment of the annual dose using factors similar to those recommended by Lian *et al.* (1995). Due to the nature of the sampling it was not possible to perform *in situ* measurement of the gamma dose-rate. The gamma dose-rate was calculated for each sample location taking into account the occurrence of sediment layers of differing radioactivity and water content and employing the fractional dose coefficients calculated by Løvborg and reproduced in Aitken (1985). No significant escape of gaseous radon was detected in TSAC measurements. The estimated values of cumulative cosmic-ray dose (Table 2) were calculated using the depth–dose-rate relationship obtained by Prescott & Hutton (1994) where the rate of overburden development was approximated on the basis of the chronological markers provide by the radiocarbon ages for the intercalated peats.

Discussion

The underlying physical principle of the luminescence dating method is the measurement of the cumulative population of trapped charge within the minerals selected for testing; a quantity that is related to the palaeodose, P. In determining P it is assumed that, prior to final burial, the sediments were sufficiently exposed to light to cause the residual trapped charge population, which participates in the production of IRSL when stimulated, to be negligible. As mentioned above, this assumption may not be correct in the case of water-lain sediment and the calculated luminescence age will be greater than the true age where this condition is not met. One approach recognized at an early stage in the development of techniques for dating sediments (Huntley 1985) is to date sediments from modern contexts of similar type to those intended for testing. Another is to employ a partial bleach methodology, which was originally developed to circumvent the problem of incomplete zeroing when P is evaluated using TL and which has been adapted to use with IRSL (Wintle 1997). However, we avoided the use of this approach because of doubts concerning the validity of its basis. In calculating luminescence ages we assumed that the zeroing process prior to burial was effective.

The value of $\dot{D}_{\alpha\beta\gamma}$ used in evaluating Equation 2 is an average for the burial period. The volume of sediment relevant to this calculation is that within a radius of *c.* 30 cm about a sampled location. It has been assumed that in estimating the contribution due to gamma radiation, the effect of the reduction in dose-rate due to the initial period of build-up of the overlying sediments is not significant. The presence of water within the pore structure of the sediment body has a moderating effect on the distribution of radiation dose within the sediment; as the moisture content increases a smaller fraction of the radiation dose is available to be absorbed by constituent luminescent minerals. The effect is most pronounced for alpha radiation and least for gamma radiation (Aitken 1985). In making corrections for moisture content to the alpha, beta and gamma components of the dose-rate we have assumed that the fraction by weight of water in the minerogenic component of the sediment is the same as that measured in the core samples. While this may be a simplistic assumption, more specific information on burial content history is lacking. An uncertainty of $\pm 5\%$ in the measured moisture content has been factored into the dose-rate calculations to reflect the possible changes in water content during burial. As indicated in Table 2, the water content in the Holocene clastic units generally increases with depth from *c.* 50% to >150% by weight in the vicinity of the basal peats. Surprisingly, however, the water content in the vicinity of the middle and upper peats is comparatively low. The extent to which changes in average water content affect the calculated luminescence age is illustrated in Fig. 4. In addition to the progressive increase in the central value of the age with water content, due to the reduction in the effective dose-rate, the error limits also increase, reflecting the incorporation of a fractional uncertainty in water content in the error analysis. It is the variation of water content during the burial period that contributes a major source of uncertainty in the calculated age, particularly for highly water-charged environments. We have assumed that any losses of water during extrusion were not significant and although the overall concordance of the luminescence and radiocarbon ages provides some support for this

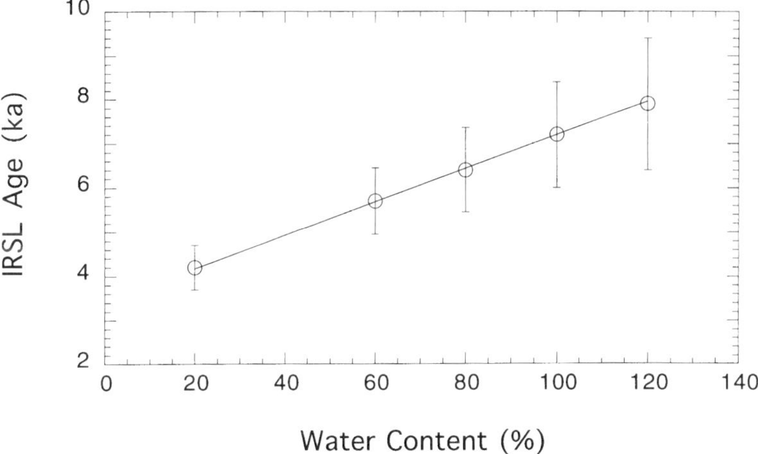

Fig. 4. Illustration of the effect of different levels of average water uptake in sediment on the calculated luminescence age.

assumption, the values of the *in situ* and the as-cored water contents are likely to differ and this remains to be investigated.

A further factor that affects the estimation of the average annual dose and thus the luminescence age is the occurrence of disequilibrium in the uranium and thorium decay chains. As discussed by Olley *et al.* (1996), departure from secular equilibrium is most likely to occur in the uranium decay chain, the extent depending on the degree of disequilibrium at deposition and whether the sediment is a chemically closed system. The presence of unsupported uranium (due to the absence of ^{230}Th and progeny) precipitated from ground-water is expected to be found within the organic sediments; the quantity is not predictable and likely to be highly variable. However, the β-TLD and TSAC measurements performed with sediment samples register contributions from such sources of uranium. A systematic overestimate of the calculated dose-rate will arise if the dose conversion coefficients applied to the TSAC results are based on the assumption of secular equilibrium where a sample contains unsupported uranium; for F21A samples the proportion of the annual dose (due to lithogenic sources), which is based on the use of such results, is in the range *c.* 35–50%. Although for the majority of samples the organic content is less than 15% (the exceptions being samples F21A-130, -775 and -6A where the organic contents are *c.* 18–25%), a nominal allowance for disequilibrium in the uranium decay chain of 3% was made in calculating those components of the annual dose derived from TSAC results (due to alpha and gamma radiation from uranium and thorium); these components contribute about 20% of the annual dose. In the case of the β-TLD measurements it has been assumed that the beta dose-rate within the sediment has not changed significantly since deposition due to either the presence of unsupported uranium within the organic component or an initial excess ^{226}Ra in the inorganic sediments (Olley *et al.* 1996). For the oldest organics, in the basal deposits of F21A the degree of accumulation of ^{230}Th during the Holocene is expected to be less than 10% of the equilibrium level for a chemically closed system on the basis of the 75 ka half-life. It has also been assumed that the activity of parent Th and progeny in the inorganic fraction of the sediments has remained unaltered during burial. The extent to which these assumptions are justified is to be examined by high-resolution gamma-ray spectrometry. However, to make an allowance for the possible effects of disequilibrium in the uranium decay chain a factor related to the proportion of water in the sediment ($\pm 0.05F_w$) was included in the assessment of uncertainty associated with the annual dose.

The cumulative contributions to the palaeodose by source type, i.e. due to combined U and Th, K and cosmic radiation as a function of core depth, can be compared in Fig. 5. In the case of the lithogenic sources, alpha, beta and gamma components of the absorbed dose are included. The main changes occur in the region underlying

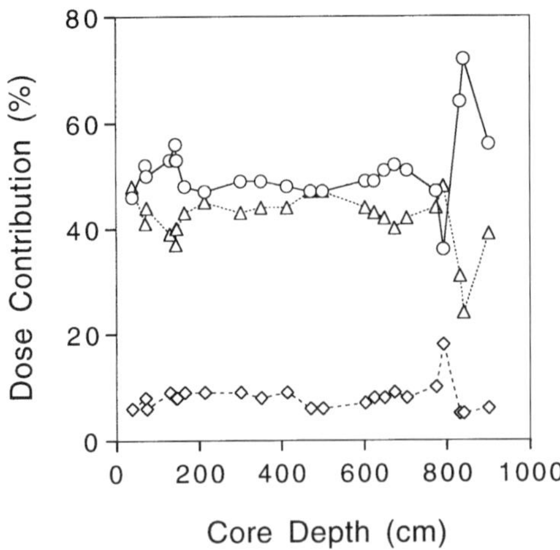

Fig. 5. Breakdown of proportion of palaeodose due to (**a**) U and Th (open circles), (**b**) K (open triangles) and (**c**) cosmic radiation (open diamonds) as a function of sample depth in core, based on the assessment of lithogenic and cosmogenic sources of dose.

the basal peat layers, where the dose from lithogenic sources is mainly carried by uranium and thorium. Although the specific radioactivity of the sediments is lower than that for the silts the effective time-averaged dose-rate is comparable because of the significantly lower water content.

Radiocarbon dating

Seven radiocarbon age determinations were performed by accelerated mass spectroscopy (AMS) at the University of Arizona following sample preparation to graphite at the Natural Environmental Research Council's (NERC) radiocarbon laboratory, East Kilbride and the results are given in Table 3. The calibrated age ranges were obtained using Oxcal v2.18 with cal 10 and cal 20 calibration data. There are also relevant radiocarbon dating results from two adjacent sites, AL-2 and AL-4, from material collected in 1976 and 1979 (Shennan 1980, 1982).

An independent relative dating technique exists using pollen analysis. West (1970) demonstrated that chronozones could be established by dating radiometrically the regional pollen assemblage zone boundaries at a pollen type-site. For England, Wales and southern Scotland the type-site that spans the past 10 ka is at Red Moss (Hibbert *et al.* 1971), and the principle of the regional parallelism of vegetation development, first expounded by von Post (1916), was confirmed when local pollen assemblage zone boundaries were dated at sites remote from Red Moss and found to have similar radiometric ages. Local pollen assemblages could be established from short, interrupted or incomplete stratigraphic sequences that had been dated

Table 3. *Radiocarbon age determinations for core F21, Adventurers' Land* [TF35770185]

Sample no.	Date of report/material sampled	Depth (cm)	Altitude (m OD)	Laboratory code & assay method	Radio carbon age (1σ, BP)	Calibrated age ranges* (cal. BP)	$\delta^{13}C$
F21-56	971208 Decomposed peat	56–57	−0.33 to −0.34	AA-26362 AMS	2435 ± 50	1σ: 2350–2710 2σ: 2350–2720	−27.6
F21-446	970530 Black amorphous peat	446–447	−4.23 to −4.24	AA-22359 AMS	4165 ± 55	1σ: 4570–4840 2σ: 4540–4860	−27.9
F21-478	970530 Black amorphous peat	478–479	−4.55 to −4.56	AA-22360 AMS	5130 ± 60	1σ: 5760–5990 2σ: 5730–6000	−26.3
F21-779	970530 Silty peat	779–780	−7.56 to −7.57	AA-22361 AMS	5925 ± 65	1σ: 6720–6890 2σ: 6620–6950	−27.0
F21-792	970521 Silty peat	792–793	−7.69 to −7.70	AA-22668 AMS	6265 ± 50	1σ: 7030–7220 2σ: 6910–7240	−27.6
F21-799	970521 Silty peat	799–800	−7.76 to −7.77	AA-22669 AMS	6255 ± 55	1σ: 7030–7210 2σ: 7000–7240	−27.8
F21-811	970530 Silty peat	811–812	−7.88 to −7.89	AA-22362 AMS	6310 ± 65	1σ: 7090–7280 2σ: 7020–7380	−21.1

* Overall range given where multiple ranges obtained.

radiometrically, and these dates confirmed or rejected by referring the local pollen assemblage zones to the regional pollen assemblage zones and chronozones. This method has been used to check radiocarbon ages for organic deposits interdigitating marine deposits in coastal lowlands in northwest England (Tooley 1978*a*, *b*). If the age of the dated material is close to a dated pollen assemblage zone boundary, corroboration of the radiometric age is assured: however, less confidence attaches to the process of corroboration if the date lies within chronozone boundaries, which, in the case of Flandrian Chronozone II, covers 2000 radiocarbon years. In the case of F21A, pollen diagrams were constructed from two adjacent sites (AL-2 and AL-4, Shennan 1980, 1982). The tree pollen assemblages from the basal peat (838–855 cm in AL-2) are characterized by *Quercus*, *Alnus*, *Ulmus* and *Tilia* and are indicative of a Flandrian II chronozone, dated from 7107 ± 120 to 5010 ± 80 BP. In the next peat (pollen analysis at 726–730 cm at AL-4, equivalent to Fig. 1 strata e–h in F21A), there is a similar pollen assemblage, and with *Ulmus*-values of 19% of the total aboreal pollen, this peat also lies within the Flandrian II chronozone. However, in the higher peat (pollen analysis at 466–444 cm at AL-4, equivalent to Fig. 1 strata o–p in F21A), whilst *Quercus* pollen remains dominant, only a single grain of *Ulmus* has been recorded and *Tilia* pollen frequencies decline significantly at the top of the peat. The presence of the pollen of ruderals, such as *Plantago lancealata* and *Cirsium*, indicates that this peat accumulated at the Flandrian II to Flandrian III chronozone boundary, i.e. 5 ka BP.

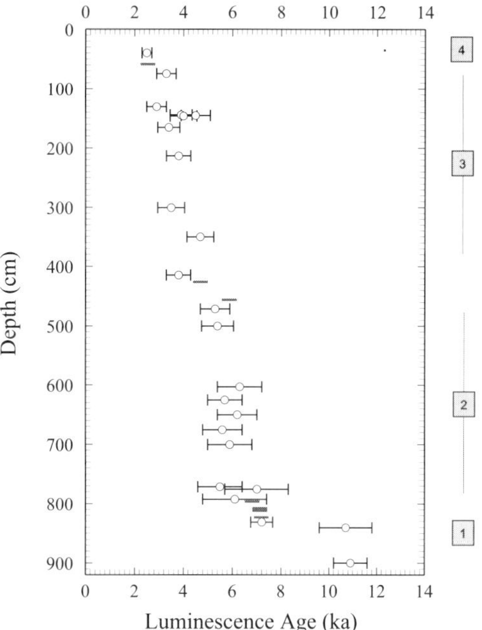

Fig. 6. Luminescence and calibrated radiocarbon ages versus depth for core F21A. The mean values of the luminescence age are indicated by open circles and the error bars represent the overall error (1σ); the calibrated radiocarbon age range (1σ) is indicated by a thick solid line. The numbered sections of the core, as discussed in the main text, are indicated to the right of the plot frame.

Comparison of luminescence and radiocarbon ages

The luminescence ages and calibrated radiocarbon age ranges are plotted as a function of core depth in Fig. 6. Both sets of results are shown at the 68% level of confidence and in the case of the luminescence ages the error bars correspond to the limits of the overall error. For the uncalibrated radiocarbon ages no additional multiplier was applied to the error term supplied; the 2σ calibrated range is also given in Table 3 to illustrate the effect on the calibrated range that would result if it were applied (i.e. equal to two).

For the purposes of this discussion the core can be divided into four sections comprising: Section 1 the predominantly sandy sediments lying below the basal Holocene peat; Sections 2 and 3, enclosed by the two layers of marine sediment between the basal and middle peats and the middle and upper peats, respectively; and finally Section 4 the sediments overlying the upper peat. In comparing the luminescence ages with the calibrated radiocarbon date ranges, which are taken to provide reliable absolute chronological markers on the basis of the foregoing discussion, we note the following.

(1) For samples immediately overlying and subjacent to the three peats, four (F21A-1, -4, -6B1 and -6B3) of the six luminescence ages overlap with the calibrated radiocarbon age ranges at 1σ limits, and the remaining two (F21A-74 and -3B) overlap at 2σ limits.
(2) The luminescence ages for the two deeper samples (F21-800 and F21-900) are consistent with their pre-Holocene origin, and the marked change in rate of sediment deposition is as expected.
(3) On the basis of the radiocarbon ages, the clastic sediments in section 2 accumulated within the age interval *c*. 5.8–6.9 ka. Visual inspection of the variation of luminescence

ages with depth in this section suggests that the ages are not distinguishable. This is confirmed by calculation of the index of homogeneity ($\sigma_w = 1.07$) and the relevant test statistic ($T = 9$ compared with $\chi^2_{8,0.5} = 15.5$) as discussed by Ward & Wilson (1978) and Wilson & Ward (1981), allowing a pooled mean age of 5.8 ± 0.3 ka (s.e., 1σ) to be obtained. Thus the luminescence ages for this section fall within the range delineated by the radiocarbon ages. The possibility that the luminescence ages are systematically younger than the true ages cannot be excluded when taking into account the caveats concerning the effect of disequilibrium on dose-rate assessment.

(4) In section 3, the boundary radiocarbon ages indicate that the clastic sediments accumulated within the age interval 2.4–4.8 ka; all ten luminescence ages fall within this range. Samples F21-2AU, -2A and -2AB were selected to provide a spatially close group (within 3 cm) as a test of experimental reproducibility. They form a coherent group ($\sigma_w = 1.05$; $T = 2$ compared with $\chi^2_{2,0.05} = 6$) with a pooled mean of 4.2 ± 0.2 ka and the 0.4 ka range of uncertainty (1σ) associated with the mean provides a gauge of sample-to-sample variability. A statistical analysis of the luminescence ages for this section suggests that a sub-group comprising the eight samples between and including F21A-130 and -3B can be considered to form a single group ($\sigma_w = 1.3$; $T = 12$ compared with $\chi^2_{7\,0.05} = 14.1$) with a pooled mean of 3.8 ± 0.3 ka. It can also be argued that all the luminescence ages within this section overlap, within 2σ limits, the interpolated radiocarbon age band drawn between the boundary radiocarbon age ranges, which is obtained by assuming a uniform rate of deposition. On the other hand the difference between the pooled mean for the 2AU–2A–2AB group and the interpolated radiocarbon age range is significant and would, if correct, tend to support the case for a more rapid rate of deposition than would be inferred from the radiocarbon ages. However, the margins of uncertainty and the consistency of seriation of the luminescence ages with depth are not yet sufficient to distinguish between these two interpretations.

Conclusions

Overall the results are encouraging, although there are issues connected with the assessment of the annual dose that require further investigation, in particular the extent of radioactive disequilibrium and the assumptions made concerning water content history. The degree to which luminescence ages vary with change in the assumed average water content sounds a note of caution when applying the method to cored samples from coastal environments of this type. Although F21A was the only core amongst the entire LOEPS coring programme to contain three datable organic horizons, cores with at least this degree of dating control are needed in further studies.

While the luminescence ages for F21A do not possess a resolution that would enable rates of sedimentation between the organic sediments to be evaluated with confidence, the indication from the results obtained for section 3 of the core, of a more rapid rate of deposition than conventionally would be inferred from the radiocarbon ages, merits further investigation. In the absence of organic chronometric markers, luminescence nonetheless provides a means of dating the Holocene minerogenic sediments within 1 ka or better, albeit with some indications of a systematic underestimate on the basis of the results from the sediments lying between the basal and middle peats.

Core F21A was taken by D. Brew at the British Geological Survey who also logged core F21. Samples for radiocarbon dating were taken from core F21 at BGS by J. B. Innes, Durham University. We thank R. J. Clark and C. Pickin for assistance in the experimental work described in this paper, and D. Hume of the Geography Department for the production of Fig. 1. We are grateful to the two reviewers of this paper, who brought to our attention a number of important issues.

References

Adamiec, G. & Aitken, M. J. 1998 Dose-rate conversion factors: update. *Ancient TL*, **16**, 37–50.

Aitken, M. J. 1985. *Thermoluminescence Dating*. Academic Press, London.

——1998. *Introduction to Optical Dating*. Oxford University Press, London.

Andrews, J. E., Boomer, I., Bailiff, I. K., Balson, P., Bristow, C., Chroston, P. N., Funnell, B. M., Harwood, G., Jones, R., Maher, B. A. & Shimmield, G. B. 1999. Sedimentary evolution of the north Norfolk barrier coastline in the context of Holocene sea-level change. *Proc. of Geology Soc. This volume.*

Bailiff, I. K. 1992. Luminescence dating of alluvial deposits. *In*: Needham, S. & Macklin, M. G. (eds) *Alluvial Archaeology in Britain*. Oxbow Monograph, **27**. Oxbow Books, Oxford, 27–35.

—— & AITKEN, M. J. 1980. The use of thermoluminescence dosimetry for evaluation of internal beta dose-rate in archaeological dating. *Nuclear Instruments and Methods*, **173**, 423–429.

GODWIN, H. & CLIFFORD, M. H. 1938. Studies of the post-glacial history of British vegetation. II. Origin and stratigraphy of deposits in southern Fenland. *Philosophical Transactions of the Royal Society B*, **229**, 363–406.

GÖKSU, H. Y., BAILIFF, I. K., BØTTER-JENSEN, L., HÜTT, G. & STONEHAM, D. 1995. Inter-laboratory beta source calibration using TL and OSL with natural quartz. *Radiation Measurements*, **24**, 479–484.

HIBBERT, F. A., SWITSUR, V. R. & WEST, R. G. 1971. Radiocarbon dating of Flandrian pollen zones at Red Moss, Lancashire. *Proceedings of the Royal Society of London*, **B177**, 161–176.

HUNTLEY, D. J. 1985. On the zeroing of the thermoluminescence of sediments. *Physics and Chemistry of Minerals*, **12**, 122–127.

LANG, A. & WAGNER, G. A. 1996. Infrared stimulated luminescence dating of archaeosediments. *Archaeometry*, **38**, 129–142.

LIAN, O. B., JINSHENG HU, HUNTLEY, D. J. & HICOCK, S. R. 1995. Optical dating studies of organic-rich sediments from southwestern British Columbia and northwestern Washington State. *Canadian Journal of Earth Science*, **32**, 1194–1207.

OLLEY, J. M., MURRAY, A. & ROBERTS, R. G. 1996. The effects of disequilibrium in the uranium and thorium decay chains on burial dose rates in fluvial sediments. *Quaternary Science Reviews (Quaternary Geochronology)*, **15**, 751–760.

ORFORD, J. D., WILSON, P., WINTLE, A. G., KNIGHT, J. & BRALEY, S. 1999. Holocene coastal dune initiation in Northumberland and Norfolk, eastern UK: climate and sea-level changes as possible forcing agents for dune initiation. *This volume*.

PARISH, R. 1994. The influence of feldspar weathering on luminescence signals and the implications for luminescence dating of sediments. *In*: ROBINSON, D. A. & WILLIAMS, R. B. G. (eds) *Rock Weathering and Landform Evolution*. John Wiley, Chichester.

PRESCOTT, J. R. & HUTTON, J. T. 1994. Cosmic ray contributions to dose rates for luminescence and ESR dating: large depths and long term variations. *Radiation Measurements*, **23** , 497–500.

—— & ROBERTSON, G. B. 1997. Sediment dating by luminescence. *Radiation Measurements*, **27**, 893–922.

REES-JONES, J. & TITE, M. S. 1997. Optical dating results for British archaeological sediments. *Archaeometry*, **39**, 177–188.

SHENNAN, I. 1980. *Flandrian Sea-level Changes in the Fenland*. PhD thesis, University of Durham.

——1982. Interpretation of Flandrian sea-level data from the Fenland, England, *Proceedings of the Geoogists' Association*, **3**(1), 53–63.

SKERTCHLY, S. B. J. 1877. *The Geology of the Fenland*. Memoirs of the Geological Survey: England and Wales. HMSO, London.

TOOLEY, M. J. 1978*a*. *Sea-level Changes: North-west England During the Flandrian Stage*. Clarendon Press, Oxford.

——1978*b*. Interpretation of Holocene sea-level changes. *Geologiska Föreningens i Stockholm Förhandlingar*, **100**, 203–212.

——1993. Long term changes in eustatic sea level. *In*: WARRICK, R. A., BARROW, E. M. & WIGLEY, T. M. L. (eds) *Climate and Sea Level Change*. Oxford University Press, Oxford, 81–107.

TROELS-SMITH, J. 1955. Karakterisering af løse jordarter. *Danmarks Geologiske Undersølgelse IV*, **3**(10), 1–73 and I-XIII Tavles.

VON POST, L. 1916. Forest tree pollen in south Swedish peat bog deposits. (Translated by DAVIS, M. B. & FAEGRI, K. 1967. *Pollen Spores*, **9**(3), 375–401.

WARD, G. K. & WILSON, S. R. 1978. Procedures for comparing and combining radiocarbon age determinations: a critique. *Archaeometry*, **20**, 19–31.

WEST, R. G. 1970. Pollen zones in the Pleistocene of Great Britain and their correlation. *New Phytologist*, **69**, 1179–1183.

WILSON, S. R. & WARD, G. K. 1981. Evaluation and clustering of radiocarbon age determinations: procedures and paradigms. *Archaeometry*, **23** 19–39.

WINTLE, A. G. 1977. Detailed study of a thermoluminescent mineral exhibiting anomalous fading. *Journal of Luminescence*, **15**, 385–393.

——1997. Luminescence dating: laboratory procedures and protocols. *Radiation Measurements*, **27**, 769–817.

The development of a methodology for luminescence dating of Holocene sediments at the land–ocean interface

M. L. CLARKE[1] & H. M. RENDELL[2]

[1] *School of Geography, University of Nottingham, University Park, Nottingham NG7 2RD, UK (e-mail: michele.clarke@nottingham.ac.uk)*
[2] *Department of Geography, University of Loughborough, Loughborough, Leicestershire LE11 3TU, UK*

Abstract: The main challenge in luminescence dating is to provide accurate ages for sediments derived from hillslope, fluvial and marine environments where grains have been transported and deposited by water. A new methodology has been devised for dating Holocene age sediments from the land–ocean interface. Alkali feldspars are recommended as the ideal dosimeter in these environments as they are: rapidly zeroed, have high sensitivity to dose, have an internal dose rate from the decay of ^{40}K, have signal intensities that are much higher than for quartz, and the equivalent doses (EDs) determined are not affected by chemical weathering. Studies of the fundamental characteristics of feldspars have been used to optimize the luminescence signal for use in dating applications. A quality assurance technique for discriminating between those samples that will give accurate dates and those that will yield inaccurate dates has been developed and is tested here on coastal-zone sediments from Lincolnshire. Sampling from sediment exposures rather than cores minimizes the uncertainties related to past water content fluctuations.

Luminescence techniques date the last exposure to light of quartz or feldspar grains in a sediment body. The techniques are based upon the fact that exposure to daylight during sediment entrainment, transport and deposition zeroes the luminescence signal, re-setting the dating clock. Once quartz or feldspar grains are buried, the luminescence signal builds up over time due to the radioactive decay of uranium, thorium and potassium isotopes present in the natural environment. The age of the sediment is derived from the determination of the radiation dose required to produce the level of 'natural' luminescence accrued in the grain since burial, divided by the dose rate to the sediment from the radioactive isotopes in the natural environment.

Luminescence techniques have traditionally been applied to sediments from a range of aeolian environments (e.g. Ollerhead *et al.* 1994; Clarke *et al.* 1996; Rendell & Sheffer 1996; Clarke & Rendell 1998; Wintle *et al.* 1998) where transport processes allow sufficient exposure to light to zero the luminescence signal completely before deposition and burial of the sediment grains. Comparison with independent age control has shown that the technique works well, giving accurate dates when applied to wind-blown sands (Stokes & Gaylord 1993; Clarke & Käyhkö 1997) and silts (Richardson *et al.* 1997).

The main challenge in luminescence dating is to provide accurate dates for sediments derived from hillslope, fluvial and marine environments where grains have been transported and deposited by water. A number of workers have attempted to provide luminescence dates for samples from water-lain contexts (Huntley 1985; Berger 1990; Berger *et al.* 1990; Berger & Easterbrook 1993; Forman *et al.* 1994; Lamothe & Auclair 1997). In these environments, transport is often rapid, occurring in a sediment-laden, turbid water body. Given that the key to providing an accurate date for deposition and burial is sufficient exposure to light to reduce the luminescence signal to zero, it is clear that sediments from these environments pose a problem due to the potential for incomplete zeroing of the luminescence signal during the erosion–transport–deposition cycle. Sediments that have been incompletely zeroed at burial are termed 'poorly-bleached' whereas full zeroing of the luminescence signal occurs in 'well-bleached' sediments.

Characteristics of the luminescence signal

Thermoluminescence (TL) methods use heat to measure the luminescence signal in the

From: SHENNAN, I. & ANDREWS, J. (eds) *Holocene Land–Ocean Interaction and Environmental Change around the North Sea*. Geological Society, London, Special Publications, **166**, 69–86. 1-86239-054-1/00/$15.00

sediment grains; whereas the newer optical luminescence methods, infra-red-stimulated luminescence (IRSL) and green-light-stimulated luminescence or optically stimulated luminescence (GLSL or OSL), use photons of a specific wavelength to measure the natural luminescence signal. Aitken (1998) refers to the newer techniques collectively as photon-stimulated luminescence. Both quartz and feldspars exhibit TL when heated and GLSL when green light (or more accurately, blue-green light) is applied to the grains. In the case of IRSL, where infra-red light is used to stimulate a luminescence signal from the grains, only feldspars respond; quartz has no IRSL signal. The reason for this difference is not known, but probably results from the different crystal structure of the two minerals. Luminescence signals result from the change in energy state of trapped charge (electrons and holes) in the crystal lattice of quartz and feldspars.

At the zeroing event, after sufficient daylight exposure to fully zero the luminescence signal, the grains are buried and, therefore, removed from light stimulation. The interaction of radioactive particles or rays (given off by uranium, thorium and potassium isotopes naturally occurring in sediments) with the crystal structure of quartz and feldspar grains provides an input of energy to charged particles in the ground state, which elevates them to higher energy levels in the crystal field where they are trapped (for further explanation see Aitken 1998). As the time period in which the quartz or feldspar grains remain buried increases, the trapped charge population increases. The sediment is then sampled, without exposing the grains to light, and taken to the laboratory where the quartz or feldspar grains undergo TL, GLSL or IRSL measurement. The heat or light applied provides sufficient energy to stimulate the crystal structure and de-trap the charge population. This charge then recombines, giving off excess energy to return to the ground state. Some of this excess energy is given off in the form of light (i.e. photons) and this is the luminescence signal. Thus, the larger the trapped charge population, the larger the number of photons produced from the crystal as the trapped charge recombines and the brighter the luminescence signal emitted.

With TL methods, heat is applied to the grains via a heating strip, which can ramp up to temperatures of around 500°C at rates of the order of 2.5°C s^{-1}. As the grains are heated, the luminescence signal emitted is measured using a photomultiplier tube (PMT). The colour of light emitted depends on the crystal structure and is linked to the presence of specific

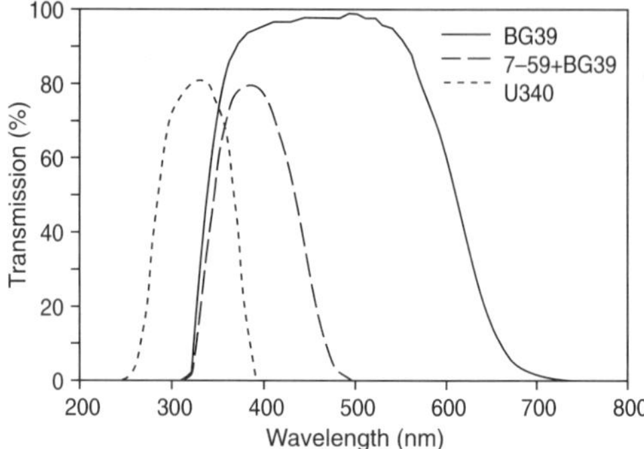

Fig. 1. Transmission characteristics of detection filters routinely used in luminescence dating.

impurities or defects. For example, in feldspars, light emitted at 550–570 nm is thought to be related to the presence of a Mn^{2+} impurity substituting on to a Ca^{2+} site in the alumino-silicate lattice (Prescott & Fox 1993). With TL, it is possible to detect emitted luminescence in any region of the spectrum from the ultraviolet (UV) to red (200–800 nm). Wavelength selection is achieved using colour filters, which are placed in front of the PMT. Examples of filters that are widely used in dating applications are shown in Fig 1. Optical and infra-red stimulation use light of a specific wavelength, or wavelength range, to stimulate luminescence from quartz and/or feldspars. Infra-red diodes, such as type TEMT484, which stimulate at 880 ± 80 nm are used during IRSL with feldspars and the luminescence emitted from the sample is detected in the dominantly blue or UV region of the spectrum (*c.* 250–500 nm). OSL (GLSL) uses light from either a filtered halogen lamp (420–550 nm), green diodes (450–600 nm) or an argon-ion laser (514.5 nm), and is more problematic for use with blue detection filters due to potential signal interference from the stimulation source. Thus, UV or violet detection filters are routinely used for OSL (GLSL) dating applications.

UV and blue detection filters have been traditionally used for TL because quartz and feldspars emit light from TL stimulation in these regions (Huntley *et al.* 1988*a*, *b*). The same filters were used in subsequent developments in IRSL and OSL dating. Research into the colour of light emitted from quartz and feldspars using photon stimulation methods has been limited to a handful of studies, often using stimulation sources not used in dating; e.g. such as a helium–neon laser that emits at 633 nm, which has been used to stimulate feldspars (Huntley *et al.* 1991; Jungner & Huntley 1991).

Photon-stimulation methods

There are two main advantages of applying photon-stimulation techniques, rather than TL, when attempting to date sediments. First, infra-red and green-light stimulation methods sample trapped charge populations, which are zeroed more rapidly in daylight exposure than the trapped charge populations sampled using TL stimulation. This difference, with respect to IRSL and TL, is shown in Fig. 2 where grains of potassium feldspar separated from a natural dune sand, were exposed to natural daylight conditions in the field for controlled lengths of time. The experiment took place on a sunny spring day in Aberystwyth, west Wales (52°25′N 04°06′W) in March 1995, and the light intensity received by the grains was measured at 5.5–5.8 mW cm^{-2} (Clarke 1996). The proportion of trapped charge sampled during IRSL measurements shows a rapid zeroing, with only 10% of the signal remaining after 60 s daylight exposure (Fig. 2). For the same time period (60 s) 85% of the trapped charge associated with TL remains unbleached. Therefore, it is more likely that sediments which have been rapidly transported and buried will be well-bleached at deposition if the IRSL rather than the TL is measured. Godfrey-Smith *et al.* (1988) have shown that a similar relationship exists between OSL and TL for both quartz and feldspar. Therefore, for both quartz and feldspar dosimeters, using light to stimulate luminescence from sediments has the potential to measure a better-bleached (i.e. more fully zeroed) signal.

The second advantage of using photon-stimulation methods is that there is no apparent residual level. When a sediment is exposed to daylight, with time, the photon-stimulated signal is bleached to a level where it is fully zeroed, i.e. there should be no IRSL or GLSL signal remaining in the grain. This is not the case for TL, which has a residual unbleachable component. This residual trapped charge, which cannot be stimulated by exposure to daylight but only by heating the sample, exists in both quartz and feldspar and provides a minimum age limit for the TL dating technique. Huntley (1985) quotes typical values of 2 ka, but with one case of <300 years, while Berger *et al.* (1990) quote a value of 300 years. The lack of a residual component with photon-stimulation methods means that there is no theoretical lower limit for OSL and IRSL dating techniques and it should therefore be possible to date samples as young as ten years old. The main obstacles to providing a date of ten years being signal-to-noise ratio with respect to luminescence measurement from the PMT, experimental artefacts such as signal recuperation (Aitken 1998) and the systematic errors in all components of the dating procedure.

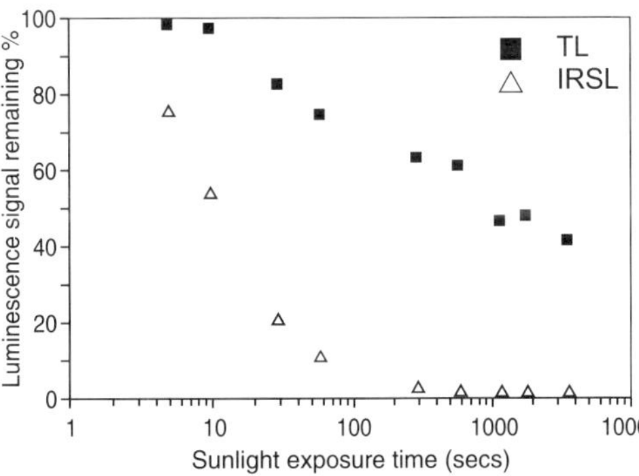

Fig. 2. The effect of sunlight on the bleaching of TL and IRSL signals from feldspar sediments (Clarke 1996).

The following sections outline the steps undertaken in the development of a rigorous but flexible methodology for accurately dating Holocene sediments from the land–ocean interface where 'poor bleaching' is a problem. Given the advantages of photon-stimulation over TL outlined above, the dating procedures developed here were undertaken using these methods, but with TL used in behavioural studies. The aim at the outset was to develop a technique that would be usable with either IRSL or GLSL and with both feldspar and quartz dosimeters, as well as polymineral fine-grained sediments containing mixtures of minerals with luminescence signals.

Characteristics of luminescence behaviour

In order to develop a dating methodology, various characteristics of luminescence behaviour need to be evaluated and procedures recommended which can be routinely adopted in dating applications. Fundamental characteristics investigated here include emission spectra, charge eviction characteristics, and the impacts of weathering and water-content variations.

Emission spectra

The first stage of the study investigated the spectral characteristics of quartz and feldspars when stimulated using photons of different wavelengths used in dating. This work was undertaken using a high-resolution spectrometer with detection in the wavelength range from near UV to near infra-red (200–800 nm) and a resolution of 3 nm (Clarke & Rendell 1997*a*). Equipment limitations with this spectrometer did not allow for the use of GLSL methods and, to date, no equipment exists worldwide that is capable of high-resolution spectral measurements of quartz and feldspar while stimulating the sample using wavelengths in the visible part of the spectrum. Given the spectral range of the equipment, it was, however, possible to use infra-red diodes, emitting at 880 ± 80 nm, without filtering the detectors to affect emission in the areas of interest for dating applications. As quartz does not exhibit IRSL, the majority of the research described here was based on characterizing feldspar behaviour.

The feldspars used in these experiments consisted of:

(a) single crystal well-characterized feldspars from across the solid-solution series, including potassium, sodium and calcium-rich samples;
(b) NIST powdered feldspar standards: NIST99a (sodium feldspar: albite) and NIST70a (potassium feldspar: microcline);
(3) sand-sized grains of feldspar routinely separated from sediments for dating purposes.

Using this range of samples, the behaviour of different feldspar types could be characterized for later use in dating applications. Measurements undertaken included TL, IRSL, radioluminescence and cathodoluminescence (Rendell & Clarke 1997). For general comparison, TL measurements were undertaken on single crystals of volcanic, granitic and hydrothermal quartz and sand-sized grains of quartz routinely separated from sediments for dating purposes.

The results of the spectra obtained using TL measurements are shown in Table 1 (Rendell & Clarke 1997). Low signal intensities were found for all of the quartzes and plagioclase feldspars compared with the sodium and potassium-rich feldspars. Not all specimens showed all emission bands and these varied in number, colour (wavelength of light, measured in nanometres) and intensity between samples and also between samples of the same mineralogy (e.g. orthoclase feldspars). In addition, both quartz and feldspars and mixed minerals showed common emission bands, probably due to comparable defect sites as has been noted by Huntley *et al.* (1988*a*, *b*) and Prescott & Fox (1993). Given these findings, it was clear that a mixed mineral fine-grained sediment consisting of quartz and feldspar grains derived from different origins will inevitably show a variable range of emission features. It is, therefore, not possible to characterize mineralogy from emission spectra alone. Furthermore, as with the existence of feldspar inclusions within quartz grains, it cannot be assumed that defining one emission wavelength region will isolate emission from a particular mineral.

Although all of the TL emission bands observed in the spectra of quartz and feldspars measured in this study appeared to be stable, the same does not apply to the infra-red-stimulated luminescence (IRSL) spectra. An unstable 290 nm (UV) IRSL peak is present in feldspars (Clarke & Rendell 1997*a*, *b*). This unstable peak is seen after laboratory irradiation (Fig. 3), in bulk potassium and sodium feldspars, both as single crystals and in mixed mineral detrital feldspars separated from sediments. The same feldspar specimen can show a stable 290 nm TL emission and an unstable 290 nm IRSL emission. The unstable IRSL emission may result from either defect transformation or an unstable trapped charge population (Clarke & Rendell 1997*a*) and is related to a sodium feldspar struc-ture, found in bulk sodium feldspars (albites) and as nanometre-scale exsolution features within perthitic potassium feldspars (orthoclases and microclines; Clarke *et al.* 1997). Decay of the unstable 290 nm IRSL peak causes charge

Table 1. *Comparison of the peak wavelength (nanometres) of TL emission bands in quartz and feldspars*

Quartz	Feldspars
290	290
330	340
360–380	400–480
400–480	530
560–580	570
620–630	600

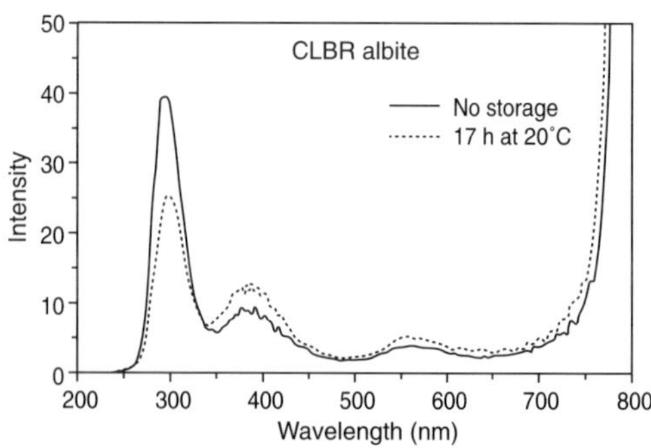

Fig. 3. The effect of room temperature storage on single crystal albite (CLBR) after 50 Gy X-ray dose (Clarke & Rendell 1997*b*).

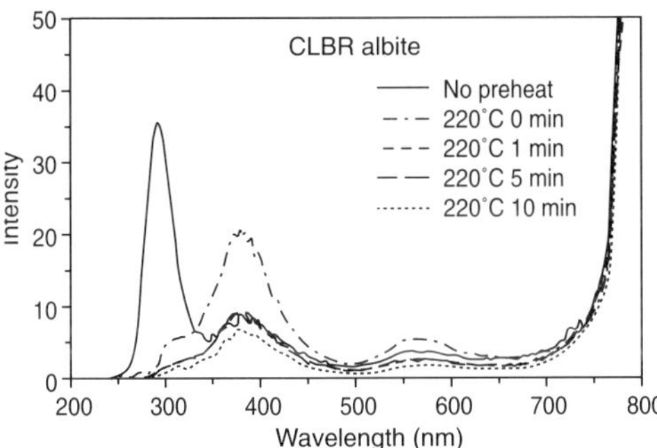

Fig. 4. The effect of sustained preheats of 1 min or more on the IRSL emission from albite, CLBR (Clarke & Rendell 1997*b*).

transfer into other (blue and green) emission peaks at room temperature and after short (<1 min) preheats at temperatures of up to 220°C (Clarke & Rendell 1997*b*). Complete removal of both the unstable peak and the transferred charge occurs only after a prolonged (>1 min) preheat at 220°C (Fig. 4). Clearly, the presence of this instability will cause problems for: (a) rapid pulse annealing routines; (b) thermo-optical routines; (c) IRSL dating unless a prolonged preheat is used.

Given the problems with separating the emissions from different minerals, there can be no definite criteria for selecting TL or photon-stimulated luminescence emission wavelengths. A filter combination centred on the blue region of the spectrum (around 400 nm) for both TL and IRSL using either quartz or feldspar would detect a range of emission bands and have the advantage of de-selecting the UV region around 290 nm, which has been demonstrated to be problematic in previous studies of feldspar dating (e.g. Balescu & Lamothe 1992, 1994). However, in the case of GLSL (OSL) of feldspars detection in the UV may be unavoidable. The identification of the unstable 290 nm emission in IRSL has necessitated the careful selection of optical filters to avoid this UV component and a blue centred filter combination, e.g. Schott BG39 + Corning 7–59, is recommended. This filter combination has been used for most of the experiments described here, including weathering tests, methodological development and dating routines. The only exception involves the use of a UV filter for GLSL measurements for the feldspar separates (Fig. 5). Other workers, while detecting the same IRSL emission peaks (Krbetschek *et al.* 1996), have come to different conclusions about the choice of detection filters (Krause *et al.* 1997).

Choice of dosimeter

Luminescence signal intensities in Holocene sediments are low and background noise may be significant. The choice of luminescence dosimeter, with which to undertake a methodological study, has thus to reflect the comparatively young age of the sediments. Feldspars were preferred to quartz for the following reasons.

(1) Potassium feldspars have an internal dose rate from the decay of ^{40}K within the feldspar lattice. In sand-size grains this internal dose may contribute as much as 40% of the total dose rate to the sample and this has two advantages. Firstly, in very young samples, such grains can accrue a significant dose above background faster than an equivalent quartz grain, meaning that feldspars have the potential to date younger samples than quartz. Secondly, water in the sediment matrix attenuates the external dose to the grains and uncertainties in past water content are minimized when a significant proportion of the total dose is internal to the grain and thus not affected by changes or uncertainties in water content.
(2) Alkali feldspars have a higher sensitivity to radiation dose than quartz. This means that the luminescence intensity derived from the grains is higher than for quartz for the same radiation dose, i.e. feldspars are 'brighter'. At low doses, for young samples, feldspars therefore have a better signal-to-noise ratio than quartz and have the potential to date younger samples than quartz.
(3) The research undertaken into IRSL emission spectra means that we can be confident that, in circumstances where a preheat routine of 220°C for 10 min is used (e.g. Duller 1991), none of the luminescence detected by the PMT is unstable.

Some feldspars exhibit a condition known as 'anomalous fading' (Wintle 1973), which has been shown to afflict feldspars stimulated by both TL (Wintle 1977) and IRSL (Spooner 1994). This results in long-term fading of the luminescence signal over time since irradiation. Within a dating programme, a feldspar that exhibits anomalous fading will give an age underestimation. The precise mechanism responsible for this anomalous behaviour is not understood and various theories have been put forward, including quantum mechanical tunnelling (Visocekas 1985), decay of luminescence centres (Wintle 1977; Debenham 1985) and localized transitions (Templer 1986). Although

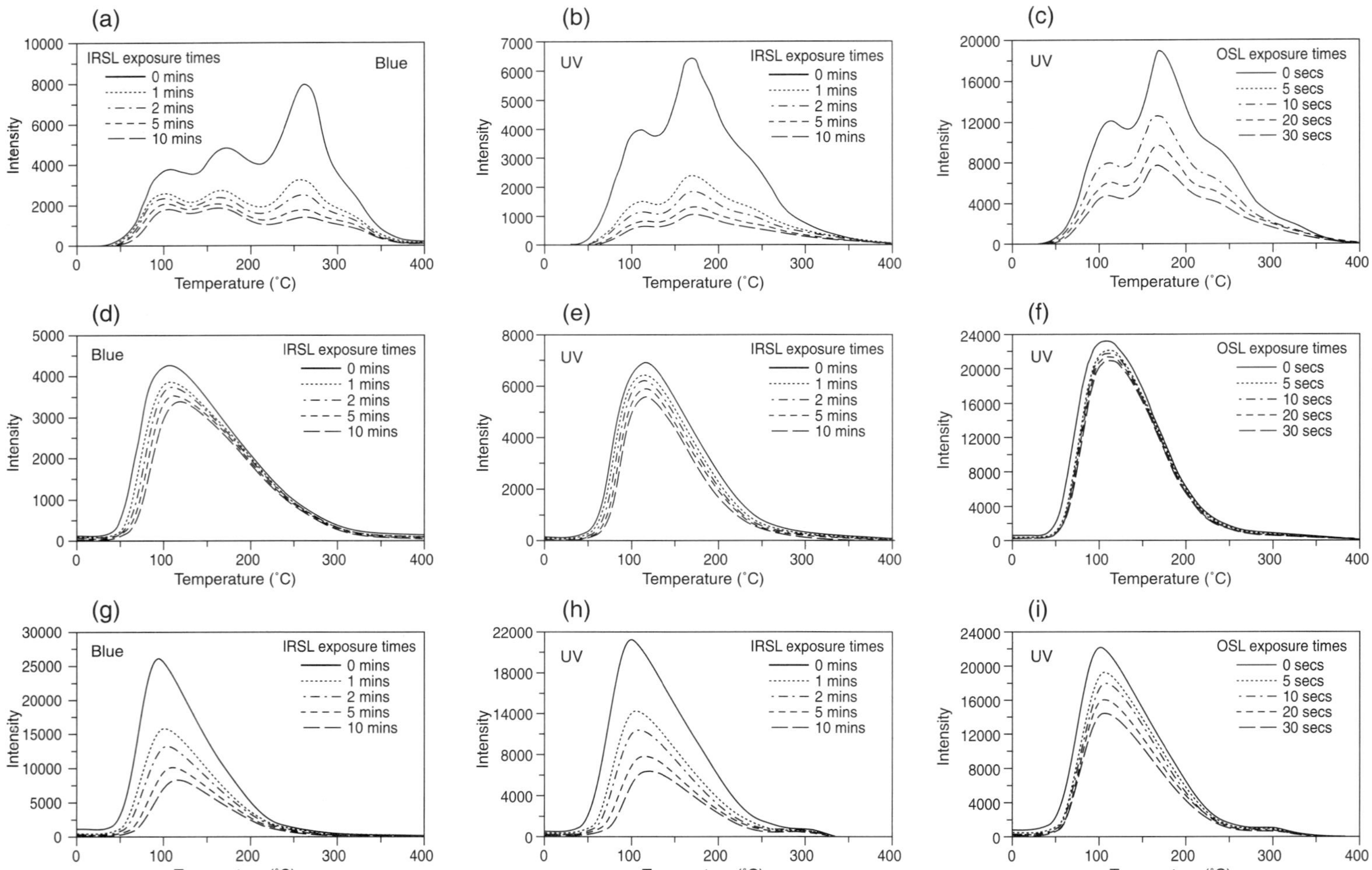

Fig. 5. The effects of infra-red and green light on the TL signals of (**a–c**) albite (CLBR), (**d–f**) microcline (FB1) and (**g–i**) ASGS adularia. Figures (**a**), (**d**) and (**g**) show the effects of infra-red exposure on TL detected in the blue region of the spectrum with a BG39 + 7–59 filter combination; (**b**), (**e**) and (**h**) show IRSL detected in the UV (U340); and (**c**), (**f**) and (**i**) show the effects of green light exposure on TL detected in the UV (U340).

feldspars from all parts of the solid-solution series have been shown to exhibit anomalous fading after laboratory irradiation (Spooner 1994), samples that show fading are comparatively rare, and most often found in volcanically derived feldspars, which exhibit a disordered lattice structure. Each sediment to be dated, should thus be tested for anomalous fading as a matter of routine; typically sample aliquots should be given a laboratory radiation dose, preheated and then stored with IRSL being read out using 0.1 s exposures at fixed time intervals during storage. Where anomalous fading is found, then the sample should be discarded as there is no agreed procedure for overcoming the problem.

Charge eviction characteristics

The loss of 290 nm IRSL emission in feldspars may result from a thermally unstable trapped charge population associated with the 290 nm recombination centre or defect transformation (Clarke & Rendell 1997*a*). Following the assumption that the cause is a thermally unstable trapped charge population, tests were undertaken on a variety of feldspars to evaluate whether it was possible to discriminate between different populations on the basis of their eviction behaviour using IRSL, GLSL and TL stimulation.

Experiments were undertaken using a Risϕ TL-OSL Reader equipped with a halogen lamp and infra red diodes. Two detection windows were used: U340 (TL, GLSL and IRSL) and a BG39 + 7–59 (TL and IRSL only; see Fig. 1). The feldspars used in the experiments included: two albites, three microclines, four orthoclases, adularia, oligoclase and two sanidines. In addition, several other plagioclase feldspars (labradorite, andesine) were measured, but had little discernible signal. The 290 nm peak is thought to be associated with sodium feldspar ($NaAlSi_3O_8$) present as either the bulk matrix, as in albites, or as exsolution features, as in perthitic microclines and orthoclases, which have a bulk potassium feldspar ($KAlSi_3O_8$) lattice structure (Clarke *et al.* 1997). Thus, the assumption is that these types of minerals will have different charge eviction behaviour than those feldspars that have no sodium phase (e.g. adularia).

The feldspars were given 300 s radiation dose from a ^{90}Sr or ^{90}Y beta source and the effect on the TL signal from exposure to infra-red and green light was measured. No preheats were given, so that in theory the thermally unstable charge component would be present. After irradiation, the samples were exposed to 0 1, 2 5 and 10 min of infra-red light or 0 5, 10, 20, 30 s of green light, followed immediately by TL to 400°C. The different exposure times with infra-red and green light relate to the fact that the two light sources have different stimulation power. The loss of TL after exposure to infra-red light is shown in Fig. 5 for an albite, a microcline and an adularia using both UV (340U) and violet-blue (BG39 + 7–59) detection windows. Both the albite and microcline showed a loss in the 290 nm peak after irradiation (Clarke & Rendell 1997*b*), therefore it was initially assumed that the lower temperature TL peaks would be preferentially 'bleached' because of the unstable charge associated with the 290 nm IRSL peak. As shown in Fig. 5, this is clearly not the case in either wavelength region. Identical tests carried out on all of the feldspars showed the same response, i.e. a loss in trapped charge population across the entire glow curve. The effect of green light on the TL signal is clearly similar to that of infra-red light, although as yet, we do not know whether the charge associated with the thermally unstable 290 nm signal in IRSL is also present in GLSL signals. However, given the similarity of charge eviction behaviour using GLSL and IRSL on feldspars, it would be reasonable to assume that an unstable 290 nm GLSL (OSL) peak is likely to be present.

What these experiments suggest is that, as there is a loss of charge from the whole TL glow curve region for all feldspars, irrespective of whether the detection window encompasses the 290 nm peak, the 290 nm emission cannot be associated with a thermally unstable trapped charge population. An implicit assumption here is that TL and IRSL do not stimulate discrete charge populations. Bearing this assumption in mind, if the 290 nm emission in IRSL was associated with a thermodynamically unstable charge population, then the lower parts of the TL peaks would be preferentially lost in the albites, microclines and orthoclases. Given that this is clearly not the case, it therefore seems likely that the 290 nm emission is related to defect transformation rather than to a specific trapped charge population feeding preferentially to the 290 nm recombination centre.

Following on from the above experiment, IRSL and GLSL shinedown curves (also known as decay curves) for the different feldspars were compared after a 300 s beta dose in order to assess whether a pattern of charge eviction exists for the different feldspar minerals. Figure 6 shows the results observed using a U340 filter for albites, microclines and orthoclases, both with and without the addition of a preheat after

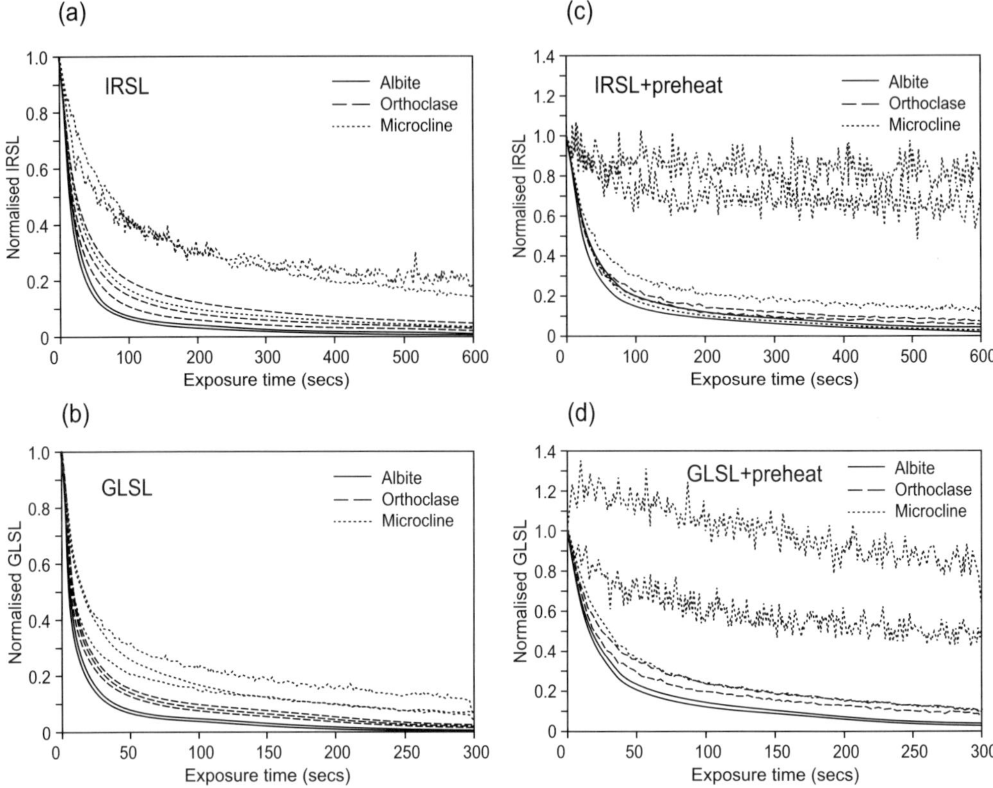

Fig. 6. A comparison of charge eviction kinetics illustrated in the shape of IRSL and GLSL shinedown (decay) curves in the UV (U340) detection window for albites, microclines and orthoclases, with and without the addition of a preheat of 220°C for 10 min.

irradiation of 220°C for 10 min. It is immediately apparent that mineralogy has a control on the shape of the shinedown curve: microclines do not evict charge as rapidly as orthoclases. However, albites are very rapidly 'bleached'. Again, this is the case with both IRSL and OSL, significantly both before and after the preheat, and for IRSL this occurs in both UV and blue wavelength regions. Therefore it may be possible to characterize the relationship between mineralogy and charge eviction in terms of the shape of the decay curve.

The results from this experiment show that sodium-rich feldspars evict trapped charge more rapidly under green and infra-red light than potassium-rich feldspars. If this were also true under daylight conditions, given that concerns about a 290 nm unstable signal are removed by addition of the recommended preheat, sodium feldspars may be a better dosimeter for poorly bleached depositional environments. However, in sediments, feldspars are derived from a range of sources and alkali feldspars are often perthitic, showing exsolved phases of sodium feldspar within a dominantly potassium feldspar matrix. Thus it is unlikely that one could differentiate between individual potassium- and sodium-rich feldspar grains, unless single grain measurements were undertaken with appropriate grain-by-grain chemical analysis. (It is also worth noting that individual grains of Holocene age are unlikely to have sufficient intensity of signal to be visible above background noise in conventional luminescence equipment.) Therefore, using standard heavy liquid grain separation to select alkali feldspars and remove plagioclases should be sufficient to select the most rapidly zeroed fraction of the feldspars present in the sediment to be dated.

Weathering

Feldspars are known to be affected by chemical weathering, undergoing dissolution and transformation into clay minerals (Berner & Holdren

1979; Holdren & Berner 1979). Concern has been expressed (Parish 1994; Käyhkö *et al.* 1999) about the potential effects of weathering on luminescence age determination and the accuracy of the dates obtained. The impact of chemical weathering by dissolution on feldspar luminescence was therefore tested using two geochemical feldspar standards (K-feldspar NIST70a and Na-feldspar NIST99a). The luminescence behaviour of both of these standards had been characterized during the spectral analysis work described above. Extreme conditions of pH were chosen, as they have been shown to determine feldspar dissolution rates (Helgeson *et al.* 1984; Holdren & Speyer 1985; Knauss & Wolery 1986; Blum 1994). The powdered standards also provide a large surface area to volume ratio for weathering attack. The impact on the IRSL was determined by giving a known radiation dose to the feldspars in the laboratory, which were then preheated at 220°C for 10 min to remove any unstable component. The feldspars were then subject to different pH environments for 48 h. A control of pH 7.0 was used, and 1 M hydrochloric acid (pH 1.0) and 1 M sodium hydroxide (pH 11.8) were applied, as solutions at these pHs have been shown to create the most rapid weathering of feldspars (Blum 1994). After 48 h, the feldspars were removed and washed in distilled water to remove all weathering products and excess charged ions. Measurement of the brightness (ln) and determination of the equivalent dose (ED) for the luminescence signal remaining after the treatment was carried out in a Risφ TL-OSL Reader (more detail on these terms is found below under dating methodology). The results showed that feldspar dissolution has no effect on either the ED or the natural luminescence intensity of the feldspar standards (within experimental error). This finding is illustrated in Table 2 for the potassium feldspar standard.

The experiment was repeated with 1 M hydrochloric acid (pH 1.0) and a control (pH 7.0), but with evaporation to dryness rather than washing in distilled water, such that the products of weathering would still be present. No significant difference in mean ED was found for either potassium or sodium feldspars when treated with pH 1.0 or pH 7.0. However, an 82% loss in signal intensity was noted for the potassium feldspar and a 63% loss in signal for the sodium feldspar after treatment with the acid. This suggests that highly charged weathering products, present on the grain surface, quench the luminescence signal. These products include clay minerals and ions such as Al^{3+}, H^+ and $Al(OH)_4^-$. Given that, during routine dating preparation, all samples are washed in distilled water and clay minerals removed, it is clear that this signal quenching would not occur. The lack of effect on the ED at the extremes of pH associated with high feldspar dissolution rates, shows that weathering of feldspars in natural environments should not have any effect on luminescence age determination. The earlier work by Parish (1994) using solutions of pH 6.0 and pH 8.0, assumed that a loss of signal intensity also represented a loss of ED, although the latter was not determined. The present study has demonstrated that there is a loss of signal intensity when weathering products are not

Table 2. *The effect of 48 h in water (pH 7.0), 1 M hydrochloric acid (pH 1.0) and 1 M sodium hydroxide (pH 11.8) on the ED and natural intensity (I_n) of potassium feldspar standard NIST70a**

Aliquot number	pH 7.0		pH 1.0		pH 11.8	
	ED (Gy)	I_n (c.p.s.)	ED (Gy)	I_n (c.p.s.)	ED (Gy)	I_n (c.p.s.)
1	0.602	10 528	0.586	8 480	0.617	14 108
2	0.625	10 936	0.620	8 152	0.639	15 206
3	0.630	13 054	0.630	15 040	0.607	10 292
4	0.616	10 104	0.617	13 976	0.608	11 574
5	0.618	12 164	0.621	11 418	0.624	13 528
6	0.605	8 328	0.610	12 376	0.620	12 502
7	0.620	10 156	0.622	12 280	0.615	8 840
8	0.626	9 736			0.611	9320
$\bar{x}$	0.618	10 626	0.615	11 675	0.618	11 796
σ_{n-1}	0.010	1 461	0.014	2 587	0.010	2 132
S_N	0.016	–	0.023	–	0.017	–

* Aliquots were given known laboratory doses of 0.6 Gy.

removed prior to measurement, but that the ED is unchanged irrespective of whether the weathering products are removed or not. Weathering processes may, however, result in the translocation of radionuclides within the sediment body and, therefore, in changes in the external natural radiation dose to sediment grains.

Water content fluctuations

Water content history is of particular importance when dating sediments from alluvial and near marine environments where fluctuations of the water-table are common. Water attenuates the external dose rate to the sample and thus can have a large effect on the calculated age. This is a particular problem for fine-grained, clay-rich sediments, which can have very high natural water contents. This is illustrated in Fig. 7 for four samples taken from a sediment core from a location near Guyhirn, Cambridgeshire (52°39′ 01°0′E; Fig. 8) which consisted of intertidal silts and clays. The effect of water content (here expressed as per cent wet weight) on the dose rate and, therefore, on the calculated age is clear, and is particularly dramatic when water contents rise above 40%. Water contents in near marine environments typically exceed this value and are thus the major uncertainty in age estimation. Bulk densities of clay-rich sediment samples can be altered by remoulding or compaction during recovery; phenomena that are well known in the oil industry. Sampling from cores may also take place days to months or even years after core recovery, which inevitably involves significant de-watering. Thus, when the core is sampled there is no indication of the ambient water content at the time of extraction and an estimated value must be used. This choice of value can significantly alter the final age (see Fig. 7). In circumstances where no other form of dating exists on the core to provide an independent age control, and given that there is no certain method for correctly assessing the past water content, then the final age calculated is a best estimate. Sampling from exposed sediment sections rather than cores has the great advantage of allowing *in situ* measurement of water content at the time of sampling and this can therefore lessen the uncertainties involved, providing a guide to water content in the ambient conditions affecting the sediment. In addition, the use of sand-sized grains of potassium feldspars as the dosimeter also decreases dose rate uncertainty since their high internal beta dose contribution is unaffected by external water content variation, compared to quartz.

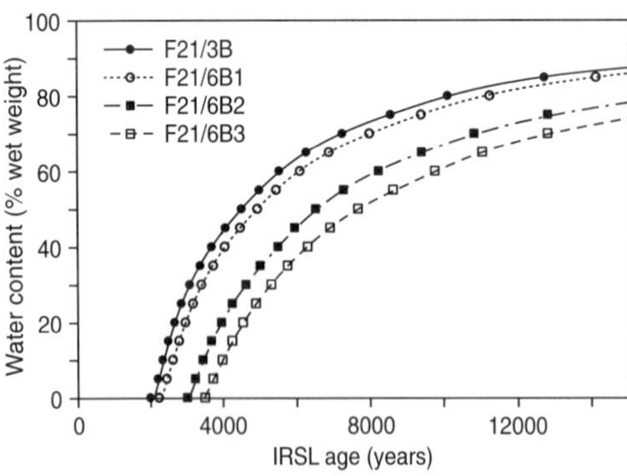

Fig. 7. Modelling the effect of water content on the calculated IRSL age of four polymineral silt samples from sediment core F21.

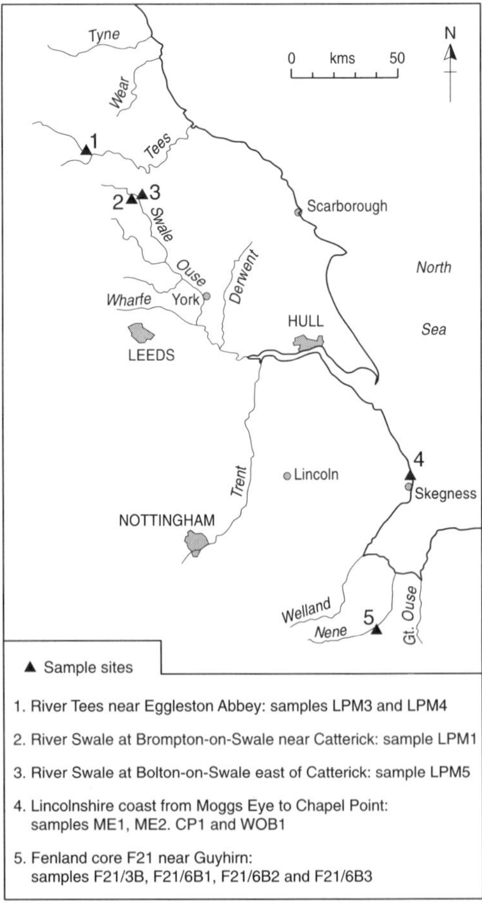

Fig. 8. Location of sample sites for data in Table 3 and Fig. 7.

Development of a dating methodology

In luminescence dating (TL, IRSL and GLSL), the age of the sample is obtained from two independent sets of measurements. The dose rate to the sample from the natural environment is obtained from dosimetry measurements undertaken both in the field and in the laboratory. For sediment grains which had their luminescence signal zeroed fully prior to burial, the equivalent dose (ED) is a measure of the accumulated luminescence signal in the sample since burial and is calibrated against known laboratory beta doses (Aitken 1998). The age of the sample (in years) = ED/dose rate. The methodological development outlined below is based on ED determination experiments.

Establishment of bleaching parameters

The use of the single aliquot approach to ED determination (Duller 1991) has several advantages, particularly in that it is possible to obtain many ED determinations for any given sample. Dating applications routinely use between ten and 18 aliquots to determine the mean ED (e.g. Clarke & Käyhkö 1997; Wintle *et al.* 1998). In well-bleached sediments, which contain grains that were fully zeroed at deposition, there will be a tight clustering of individual ED values and a small standard deviation from the mean ED. Thus, the true ED is known with a high degree of precision. In poorly bleached sediments, the single aliquot approach will show a wide scattering of EDs between aliquots, as a result of different residual levels remaining in the grains at burial due to incomplete zeroing, and the standard deviation from the mean ED will be large. This approach was used by Clarke (1996), who defined a set of criteria for characterizing samples that are poorly bleached and would be a problem for dating, giving an inaccurate age for burial of the sediment. This approach differs from earlier attempts to examine ED scatter by Li (1994), who used a method for extrapolating a minimum ED from a single aliquot data set; and Duller (1994), who assumed that some of the grains in a particular data set would be well-bleached. More recently, Lamothe & Auclair (1997) developed an approach based on the statistical analysis of the brightness of single grains in order to discriminate between unbleached and well-bleached samples. Their approach also differs from that of Clarke (1996), in that she suggested that using the standard deviation (σ_{n-1}) about the mean ED ($\bar{x}$) provides a measure of the existing trapped charge at deposition. For example, a sample with an ED of 12.3 ± 0.2 Gy could be classified as well-bleached, whereas a sample with 12.3 ± 12 Gy would be classified as poorly bleached (Clarke 1996, p. 613). A threshold was set at 5 Gy to provide a cut-off between well-bleached and poorly bleached samples based on analysis of over 70 samples. However, as charge builds up in the lattice over time since burial, the residual trapped charge, which was present from the previous depositional phase, becomes a smaller proportion of the overall trapped charge component in the grain. Thus, at 12.3 Gy a residual component of 12 Gy is significant; but once the total trapped charge has reached 100 Gy, the residual is less of a problem. Thus, the sole use of the standard deviation as a bleaching parameter works well for young samples but is inadequate when applied to older samples. Clarke (1996) recommended the use of an additional parameter, S_N, which provides a measurement of the overall trend in scatter on the ED. S_N is defined as the standard deviation divided by the mean. Unlike the standard deviation, which will remain constant through time as charge builds up in the lattice, the S_N value will decrease over time. Thus if $\sigma_{n-1} < 5$ Gy and $S_N < 0.1$, the sample is well-bleached; where $\sigma_{n-1} > 5$ Gy and $S_N > 0.1$, the sample is poorly bleached at deposition and would give an inaccurate luminescence age. A sample has to fail both of the parameters to classify as poorly bleached. Normalized scatter plots of aliquot EDs and natural intensities (both normalized to their respective means) were proposed by Clarke (1996) for graphical representation of bleaching behaviour.

In a demonstration of the applicability of the Clarke (1996) criteria, Table 3 shows the individual ED values, the mean and bleaching parameters for a range of samples taken from sedimentary environments at the Lincolnshire land–ocean interface and in two UK rivers (see Fig. 8). From Table 3 it is clear that only one of the samples, CP1, which is a freshwater marsh clay (see below), is well-bleached and would provide an accurate luminescence age for deposition and burial. The remaining four river sediments and sample ME1, which is an intertidal silt, are poorly bleached and would give inaccurate dates. These samples are represented in scatter plots in Fig. 9, which shows the distribution of normalized ED and normalized natural brightness (I_n) for the samples. Well-bleached samples, which would give accurate luminescence ages, provide data that cluster tightly around the centre of the graph (Fig. 9b); whereas poorly bleached samples, which would give an inaccurate age, give data that are more

Table 3. *Luminescence dating samples* taken from the coastal zone and fluvial environments: equivalent dose (ED) and natural brightness (l_n) measurements (in counts per second, c.p.s.) for each aliquot including mean ($\bar{x}$), standard deviation (σ_{n-1}) and bleaching parameters (S_N)*

Aliquot number	Lincolnshire coast* ME1		Lincolnshire coast* CP1		River Swale† LPM1		River Swale† LPM5		River Tees† LPM3		River Tees† LPM4	
	ED (Gy)	l_n (c.p.s.)	ED (Gy)	l_n (c.p.s.)	ED (Gy)	l_n (c.p.s.)	ED (Gy)	l_n (c.p.s.)	ED (Gy)	l_n (c.p.s.)	ED (Gy)	l_n (c.p.s.)
1	69.25	6344	10.30	1022	56.00	1466	154.65	3254	43.48	601	34.31	2 182
2	47.13	4674	10.30	836	1125.51	5641	102.98	2326	126.15	3984	13.94	1 014
3	17.21	716	9.67	752	53.44	2618	374.84	7000	37.03	1907	27.54	1 886
4	16.25	1008	9.13	810	64.06	1167	127.91	3606	36.38	1316	18.52	1 921
5	33.54	2124	9.57	866	30.23	773	210.32	5103	81.15	5922	32.98	1 810
6	8.69	7156	9.60	764	158.90	1688	202.51	5729	100.86	4674	175.78	21 528
7	11.90	9556	10.77	810	57.68	1665	147.88	3028	123.42	6942	70.35	5 139
8	9.60	8126	11.20	860	194.99	4637	290.49	7617	37.96	1732	113.87	11 820
9	11.14	3350	12.61	1174	88.60	2759	174.59	3390	45.89	2423	67.22	4 659
10	8.51	7024	11.00	914	64.97	802	229.14	8592	44.61	836	64.81	1 369
11	11.34	4786	9.29	590	83.29	698	135.12	3443	185.86	6128	74.84	5 718
12	8.18	4218	10.28	662	81.21	1630	387.32	4206	166.84	6156	59.79	7 643
13					160.39	1992	353.58	7274	18.10	378	81.54	6 441
14					28.33	1417	170.03	4714	30.91	1183	21.97	846
15					187.01	615	158.42	3623	60.22	2792	102.52	6 442
16							145.24	3269				
17							209.24	4670				
18							165.34	4989				
$\bar{x}$	21.06	4924	10.31	838	162.31	1971	207.87	4769	75.92	3132	64.00	5 361
σ_{n-1}	19.26	2813	0.985	154	272.07	1449	86.86	1803	52.68	2298	43.49	5 432
S_N	0.914	–	0.096	–	1.676	–	0.418	–	0.694	–	0.679	–

* The Lincolnshire coast samples (located 53°11′N, 0°21′E) consist of ME1, an intertidal silt, and CP1, a freshwater marsh clay; both are polymineral fine grains (4–11 μm).
† The locations of the river samples are described in Clarke *et al* (1999) and were taken from fresh exposures on terrace deposits. The samples are sand-size (180–212 μm) potassium-rich feldspar grains.

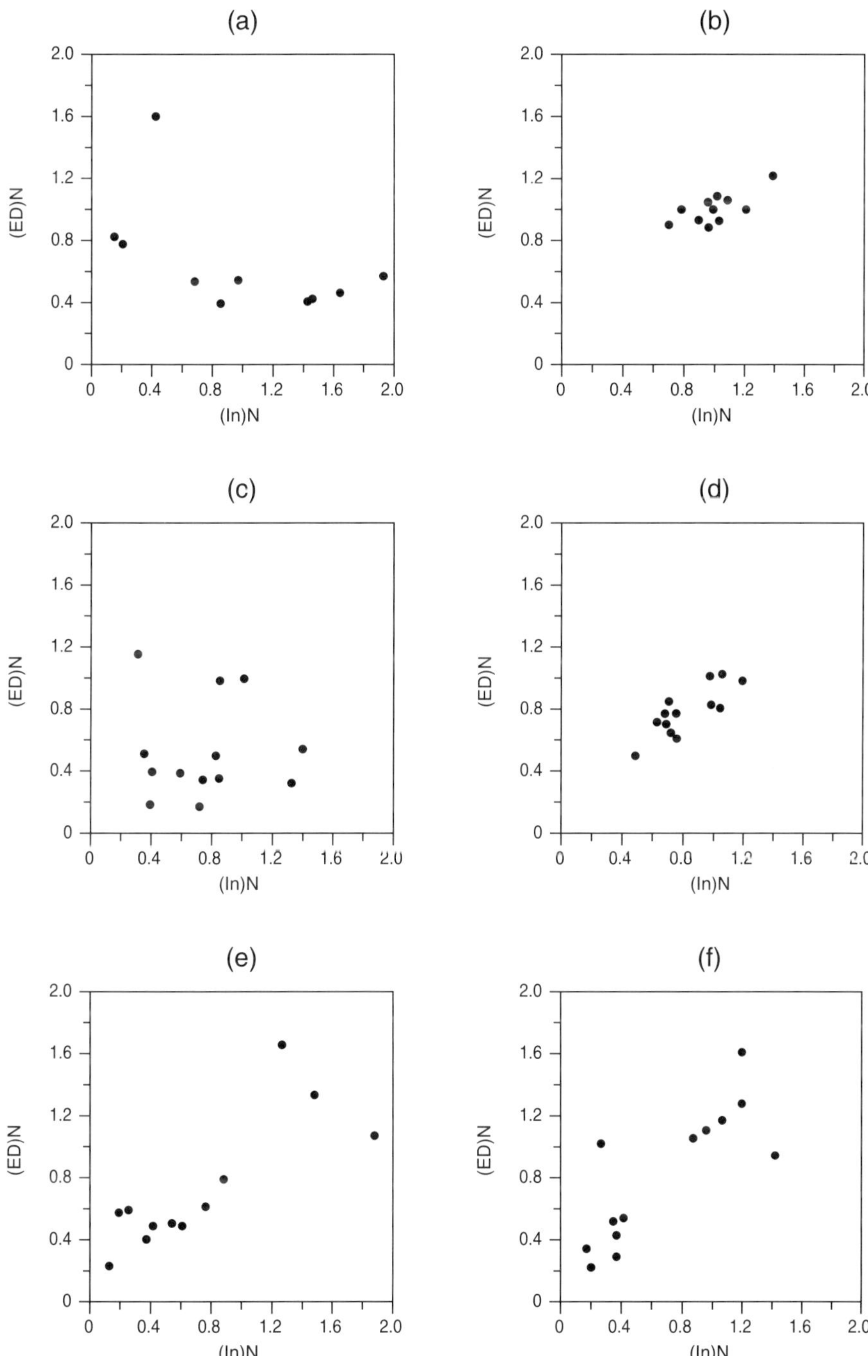

Fig. 9. Scatter plots of normalized equivalent dose $(ED)_N$ against normalized natural intensity $(I_n)_N$ for the coastal zone and river sediments: (**a**) ME1, (**b**) CP1, (**c**) LPM1, (**d**) LPM5, (**e**) LPM3, (**f**) LPM4 (Table 3).

widely scattered. Using the dose rate obtained from the samples it is possible to illustrate the nature of the scatter by using the range of ED values to calculate the equivalent range in age represented by aliquots of the sample. Well-bleached aliquots of CP1 show a range in age from 3140 to 4340 years compared with poorly bleached samples LPM5: 12 180–483 680 years and LPM4: 6590–83 100 years (which are described in detail in Clarke *et al.* 1999). It is clear from these examples of poorly bleached sediments that mean values of ED used in calculation of the age would be arbitrary. It must be stressed here that each sediment sample needs to be assessed separately. Not all fluvial sediments may be poorly bleached; however, the fluvial samples that we have analysed so far (Clarke *et al.* 1999) all fall into this category and would therefore give inaccurate dates. A cautious note with regard to water-lain sediments is also given by Forman *et al.* (1994). Samples that fail the criteria defined by Clarke (1996), due to the fact that they are poorly bleached, should be discarded.

A partial bleach approach

One of the aims of this project, at the outset, was to develop a methodology for application to poorly bleached samples (as defined above) with the objective of obtaining the 'true' ED for deposition and burial after the last transport phase, leading to an accurate luminescence age. As a result, a single aliquot partial bleach approach was devised, based on the methodology initially developed for TL (Wintle & Huntley 1980). The partial bleach approach is used to ascertain the proportion of the remnant signal remaining in the sediment grain at burial due to incomplete zeroing, i.e. where poorly bleached samples have a remnant signal plus a depositional signal, which has built up since the most recent burial event. An accurate identification of the remnant dose component of the ED, would allow an accurate age for the most recent burial event to be determined (i.e. 'true' ED, measured in grays = depositional dose − remnant dose).

For this approach to work, the remnant component must form a discrete population, which can remain distinct from the additional trapped charge that subsequently builds up after burial (the depositional dose). The methodology devised here was used on poorly bleached samples described in Clarke (1996). The aliquots were treated in the same way as those for additive dose single aliquot IRSL (using a preheat of 220°C for 10 min). However, the chosen stimulation time (10 s) is greater than for the additive dose (0.5 or 1 s), and thus the preheat and signal loss calibration must be increased. The experiment was set up using a 10 s infra-red exposure, and the photon count data (decay curve), once collected, was split into 1 s timeslices. The preheat and signal loss calibration was undertaken using 10 s exposures,

Table 4. *Single aliquot IRSL additive dose and partial bleach EDs (Gy) for two poorly bleached samples of colluvium (C3) and alluvium (F3)*

Aliquot number	Colluvium: C3*			River Tyne: F3*		
	Additive dose (1.0 s)	Additive dose (10.0s)	Partial bleach (1.0 & 10.0 s)	Additive dose (1.0 s)	Additive dose (10.0 s)	Partial bleach (1.0 & 10.0 s)
1	33.31	37.45	24.65	29.55	32.39	23.57
2	27.62	30.65	21.66	54.06	51.24	60.41
3	41.96	47.79	30.99	48.78	51.85	41.87
4	35.28	40.15	25.98	56.13	57.75	52.74
5	29.04	32.54	21.96	62.87	63.13	62.19
6	36.93	40.50	29.98	74.69	81.59	55.76
7	27.79	31.25	20.34	79.40	86.39	61.86
8	33.78	37.74	25.33	60.43	63.38	56.75
9	32.13	35.37	25.39	44.25	47.22	36.72
10	30.78	35.07	22.22	50.52	51.58	47.45
11	30.96	34.35	23.82	35.48	37.33	29.99
12	30.30	33.64	23.25	55.28	59.14	45.56
$\bar{x}$	32.49	36.38	24.63	54.29	56.92	47.91
σ_{n-1}	4.13	4.79	3.23	14.38	15.75	12.79
S_N	0.127	0.132	0.131	0.265	0.276	0.267

* The locations of the colluvium and River Tyne sample are described in Clarke (1996).

and this was also split into 1 s integrals and the appropriate integrals used to calibrate the equivalent photon count timeslices (e.g. a 2–3 s preheat integral would correct the 2–3 s photon count data). This procedure does not substantially increase measurement time above that of a routine single aliquot additive dose measurement, and allows, in this case, up to 11 ED determinations to be obtained upon each aliquot. The only obvious disadvantage of this methodology was dealing with the vast quantities of data produced from each run.

This method was tested on a range of poorly bleached samples and is illustrated here for C3, a sandy colluvium from South Africa, and F3, a sandy alluvium from the River Tyne (see Clarke 1996). The Tyne sample, in particular, showed a high degree of inter-aliquot scatter, but the dose response from individual aliquots showed that the material was well behaved (see Clarke *et al.* 1999). The 0.5, 1 and 10 s calibrated integrals were used to obtain additive dose growth curves (and EDs), and the 1 and 10 s curves were used to determine a partial bleach ED (see Table 4). It was hoped that, if successful, the partial bleach approach, applied to all of the aliquots, would decrease the scatter on the mean ED as the 'true' residual level in the last depositional cycle was approached.

Table 4 shows the EDs for each aliquot at these exposure times and the bleaching parameters for the sample as a whole. The results show that the partial bleach approach decreases the overall ED (as would be expected given that a bleaching residual level which is above zero is being defined), but does not significantly decrease the scatter (Table 4). Therefore, although the procedure itself was shown to work well, the results show that no remnant component could be identified for any of the samples tested. This implies that there is no distinct remnant charge component and confirms that the only thing that can be done with poorly bleached samples, identified using the Clarke (1996) criteria, is to continue to discard them.

A new methodology with quality assurance capability

Samples with independent age control have been used to test the rigour of the Clarke (1996) dual bleaching criteria (outlined above). The success of this methodology, when compared with independent age control, has resulted in the proposal that the single aliquot method and the Clarke (1996) criteria be used as a universal quality assurance method for luminescence dating (Clarke *et al.* 1999). This is because the approach is equally applicable to both feldspar and quartz dosimeters, to a range of grain sizes from sands to silts, and has been shown to be rigorous in comparison with independent age control (Clarke *et al.* 1999).

The single aliquot procedure used by Clarke *et al.* (1999) has been applied to a sequence of sediments that are exposed along the Lincolnshire coast. The exposures provide a range of sedimentary environments (dune sand, intertidal silts, freshwater marsh and saltmarsh) and different grain sizes (allowing the use of sand-sized feldspars and polymineral fine silt grains). Independent age control exists for the sequence in the form of radiocarbon dates on wood and peat. The Lincolnshire sites were chosen, as using field sections is advantageous in that it provides the opportunity to take *in situ* measurements of water content, *in situ* gamma dose rate measurements and allows an abundance of sediment to be sampled for laboratory analysis.

The Lincolnshire coastal sequence described below occurs north of Chapel St Leonards and at Ingoldmells Point (53°11′N 0°21′E). Shoreface exposures of the lower part of the sequence, found above MLWS at Wolla Bank [557 750] and Chapel Six Marshes [560 743], were sampled in February 1996. These exposures are now buried by beach recharge, undertaken by the Environment Agency to improve sea defences, involving onshore pumping of millions of tonnes of sand and gravel on to the beaches between Mablethorpe (53°21′N 0°14′E) and Skegness (53°09′N 0°20′E). The base of the sequence consists of a mid-Holocene forest bed, rooted in Pleistocene till (Fig. 10). Whole oak stumps were exposed, along with associated peat, at low tide on the beach at Wolla Bank before the recharge work. A rapid marine transgression killed off the forest, allowing insufficient time for a build up of thick brackish or freshwater swamp deposits. This period was followed by a still stand or slight relative fall in sea-level when freshwater marsh clay and Iron Age peat accumulated. Radiocarbon dates were obtained from the Iron Age peat (Pye pers. comm.) and are shown as calibrated dates in Fig. 10. Intertidal mud deposition, which buried the Iron Age peat, probably occurred in a back-barrier environment. The present coastline was overrun by the receding dune-capped barrier around the seventeenth or eighteenth century (Pye pers. comm.), exposing the older sediments on the foreshore.

Two samples were taken from a trench excavated at Marsh Yard, behind the artificial dune barrier created at the beach top at Moggs Eye [545 776], which lies seaward of the roman

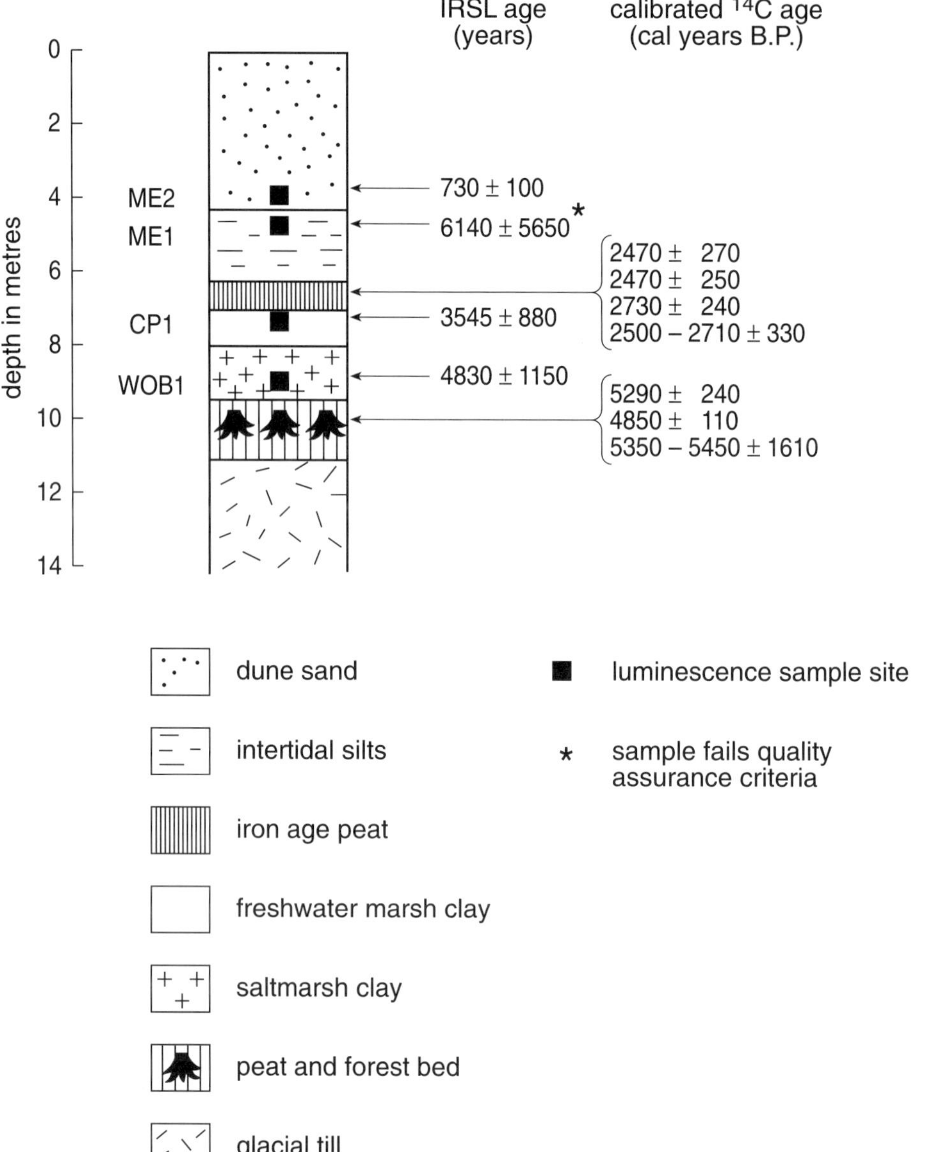

Fig. 10. IRSL ages for coastal zone sediments from Lincolnshire.

bank at Bank House. ME1 was located in the intertidal silts and ME2 was taken from the overlying dune sand (Fig. 10). The samples were carved out of the face using a knife and a trowel and placed into an opaque black sample bag, whilst the trench was covered over using a black tarpaulin. In this way, the samples were taken without being exposed to light.

The freshwater marsh clay was sampled from a location on the beach at Chapel Six Marshes immediately beneath the Iron Age peat. Blocks of the sediment (CP1) were cut with a spade, and these were transferred into an opaque black bag. The outer faces, which were exposed to light, were then trimmed off under subdued lighting in the laboratory. The final sample (WOB1), taken from the base of the saltmarsh deposit, was taken from an exposure on the beach at Wolla Bank immediately above the forest bed, using the same procedure as for CP1.

The IRSL ages obtained from the sequence are shown in Fig. 10. They are calculated in each case using the measured field water content. The IRSL ages show good agreement with the radiocarbon dates, with the exception of sample ME1, which is the only sample that failed the Clarke (1996) criteria, and was thus expected to give an inaccurate age and would have been discarded. This dating application is a useful illustration of the sensitivity of the quality assurance technique and shows that IRSL techniques can provide accurate ages from freshwater and saltmarsh environments. It is also clear from this work, and that of others (e.g. Berger *et al.* 1990; Berger & Easterbrook 1993; Forman *et al.* 1994), that samples from a range of sedimentary coastal zone environments will provide a luminescence date. The new methodology outlined above shows that not all samples will provide dates that are accurate.

Conclusion

The development of a methodology for luminescence dating Holocene sediments at the land–ocean interface has involved detailed analysis of the fundamental characteristics of luminescence behaviour to identify a suitable signal of sufficient long-term stability to allow accurate ages to be obtained. An unstable IRSL signal has been identified and a preheat procedure proposed to eliminate the problem. The use of feldspars as the dosimeter in a dating regime is advantageous for young sediments derived from facies where significant water content fluctuations are likely. In these conditions, sampling from *in situ* exposures is preferable to sampling from sediment cores.

Alkali feldspars bleach rapidly in daylight and chemical weathering does not affect the equivalent dose derived from feldspar grains. Given these fundamental characteristics, a quality assurance methodology has been developed for application to all types of sedimentary environment. Application of this method, using IRSL techniques to date the alkali feldspar fraction of coastal zone sediments has shown that accurate ages can be obtained from freshwater and saltmarsh environments.

This work was funded by Natural Environment Research Council Award GST/02/755. We would like to thank members of the Aberystwyth Laboratory: A. G. Wintle, C. A. Richardson and F. M. Musson for discussions during the course of this project, K. Pye for collaboration on dating the Wolla Bank sections, M. G. Macklin for collaboration on sampling the Rivers Tees and Swale, and H. Glaves for help in sampling core F21 at the British Geological Survey, Keyworth. M. L. Clarke would like to acknowledge funding from NERC Award GST/02/762 during development of the single aliquot partial bleach approach. This is LOIS publication number 548.

References

AITKEN, M. J. 1998. *An Introduction to Optical Dating: the Dating of Quaternary Sediments by the Use of Photon-Stimulated Luminescence*. Oxford University Press, Oxford.

BALESCU, S. & LAMOTHE, M. 1992. The blue emission of K-feldspar coarse grains and its potential for overcoming TL age underestimates. *Quaternary Science Reviews*, **11**, 45–51.

—— & ——1994. Comparison of TL and IRSL age estimates of feldspar coarse grains from waterlain sediments. *Quaternary Geochronology*, **13**, 437–444.

BERGER, G. W. 1990. Effectiveness of natural zeroing of the thermoluminescence in sediments. *Journal of Geophysical Research*, **95**, 12 375–12 397.

—— & EASTERBROOK, D. J. 1993. Thermoluminescence dating tests for lacustrine, glaciomarine, and floodplain sediments from Western Washington and British Columbia. *Canadian Journal of Earth Sciences*, **30**, 1815–1828.

——, LUTERNAUER, J. L. & CLAGUE, J. J. 1990. Zeroing tests and application of thermoluminescence dating to Fraser River delta sediments. *Canadian Journal of Earth Sciences*, **27**, 1737–1745.

BERNER, R. A. & HOLDREN, G. R., JR. 1979. Mechanism of feldspar weathering. II. Observations of feldspars from soils. *Geochimica et Cosmochimica Acta*, **43**, 1173–1186.

BLUM, A. E. 1994. Feldspars in weathering. *In*: PARSONS, I. (ed.) *Feldspars and their Reactions*. Kluwer Academic Publishers, Dordrecht, 595–630.

CLARKE, M. L. 1996. IRSL dating of sands: bleaching characteristics at deposition using single aliquots. *Radiation Measurements*, **26**, 611–620.

—— & KÄYHKÖ, J. A. 1997. Evidence of Holocene aeolian activity in sand dunes from Lapland. *Quaternary Geochronology*, **16**, 341–348.

—— & RENDELL, H. M. 1997*a*. Stability of the IRSL spectra of alkali feldspars. *Physica Status Solidi B*, **199**, 597–604.

—— & ——1997*b*. Infra-red luminescence spectra of alkali feldspars. *Radiation Measurements*, **27**, 221–136.

—— & ——1998. Climate change impacts on sand supply and the formation of desert dunes in the south-west USA. *Journal of Arid Environments*, **39**, 517–531.

——, ——, SANCHEZ-MUÑOZ, L. & GARCIA-GUINEA, J. 1997. A comparison of luminescence spectra and structural composition of perthitic feldspars. *Radiation Measurements*, **27**, 137–144.

——, —— & WINTLE, A. G. 1999. Quality assurance in luminescence dating. *Geomorphology*, **29**, 173–185.

——, WINTLE, A. G. & LANCASTER, N. 1996. Infra-red stimulated luminescence dating of sands from the Cronese Basins, Mojave Desert. *Geomorphology*, **17**, 199–205.

Debenham, N. C. 1985. Use of UV emissions in TL dating of sediments. *Nuclear Tracks and Radiation Meaurements*, **10**, 717–724.

Duller, G. A. T. 1991. Equivalent dose determination using single aliquots. *Nuclear Tracks and Radiation Measurements*, **18**, 371–378.

——1994. A new method for the analysis of infrared stimulated luminescence data from potassium feldspars. *Radiation Measurements*, **23**, 281–285.

Forman, S. L., Lepper, K. & Pierson, J. 1994. Limitations of infra-red stimulated luminescence in dating high arctic marine sediments. *Quaternary Geochronology*, **13**, 545–550.

Godfrey-Smith, D. I., Huntley, D. J. & Chen, W.-H. 1988. Optical dating studies of quartz and feldspar sediment extracts. *Quaternary Science Reviews*, **7**, 373–380.

Helgeson, H. C., Murphy, W. M. & Aagaard, P. 1984. Thermodynamic and kinetic constraints on reaction rates among minerals and aqueous solution. II. Rate constants, effective surface area and the hydrolosis of feldspar. *Geochimica et Cosmochimica Acta*, **48**, 2405–2432.

Holdren, G. R., Jr & Berner, R. A. 1979. Mechanism of feldspar weathering. I. Experimental studies. *Geochimica et Cosmochimica Acta*, **43**, 1161–1171.

—— & Speyer, P. M. 1985. pH dependent changes in the rates and stoichiometry of dissolution of an alkali feldspar at room temperature. *American Journal of Science*, **285**, 994–1026.

Huntley, D. J. 1985. On the zeroing of the thermoluminescence of sediments. *Physics and Chemistry of Minerals*, **12**, 122–127.

——, Godfrey-Smith, D. I. & Haskell, E. H. 1991. Light-induced emission spectra from some quartz and feldspars. *Nuclear Tracks and Radiation Measurements*, **18**, 127–131.

——, ——, Thewalt, M. L. W. & Berger, G. W. 1988*a*. Thermoluminescence spectra of some mineral samples relevent to thermoluminescence dating. *Journal of Luminescence*, **44**, 41–46.

——, ——, ——, Prescott, J. D. & Hutton, J. T. 1988*b*. Some quartz thermoluminescence spectra relevant to thermoluminescence dating. *Nuclear Tracks and Radiation Measurements*, **14**, 27–33.

Jungner, H. & Huntley, D. J. 1991. Emission spectra of some potassium feldspars under 633 nm stimulation. *Nuclear Tracks and Radiation Measurements*, **18**, 125–126.

Käyhkö, J. A., Worsley, P., Pye, K. & Clarke, M. L. 1999. A revised chronology for aeolian activity in subarctic Fennoscandia during the Holocene. *The Holocene*, **9**, 195–205.

Knauss, K. G. & Wolery, T. J. 1986. Dependence of albite dissolution kinetics on pH and time at 25 and 70°C. *Geochimica et Cosmochimica Acta*, **50**, 2481–2497.

Krause, W. E., Krbetschek, M. R. & Stolz, W. 1997. Dating of Quaternary lake sediments from the Schirmacher Oasis (East Antarctica) by infra-red stimulated luminescence (IRSL) detected at the wavelength of 560 nm. *Quaternary Geochronology*, **16**, 387–392.

Krbetschek, M. R., Rieser, U. & Stolz, W. 1996. Optical dating: some luminescence properties of natural feldspars. *Radiation Protection Dosimetry*, **66**, 407–412.

Lamothe, M. & Auclair, M. 1997. Assessing the datability of young sediments by IRSL using an intrinsic laboratory protocol. *Radiation Measurements*, **27**, 107–117.

Li, S.-H. 1994. Optical dating: insufficiently bleached sediments. *Radiation Measurements*, **23**, 563–567.

Ollerhead, J., Huntley, D. J. & Berger, G. W. 1994. Luminescence dating of sediments from Buctouche Spit, New Brunswick. *Canadian Journal of Earth Sciences*, **31**, 523–531.

Parish, R. 1994. The influence of feldspar weathering on luminescence signals and the implications for luminescence dating of sediments. *In*: Robinson, D. A. & Williams, R. B. G. (eds) *Rock Weathering and Landform Evolution*. Wiley, Chichester, 243–258.

Prescott, J. R. & Fox, P. J. 1993. Three dimensional thermoluminescence spectra of feldspars. *Journal of Physics, D: Applied Physics*, **26**, 2245–2254.

Rendell, H. M. & Clarke, M. L. 1997. Thermoluminescence, radioluminescence and cathodoluminescence spectra of alkali feldspars. *Radiation Measurements*, **27**, 263–272.

—— & Sheffer, N. L. 1996. Luminescence dating of sand ramps in the eastern Mojave Desert. *Geomorphology*, **17**, 187–197.

Richardson, C. A., McDonald, E. V. & Bussacca, A. J. 1997. Luminescence dating of loess from the Northwest United States. *Quaternary Geochronology*, **16**, 403–415.

Spooner, N. A. 1994. The anomalous fading of infra-red stimulated luminescence from feldspars. *Radiation Measurements*, **23**, 625–632.

Stokes, S. & Gaylord, D. R. 1993. Optical dating of Holocene dune sands in the Ferris Dune Field, Wyoming. *Quaternary Research*, **39**, 274–281.

Templer, R. H. 1986. The localised transition model of anomalous fading. *Radiation Protection Dosimetry*, **17**, 495–497.

Visocekas, R. 1985. Tunnelling radiative recombination in labradorite: its association with anomalous fading. *PACT*, **3**, 258–265.

Wintle, A. G. 1973. Anomalous fading of thermoluminescence in mineral samples. *Nature*, **245**, 143–144.

——1977. Detailed study of a thermoluminescent mineral exhibiting anomalous fading. *Journal of Luminescence*, **15**, 385–397.

—— & Huntley, D. J. 1980. Thermoluminescence dating of ocean sediments. *Canadian Journal of Earth Sciences*, **17**, 348–360.

——, Clarke, M. L., Musson, F., Orford, J. & Devoy, R. 1998. Luminescence dating of recent dunes on Inch Spit, Dingle Bay, southwest Ireland. *The Holocene*, **8**, 331–339.

Holocene environmental change in the Yorkshire Ouse basin and its influence on river dynamics and sediment fluxes to the coastal zone

M. G. MACKLIN,[1] M. P. TAYLOR,[2] K. A. HUDSON-EDWARDS[3] & A. J. HOWARD[4]

[1] *Institute of Geography and Earth Sciences, The University of Wales, Aberystwyth, Aberystwyth, Ceredigion SY23 3DB, UK (e-mail: mvm@aber.ac.uk)*
[2] *School of Geography, University of Oxford, Mansfield Road, Oxford OX1 3TB, UK*
[3] *Department of Geology, Birkbeck College, University of London, London WC1E 7HX, UK*
[4] *The School of Geography, The University of Leeds, Leeds LS2 9JT, UK*

Abstract: Geomorphological, geochemical and geochronological investigations of Holocene fluvial sedimentary sequences have been undertaken within a range of upland, piedmont and lowland valley floor reaches in the Yorkshire Ouse catchment, northern England. The aims of these studies have been to: (a) evaluate the effects of prehistoric and historic land-use change on catchment erosion and sediment delivery to river channels and floodplains; (b) establish the degree to which episodes of river erosion and sedimentation are controlled by climate-related variations in flood regime; and (c) assess the spatial heterogeneity of river response to environmental change and how this is likely to influence short- and long-term sediment storage, as well as sediment transfer to the Humber Estuary. Similar discontinuities in the Holocene alluvial record are evident at many sites in the Yorkshire Ouse catchment, though local differences in river sensitivity to externally imposed change have resulted in a complicated and often unique relationship between river behaviour and environmental change. The large proportion of particulate-borne contaminant metals (resulting predominantly from historical mining) stored in the Vale of York strongly indicates that sediment delivery from the Ouse catchment to the Humber Estuary during the Holocene may have been relatively low. This suggests that the degree of connectivity between river, estuarine and coastal transport systems, as well as spatial and temporal variations in fluvial sediment storage, are the key controls of long-term land–ocean sediment fluxes.

One of the central research issues in Holocene river studies in the UK, and elsewhere in the world, is the effect of climate and land-use change on river dynamics, sediment fluxes and sediment delivery in drainage basins. Over the last decade or so, there has been a shift in the UK away from the view that human activity was the sole agent controlling Holocene river sedimentation (e.g. Bell & Walker 1992; Robinson 1992; Tipping 1992), towards an appreciation that climate has changed significantly over the past 10 ka and that rivers have periodically entrenched or alluviated their channels and floodplains in response to climate-related variations in runoff and flow regime (e.g. Macklin *et al.* 1992; Macklin & Lewin 1993; Passmore *et al.* 1993; Macklin 1999). The questions that are now being asked is the *degree* to which a catchment (or reach within a catchment) has been influenced by environmental change resulting from inadvertent or deliberate anthropogenic interference, or from natural processes such as variations in climate, and how these factors have varied in importance over time and geographically during the Holocene.

Geomorphological, geochemical and geoarchaeological investigations of Holocene fluvial sedimentary sequences in the Yorkshire Ouse basin, northern England, have been undertaken to address this important topic. In particular, to assess the spatial heterogeneity of river response to environmental change and how this influences short- and long-term sediment storage and transfer to the coastal zone. This forms part of the Natural Environmental Research Council's (NERC) Land–Ocean Interaction Study (LOIS) community research programme in which the Yorkshire Ouse basin has been the principal focus of hydrological, geochemical and geological investigations. Many of these studies have already been published (Macklin *et al.* 1997; Taylor & Macklin 1997; Dawson & Macklin

From: SHENNAN, I. & ANDREWS, J. (eds) *Holocene Land–Ocean Interaction and Environmental Change around the North Sea*. Geological Society, London, Special Publications, **166**, 87–96. 1-86239-054-1/00/$15.00

1998; Howard & Macklin 1998; Hudson-Edwards *et al.* 1999*a*, *b*; Longfield & Macklin 1999) and in this paper a synthesis of this work is presented. Additionally, an updated interpretation of Holocene river activity in the Yorkshire Ouse basin is proposed, which utilizes new data from the Yorkshire Dales and, for the first time, compares the regional fluvial record with climatic reconstructions from peat bogs in the Humberhead Levels (Smith 1985) and the North York Moors (Chiverrell 1998).

Study area

Geomorphological and geochemical investigations have been undertaken in all of the major river systems in the Ouse basin that drain the Yorkshire Dales (Fig. 1). These are the Rivers Swale, Ure, Nidd, Wharfe and Aire, which have a total catchment area just exceeding 8000 km^2. Topographically, the Carboniferous limestone and sandstone Pennine Hills in the west, which rise to over 700 m (Fig. 2), dominate these catchments. Beyond the Pennine escarpment lies the relatively flat, low-lying Vale of York, developed in Triassic and Permian bedrock. The eastern side of the Ouse catchment is formed by the Jurassic uplands of the North York Moors and is drained by the River Derwent. To establish a basin-wide picture of Holocene river development, study reaches were selected in the Pennine uplands of the Ouse catchment (River Wharfe at Kettlewell), in the piedmont zone at the margin of the Pennines (River Swale at Catterick) and in the Vale of York (River Swale at Myton-on-Swale, River Nidd at Kirk Hammerton, River Ouse at York, River Wharfe at Tadcaster and River Aire at Beal, Fig. 1).

Holocene river evolution, channel change and sedimentation styles have varied considerably between these areas. In the higher-relief upland and piedmont zones, including upper Swaledale and Wharfedale, Pleistocene and Holocene age river terraces testify to laterally and vertically mobile channels that have episodically incised their beds and eroded their banks. At the

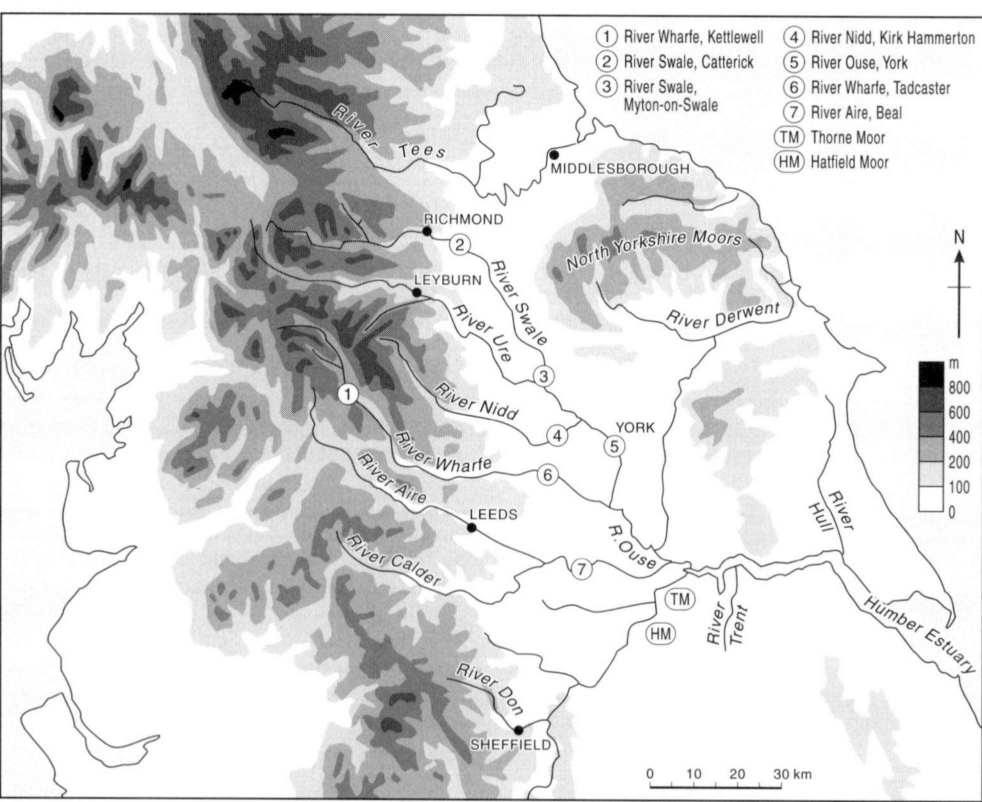

Fig. 1. Relief and drainage network in the Yorkshire Ouse basin and adjoining region. The location of study reaches (marked 1–7) is also shown.

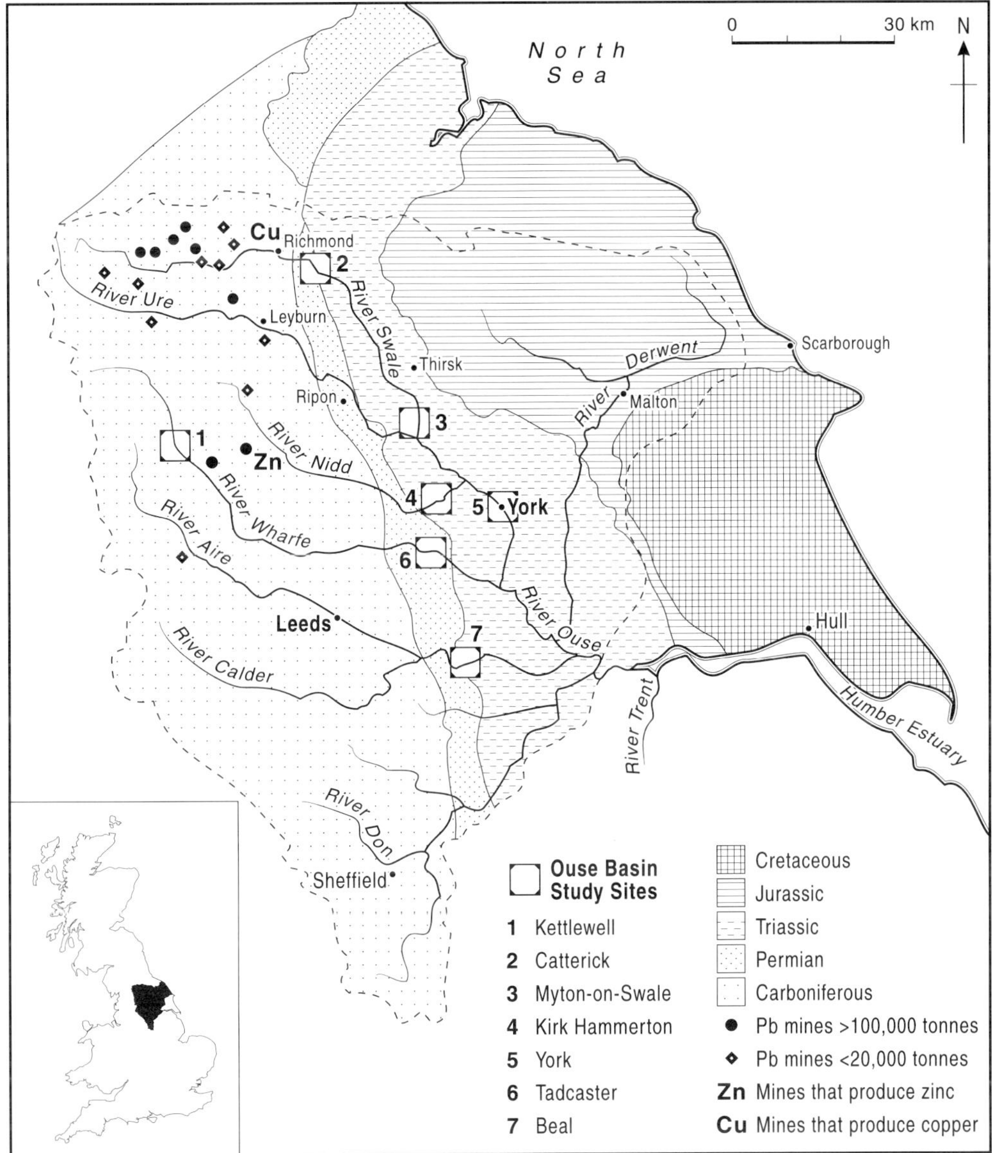

Fig. 2. Geology of the Ouse catchment and location of principal metal mines.

beginning of the Holocene this was a consequence of reduced sediment supply following the end of the Late Devensian glaciation (Taylor & Macklin 1997) and more recently, in historical times, as a result of increased runoff following changes in river basin hydrology and the effects of major floods (Merrett & Macklin 1999). At both upland study reaches on the Wharfe and Swale, map evidence shows that there has been no appreciable channel movement in the last 200 years and that the last major period of channel change occurred before the end of the eighteenth century. Most of the river channels in the Vale of York are presently inactive and have low-sinuosity meandering patterns. Valley floor relief is generally low (<2m), except where channels have been embanked and artificially raised. Before large-scale drainage and channelization, which began in the early seventeenth century, documentary sources and maps show that many channels in lowland areas were divided, anastomosed systems with adjoining riparian wetlands

(Dinnin 1997). Borehole records at these sites show Holocene sediment fills commonly exceeding 10 m in thickness, comprising multistorey, sandy–gravel channels and organic-rich silty floodplain units that formed in response to long-term valley floor aggradation.

Field and laboratory analytical methods

At each study reach, valley floor morphology was mapped at a scale of 1:10 000. The height of river terrace surfaces and palaeochannels were levelled to an accuracy of 0.01 m using an electronic distance meter and, where possible, tied into Ordnance datum (OD) using the nearest benchmark. River terrace and floodplain units at Kettlewell and Catterick were logged and sampled for geochemical analyses and ^{14}C dating using machine-cut trenches and eroding riverbank sections. In the Vale of York, due to the absence of riverbank exposures and high water tables, sediments were sampled using percussion drilling. In total 45 boreholes, 14 trenches and ten river-bank-sections were studied. Geochemical analyses were carried out in all reaches to establish first, the effects of historic base metal mining on rivers draining the Yorkshire Dales orefield; and secondly, to determine the principal sources of alluvial sediment and whether these had varied during the Holocene. All samples (249 total) were analysed by X-ray fluorescence (XRF) spectrometry at the British Geological Survey for Al_2O_3, CaO, Fe_2O_3, K_2O, MgO, MnO, SiO_2, TiO_2, As, Ba, Co, Cr, Cu, Ni, Pb, Rb, Sr, Y Zn and Zr, and full details of the analytical procedures can be found in Hudson-Edwards *et al.* (1999*b*). Age control was provided by 52 ^{14}C dates on a variety of organic materials (wood, bone and peat) incorporated within alluvial deposits. These have been calibrated using the Oxcal Program v2.18 (Bronk Ramsey 1995) and are reported as calibrated years (calibrated BC or calibrated AD).

Holocene river sedimentation and erosion in the Ouse catchment: evaluating the impact of climate and sea-level change

In Fig. 3 Holocene alluvial units at the Kettlewell, Catterick, Myton-on-Swale and Beal study reaches (where ^{14}C dating control is most complete) are plotted on height–age diagrams, along with the principal recurrence surfaces in the Humberhead Levels identified by Smith (1985). The recurrence surfaces are interpreted as representing wet shifts in climate and have been dated to 1260–820 and 820–400 cal. BC, and to cal. AD 220–500, cal. AD 640–890 and cal. AD 1250–1410. More recent work on raised and blanket mires, both locally in the North York Moors (Chiverrell 1998) and elsewhere in northern Britain (Barber *et al.* 1994; Chambers *et al.* 1997), have also demonstrated major changes in peat humification at these times and a shift towards a wetter and/or cooler climate.

In terms of general river behaviour, the last 5 ka in the upland and piedmont reaches of the Ouse basin have been characterized by episodic valley floor sedimentation and channel bed incision, which increased in both amplitude and frequency at *c.* 900–430 cal. BC, cal. AD 725–1015 and again at cal. AD a 1205–1450 (Fig. 3). By contrast in the Vale of York, and in the perimarine parts of the Ouse catchment, there has been progressive valley floor alluviation with accelerated phases of sedimentation beginning at *c.* 3700–3380 cal. BC and *c.* cal. AD 1020–1220 at Myton-on-Swale, and at 2320–1970 cal. BC and cal. AD 1040–1290 on the River Aire at Beal (Fig. 2). In the last 6 ka, nine major phases of river alluviation and incision can be recognized in the Ouse catchment and these are summarized in Table 1. Many of these episodes coincide with inferred climate shifts identified by Smith (1985) in the Humberhead Levels and by Chiverrell (1998) in the North York Moors.

Alluviation of the entrenched, early Holocene valley floor at Myton-on-Swale in the Vale of York, and a switch from deposition of gravel to fine-grained sands and silts, began at *c.* 3700–3380 cal. BC. There was also alluviation upstream in the piedmont reach at Catterick at this time, which continued until *c.* 1850 cal. BC. Between 3510–2930 cal. BC paludification and peat development began in the northern part of the Humberhead Levels at Thorne Moors (Smith 1985), which has been linked to climate and sea-level change. A shift to wetter climatic conditions at 3550 cal. BC is recorded at many mire and lake sites in northern Britain (Barber *et al.* 1994; Anderson *et al.* 1998) and in the Humber Estuary there is evidence of a marine transgression and shoreline retreat at this time (Long *et al.* 1998). Myton-on-Swale, however, is located too far upstream to have been directly affected by base-level change associated with sea-level rise and, it is more likely, a climate-related increase in valley floor water-tables and the frequency of floodplain inundation were the main factors controlling sedimentation. Anthropogenic interference is considered unlikely, as pollen records in the region show very little

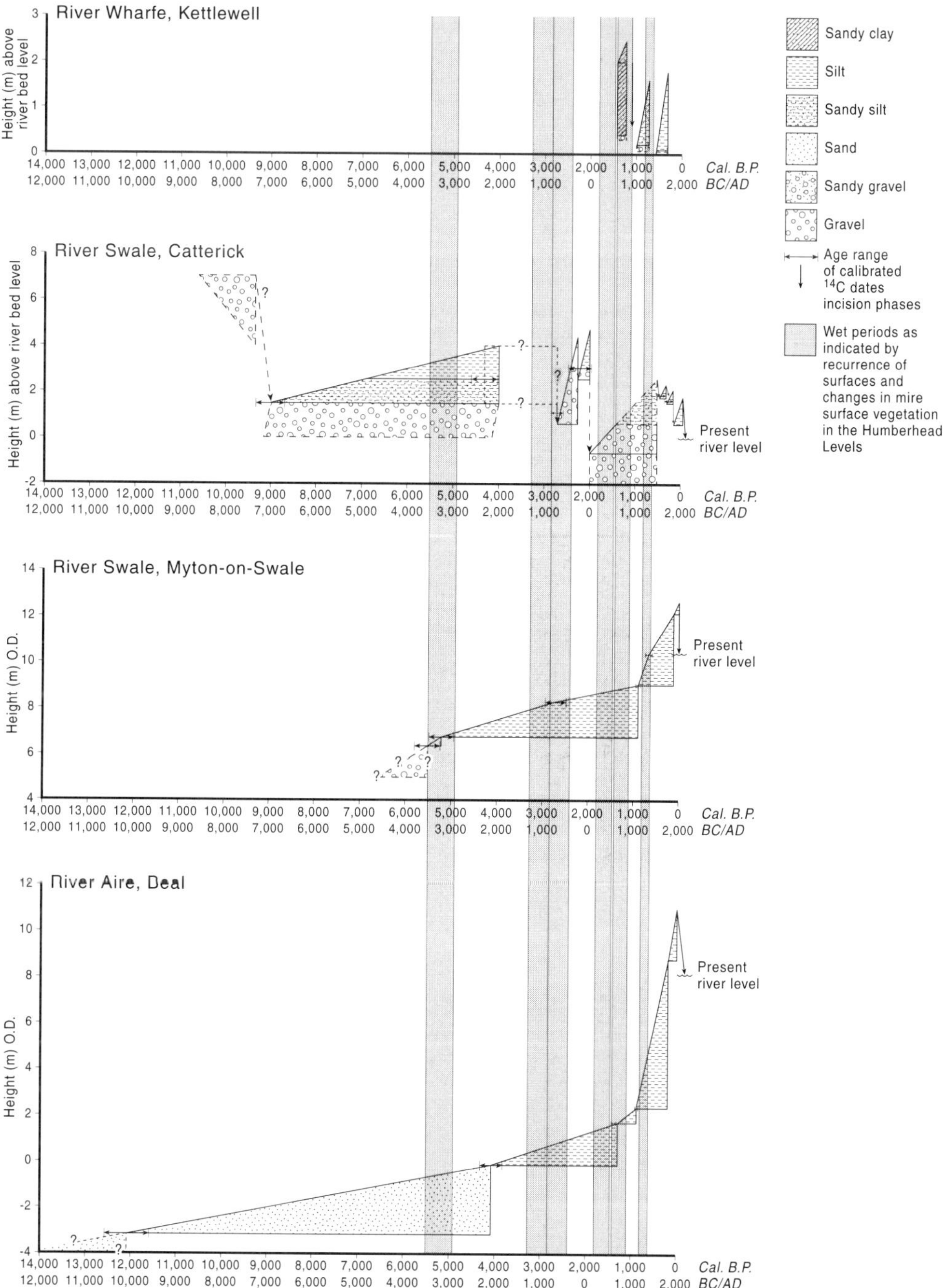

Fig. 3. Height–age diagram of Holocene alluvial deposits in the Rivers Wharfe, Swale and Aire. Wet periods of climate (after Smith 1985) identified from mire stratigraphies in the Humberhead Levels are also shown.

human impact on vegetation during the Neolithic. The period 3700–3380 cal. BC, appears to have been a phase of sediment storage in the piedmont and lowland parts of the Ouse catchment, with sedimentation rates at Beal (Fig. 3) remaining constant, suggesting that sediment delivery from the Ouse catchment to the Humber Estuary was quite limited. The first probable

Table 1. *Major periods of Holocene river activity in the Yorkshire Ouse catchment*

Calibrated date (years)	River activity	Description
AD 1420–1645	Alluviation (fine grained)	In upland, piedmont, lowland and perimarine reaches coinciding with recurrence surface in the North York Moors (Chiverrell 1998) at cal. AD 1400–1450
AD 1205–1450	Incision	In upland and piedmont reaches coinciding with recurrence surface in the Humberhead Levels (Smith 1985) at cal. AD 1250–1410
AD 1015–1290	Alluviation (fine grained)	In upland, lowland and perimarine reaches coinciding with recurrence surface in the North York Moors (Chiverrell 1998) at cal. AD 1250–1300
AD 775–1015	Incision	In uplands coinciding with recurrence surfaces in the Humberhead Levels (Smith 1985) at cal. AD 640–890 and in the North York Moors (Chiverrell 1998) at cal. AD 800–900
AD 645–775	Alluviation (fine grained)	In uplands coinciding with recurrence surfaces in the Humberhead Levels (Smith 1985) at cal. AD 640–890 and in the North York Moors (Chiverrell 1998) at cal. AD 800–900
900–430 BC	Alluviation (coarse grained)	In piedmont, lowland and perimarine reaches coinciding with recurrence surfaces in the Humberhead Levels (Smith 1985) at 820–400 cal. BC and in the North York Moors (Chiverrell 1998) at 500–400 cal. BC
1850–900 BC	Incision	In piedmont reach coinciding with recurrence surface in the Humberhead Levels (Smith 1985) at 1260–820 cal. BC
2320–1850 BC	Alluviation (fine grained)	In piedmont, lowland and perimarine reaches
3700–3380 BC	Alluviation (fine grained)	In piedmont and lowland reaches coinciding with paludification and peat development in the Humberhead Levels (Smith 1985) at *c.* 3510–2930 cal. BC

evidence for sea-level change directly affecting river sedimentation occurs at Beal at 2320–1970 cal. BC when there is a change from sandy to silty alluvium, coinciding with a positive sea-level tendency in the Humber Estuary dated to 2318–1882 cal. BC (Long *et al.* 1998), and sedimentation rates rise.

Sometime between 1850–900 cal. BC the first major episode of Holocene channel bed incision occurred in the piedmont reach of the River Swale at Catterick. Lowland and perimarine parts of the catchment, however, show no major changes until *c.* 900–430 cal. a BC when deposition of gravel and pebbly sand was accompanied by refilling of the valley floor at Catterick. The coarse nature of this sedimentation phase, and the fact that it is recorded at nearly all of the sites investigated in the Ouse catchment, indicates that it represents a significant geomorphic event, very probably a single or series of major floods. A change to wetter conditions documented by prominent recurrence surfaces in the Humberhead Levels (Smith 1985) and in the North York Moors (Chiverrell 1998), together with conspicuous peaks in the number of dated alluvial units elsewhere in Britain (Macklin 1999), indicate that in many parts of the country this was an important period of river channel and valley floor change. The effects of climatic deterioration and larger floods in the region are likely to have been exacerbated by woodland clearance from around 1850 cal. BC until 100 cal. BC, when much of the mid-Holocene woodland had been cleared (Tinsley 1975; Atherden 1976). Bronze Age and Iron Age deforestation, while very probably increasing runoff and flood peaks, does not appear to have significantly increased sediment supply to the lower reaches of the Ouse catchment. Indeed, accelerated deposition of fine-grained sediment on floodplains in the Vale of York, north of the Humber Estuary, which could have originated from land use-related catchment disturbance, occurs much later in the eleventh century AD.

From the middle of the first millennium AD until the present day, river instability and incision episodes in the upland and piedmont parts of the Ouse catchment precede, and may have been precursors to, alluviation in downstream lowland and perimarine reaches. Alluviation at *c.* cal. AD 645–775 in upper Wharfedale was very shortly followed by incision sometime before cal. AD 775–1015 (Fig. 3). This represents a major change in the vertical behaviour of river systems in the Yorkshire Dales and occurred during a regionally significant wet shift, which has been dated to *c.* cal. AD 640–890 in the Humberhead Levels (Smith 1985) and to *c.* cal. AD 800–900 in the North York Moors (Chiverrell 1998).

The High Middle Ages marks the most extensive and one of the most rapid phases of alluviation in the Ouse basin that has occurred in the last 5 ka. Accelerated sedimentation is recorded between cal. AD 1015 and 1290 with rates of deposition at Myton-on-Swale and at Beal, especially, increasing by more than tenfold. The most obvious explanation for this dramatic increase in floodplain deposition was the massive expansion of tillage into areas that had not been cultivated until the Medieval Warm Period (Lamb 1995), although dating control is not precise enough to determine when large-scale alluviation began. It is possible, however, that alluviation may have started towards the end of the thirteenth century AD when climatic deterioration at *c.* cal. AD 1250–1300 (Chiverrell 1998), combined with anthropogenically enhanced sediment delivery, would have resulted in very high sedimentation rates. By cal. AD 1205–1450 upland and piedmont parts of the Ouse catchment were once again incising, coinciding with a shift to wetter climatic conditions recorded by a recurrence surface in the Humberhead Levels dated to cal. AD 1250–1410.

The last significant alluviation episode in the Yorkshire Ouse catchment occurred at cal. AD 1420–1645 and, while not of the same magnitude as the High Middle Ages alluviation event, deposition of fine-grained sediment is recorded throughout the Ouse basin, from the Yorkshire Dales downstream to the limit of tidal flows. Although peat cutting has destroyed the recent palaeoenvironmental record in the Humberhead Levels, major changes in mire surface wetness are recorded in many parts of Britain (Chambers *et al.* 1997) at this time, including the North York Moors at cal. AD 1400–1450 (Chiverrell 1998), marking the beginning of the Little Ice Age.

The coincidence between major episodes of river erosion and alluviation in the Yorkshire Ouse catchment and wetter periods of climate, indicates that increases in flood frequency and magnitude (if there was a large snowmelt component, Rumsby & Macklin 1994), were the primary controls of river behaviour during the Holocene. Variations in reach response to changes in flood regime are most likely to relate to upstream sediment supply and storage, and the position of a reach within the catchment. Thus, until *c.* 2 cal. ka BC valley floors at all of the study reaches were alluviating at a relatively slow rate. Thereafter, a divergence in river response emerges, especially in the early Medieval period, with incision in upland and piedmont parts of the catchment generally preceding alluviation in downstream lowland and peri-marine reaches. It appears that material eroded from headwater catchments through stream incision and network enlargement (cf. Coulthard *et al.* 1998) was temporarily stored and then transported downstream during the next period of cooler and wetter climate resulting in sedimentation in the lower part of the catchment.

The effects of prehistoric and historic land-use change on alluvial sediment sources and sediment delivery

When considering the abundance of archaeological remains in the Yorkshire Ouse catchment, and the long history of palynological research in the area, surprisingly little is known of the development of agriculture in the region. The history of base metal mining, that in the Yorkshire Dales may have begun as early as the Roman occupation, is much better documented. The effect of metal mining on the elemental composition of alluvial sediments, and on sediment delivery in the Ouse catchment, is relatively easy to measure. Contaminant metals, such as Pb and Zn, represent distinctive tracers and when they are present in elevated concentrations within alluvial sediments are usually an unambiguous indicator of mining activity within a river catchment (Macklin 1985).

In Fig. 4 rates of Cu, Pb and Zn deposition at Catterick, Myton-on-Swale, York, Kirk Hammerton, Tadcaster and Beal are plotted over the last 2 ka in 250-year time slices. Metal delivery starts to rise between cal. AD 1000 and 1250, which at York can be directly related to eleventh century mining activity dated by archaeological evidence (Hudson-Edwards *et al.* 1999*a*). Unexpectedly, there is no evidence for metal pollution at York, or in any other of the regions' rivers, during the Roman occupation. The highest rates of contaminant metal deposition have occurred in the last 250 years, with most of the mining waste (up to 94% of Pb at Myton-on-Swale, Hudson-Edwards *et al.* 1999*b*) being stored in lowland river reaches within the Vale of York. This would suggest that contaminant metal delivery to the Humber Estuary from historic mining in the northern Dales rivers might have been relatively low since AD 1750 and possibly in earlier periods.

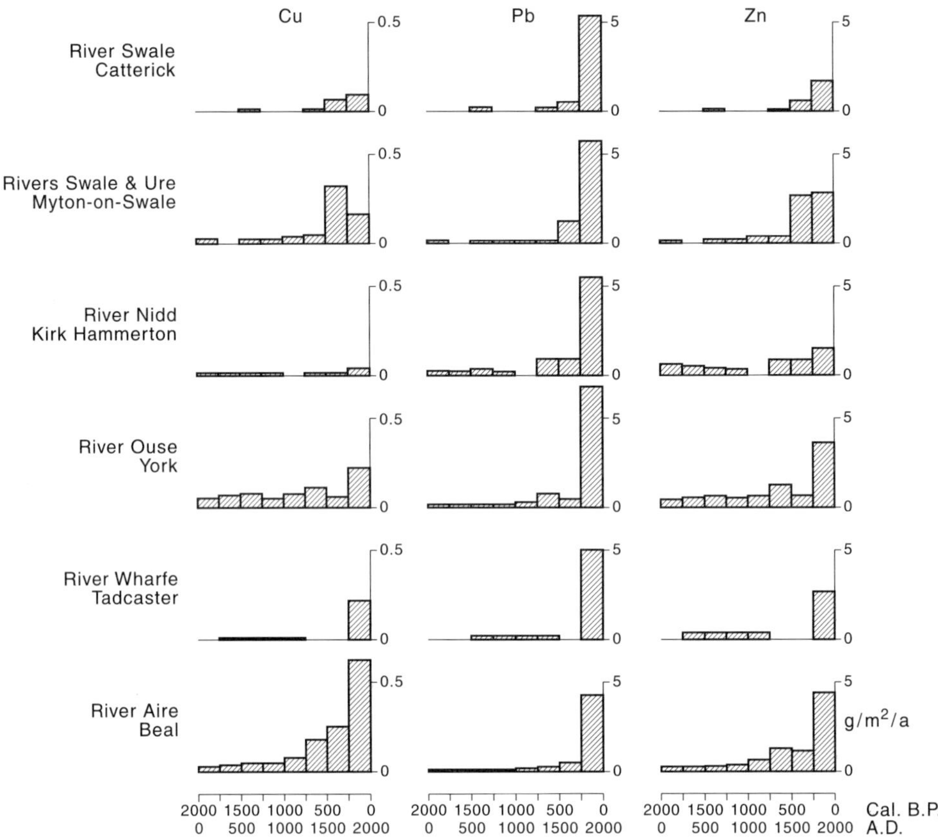

Fig. 4. Rates of Cu, Pb and Zn deposition in the Ouse catchment over the last 2 ka.

The impact of prehistoric and historic deforestation, changing land-use and agricultural practices on river sedimentation in the Yorkshire Ouse catchment has proved to be more difficult to quantify. This is because firstly, catchment erosion linked with farming and cultivation represents a diffuse rather than point source of sediment; and secondly, because the timing and spatial pattern of land-use change (e.g. tree clearance) is not well constrained. Most pollen records in the region (e.g. Tinsley 1975; Atherden 1976) do, however, indicate woodland clearance from the Middle Bronze Age (*c.* 1850 cal. BC) onwards, with very rapid and widespread deforestation in the Late Iron Age (*c.* 200–100 cal. BC) to levels approaching those found at the present day. It would be anticipated that large-scale woodland removal, particularly in the upland parts of the Ouse catchment, would have increased slope erosion and accelerated floodplain sedimentation.

All of the study reaches (with the exception of Kettlewell) are underlain by Triassic or Permian bedrock, and the impact of erosion from the upland part of the catchment would be expected to be represented by a change in alluvial sediment composition from local Pleistocene, or Triassic, to non-local Carboniferous sources. This indeed appears to be the pattern, with the Rivers Nidd, Ouse, Swale and Wharfe all showing a change in sediment type before the first millennium AD (Fig. 5). Dates for the switch from local (Permo-Triassic–Pleistocene) to non-local (Carboniferous) sediment sources do vary considerably between river systems, most likely reflecting differences when significant erosion began in each upland basin and drainage network configuration. In the Rivers Ure, Swale and Nidd there is an early change in sediment sources at 3.5, 2 and 1.5 cal. ka BC, respectively, which pre-date large-scale tree clearance and most likely reflect natural increases in sediment delivery from the Dales resulting from gullying and channel incision. In the Ouse and Wharfe systems, a change to Carboniferous sediment sources occurs significantly later between *c.* 500

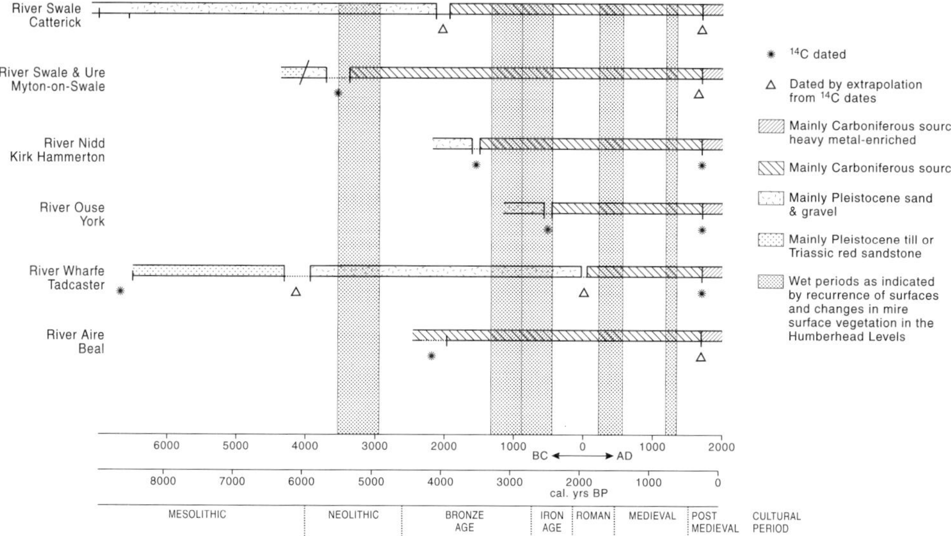

Fig. 5. Changes in alluvial sediment sources in the Ouse catchment during the Holocene.

and 50 cal. BC and may be related to prehistoric land-use change. At present, however, it is not possible to test linkages between deforestation and enhanced sediment delivery from particular tributaries of the Ouse, as land-use histories of individual sub-catchments are not well-documented.

Conclusion

This basin-wide study of Holocene fluvial sedimentary sequences in the Ouse catchment has revealed a close correspondence between periods of cooler and/or wetter climate and phases of river instability marked either by accelerated sedimentation or erosion. Land-use change, most notably in the High Middle Ages, appears to have significantly increased fine-grained sediment supply to river systems in the region, resulting in higher rates of floodplain deposition and widespread alluviation in the last 1 ka. Changes in river behaviour in the Ouse catchment at *c.* 3500, *c.* 2000 and 900–430 cal. BC, and at cal. AD 645–775, cal. AD 1015–1290 and cal. AD 1420–1645 were very probably accompanied by altered sediment fluxes to the Humber estuary. The large proportion of particulate-borne contaminant metals (resulting predominantly from historical mining) stored in the Vale of York, however, strongly indicates that sediment delivery from the Ouse catchment to the Humber Estuary during the Holocene may have been relatively low. This suggests that the degree of connectivity between river, estuarine and coastal transport systems, as well as spatial and temporal variations in fluvial sediment storage, are the key controls of long-term land to ocean sediment fluxes.

We are most grateful to NERC for supporting investigations in the Yorkshire Ouse catchment through a research grant (GST/02/0758) to M. G. Macklin. M. G. Macklin would like to especially thank the referees, T. Burt and M. Evans, for their most helpful comments on this paper.This is LOIS publication number 585.

References

ANDERSON, D. E., BINNEY, H. A. & SMITH, M. A. 1998. Evidence for abrupt climatic change in northern Scotland between 3900 and 3500 calendar years BP. *The Holocene*, **8**, 97–103.

ATHERDEN, M. A. 1976. The impact of late prehistoric cultures on the vegetation of the North York Moors. *Transactions of the Institute of the British Geographers, New Series*, **1**, 284–300.

BARBER, K. E., CHAMBERS, F. M., MADDY, D., STONEMAN, R. E. & BREW, J. S. 1994. A sensitive high-resolution record of Late Holocene climatic change from a raised bog in northern England. *The Holocene*, **4**, 198–205.

BELL, M. & WALKER, M. J. C. 1992. *Late Quaternary Environmental Change: Physical and Human Perspectives*. Longman, Harlow.

BRONK RAMSEY, C. 1995. *OXCAL Program v2.18*. Oxford Radiocarbon Accelerator Unit: University of Oxford.

CHAMBERS, F. M., BARBER, K. E., MADDY, D. & BREW, J. 1997. A 5500-year proxy-climate and vegetation record from blanket mire at Talla Moss, Borders, Scotland. *The Holocene*, **7**, 391–399.

CHIVERRELL, R. C. 1998. *Moorland Vegetation History and Climate Change on the North York Moors During the Last 2000 Years*. PhD thesis, University of Leeds.

COULTHARD, T. J., KIRKBY, M. J. & MACKLIN, M. G. 1998. Non-linearity and spatial resolution in a cellular automaton model of a small upland basin. *Hydrology and Earth System Science*, **2**(2–3), 257–264.

DAWSON, E. J. & MACKLIN, M. G. 1998. Speciation of heavy metals on suspended sediment under high flow conditions in the River Aire, west Yorkshire. *Hydrological Processes*, **12**, 1483–1494.

DINNIN, M. 1997. The drainage history of the Humberhead Levels. *In*: VAN DE NOORT, R. & ELLIS, S. (eds) *Wetland Heritage of the Humberhead Levels*. Humber Wetlands Project, Kingston upon Hull 19–30.

HOWARD, A. J. & MACKLIN, M. G. (eds) 1998. *The Quaternary of the Eastern Yorkshire Dales: Field Guide. The Holocene Alluvial Record*. Quaternary Research Association, London.

HUDSON-EDWARDS, K. A., MACKLIN, M. G., FINLAYSON, R. & PASSMORE, D. G. 1999*a*. Medieval lead pollution in the River Ouse at York, England. *Journal of Archaeological Science*, **26**, 809–819.

——, —— & TAYLOR, M. P. 1999*b*. 2000 years of sediment-borne heavy metal storage in the Yorkshire Ouse Basin, NE England, UK. *Hydrological Processes*, **13**, 1087–1102.

LAMB, H. H. 1995. *Climate History and the Modern World*. 2nd edition. Routledge, London.

LONG, A. J., INNES, J. B., KIRBY, J. R., LLOYD, J. M., RUTHERFORD, M. M., SHENNAN, I. & TOOLEY, M. J. 1998. Holocene sea-level change and coastal evolution in the Humber estuary, eastern England: an assessment of rapid coastal change. *The Holocene*, **8**, 229–247.

LONGFIELD, S. A. & MACKLIN, M. G. 1999. The influence of recent environmental change on flooding and sediment fluxes in the Yorkshire Ouse basin, northeast England. *Hydrological Processes*, **13**, 1051–1066.

MACKLIN, M. G. 1985. Flood-plain sedimentation in the upper Axe Valley, Mendip, England. *Transactions of the Institute of British Geographers*, **10**(2), 235–244.

——1999. Holocene river environments in prehistoric Britain: human interaction and impact. *Quaternary Proceedings, Journal of Quaternary Science*, **14**, in press.

—— & LEWIN, J. 1993. Holocene river alluviation in Britain. *Zeitschrift für Geomorphologie (Supplement)*, **88**, 109–122.

——, HUDSON-EDWARDS, K. A. & DAWSON, E. J. 1997. The significance of pollution from historic metal mining in the Pennine orefields on river sediment contaminant fluxes to the North Sea. *The Science of the Total Environment*, **194/195**, 391–397.

——, RUMSBY, B. T. & HEAP, T. 1992. Flood alluviation and entrenchment: Holocene valley floor development and transformation in the British uplands. *Geological Society of America Bulletin*, **104**, 631–643.

MERRETT, S. P. & MACKLIN, M. G. 1999. Historic river response to extreme flooding in the Yorkshire Dales, northern England. *In*: BROWN, A. G. & QUINE, T. M. (eds) *Fluvial Processes and Environmental Change*. Wiley, Chichester, 345–360.

PASSMORE, D. G., MACKLIN, M. G., BREWER, P. A., LEWIN, J. RUMSBY, B. T. & NEWSON, M. D. 1993. Variability of Late Holocene braiding in Britain. *In*: BEST, J. L. & BRISTOW, C. S. (eds) *Braided Rivers*. Geological Society Special Publications, **75**, 205–229.

ROBINSON, M. 1992. Environment, archaeology and alluvium on the river gravels of the South Midlands. *In*: NEEDHAM, S. & MACKLIN, M . G. (eds) *Alluvial Archaeology in Britain*. Oxbow Press, Oxford, 197–208.

RUMSBY, B. T. & MACKLIN, M. G. 1994. Channel and floodplain response to recent abrupt climate change: the Tyne basin, northern England. *Earth Surface Processes and Landforms*, **19**, 499–515.

SMITH, B. 1985. *A Palaeoecological Study of Raised Mires in the Humberhead Levels*. PhD thesis, University of Wales, Cardiff.

TAYLOR, M. P. & MACKLIN, M. G. 1997. Holocene alluvial architecture and valley floor development on the River Swale, Catterick, North Yorkshire, UK. *Proceedings of the Yorkshire Geological Society*, **51**, 317–327.

TINSLEY, H. M. 1975. The former woodland of the Nidderdale Moors (Yorkshire) and the role of early man in its decline. *Journal of Ecology*, **63**, 1–26.

TIPPING, R. 1992. The determination of cause in the generation of major prehistoric valley fills in the Cheviot Hills, Anglo-Scottish border. *In*: NEEDHAM, S. & MACKLIN, M. G. (eds) *Alluvial Archaeology in Britain*. Oxbow Press, Oxford, 111–121.

The Holocene evolution of the Humber Estuary: reconstructing change in a dynamic environment

S. E. METCALFE,[1] S. ELLIS,[2] B. P. HORTON,[3] J. B. INNES,[3]
J. McARTHUR,[3] A. MITLEHNER,[1] A. PARKES,[2] J. S. PETHICK,[4] J. REES,[5]
J. RIDGWAY,[5] M. M. RUTHERFORD,[3] I. SHENNAN[3] & M. J. TOOLEY[6]

[1] *Department of Geography, University of Edinburgh, Edinburgh EH8 9XP, UK (e-mail: sem@geo.ed.ac.uk)*
[2] *Department of Geography, University of Hull, Hull HU6 7RX, UK*
[3] *Department of Geography, University of Durham, Durham DH1 3LE, UK*
[4] *Department of Marine Sciences and Coastal Management, University of Newcastle, Newcastle upon Tyne NE1 7RU, UK*
[5] *British Geological Survey, Keyworth, Nottingham NG12 5GG, UK*
[6] *School of Geography, Kingston University, Kingston-upon-Thames KT1 2EE, UK*

Abstract: The Holocene sequence of the Humber Estuary displays a wide range of sediment types within which the preservation of microfossils is highly variable. Its evolution has been reconstructed using a range of environmental proxies with chronological control provided by more than 90 radiocarbon dates. Results are presented of diatom analyses from three cores typical of the inner, middle and outer estuary (HMB20, HMB7 and HMB12) and of three cores that illustrate the role of organic deposits (peats) and their associated pollen (HMB13, HMB12 and the Ancholme Valley) in the definition of sea-level index points. The reconstruction of relative sea-level change shows a rapid rise in the early Holocene, followed by a reduced rate of rise in the mid–late Holocene. This reconstruction, together with information on the pre-Holocene surface and the different palaeoenvironments from the cores have been integrated within a geographical information system and then interpreted to yield a series of palaeogeographical maps of the Humber at 1000-year time slices between 8 and 3 cal. ka BP. The marine transgression progressed up the estuary after 8 cal. ka BP, reaching the inner estuary by 6 cal. ka BP. The expansion of intertidal environments probably reached its maximum around 3 cal. ka BP. Changes since 3 cal. ka BP are described using archaeological and historical records. Tidal asymmetry is a major controlling factor on the balance of sediment accretion and erosion in the estuary. Sedimentary and bathymetric evidence suggests a damped oscillation between flood and ebb asymmetry in the Humber over the Holocene period. Such a conclusion would be of great importance to estuarine managers and users since it could be used to predict the future development of the estuary.

The Humber is one of the largest estuaries in the UK draining an inland area of more than 24 000 km^2 before entering the North Sea at Spurn Point (Fig. 1). Five major rivers (Derwent, Ouse, Aire, Don and Trent) flow in to the estuary above Trent Falls. The Rivers Ancholme and Hull enter further to the east. In spite of this large inland catchment, the flows of the modern estuary are dominated by the tides. The Humber is macrotidal (7.2 m) and the tidal wave is highly asymmetrical, with strong flood currents bringing sediments into the estuary from the North Sea (Pethick 1990). Suspended sediment loads are high, with the maximum occurring about 45 km up-stream of the mouth near the Humber Bridge. The estuary is well mixed and dynamic (Millward & Glegg 1997). Pethick (1990) suggested that the estuary can be divided into two sections: a mature inner estuary and an immature outer estuary separated by the axis of the chalk Yorkshire and Lincolnshire Wolds, the Humber Gap. A buried cliff cut into the chalk marks an interglacial coastline (Gaunt *et al.* 1992).

The morphology and hydrology of the Humber have been profoundly affected by glaciation: both directly through the advance and retreat of ice sheets and indirectly through

From: SHENNAN, I. & ANDREWS, J. (eds) *Holocene Land–Ocean Interaction and Environmental Change around the North Sea*. Geological Society, London, Special Publications, **166**, 97–118. 1-86239-054-1/00/$15.00

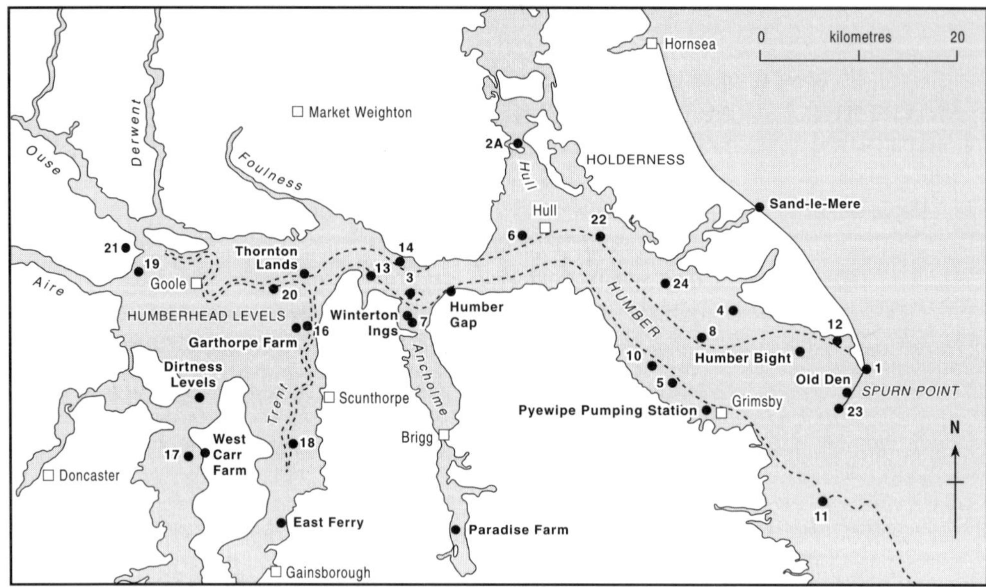

Fig. 1. Location map of the Humber Estuary shows the sites referred to in the text, extent of Holocene fluviatile–estuarine–wetland sediments, and outline of the present estuary (dashed line).

changes in relative sea-level and, therefore, in base level. In the Late Devensian (Dimlington stadial *c.* 18–13 ^{14}C ka BP), ice advanced westwards from the North Sea basin as far as the Humber Gap and blocked the drainage of the Humber into the North Sea creating pro-glacial Lake Humber. A minimum age for regional deglaciation of 13 045 ± 270 BP (15 508 cal. a BP), comes from a kettlehole in the Roos Valley in Holderness. Lake Humber probably drained sometime between 12.5 and 11.1 ka BP (14 648 and 13 011 cal. a BP) (Dinnin 1997*b*). The Late Glacial and very early Holocene were marked by aeolian activity. Once the Humber Valley was ice free, lower sea-levels resulted in major valley incision. In Holderness, valley systems are found more than 9 m below the modern ground surface (Dinnin & Lillie 1995). This phase of incision probably lasted until about 8.5 ka BP (*c.* 9470 cal. a BP), by which time sea-level had risen and channels and valley systems began to aggrade.

The Humber wetlands (defined as occurring below 10 m Ordnance datum, OD) cover some 330 000 ha (3300 km^2) and are of considerable archaeological significance (Van de Noort & Davies 1993; Van de Noort & Ellis 1995*a*). The lowlands have been a focus for settlement and for agricultural and industrial development, and in many cases are now highly disturbed. Drainage and the practice of warping (see below) will have had a significant impact on recent sediments. In the Humberhead Levels alone, it has been estimated that warp 'masks' the palaeosurface over some 7000 ha (Lillie & Weir 1997). It is clear, therefore, that the reconstruction of past environments in the Humber Estuary is challenging, with large amounts of sediment being moved around the system, stored and reworked over different time-scales. In the recent past, human impact has been significant.

Background to the study

In spite of its geographical and economic importance surprisingly little was known about the post-glacial evolution of the Humber prior to the Land–Ocean Interaction Study (LOIS) research initiative (e.g. Smith 1958). A review of palaeoecological studies in the Humber lowlands has been provided by the Humber Wetlands Project (Van de Noort & Davies 1993; Van de Noort & Ellis 1995*a*, 1997). The greatest emphasis has been on pollen-based studies, which have recorded vegetation change since the last glacial event. The history of Holocene changes in relative sea-level was also poorly known (Gaunt & Tooley 1974; Long *et al.* 1998). Given the scarcity of data, there have been few attempts to reconstruct the evolution of the estuary through the Holocene, although Berridge & Pattison (1994) presented maps for 7.5, 3.5, 2 and 1 ka BP (uncalibrated).

The overall aim of the present study was to compile a database of changes in the Humber Estuary over the last 10 ka. Changes in sedimentary environments at different locations through time would reflect changes in the shape of the estuary, the different locations of the land–ocean boundary and the balance between fresh- and seawater inputs. These palaeoenvironmental data, within a chronological framework provided by radiocarbon dating, were then used both to improve the relative sea-level curve for the Humber and to produce a series of palaeogeographic maps of the estuary as it changed through the Holocene.

Approach

The basis for the environmental reconstruction was a series of cores taken around the estuary (see Ridgway *et al.* this volume). These were both the Land–Ocean Evolution Perspective Study (LOEPS) HMB cores and additional cores taken by the Universities of Durham and Hull. Our approach was to use a range of environmental proxies, including sediment texture, sediment geochemistry and pollen and diatom analyses, to identify a suite of environments that occurred in the Humber in different places and at different times (e.g. high and low saltmarsh, mudflat, alder carr). In the estuarine environment, the preservation of microfossils is highly variable and any single indicator is unlikely to provide a continuous record throughout all cores. A combination of indicators is much more likely to be successful. Details of the methods used are given in Ridgway *et al.* (this volume). The results of the physical and geochemical analyses of the Holocene sediments are described in Rees *et al.* (this volume) and Andrews *et al.* (this volume). Here we focus on the palaeoecological record, the reconstruction of relative sea-level and the production of the palaeogeographic maps. For the period after 3 ka BP purely geological methods do not provide adequate data for environmental reconstructions and we have used the available archaeological and historical information to reconstruct change in the estuary.

Dating

Ninety-five radiocarbon dated samples provide the chronostratigraphic framework. These comprise 63 samples that have been validated as sea-level index points (Table 1). This means that they can be related to a past tide level (the reference water-level, RWL) and a tendency of sea-level. A positive sea-level tendency is an increase in marine influence, and a negative tendency is a decrease (e.g. Shennan 1982). Each date is calibrated following Stuiver & Reimer (1993). A further 32 index points are limiting dates (Table 2). These are dates mainly from peats for which there is no biostratigraphic indicator of tidal conditions. They therefore provide a limiting age on the later tidal sediments at the site, and a spatial limit on the extent of tidal sedimentation.

Results

The changes in the Humber were identified with reference to a variety of modern environments (see below), based on a model of coastal sedimentary environments devised by Shennan (1986, 1992) initially for the Fenland of East Anglia. These environments have a relative relationship based primarily upon altitudinal position within, or above, the tidal cycle, which correspond to particular floral and faunal communities and depositional conditions. Although some of these environments could be identified based on one proxy, it was often necessary to combine the information derived from diatoms, pollen, molluscs and the physical and geochemical properties of the sediments. Hence, the original model has been expanded to include several non-biological criteria to cope with the highly complex situation of the evolving Humber Estuary. The combination of the environmental interpretations and the chronology based on radiocarbon dates and pollen stratigraphic data, formed the basis for both the Humber sea-level curve and the palaeogeographical maps (see below).

Diatom records

Samples for diatom analyses were taken from all the Humber cores and prepared using standard techniques (see Ridgway *et al.* this volume). Diatoms were recovered from all the HMB cores, but the degree of retrieval and preservation varied markedly both within and between cores. Both contemporaneous and post-depositional effects account for this variability. Despite these difficulties, some cores have remarkably complete diatom records (e.g. HMB10, HBM16 and HBM24). Some taxa are especially resistant to post-depositional dissolution (e.g. *Paralia sulcata* (Ehrenberg) Cleve, *Actinoptychus senarius* (Ehrenberg) Ehrenberg, *Scoloneis tumida* (Brebisson ex Kutzing) D. G. Mann) and dominate the flora of some outer-estuary cores in medium

Table 1. *Sea-level index points from the Humber Estuary*

Site	Laboratory code	^{14}C age $\pm 1\sigma$	Calibrated age cal. a BP			Altitude (m OD)	RWL (m OD)	Chane in MSL from present	Area	Type	Tendency	Reference
			Max.	Mean	Min.							
Barrow Haven	SRR1374	2325 ± 60	2466	2341	2153	1.43	3.75	−2.32 ± 0.42	OUTER	I	±	Long *et al.* 1988
Barrow Haven	SRR4897	2040 ± 40	2107	1985	1882	1.77	3.75	−2.18 ± 0.42	OUTER	I	−	Long *et al.* 1998
Barrow Haven	SRR1373	1080 ± 40	1063	966	927	2.07	3.75	−1.48 ± 0.23	OUTER	I	+	Long *et al.* 1998
Dirtness Levels, DL961	AA26377	3880 ± 55	4427	4321	4095	−1.84	4.58	−6.22 ± 0.22	INNER	I	+	
Dirtness Levels, DL961	AA26375	3020 ± 60	3361	3211	2993	−0.12	4.58	−4.88 ± 0.22	INNER	I	−	
Dirtness Levels, DL961B	AA26376	4610 ± 60	5561	5309	5053	−2.94	4.58	−7.52 ± 0.41	INNER	I	+	
Dunswell, HMB2A	AA25592	5000 ± 55	5901	5731	5611	−0.65	3.55	−4.20 ± 0.22	OUTER	BB	+	
East Clough	SRR4748	3640 ± 45	4085	3948	3832	−0.35	3.67	−3.82 ± 0.23	INNER	B	+	Long *et al.* 1998
East Clough	SRR4749	3770 ± 45	4268	4114	3983	−0.56	3.67	−4.03 ± 0.23	INNER	BB	+	Long *et al.* 1998
East Ferry, EF95-2	AA27615	4890 ± 55	5734	5612	5488	−2.01	4.55	−6.36 ± 0.22	INNER	I	+	
East Ferry, EF95-2	AA23438	3090 ± 50	3386	3299	3160	0.29	4.55	−4.44 ± 0.22	INNER	I	−	
East Ferry, EF95-2	AA22678	1880 ± 45	1892	1820	1705	0.93	4.55	−3.42 ± 0.22	INNER	I	+	
Garthorpe Farm, GF96-1B	AA27584	5870 ± 55	6849	6719	6539	−4.38	4.14	−8.92 ± 0.41	INNER	BB	+	
Garthorpe Farm, GF96-1B	AA24138	4845 ± 55	5702	5592	5340	−2.91	4.14	−6.85 ± 0.22	INNER	B	+	
Garthorpe Farm, GF96-1B	AA23888	4365 ± 65	5246	4872	4830	−2.6	4.14	−6.92 ± 0.22	INNER	I	−	
Garthorpe Farm, GF96-1B	AA23887	4060 ± 60	4819	4526	4409	−1.89	4.14	−5.83 ± 0.22	INNER	I	+	
Garthorpe Farm, GF96-1B	AA23886	3920 ± 75	4532	4374	4095	−1.05	4.14	−4.99 ± 0.22	INNER	I	+	
Garthorpe, HMB16	AA23437	3425 ± 65	3836	3664	3475	−2.56	4.14	−6.50 ± 0.22	INNER	I	+	
Halsham Carrs	GU5475	4260 ± 50	4867	4830	4577	−1.34	3.63	−5.10 ± 0.25	OUTER	I	−	Dinnin & Lillie 1995
Halsham Carrs	GU5476	4150 ± 50	4833	4642	4456	−1.34	3.63	−5.10 ± 0.25	OUTER	I	−	Dinnin & Lillie 1995
Hasholme	HAR7007	5710 ± 100	6739	6484	6298	−4.41	4.34	−8.95 ± 0.23	INNER	I	−	Millett & McGrail 1987
Humber Bight, HI1	AA26378	6140 ± 65	7181	7008	6858	−7.23	3.18	−10.21 ± 0.22	OUTER	B	+	
Immingham, HMB10	AA23433	6520 ± 75	7526	7387	7233	−6.77	3.35	−10.12 ± 0.22	OUTER	BB	+	
Immingham, HMB10	AA23432	6245 ± 80	7275	7170	6905	−6.66	3.35	−9.81 ± 0.22	OUTER	B	+	
Kilnsea Waaren	UB3900	4562 ± 59	5449	5292	4994	−1.29	3.08	−4.37 ± 0.81	OUTER	B	+	Long *et al.* 1998
Kilnsea Warren	UB3901	4384 ± 54	5245	4928	4838	−1.19	3.08	−4.07 ± 0.23	OUTER	BB	+	Long *et al.* 1988
Lockham, HMB12	AA23434	5425 ± 70	6391	6238	5997	−3.45	3.08	−6.53 ± 0.22	OUTER	I	+	
Lockham, HMB12	AA22672	5325 ± 50	6271	6138	5944	−3.3	3.08	−6.18 ± 0.22	OUTER	I	+	
Lockham, HMB12	AA23891	4235 ± 60	4871	4829	4569	−1.27	3.08	−3.65 ± 0.71	OUTER	I	−	
Lockham, HMB12	AA23890	4040 ± 65	4816	4479	4354	−1.16	3.08	−3.54 ± 0.71	OUTER	I	−	
Newlands, HMB19	AA23440	5075 ± 55	5930	5821	5670	−3.25	4.60	−8.03 ± 0.22	INNER	I	−	

Newlands, HMB19	AA27141	3395 ± 75	3833	3631	3464	−0.76	4.60	−5.36 ± 0.41	INNER	I	±	
Newlands, HMB19	AA23439	4800 ± 55	5646	5527	5330	−2.84	4.60	−7.24 ± 0.22	INNER	I	+	
Old Den, HI3	AA27583	6875 ± 60	7782	7650	7548	−10.31	3.18	−13.29 ± 0.30	OUTER	B	+	
Paradise Farm, PA94-12	CAM41320	3490 ± 50	3879	3775	3628	0.22	4.54	−4.32 ± 0.41	INNER	BB	+	
Paradise Farm, PA94-12	CAM41317	3170 ± 50	3470	3373	3264	0.49	4.54	−3.85 ± 0.22	INNER	B	+	
Pyewipe Pumping Station	AA27586	4985 ± 55	5895	5724	5600	−2.38	3.18	−5.56 ± 0.22	OUTER	B	+	
Pyewipe Pumping Station	AA27585	5065 ± 60	5930	5821	5656	−2.22	3.18	−5.20 ± 0.22	OUTER	BB	+	
Roos Drain	GU5477	5080 ± 60	5936	5807	5664	−2.61	3.66	−6.40 ± 0.25	OUTER	I	−	Dinnin & Lillie 1995
Roos Drain	GU5478	5200 ± 60	6171	5935	5767	−2.61	3.66	−6.40 ± 0.25	OUTER	I	−	Dinnin & Lillie 1995
Roos Drain	GU5483	5010 ± 70	5916	5736	5599	−2.46	3.66	−6.25 ± 0.25	OUTER	I	−	Dinnin & Lillie 1995
Roos Drain	GU5484	5140 ± 90	6170	5911	5665	−2.46	3.66	−6.25 ± 0.25	OUTER	I	−	Dinnin & Lillie 1995
Sand-le-Mere, SM95-5	AA23821	2630 ± 60	2849	2752	2550	−0.81	3.74	−4.71 ± 0.22	OUTER	I	−	
Sandholme Lodge	SRR4894	5615 ± 45	6484	6408	6301	−3.35	4.33	−7.48 ± 0.23	INNER	I	+	Long *et al.* 1998
Sandholme Lodge	SRR4743	4170 ± 45	4835	4719	4534	−1.78	4.33	−6.31 ± 0.23	INNER	I	−	Long *et al. 1998*
South Ferriby, HMB3	AA25578	5985 ± 55	6935	6825	6721	−4.76	3.81	−8.57 ± 0.22	INNER	BB	+	
South Ferriby, HMB3	AA25577	6000 ± 55	6996	6830	6727	−4.66	3.81	−8.27 ± 0.22	INNER	B	+	
South Marsh, HMB5	AA23431	6050 ± 70	7153	6887	6736	−6.48	3.31	−9.59 ± 0.22	OUTER	BB	+	
South Marsh, HMB5	AA23430	6135 ± 75	7202	7006	6798	−6.37	3.31	−9.48 ± 0.22	OUTER	B	+	
Thirtle Bridge	GU5480	2370 ± 50	2703	2350	2321	0.66	3.65	−3.12 ± 0.25	OUTER	I	−	Dinnin & Lillie 1995
Thirtle Bridge	GU5479	2250 ± 50	2345	2250	2136	0.66	3.65	−3.12 ± 0.25	OUTER	I	−	Dinnin & Lillie 1995
Thirtle Bridge	GU5490	2590 ± 50	2773	2744	2495	0.79	3.65	−2.99 ± 0.25	OUTER	I	−	Dinnin & Lillie 1995
Thirtle Bridge	GU5489	2420 ± 50	2716	2362	2340	0.79	3.65	−2.99 ± 0.25	OUTER	I	−	Dinnin & Lillie 1995
Thornton Lands, TL96-1	AA26380	5560 ± 65	6473	6357	6211	−3.8	4.14	−8.12 ± 0.22	INNER	I	−	
Thornton Lands, TL96-1	AA26379	3760 ± 80	4405	4111	3884	−2.86	4.14	−6.80 ± 0.22	INNER	I	+	
Union Dock	SRR4744	5665 ± 45	6543	6430	6316	−7.68	3.22	−10.70 ± 0.23	OUTER	I	+	Long *et al.* 1998
Union Dock	SRR4745	5900 ± 45	6852	6752	6639	−7.78	3.22	−11.17 ± 0.23	OUTER	I	−	Long *et al.* 1968
Union Dock	SRR4746	6645 ± 45	7544	7496	7392	−8.78	3.22	−11.80 ± 0.23	OUTER	BB	+	Long *et al.* 1998
Winterton Ings, WTI95-3	CAM41319	3560 ± 60	3986	3836	3689	−0.6	3.94	−4.34 ± 0.22	INNER	B	+	
Winterton Ings, WTI95-6	CAM41321	4300 ± 50	4979	4855	4658	−2.99	3.94	−6.73 ± 0.22	INNER	B	+	
Brigg	OXA7137	4990 ± 75	5912	5726	5591	−6.34	3.81	−10.33 ± 0.23	INNER	I	−	Neumann 1998
South Ferriby	OXA7052	3670 ± 40	4120	3940	3868	−0.4	3.81	−4.21 ± 0.42	INNER	B	+	Neumann 1998
South Ferriby	GU5704	1840 ± 60	1885	1735	1607	1.2	3.81	−2.61 ± 0.42	INNER	I	−	Neumann 1998

The calibrated ages shown are the age ranges which contain 95.4% of the area under the probability curve. RWL, present altitude of reference water level. Change in MSL from present includes error estimates for reference water level and measurement of altitude. 'Area' indicates whether a site is inner or outer estuary (boundary is the Humber Gap). 'Type' indicates whether an index point is from an intercalated (I), basal (B) or base of basal (BB) sedimentary environment.

Table 2. *Limiting data from the Humber Estuary*

Site	Laboratory Code	^{14}C age $\pm 1\sigma$	Calibrated age cal. a BP			Altitude (m OD)	RWL (m OD)	Change in MSL from present	Reference
			Max.	Mean	Min.				
Brigg	OXA7091	3940 ± 45	4513	4408	4236	−0.92	4.68	−5.22 ± 1.58	Neumann 1988
Brigg	OXA7090	4300 ± 50	4979	4855	4658	−1.47	4.68	−5.77 ± 1.58	Neumann 1998
Brigg	GU5700	4730 ± 100	5652	5341	5086	−2.46	4.68	−6.76 ± 1.58	Neumann 1998
Brigg	GU5699	5040 ± 170	6188	5834	5333	−2.67	4.68	−6.97 ± 1.58	Neumann 1998
Brigg	OXA7136	6170 ± 90	7225	7079	6802	−4.61	4.68	−8.91 ± 1.58	Neumann 1998
Garthorpe, HMB16	AA25585	7265 ± 60	8135	8034	7922	−11.43	4.14	−15.19 ± 1.58	
Garthorpe, HMB16	AA25586	7745 ± 60	8580	8464	8374	−11.65	4.14	−15.41 ± 1.58	
Hull Market Place	IGS100	6890 ± 100	7904	7659	7529	−9.73	3.64	−12.99 ± 1.70	Gaunt & Tooley 1974
Hull Market Place	IGS99	6970 ± 100	7937	7729	7559	−11.55	3.64	−14.81 ± 1.70	Gaunt & Tooley 1974
Kilnsea, HMB1	AA25575	4130 ± 50	4829	4669	4448	−0.87	3.46	−3.96 ± 1.43	
Ousefleet, HMB20	AA25590	7190 ± 60	8111	7945	7837	−12.85	4.07	−16.54 ± 1.58	
Ousefleet, HMB20	AA25591	8240 ± 110	9448	9214	8775	−12.92	4.07	−16.61 ± 1.58	
Redbourne Hayes	GU5703	3590 ± 50	4063	3875	3720	−0.05	4.21	−3.88 ± 1.58	Neumann 1998
Redbourne Hayes	OXA7065	3595 ± 50	4068	3881	3722	−0.06	4.21	−3.77 ± 1.58	Neumann 1998
Redbourne Hayes	GU5701	3940 ± 70	4537	4408	4150	−0.26	4.21	−4.09 ± 1.58	Neumann 1998
Redbourne Hayes	OXA7064	4494 ± 31	5292	5218	4988	−1.92	4.21	−5.75 ± 158	Neumann 1998
South Farm, HMB8	AA25581	7145 ± 60	8063	7923	7806	−11.67	3.32	−14.61 ± 1.58	
South Farm, HMB8	AA25582	8555 ± 65	9646	9491	9397	−11.95	3.32	−14.89 ± 1.58	
South Ferriby	OXA7066	2635 ± 45	2791	2753	2720	1.33	3.81	−2.48 ± 1.58	Neumann 1998
South Ferriby	OXA7067	2690 ± 70	2939	2771	2726	2.01	3.81	−1.80 ± 1.58	Neumann 1998
South Ferriby	OXA7053	3310 ± 40	3630	3517	3468	0.53	3.81	−3.28 ± 1.58	Neumann 1998
South Ferriby	GU5706	3890 ± 60	4504	4328	4095	−1.12	3.81	−4.93 ± 1.58	Neumann 1998
South Ferriby	OXA7055	4650 ± 50	5566	5432	5287	−2.07	3.81	−5.88 ± 1.58	Neumann 1998
South Ferriby	GU5707	4700 ± 80	5597	5365	5099	−2.12	3.81	−5.55 ± 1.58	Neumann 1998
South Ferriby	OXA7054	4960 ± 40	5849	5693	5602	−2.99	3.81	−6.42 ± 1.58	Neumann 1968
South Ferriby	OXA7056	5440 ± 45	6304	6224	6110	−5.87	3.81	−9.30 ± 1.58	Neumann 1998
South Ferriby	OXA7057	6000 ± 50	6983	6821	6730	−3.32	3.81	−6.75 ± 1.58	Neumann 1998
Union Dock	SRR4747	8170 ± 45	9251	9088	8981	−8.91	3.22	−11.75 ± 1.48	Long *et al.* 1998
Whitton Ness, HMB13	AA22674	3910 ± 45	4439	4376	4155	−1.88	3.91	−5.41 ± 1.58	
Whitton Ness, HMB13	AA23436	4830 ± 70	5717	5588	5330	−2.6	3.91	−6.13 ± 1.58	
Winterton Ings, WTI94-3	CAM41322	4100 ± 50	4823	4564	4432	−1.36	3.94	−4.92 ± 1.58	
Winterton Ings, WTI95-6	CAM41318	5990 ± 60	6995	6826	6679	−3.84	3.94	−7.40 ± 1.69	

The calibrated ages shown are the age ranges which contain 95.4% of the area under the probability curve. RWL, present altitude of reference water level. Change in MSL from present includes error estimates for reference water level and measurement of altitude.

sands and silts. The environmental interpretation of samples containing only such taxa was, therefore, made with caution and in conjunction with the results of other proxies from the same cores. The ecological groupings of the Humber diatoms are based primarily on the work of de Wolf (1982), Denys (1991) and Vos & de Wolf (1993), who grouped taxa found in Holocene cores from coastal wetlands in the Netherlands and Belgium. Their groupings reflect ecological conditions found mainly in microtidal environments, so their applicability to the macrotidal Humber has limitations. There also appear to be errors in the environmental preferences assigned to some taxa. These authors' categorizations, however, remain the most comprehensive to

Table 3. *Diatom associations from palaeoenvironmental facies*

	Environment	Diatom association	Example
1	River (above tidal limit)	*Eunotia nodosa, E. pectinalis, Gomphonema parvulum, Pinnularia gentilis, P. nobilis, P. viridis*	HMB13 625–640 cm
2	Raised bog	*Eunotia monodon, E. valida*	HMB17 0–165 cm
3	Oak/hazel fenwood	*P. viridis, Amphora ovalis, Cymbella cistula, C. sinuata, Rhopalodia gibba, Cyclotella striata*	HMB20 1640–1670 cm
4	Alder carr	*C. cistula, Navicula peregrina, P. viridis, Stauroneis phoenicenteron, Fragilaria pinnata, Eunotia curvata, E. monodon*	HMB20 1590–1638 cm
5	Sedge fen	*Epithemia adnata, E. turgida, Gomphonema acuminatum, Navicula cincta, Stephanodicus rotula*	HMB18 1086–1100 cm
6	Coastal reedswamp	*N. peregrina, N. digitoradiata, N. elegans, Nitzschia scalaris*	HMB12 755–765 cm
7	Intercreek areas, flooded by high spring tides	*Paralia sulcata, Rhaphoneis amphiceros, Actinoptychus senarius, Tryblionella navicularis, T. punctata*	HMB24 220–950 cm
8	Creeks and creek levees (intertidal channels)	*P. sulcata, C. striata, R. amphiceros, A. senarius, Podosira stelligera, Delphineis surirella, T. navicularis, Scolioneis tumida*	HMB5 400–730 cm
9.1	High saltmarsh	*Diploneis interrupta, Nitzschia vitrea, D. ovalis, T. navicularis*	HMB12 450–540 cm
9.2	Low saltmarsh	*Campylodiscus echeneis, Diploneis didyma, D. smithii, Triblionella marginulata*	HI1 540–680 cm
10	Mudflat	*Scolioneis tumida, T. navicularis, Gyrosigma acuminatum, G. balticum, Pleurosigma angulatum, P. sulcata*	HI1 20–300 cm
11	Sandflat	*P. sulcata, R. amphiceros, Achnanthes delicatula, Lyrella lyroides, Petroneis humerosa, Delphineis surirella, Coscinodiscus* sp.	HI1 5–250 cm
12	Subtidal channel	*P. sulcata, A. senarius, Coscinodiscus* spp., *R. amphiceros, Triceratium* spp., *D. surirella*	HMB14 220–890 cm
13	Fresh/brackish marsh with standing water	*A. ovalis, Ellerbeckia arenaria, Cyclotella meneghiniana, Cymbella aspera, C. cistula, E. adnata, Martyana martii, F. pinnata, Surirella striatula, R. gibba, Synedra ulna*	HMB16 1300–1470 cm
14	Beach sands and gravels, dunes	No diatom data	HMB23 15–1334 cm
15	Pre-Holocene (pleistocene till, bedrock)	Limited diatom data	Base of most cores
16	Eroded contact points	No diatom data	

date for North Sea Quaternary coastal environments. Some taxa found in the Humber cores were not reported in these publications and their environmental interpretation was based on information from Hendey (1964), Krammer & Lange-Bertolot (1986, 1988, 1991*a*, *b*), Round *et al.* (1990), Snoeijs *et al.* (1991–1997) and Juggins (1992). A few fossil taxa are rare or unknown in modern samples (e.g. *Eupodiscus radiatus*) and their ecology is uncertain. Other taxa, such as *Actinoptycus senarius* and *Paralia sulcata*, were much more common in older sediments than in younger ones. A similar trend has been observed in other estuaries and may reflect habitat loss due to anthropogenic disturbance. Quantitative reconstructions of environmental parameters (e.g. salinity, nutrient availability) are not well-developed for estuarine and coastal diatoms in the way that they are for many freshwater systems (e.g. Birks *et al.* 1990; Gasse *et al.* 1997; Horton *et al.* 1999*a*, *b*). Juggins (1992) established the viability of such an approach for salinity in the Thames Estuary and further studies for the UK are in progress (Zong & Horton 1998, 1999). Clearly there is great scope for applying transfer functions to the diatoms from the Humber cores once they become available.

A total of 16 environmental facies were identified for the Humber, of which 13 were distinguishable from the diatom evidence. Table 3 lists the environments identified, their typical diatom species associations and examples of their occurrence in the HMB cores.

Details of the results from each of the Humber cores (Fig. 1) will be available on the LOEPS CD-ROM. Data from three cores felt to typify the sequences from the inner (HMB16) and outer (HMB10, HMB12) parts of the estuary are described Ridgway *et al.* (1999). Here, the diatom results from three cores (HMB20, HMB7 and HMB12) from the inner, middle and outer estuary (Fig. 2) are summarized to illustrate both the range of environments recorded and the variable diatom preservation in the cores.

Ousefleet, core HMB20 (top at +2.88 m OD)

HMB20 (length 18 m) from Ousefleet [482250 423150], is the longest core recovered from the inner part of the Humber Estuary (Fig. 1) and has a chronology provided by five radiocarbon dates. Diatoms were well preserved, but sparse, in the lower part of the core (950–1650 cm, −6.62 to −13.62 m OD) and form the basis of the environmental interpretation (Fig. 2). Above 950 cm (−6.62 m OD) valves were too poorly preserved to count and the environmental reconstruction is based on geochemistry and lithology (Rees *et al.* this volume). The basal unit of the core is glaciofluvial sand and gravel overlain by a peaty silt with abundant root and stem debris. Diatoms were found in one level and were entirely freshwater taxa, dominated by epiphytic and epipelic species. Standing freshwater with abundant macrophytes is indicated. Similar conditions were recorded in the overlying peat which has limiting dates of 8240 ± 110 BP (about 9214 cal. a BP) and 7190 ± 60 BP (7945 cal. a BP) (1580–1573 cm, −12.92 to −12.85 m OD). Environmental facies 3 (oak–hazel fenwood) and 4 (alder carr) were assigned. Between this and the overlying peat layer at 1496 cm (−12.08 m OD), the diatoms indicate a spike of increased salinity, but this is not reflected in the pollen record. This upper peat was dated at 7205 ± 85 BP (7954 cal. a BP). The depositional environment between 1500 and 1400 cm (−12.12 to −11.12 m OD) was clearly complex, with peat overlain by sandy silt and sand, and rapid changes in the diatom flora from marine to freshwater to marine in rapid succession. Truly marine conditions do not appear to have been established until about 1408 cm (−11.20 m OD), at the base of the silt unit (environmental facies 10, mudflat) overlying the sand. The whole sequence up to the top of the sand layer shows a predominantly freshwater environment with occasional spikes of brackish and marine diatoms. It is probable that the latter represent sediment reworking and transport. Up-core, the sediments become more sandy and valves are generally poorly preserved (environmental facies 8 creeks and creek levees/intertidal channels). The final level with well-preserved diatoms is in a silty unit with abundant stems and roots at 990 cm (−7.02 m OD) (environmental facies 6 coastal reedswamp). A radiocarbon date from 715 cm (−4.27 m OD) yielded a date of 4735 ± 55 BP (average 5314 cal. a BP). The fine sands and silts above 703 cm (−4.15 m OD) preserved only occasional marine diatoms and were assigned to the environmental facies 12 (subtidal channel) and 10 (mudflat). This core illustrates the transition of the inner estuary from being one of freshwater habitats (pools and rivers), to increasing tidal influence and then fully estuarine.

Winterton Carrs, core HMB7 (top at +1.9 m OD)

Core HMB7 (8 m) comes from Winterton Carrs [496430 418920] on the south side of the middle

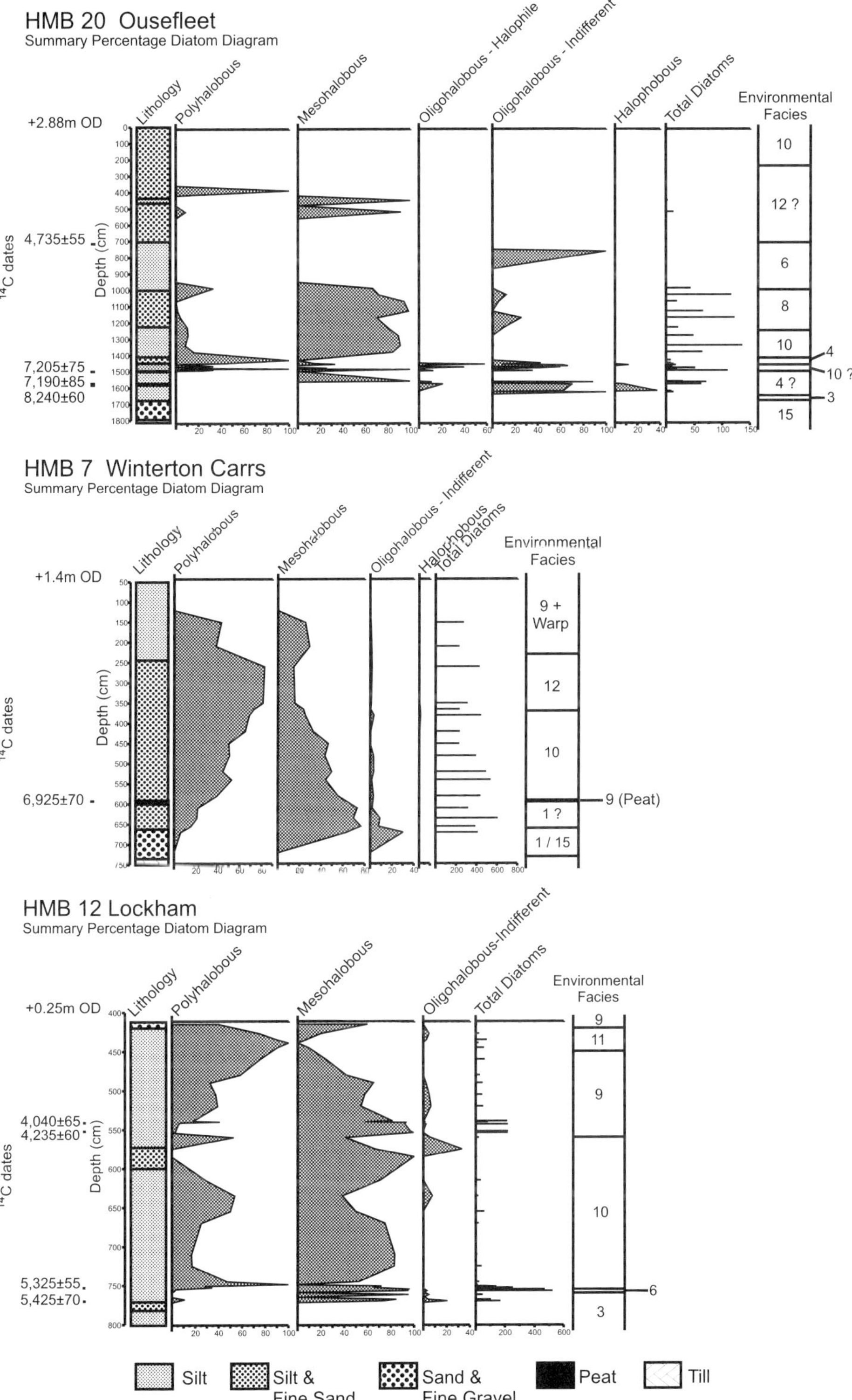

Fig. 2. Summary diagram of the diatom records from HMB20, HMB7 and HMB12 showing salinity variations. Environmental facies are described in Table 3.

part of the estuary (upstream of the Humber gap) (Fig. 1). Chronological control is provided by two radiocarbon dates, one accelerator mass spectrometry (AMS) date and one conventional. Diatom preservation was generally good in this core and a wide range of environments are recorded (Fig. 2). The base of the core is till, overlain by a very mixed assemblage, possibly a flood deposit. Abundant and well-preserved diatoms in silts above a thin saltmarsh silty peat (6925 ± 70 BP, giving a mean calibrated age of 7686 cal. a BP, and is therefore possibly reworked) are believed to represent a mudflat (environmental facies 10). Mudflat species decline in abundance up the core and by 360 cm (−1.70 m OD) marine planktonic and tychoplanktonic species dominate. Together with the lithology of sand silts, this has been interpreted as a subtidal channel (environmental facies 12). In the upper 220 cm (−0.30 m OD) of the core, the diatoms suggest a high saltmarsh developing (more aerophilous taxa), but the soil textures and the presence of high levels of metals of anthropogenic origin indicate that this may be reclaimed land, with nearby saltmarsh acting as the source of the diatoms.

Lockham, core HMB12 (top at +4.25 m OD)

HMB12 is a relatively short (8.11 m) core from Lockham [539000 417200] on the north side of the outer estuary (Fig. 1). There are 11 radiocarbon dates from the core as a whole, of which four are shown on Fig. 2. Diatom preservation is highly sporadic, but the core illustrates the dynamic nature of the estuary. Other aspects of this core are discussed in Ridgway *et al.* (1999). Diatoms are well preserved in a peaty silt unit (771–755 cm, −3.46 to −3.30 m OD) with both freshwater and brackish-water taxa. Taken with the results of pollen analysis, the base of the core has been interpreted as indicating oak fenwood (environmental facies 3) (bracketing date ranges 5425 ± 35 and 5325 ± 55 BP, 6391–5997 and 6271–5944 cal. a BP). Up-core, the diatoms indicate increasing salinities and a reedswamp (environmental facies 6), although the counts are low. Above 754 cm (−3.29 m OD), the diatoms indicate the development of a saltmarsh, followed by a mudflat (600–573 cm, −1.75 to −1.48 m OD) from which only a few robust diatoms were preserved. A rare example of a regressive contact in the Humber occurs at 573 cm (−1.48 m OD), with the return of saltmarsh conditions. Diatoms are quite abundant and well preserved. Both the lower and upper parts of the saltmarsh unit have been dated to 4235 ± 60 (4871–4569) and 4040 ± 65 BP (4816–4354 cal. a BP), respectively. As in other cores, diatom preservation in sandy sediments is poor. Sands occur between 450 and 426 cm (−0.25 and −0.10 m OD) and the interpretation of this part of the sequence as a sandflat (environmental facies 11) is based largely on the laminated texture. Resistant marine planktonic taxa are present, but the significance of these is uncertain (see above). Silts between 426 and 285 cm (−0.10 and +1.40 m OD) are believed to record further saltmarsh development at this site, although diatoms are low in numbers and poorly preserved. Radiocarbon dates in this section yielded a wide variety of ages. Few diatoms were recovered in the sandy sediments above 275 cm (+1.50 m OD). Sandflat sands occur towards the top of the core and are overlain by made-ground. This site clearly experienced highly variable conditions during sea-level rise, with mudflats, saltmarsh and sandflats occurring at different times.

Organic deposits as indicators of former sea-level

Organic deposits and their associated pollen play an important role in reconstructing former sea levels. Evidence for change comes in three forms: deposits that have only limiting value; those deposits containing indicators of intertidal influence, which rest on non-marine units and are overlain by clastic marine sediments, and peat layers intercalated within marine clastic units. In this section we present examples from the inner (HMB13) and outer parts (HMB12) of the main estuary and from the Ancholme Valley (see Fig. 1). Long pollen diagrams through freshwater peats were not completed under LOIS, but enough previously published research exists to enable the construction of the palaeogeographic maps between the locations of the LOIS cores.

Whitton Ness, core HMB13 (top at +3.8 m OD)

Organic deposits that have only a limiting value represent a significant class of palaeoenvironmental evidence. Pollen analysis shows that these were formed under freshwater conditions and contain no indicators of marine influence, but are overlain unconformably by marine clastic sediments. A sharp peat–clastic contact is usual and indicates a degree of erosion of the peat, or at least an hiatus in deposition before the later introduction of estuarine sediments.

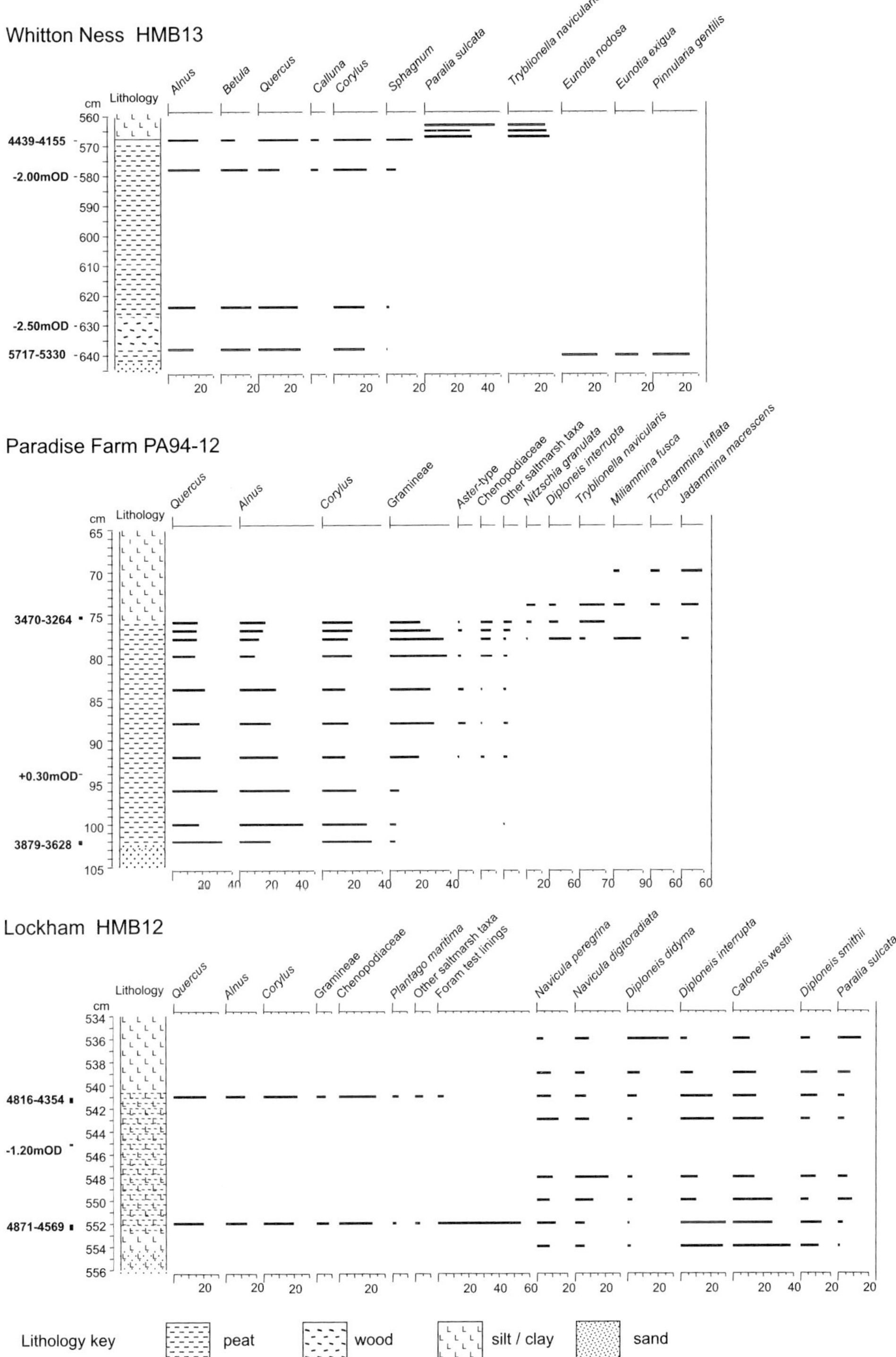

Fig. 3. Microfossil diagrams: (**a**) Whitton Ness HMB13, showing a limiting value; (**b**) Paradise Farm PA9412, showing a basal peat with sea-level index points; (**c**) Lockham HMB12, showing an intercalated peat with sea-level index points. Pollen expressed as per cent total land pollen; spores as per cent Σ(land pollen + spores); diatoms as per cent. Calibrated radiocarbon ages, altitude (metres above OD) and depth (cm) down-core shown on the left of the lithology column. The stratigraphy is drawn according to Troels-Smith (1955).

The time gap between the peat and clastic units cannot be estimated, and so the age of the top of the peat cannot be used to date the later marine event. Such a truncated freshwater peat is still very useful in providing limiting age and altitude values, which constrain the possible position of sea-level. The type of freshwater community represented by the pollen assemblage in the top of the peat may provide a measure of its altitudinal distance above mean high water spring tide (MHWST) by its relative position in the coastal vegetation succession, and therefore an estimate of the degree of erosion of the contact. Limiting values have also proved very useful in the reconstruction of palaeogeographic maps of coastal zone environments. Figure 3a shows an example of limiting data from Whitton Ness (HMB13) in the inner Humber Estuary. A well-humified peat with wood in its lower levels rests upon a sand over gravel (640 cm, −2.16 m OD), and is overlain via a sharp contact (567 cm, −1.87 m OD) by a laminated sandy silt. This silt contains peat fragments, supporting the view that the sharp contact may be erosive, and a rich diatom flora which indicates a marine/estuarine origin. The assemblage is almost entirely poly- and mesohalobous, the former dominated by *Paralia sulcata* and the latter by *Tryblionella navicularis*. The peat only preserves diatoms in its lower levels, where an entirely freshwater, oligohalobous, flora occurs with *Eunotia nodosa*, *E. exigua* and *Pinnularia gentilis* prominent. Tree and shrub pollen *Quercus*, *Alnus*, *Betula* and *Corylus* dominate the uppermost peat level with few taxa of freshwater aquatic successions. *Calluna* and *Sphagnum* suggest an acid, mainly terrestrial bog surface, and so a significant degree of erosion may have occurred here. If so, then marine inundation may have been substantially later than the limiting date on the upper contact of 3910 ± 45 BP (4439–4155 cal. a BP). The date of 4830 ± 70 BP (5717–5330 cal. a BP) on the base of the peat also forms a limiting value at a lower altitude, recording peat inception and environmental change, which is important for palaeogeographic reconstruction. Both dates are supported by the post-elm decline pollen data and constrain the altitudinal limits of possible sea-level curves for the inner Humber Estuary.

Paradise Farm (top at 1.24 m OD)

A second class of data, which provides a more precise relationship with past sea-level, is formed by basal organic deposits that are overlain by clastic marine sediments and rest upon non-marine units, but which themselves include indicators of intertidal influence. These organic deposits were, therefore, formed at least partly under saltmarsh conditions or very close to them. In some cases the whole of the peat profile contains microfossils of saltmarsh or estuarine origin, so that peat inception took place within direct tidal influence; while in others the base of the peat formed within freshwater environments and intertidal indicators become present in the peat further up the profile. In both cases these usually increase in frequency towards the contact with the overlying marine sediment. Usually this upper-peat contact forms a gradual transition to the marine facies and, even where some erosion may be suspected, the presence of saltmarsh indicators in the peat shows this cannot have been great and so the significance of the contact is not devalued. Such transitional transgressive peat–clastic contacts have a measurable altitudinal relationship to a past sea-level and when radiocarbon dated form reliable index points from which a sea-level curve may be constructed. If peat inception began under freshwater conditions in such a profile, the dated base of the peat may still be considered an index point recording a positive sea-level tendency. The meaning, however, is less precise as organic accumulation may well have been caused by rising ground-water tables driven by rising sea-level. Such data are also very important for palaeogeographic reconstruction. Figure 3b shows an example of a basal peat profile with sea-level index points from Paradise Farm in the valley of the River Ancholme (Fig. 1). Peat inception took place at 3490 ± 50 BP (3879–3628 cal. a BP) (+0.29 m OD) over a basal sand and the pollen data are compatible with the radiocarbon age, dominated by trees and shrubs, *Corylus*, *Alnus* and *Quercus*. For the lower third of the peat there are no saltmarsh taxa recorded and Gramineae and freshwater marsh pollen are very low. Rising ground-water and waterlogging initiated peat growth, a process likely to have resulted from approaching estuarine conditions. In mid-profile, rising Gramineae values are accompanied by the introduction of saltmarsh taxa, primarily Chenopodiaceae and *Aster*-type, which increase greatly in diversity and frequency towards the top of the peat. A succession through reedswamp and saltmarsh environments is indicated, supported by the appearance in the upper part of the peat of high numbers of poly- and mesohalobous saltmarsh diatoms, primarily *Tryblionella navicularis* and *Diploneis interrupta*, and saltmarsh foraminifera *Miliammina fusca*, *Trochammina inflata* and *Jadammina macrescens*. These continue into the overlying clastic unit. The upper transitional peat–clastic contact at

+0.51 m OD supports the biostratigraphic evidence of an unbroken environmental transition during the introduction of estuarine conditions and it is a reliable sea-level index point dated 3170 ± 50 BP (3470–3264 cal. a BP). The date on the base of the peat is an index point during a positive sea-level tendency.

Lockham, core HMB12 (top at +4.25 m OD)

The third type of sea-level data is from peat layers intercalated within marine clastic units, and these are widespread within all parts of the Humber Estuary. Peat deposition represents a temporary period of relative sea-level fall, or at least a reduced rate of rise, so that organic sedimentation under saltmarsh conditions replaces clastic estuarine sedimentation. Sometimes this process proceeds far enough to allow freshwater deposition under reedswamp, fen and even carr environments before renewed positive relative sea-level movement reverses the succession and reintroduces intertidal clastic sedimentation via saltmarsh facies. The regressive and transgressive contacts to these intercalated peats may be gradual or erosive and sharp, and require individual evaluation as index points. Intercalated peats are much more prone to consolidation under the weight of later sediment than are basal peats, which rest on a hard stratum, and so their observed altitude can be much lower than their original, making them more difficult to interpret. Figure 3c shows an example of an intercalated peat from the outer Humber Estuary at Lockham (HMB12), where a thin clayey amorphous peat between −1.28 and −1.16 m OD interrupts a long silt and clay sequence. The pollen assemblage is very similar throughout this organic layer and reflects a post-elm decline (i.e. post-5 ka BP) *Quercus*, *Alnus* and *Corylus* woodland. Very high frequencies of saltmarsh pollen occur, particularly Chenopodiaceae and *Plantago maritima*, but including other types like *Armeria* and *Spergularia* as well as abundant foraminiferal test linings. This saltmarsh component reaches almost 30% of total land pollen and suggests deposition under lower to mid-saltmarsh conditions. This is confirmed by the high frequencies of poly- and mesohalobous diatoms within the organic layer and in the clay above and below. Saltmarsh diatoms *Navicula peregrina*, *N. digitoradiata*, *Diploneis interrupta*, *D. didyma* and *Caloneis westii* dominate and *Diploneis smithii* and *Paralia sulcata* are the most common more-marine forms. This succession records the temporary replacement of intertidal mudflat by lower saltmarsh environments. Both contacts are gradual and are reliable sea-level index points, the lower regressive contact dated 4235 ± 60 BP (4871–4569 cal. a BP) and the upper transgressive contact dated 4040 ± 65 BP (4816–4354 cal. a BP), which agrees well with the relative pollen age.

Holocene relative sea-level changes in the Humber Estuary

Sea-level history

Although previous studies established a data resource that could be used to reconstruct the sea-level history of the Humber region (Gaunt & Tooley 1974; Tooley 1978, Dinnin & Lillie 1995; Long *et al.* 1998), the systematic collection and integration of the great volume of new LOIS data now enables us to attempt a more holistic appraisal of the nature and effects of sea-level change within the Humber Estuary. The increased number of radiocarbon-dated sea-level index points and other palaeoenvironmental analyses, such as the examples shown above, provide an enhanced data set (Tables 1 and 2). The spatial and temporal distribution of the data points is greatly improved, so that differences between various parts of the estuarine system may be investigated. The distribution of data has also been extended into the earlier and later Holocene. The Holocene sea-level data from the Humber is presented in the form of a regional sea-level curve (Fig. 4). The upward trend of relative sea level displayed by the shape of the curve, rapid in the early Holocene then at a much reduced rate in the mid–late Holocene, conforms with the modelled sea-level predictions of Lambeck (1995) for an area not far beyond the limits of the maximum Devensian ice advance. Local-scale processes that operate within the Humber Estuary influence individual sea-level index points. These include sedimentation rates, including influences from variations in catchment discharge, changes in tidal range and the tide level at which indicators form, and post-depositional sediment consolidation. The scatter of index points on the graph of relative sea-level change includes the total influence of these local-scale processes and also any differential isostatic movements. Shennan *et al.* (this volume) illustrate these influences in explaining the differences between the altitudes of individual index points and the summary curve. For example, the curve lies close to the index points from basal peats; whereas most of the index points from intercalated peats lie below the

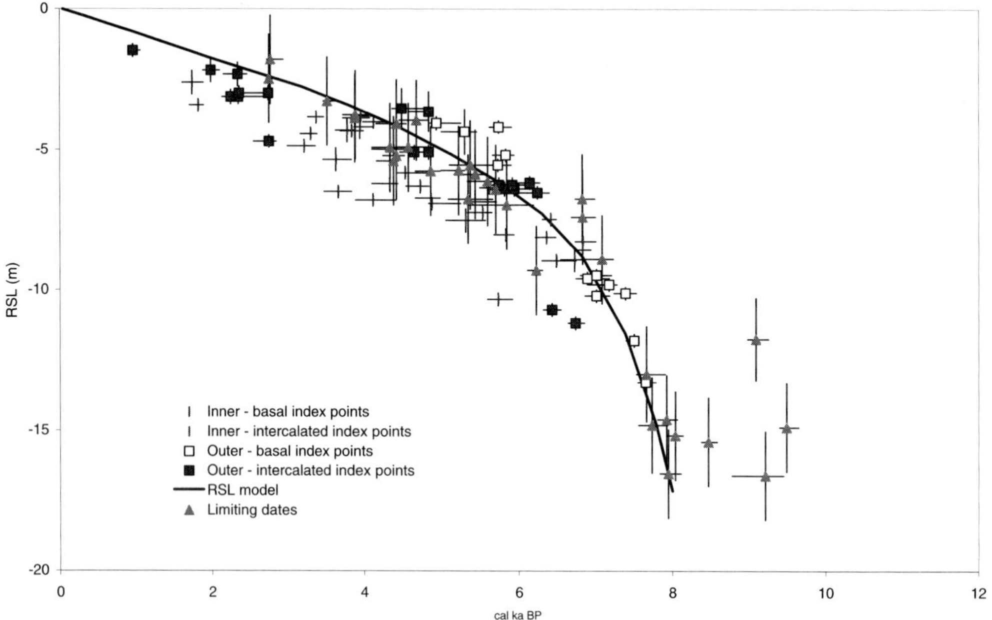

Fig. 4. Humber Estuary relative sea-level (RSL) index points. Error terms are calculated as described in the text for each index point, but only appear when they extend beyond the size of the symbol (see also Shennan *et al.* this volume).

curve, indicating the likely effect of sediment consolidation. The pattern of residuals for the basal-peat index points has a spatial component, indicative of tidal-range changes through time and/or differential isostatic movements (Shennan *et al.* this volume).

Palaeogeographies of the Humber Estuary 8–3 ka BP

With less than 100 radiocarbon-dated samples available, any attempt to describe the palaeogeographic evolution of the Humber Estuary and associated lowlands (*c.* 3300 km^2) through the Holocene is clearly constrained by the assumptions behind the reconstructions that are used to interpolate between dated samples through both time and space. The approach adopted here is to use: the relative sea-level reconstruction (Fig. 4); the reconstructions of past sedimentary environments from sediment cores and their relationships to tide levels to classify (based on altitude) a digital terrain model (DTM) for each of the main sediment units in the Humber sequences derived from the analysis of the borehole archive at the British Geological Survey (BGS; Rees *et al.* this volume). Thus the relative sea-level reconstruction, including a spatially variable parameter for an increasing high tide level up-estuary, similar to present, is overlaid within a geographic information system (GIS) on a DTM to predict palaeogeographic reconstructions for 8–3 cal. ka BP at 1 ka intervals. The number of reference points to test these predictions in terms of detailed age and biostratigraphic control is small (Tables 1 and 2) and it is unrealistic to consider the spatial patterns of the 16 environments discussed earlier (Table 3). In the following reconstructions we consider the five or six major environments.

For each palaeogeographic reconstruction the GIS model produced a classification from which we interpreted the final version. The example of the GIS classification shows both the potential and limitation of the approach (Fig. 5). Where the density of original data points is high with respect to the height variations of the DTM surface, the model produces an easily testable reconstruction. For example at 5 cal. ka BP the tidal channel in the eastern part of the estuary is flanked to the north and south by narrow zones of tidal flat and saltmarsh. However, for the same period in the large embayment in the western part, the density of boreholes is low, with angular shaped areas resulting from the coarse resolution of the DTM. The DTM is further limited because it does not include the post-depositional effects of sediment erosion or

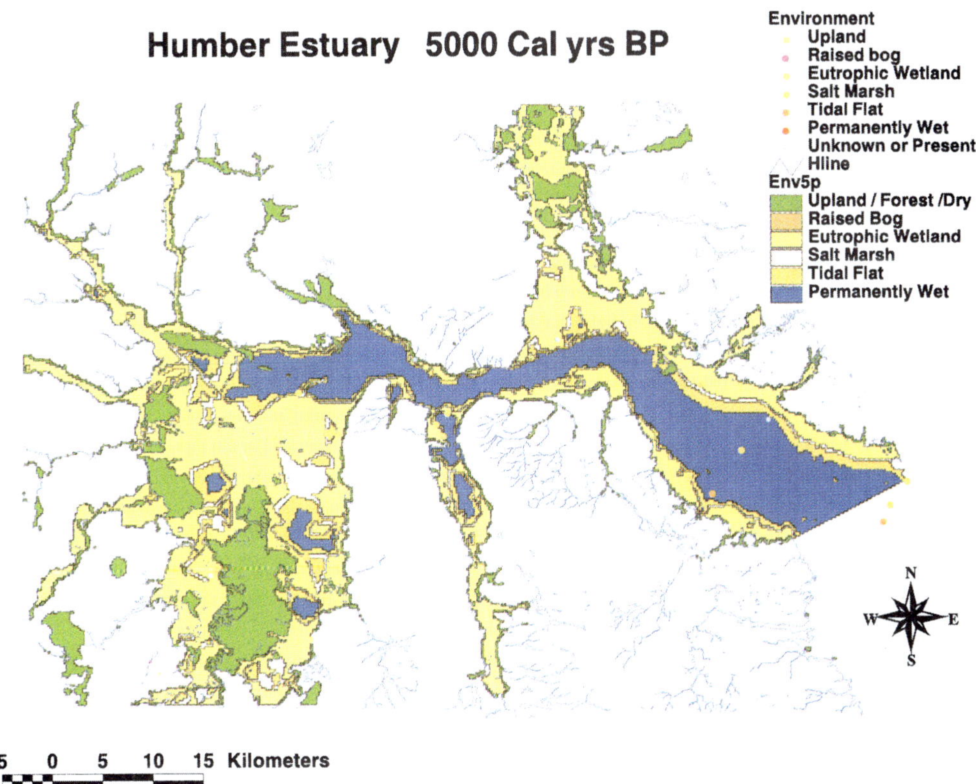

Fig. 5. GIS classification of palaeoenvironments in the Humber Estuary at 5 cal. ka BP.

sediment consolidation. This results in the low-altitude locations within the upper reaches of different arms of the estuary being unconnected to the main subtidal channel of the estuary. Therefore we use these GIS models as an interim stage, which we test against all the available biostratigraphic and chronostratigraphic information, and apply a contiguity constraint whereby the palaeogeographic reconstructions must show environmental gradients compatible with present day situations. This interpretation is carried out outside of the GIS.

For the reconstructions illustrated in Fig. 6 we show five major environments (Table 4), no longer distinguishing a separate saltmarsh class. The intricacies of both the subdivisions within a class and the transitions at the zone boundaries are not shown and are not justified given the data available to test the reconstructions.

8 cal. ka BP reconstruction

Estuarine subtidal and intertidal environments were restricted to the outer estuary (Fig. 6), when sea-level was about 17 m below present (Fig. 4). The mainly marine Garthorpe sediment suite described by Rees *et al.* (this volume) represents these environments. Numerous dates on freshwater peats (Table 2) and the microfossil assemblages from them, reveal the widespread occurrence of eutrophic wetland with a mosaic of oak–hazel fenwood, open standing water and sedge fen. Currently, it is not possible to provide a detailed reconstruction of the river channels, although it is clear that the main channel followed a meandering course around the line of the present estuary (Rees *et al.* this volume). Previous boreholes (e.g. Gaunt 1994; Van de Noort & Ellis 1997) have located older, deeply incised channels of the main rivers, beyond tidal influence. HMB20 (Fig. 1) lies well inland of the intertidal environments shown in the reconstruction, but the spikes in marine–brackish-water diatom taxa indicate at least temporary incursions of saline water. The samples shown as limiting dates on the sea-level graph fit the regional sea-level curve and may, therefore, represent extreme water levels penetrating up the estuary. It is possible that they represent the

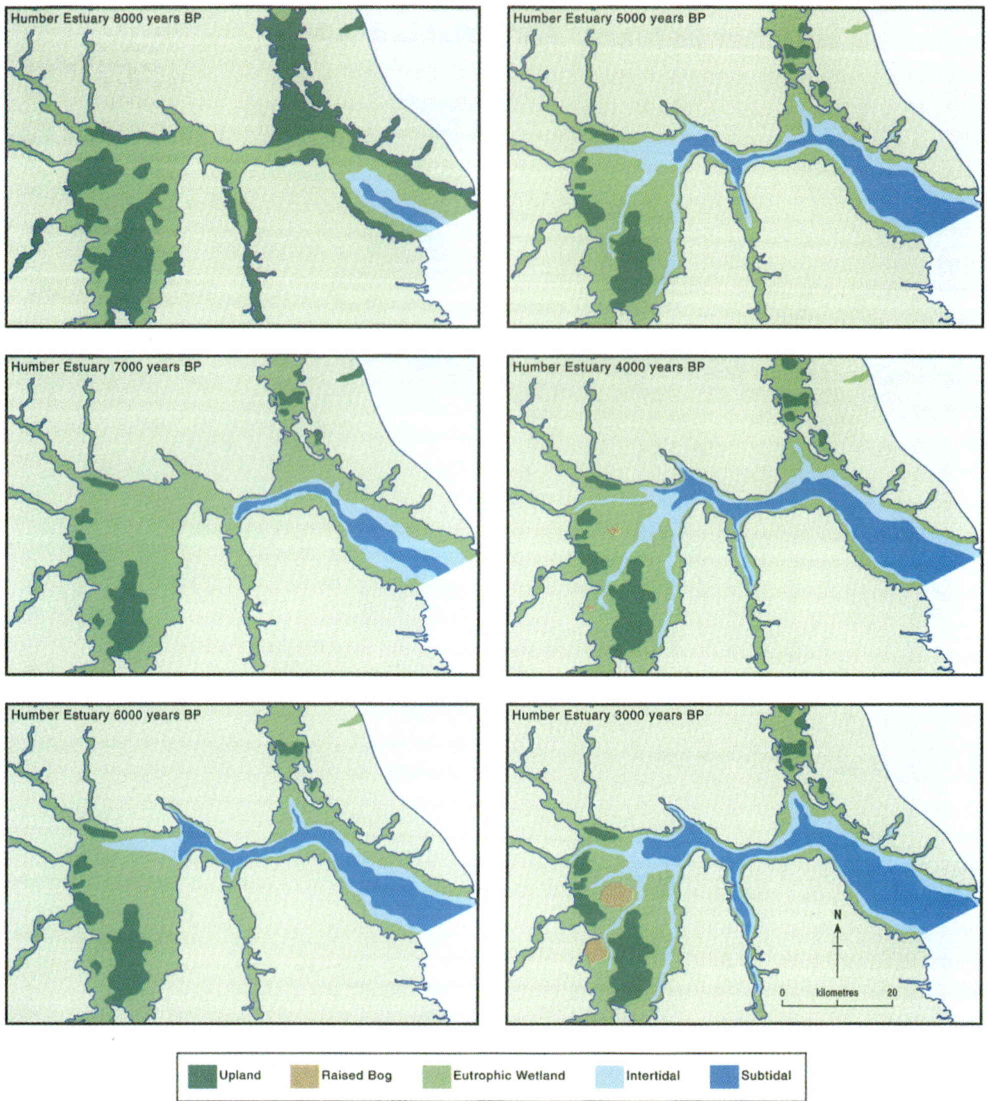

Fig. 6. Palaeogeographical maps of the Humber Estuary from 8 to 3 cal. ka BP. Changes in the configuration of the coastline to the north and south of the estuary are not shown.

effects of the tsunami recorded widely in north-east Scotland (Dawson *et al.* 1988).

7 cal. ka BP reconstruction

With sea-level around 10 m below present numerous cores from locations in the outer estuary record the transgression of intertidal environments across previously eutrophic wetland (Fig. 6). The limit of intertidal sedimentation occurs around the confluence of the Ancholme with the main channel, indicated by the transgressive overlap at South Ferriby (HMB3, Table 1) at 6727–6996 cal. a BP. The rest of the inner estuary, the Ancholme and the valley of the Hull remain dominated by eutrophic wetland (Newland and Butterwick sediment suites of Rees *et al.* this volume). There are indications of very limited incursions of saline water into the fresh aquatic systems in the lower Ouse–Trent area.

Table 4. *Environments identified on the palaeogeographic maps*

	Palaeogeographic environment class		Biostratigraphic and lithostratigraphic environment class (number from Table 3)
1	Upland	15	Pre-Holocene, till, bedrock
2	Raised bog	2	Raised bog
3	Eutrophic wetland and freshwater aquatic	1	River above tidal limit
		3	Oak/hazel fenwood
		4	Alder carr
		5	Sedge fen
		13	Fresh/brackish marsh with standing water
4	Intertidal	6	Coastal reedswamp
		7	Intercreek areas flooded by highest tides
		8	Creeks and creek levees
		9	Saltmarsh
		10	Mudflat
		11	Sandflat
		14	Beach sand and gravel
5	Estuarine subtidal	12	Subtidal channel

6 cal. ka BP reconstruction

The transgression continues up-estuary (Fig. 6) with intertidal environments encroaching into the lower reaches of the Trent (e.g. Garthorpe Farm, Table 1), Hull, Ouse, Aire (e.g. Newlands, Table 1) and Foulness Valleys (e.g. Sandholme Lodge; Long *et al.* 1998; Table 1). Intercalated peats within some of the valley sequences slightly after 6 cal. ka BP indicate a fine balance between the intertidal and freshwater sedimentary environments.

5 cal. ka BP reconstruction

By 5 cal. ka BP intertidal sedimentation occurs in all of the valleys draining into the estuary, but the density of boreholes with chronostratigraphic data makes it very difficult to define the extent of different environments in detail (Fig. 6). For example, the index points from East Ferry and Garthorpe Farm (Table 1) record the extent of intertidal environments along the Trent. A single index point (Van de Noort & Ellis 1998) from the Ancholme Valley (Brigg, 5591–5912 BP, Table 1) plots as an outlier on the relative sea-level graph (Fig. 4) and the other index points from the Ancholme are much younger and at higher altitudes (e.g. Winterton Ings, Table 1) or limiting values (Brigg, Table 2). The palaeogeographic maps show the significant areas of intertidal environments but not the narrow tidal reaches of individual rivers, such as that probably recorded by the Brigg index point.

4 cal. ka BP reconstruction

Intertidal environments continue to expand up the valley systems (Fig. 6). The limited number of radiocarbon dates from the late Holocene precludes a more detailed reconstruction.

3 cal. ka BP reconstruction

This reconstruction represents the probable maximum extent of intertidal expansion (Fig. 6). There are too few index points to define the limit in each valley system. Negative tendencies on intercalated peats occur before 3 cal. ka BP in the Old Don and Trent (Dirtness Levels and East Ferry, respectively, Table 1), but the removal of any surface peats due to drainage or agriculture eliminates the evidence of the final regression. Numerous other sites record the removal of intertidal sedimentation after 3 cal. ka BP (Table 1 also see discussion in Long *et al.* 1998).

The Humber and its wetland environments over the last 3 ka

The last 3 ka have seen major changes to the Humber and its wetlands, partly due to sea-level change but primarily due to anthropogenic activity. Any changes in the rate of sea-level rise must be slight since there are no discernible departures from the average trend that could not be attributed to other factors such as sediment consolidation. At some sites archaeological data indicate local changes. At around 3 ka, sea-level continued to rise (see Fig. 4), leading to sediment accretion around the estuary margins and a transition from coastal reedswamp to saltmarsh at the Bronze Age–Iron Age transition (Neumann 1998). Archaeological evidence suggests that drier conditions occurred during the subsequent Roman period, indicating falling sea-level or lower amplitude tidal regimes, which allowed the expansion of settlement into previously wet areas, e.g. in the lower Hull Valley (Didsbury 1990). The nature of this change is not clarified by the sea-level curve (Fig. 4). A Roman road was built northwards from South Ferriby towards the Humber during the first or second century AD, but this became buried by estuarine sediment within around 200 years of its construction, indicating rising sea-level (Neumann 1998). Similar situations can be seen on the north side of the estuary at Faxfleet, where a Roman site was probably abandoned in the fourth century AD as a result of rising water-levels (Sitch 1990), and at Kilnsea, in southern Holderness, where a possible Roman settlement is considered to have become buried by subsequent sediment accretion (Van de Noort & Ellis 1995*b*). Delivery of sediment to the Humber system due to human deforestation and colluviation in the catchment will have been a minor factor before the later Holocene (Long *et al.* 1998). It was greatly increased in Roman and later times as more intensive forest clearance and land-use caused extensive soil erosion, either locally or from the wider catchment, and a major influx of clastic sediment (Buckland & Sadler 1985; Dinnin 1997*b*).

Until the Medieval period, the low-lying areas adjacent to the Humber and the lower reaches of its tributaries were essentially wetlands on account of the proximity of the water-table to the surface and/or their susceptibility to inundation by estuarine or riverine flooding, but from that time onwards, widespread changes occurred due to artificial drainage, land reclamation and land-use changes (Van de Noort & Davies 1993). Some of the earliest drainage activity occurred in the lower Hull Valley and at Walling Fen in the southern Vale of York during the twelfth and thirteenth centuries, with embankment of the Humber and excavation of artificial drainage channels (Sheppard 1958, 1966). Drainage activity in the region increased in the seventeenth century, in particular with the large-scale diversion of existing watercourses and the excavation of new channels in the Humberhead Levels, under the direction of Cornelius Vermuyden (Dinnin 1997*a*). Geochemical analysis of the estuarine sediments indicates the presence of deposits enriched with anthropogenic metals, especially lead, in the post-Medieval period (Rees *et al.* this volume).

Land reclamation was also undertaken from the eighteenth century onwards by the process of 'warping', in which the ground-level was raised by the addition of sediment in order to reduce the effects of waterlogging. 'Dry' or 'cart' warping involved transporting material by carts and dumping it on the land, while the much more extensive technique of 'flood' warping involved the embankment of areas that were then flooded with sediment-rich estuarine waters fed on to the land at high tide via a series of artificial channels. The sediment was then allowed to settle out as the water drained away on the falling tide, leaving a thin layer of material overlying the previous ground surface (Ellis 1990). By repeating this cycle, the ground-level could be raised significantly, by up to $1\,\mathrm{m\,a^{-1}}$ in some cases. Large areas of warped land occur in the Humberhead Levels and lower Trent Valley (Lillie & Weir 1997; Lillie 1998). In the area around Sunk Island, in southern Holderness (Fig. 1), a slightly different procedure was followed, whereby the area between the island and the Holderness shore was progressively reclaimed from the estuary during the eighteenth and nineteenth centuries by the construction of a series of embankments, which allowed sediments to accumulate by natural accretion (Dinnin 1995). Of the Humber cores, HMB16 and HMB18 are probably associated with areas of warp, and HMB4 and HMB8 contain accreted sediments associated with the extension of Sunk Island.

More recently, the wetland areas associated with the peat deposits of Thorne and Hatfield Moors in the Humberhead Levels have become altered as a result of drainage and commercial peat extraction (Dinnin 1997*a*). The peat has consequently become de-watered and has been reduced to a depth of less than 2 m over much of these areas, although recent collaboration between the peat extractors and English Nature has led to the re-wetting of some areas with a view to re-establishing wetland habitats (Van de Noort & Davies 1993).

In addition to artificial drainage, the conversion of much of the low-lying land around the Humber from pastoral to arable farming has led to enhanced desiccation due to increased soil aeration as a result of ploughing and related cultivation practices. In the former area of Humberside, for example, 46% of grassland was converted to arable between 1974 and 1985, mainly for the production of cereals (Symes 1987). Mineral extraction and urban and industrial development have also led to the deterioration of wetland areas either by lowering of the water-table or the physical removal of material. The desiccation of former wetland areas has lead to the oxidation of organic materials, and this poses a continuing threat to the preservation of wetland palaeoecological and archaeological remains (Van de Noort & Davies 1993).

Progressive artificial drainage, warping, river embankment, conversion from pastoral to arable farming, mineral extraction, and urban and industrial development have all led to a situation in which few natural wetland areas remain. Although some of these are now officially protected, e.g. Derwent Ings on the Derwent floodplain, Pulfin in the Hull Valley, and Allerthorpe and Skipwith Commons in the Vale of York, an indirect threat to their preservation is still often present via de-watering activities in adjacent, unprotected areas (Ellis *et al.* in press). Artificial drainage activities will also have led to the more efficient movement of water into the estuary, while artificial embankment has constrained its natural flow dynamics. In combination, these factors are likely to have altered the patterns of sediment erosion and deposition, and thus will produce an additional complication when taken along with sea-level change, in the prediction of the future evolution of the estuary (Pethick 1990). They have also had a profound effect on the biogeochemical 'metabolism' of the estuary (Andrews *et al.* this volume).

Discussion and conclusions

A range of techniques has been used to reconstruct the Holocene evolution of the Humber Estuary (see also Andrews *et al.*; Rees *et al.*; Ridgway *et al.* this volume). The Holocene sequence displays a wide range of sediment types within which the preservation of microfossils is highly variable. As had been anticipated, no single proxy would have provided a continuous record of change, and a combination of data has been used to reconstruct the palaeoenvironments. Where present, the diatom flora provides a valuable record of salinity variations, reflecting the changing balance of marine and freshwater inputs. The diatoms suggest at least occasional marine incursions into the inner estuary even in the early Holocene when most of the sediments were originating from the catchment of the Trent or the Vale of York drift. The environmental interpretations based on biological proxies are generally supported by the geochemical data. Very variable geochemical signatures for sediments in the lower parts of many cores suggest derivation from local sources in a fluvial environment, while more uniform and widespread signatures higher in the cores indicate sediment well-mixed under estuarine conditions (Rees *et al.* and Ridgway *et al.* this volume). The detail of the environmental reconstruction and hence of both the sea-level curve (Fig 4. and Shennan *et al.* this volume) and the palaeogeographical maps (Fig. 6), is constrained by the spatial distribution of the cores and the number of radiocarbon dates. The data show the marine transgression progressing up the estuary after 8 cal. ka BP, reaching the inner estuary by 6 cal. ka BP. The possible role of a till or bedrock sill, crossing the estuary just east of the Humber Gap, in limiting marine influence in the inner estuary before *c.* 7.4 cal. ka BP is discussed in Rees *et al.* (this volume). The detail of the records from individual sites, particularly in the valley systems entering the main estuary, reflects the variable balance of forces within the highly dynamic estuarine system. The palaeogeographic maps mark a major advance on our previous understanding of the history of the Humber Estuary, but can only provide a framework for future studies.

Tidal asymmetry, i.e. the dominance of either the flood or ebb tide current velocity, is a major controlling factor of estuarine process and morphology and it is interesting to apply theoretical models of asymmetry (e.g. Dronkers 1998) to the Holocene palaeogeography of the Humber as described in this chapter, and to the Holocene sedimentary sequences described by Rees *et al.* (this volume).

As sea level rose in the outer estuary, from *c.* 8 cal. ka BP, the ratio of mean water depth to tidal amplitude was relatively high, resulting in a marked flood-tide asymmetry. Assuming a plentiful supply of suspended sediment was available at that time from marine sources, such a flood dominance would result in a net input of fine-grained sediment to the estuary and would account for the rapid development of intertidal sediment shown in borehole sequences and in the accommodation-space infill described by Rees *et al.* (this volume). As intertidal elevations increased, however, so the ratio of water depth to tidal range decreased leading to a gradual reduction in the flood-tide dominance until the net landward sediment movement was replaced by a net seaward movement. Dronkers (1998) describes how such a transition in sediment transport is brought about by a change from wide, low elevation intertidal areas, often with a high sand content, in which shallow channels migrate across the estuary width, to one in which high elevation mudflats are intersected by deep subtidal channels. The evolving morphology results in tidal ebb dominance, which leads to net erosion of the intertidal areas, via a morphological hysteresis, so that a reversion to the former morphology is initiated and flood dominance with net sediment input is restored. It can be argued (Pethick 1994) that such oscillatory behaviour is then repeated with a gradual tendency towards tidal symmetry so that the estuarine system eventually attains a form of dynamic equilibrium. The evidence from the Humber suggests that such an oscillatory development did take place, interrupted by a rapid transitional period when the possible glacial morainic sill was breached (see above). Until the breaching of the sill, the evidence from the outer estuary suggests a gradual transition from a low intertidal/shallow subtidal channel morphology to high intertidal flats with extensive saltmarsh and deep subtidal channels, giving rise to the Garthorpe sedimentary suite. Such morphology would have led to increasing ebb-dominance and intertidal erosion. However, the breaching of the morainic sill led to a major increase both in tidal prism and in sediment inputs to the estuary so offsetting the rate of erosion and extending the Garthorpe phase until <3.8 cal. ka BP. At this time the sediment record indicates the onset of a channel migration phase: the Saltend Suite, indicating flood dominance, which persisted until *c*. 1.5 cal. ka BP. This was superseded in the sedimentary record by typical ebb-dominant morphology: high intertidal flats and extensive saltmarsh with deep subtidal channels and referred to as the Sunk Island Suite (see Rees *et al.* this volume).

Sedimentary evidence for morphological change over the past 500 years is more equivocal perhaps owing to an increase in the periodicity of oscillations. Detailed bathymetric evidence available over the past 150 years has demonstrated that a period of accretion ended in the 1950s and has been superseded by net erosion in the estuary. Whether this transition marks the latest oscillation in tidal asymmetry or is the result of anthropogenic interference is the subject of ongoing research.

Although much of this evidence is fragmentary it may be interpreted as showing a damped oscillation between flood and ebb asymmetry in the Humber Estuary over the Holocene period. The oscillations occur with increasing frequency towards the present, suggesting a movement towards a tidal symmetrical equilibrium, a tendency supported by recent research, which shows that the Humber tides are close to such a symmetrical end point (Townend 1999). Such a conclusion would be of great importance to estuarine managers and users since it could be used to predict the future development of the estuary.

This study was funded as LOIS/LOEPS 348. We would like to thank H. Glaves (BGS), G. Samways (UEA), J. Andrews (UEA), J. Lloyd (University of Durham) and R. McCulloch (University of Edinburgh) for their contributions.

References

ANDREWS, J. E., SAMWAYS, G., DENNIS, P. F. & MAHER, B. A. 2000. Origin, abundance and storage of organic carbon and sulphur in the Holocene Humber Estuary: emphasizing Lunar impact on storage changes. *This volume*.

BERRIDGE, N. G. & PATTISON, J. 1994. *Geology of the Country around Grimsby and Patrington*. HMSO, London

BIRKS, H. J., LINE, J. M., JUGGINS, S., STEVENSON, A. C. & TER BRAAK, C. J. 1990. Diatoms and pH reconstruction. *Philosophical Transactions of the Royal Society of London*, **B327**, 263–278.

BUCKLAND, P. C. & SADLER, J. 1985. The nature of late Flandrian alluviation in the Humberhead Levels. *East Midland Geographer*, **8**, 239–251.

DAWSON, A. G., LONG, D. & SMITH, D. E. 1988. The Storegga Slides. Evidence from eastern Scotland for a possible tsunami. *Marine Geology*, **82**, 271–276.

DE WOLF, H. 1982. Method of coding of ecological data from diatoms in computer utilisation. *Mededelingen Rijks Geologische Dienst*, **36**, 95–99.

DENYS, L. 1991. A check-list of diatoms in the Holocene deposits of the Western Belgian coastal plain with a survey of their apparent ecological requirements. I. Introduction, ecological code and complete list. *Belgische Geologische Dienst, Professional Paper 1991/2*, **246**, 1–41.

DIDSBURY, P. 1990. Exploitation of the alluvium of the lower Hull Valley in the Roman period. *In*: ELLIS, S. & CROWTHER, D. R. (eds) *Humber Perspectives: a Region Through the Ages*. Hull University Press, Hull, 199–210.

DINNIN, M. 1995. Introduction to the palaeoenvironmental survey. *In*: VAN DE NOORT, R. & ELLIS, S. (eds) *Wetland Heritage of Holderness*. Humber Wetlands Project, School of Geography & Earth Resources, University of Hull, Hull, 27–48.

—— 1997*a*. The drainage history of the Humberhead Levels. *In*: VAN DE NOORT, R. & ELLIS, S. (eds) *Wetland Heritage of the Humberhead Levels*. Humber Wetlands Project, School of Geography & Earth Resources, University of Hull, Hull, 19–30.

—— 1997*b*. Introduction to the palaeoenvironmental survey. *In*: VAN DE NOORT, R. & ELLIS, S. (eds) *Wetland Heritage of the Humberhead Levels*. Humber Wetlands Project, School of Geography & Earth Resources, University of Hull, Hull, 31–45.

—— & LILLIE, M. 1995. The palaeoenvironmental survey of southern Holderness and evidence for sea-level change. *In*: VAN DE NOORT, R. & ELLIS, S.

(eds) *Wetland Heritage of Holderness: an Archaeological Survey*. Humber Wetlands Project, School of Geography & Earth Resources, University of Hull, Hull, 87–120.

DRONKERS, J. 1996. Tidal asymmetry and estuarine morphology. *Netherlands Journal of Sea Research*, **20**, 117–131.

ELLIS, S. 1990. Soils. *In*: ELLIS, S. & CROWTHER, D. R. (eds) *Humber Perspectives: a Region Through the Ages*. Hull University Press, Hull, 29–42.

——, VAN DE NOORT, R. & MIDDLETON, R. in press. Environmental degradation and preservation: the case of the Humber wetlands. *Geography*.

GASSE, F., BARKER, P., GELL, P., FRITZ, S. & CHALIE, F. 1997. Diatom-inferred salinity in palaeolakes: an indirect tracer of climate change. *Quaternary Science Reviews*, **16**, 547–563.

GAUNT, G. D. 1994. *Geology of the Country around Goole, Doncaster and the Isle of Axholme*. Memoirs of the Geological Survey of Great Britain, Sheets 79 and 88, British Geological Survey, London, HMSO.

—— & TOOLEY, M. J. 1974. Evidence for Flandrian sea-level changes in the Humber estuary and adjacent areas. *Bulletin of the Institute of Geological Sciences*, **48**, 25–41.

——, FLETCHER, T. P. & WOOD, C. J. 1992. *Geology of the Country around Kingston upon Hull and Brigg*. Memoirs of the Geological Survey of Great Britain. Sheets 80 and 89, British Geological Survey, London, HMSO.

HENDEY, N. I. 1964. *An Introductory Account to the Smaller Algae of British Coastal Waters. Part V Bacillariophyceae*. HMSO, London.

HORTON, B. P., EDWARDS, R. J. & LLOYD, J. M. 1999*a*. Reconstruction of former sea-level using a foraminiferal-based transfer function. *Journal of Foraminiferal Research*, **29**, 117–129.

——, —— & ——2000*b*. Implications of a microfossil transfer function in Holocene sea-level studies. *This volume*.

HUGHEN, K. A., OVERPECK, J. T., LEHMAN, S. J., KASHGARIAN, M., SOUTHON, J. R. & PETERSON, L. C. 1998. A new ^{14}C calibration data set for the last deglaciation based on marine varves. *Radiocarbon*, **40**, 483–494.

JUGGINS, S. 1992. *Diatoms in the Thames Estuary, England: Ecology, Palaeoecology and Salinity Transfer Function*. Bibliotecha Diatomologica Band 25. Berlin, Cramer.

KRAMMER, K. & LANGE-BERTOLOT, H. 1986. *Bacillariophyceae. Susswasserflora von Mitteleuropa, Band 2 Teil 1: Naviculaceae*. Gustav Fischer Verlag, Jena.

—— & ——1988. *Bacillariophyceae. Susswasserflora von Mitteleuropa, Band 2 Teil 2: Bacillariaceae, Epithemiaceae, Surirellaceae*. Gustav Fischer Verlag, Stuttgart.

—— & ——1991*a*. *Bacillariophyceae. Susswasserflora von Mitteleuropa, Band 2 Teil 3: Centrales, Fragilariaceae, Eunotiaceae*. Gustav Fischer Verlag, Jena/Stuttgart.

—— & ——1991*b*. *Bacillariophyceae. Susswasserflora von Mitteleuropa, Band 2 Teil 4: Achnanthaceae, kritische erganzungen zu Navicula (lineolatae) und Gomphonema*. Gustav Fischer Verlag, Stuttgart/Jena.

LAMBECK, K. 1995. Late Devensian and Holocene shorelines of the British Isles and North Sea from models of glacio–hydro-isostatic rebound. *Journal of the Geological Society of London*, **152**, 437–448.

LILLIE, M. C. 1998. Alluvium and warping in the lower Trent Valley. *In*: VAN DE NOORT, R. & ELLIS, S. (eds) *Wetland Heritage of the Ancholme and lower Trent Valleys*. Humber Wetlands Project, Centre for Wetland Archaeology, University of Hull, Hull, 103–122.

—— & WEIR, D. 1997. Alluvium and warping in the Humberhead Levels: the identification of factors obscuring palaeo-landsurfaces and the archaeological record. *In*: VAN DE NOORT, R. & ELLIS, S. (eds) *Wetland Heritage of the Humberhead Levels*. Humber Wetlands Project, School of Geography & Earth Resources, University of Hull, Hull, 191–218.

LONG, A. J., INNES, J. B., KIRBY, J. R., LLOYD, J. M., RUTHERFORD, M. M., SHENNAN, I. & TOOLEY, M. 1998. Holocene sea-level change and coastal evolution in the Humber Estuary, eastern England: an assessment of rapid coastal change. *The Holocene*, **8**, 229–247.

MILLWARD, G. E. & GLEGG, G. A. 1997. Fluxes and retention of trace metals in the Humber estuary. *Estuarine, Coastal and Shelf Science* (supplement A), **44**, 97–105.

NEUMANN, H. 1998. The palaeoenvironmental survey of the Ancholme valley. *In*: VAN DE NOORT, R. & ELLIS, S. (eds) *Wetland Heritage of the Ancholme and lower Trent Valleys*. Humber Wetlands Project, Centre for Wetland Archaeology, University of Hull, Hull, 75–101.

PETHICK, J. S. 1990. The Humber Estuary. *In*: ELLIS, S. & CROWTHER, D. R. (eds) *Humber Perspectives, a Region Through the Ages*. Hull University Press, Hull, 54–67.

——1994. Estuarine processes. *In*: TOFT, A. (ed.) *A Saltmarsh Management Guide*. Environment Agency, R&D Report, **444**.

REES, J., RIDGWAY, J., ELLIS, S., KNOX, R. W. O'B., NEWSHAM, R. & PARKES, A. 1999. Holocene sediment storage in the Humber Estuary. *This volume*.

RIDGWAY, J., ANDREWS, J. E., ELLIS, S., HORTON, B. P., INNES, J. M., KNOX, R. W. O'B., MCARTHUR, J. J., MAHER, B. A., METCALFE, S. E., MITLEHNER, A., PARKES, A., REES, J. G., SAMWAYS, G. & SHENNAN, I. 2000. Analysis and interpretation of Holocene sedimentary sequences: techniques applied in the Humber Estuary. *This volume*.

ROUND, F., MANN, D. G. & CRAWFORD, R. M. 1990. *The Diatoms: Biology and Morphology of the Genera*. Cambridge University Press, Cambridge.

SHENNAN, I. 1982. Interpretation of Flandrian sea-level data from the Fenland, England. *Proceedings of the Geologists' Association*, **93**, 53–63.

——1986. Flandrian sea-level changes in the Fenland II. Tendencies of sea-level movement, altitudinal changes, and local and regional factors. *Journal of Quaternary Science*, **1**, 155–179.

——1992. Late Quaternary sea-level changes and crustal movements in eastern England and eastern Scotland: an assessment of models of coastal evolution. *Quaternary International*, **15/16**, 161–173.

——, LAMBECK, K., FLATHER, R., HORTON, B. P., INNES, J. B., MCARTHUR, J. J., LLOYD, J. M., RUTHERFORD, M. M. & WINGFIELD, R. 2000. Modelling western North Sea palaeogeographies and tidal changes during the Holocene. *This volume*.

SHEPPARD, J. 1958. *The Draining of the Hull Valley*. East Yorkshire Local History Society, York.

——1966. *The Draining of the Marshlands of South Holderness and the Vale of York*. East Yorkshire Local History Society, York.

SITCH, B. 1990. Faxfleet 'B', a Romano-British site near Broomfleet. *In*: ELLIS, S. & CROWTHER, D. R. (eds) *Humber Perspectives: a Region Through the Ages*. Hull University Press, Hull, 158–171.

SMITH, 1958. Post-glacial deposits in south Yorkshire and North Lincolnshire. *New Phytologist*, **57**, 19–49.

SNOEIJS, P. (ed.) 1991–1997. *Intercalibration and Distribution of Diatom Species in the Baltic Sea*. Opulus Press, Uppsala.

STUIVER, M. & REIMER, P. J. 1993. Extended ^{14}C data base and revised CALIB 3.0 ^{14}C age calibration program. *Radiocarbon*, **35**, 215–230.

SYMES, D. G. 1987. Agriculture. *In*: SYMES, D. G. (ed.) *Humberside in the Eighties*, Department of Geography, University of Hull, Hull, 9–22.

TOOLEY, M. J. 1978. *Sea-level Changes in North-west England During the Flandrian Stage*. Clarendon Press, Oxford.

TOWNEND, I. 1999. *Humber Estuary Geomorphological Studies*. Interim report to the Environment Agency.

TROELS-SMITH, J. 1955. Characterization of unconsolidated sediments. *Danmarks Geologiske Undersogelse, Series IV*, **3**, 38–73.

VAN DE NOORT, R. & DAVIES, P. 1993. *Wetland Heritage. An Archaeological Assessment of the Humber Wetlands*. School of Geography & Earth Resources, University of Hull, Hull.

—— & ELLIS, S. (eds) 1995*a*. *Wetland Heritage of Holderness. An Archaeological Survey*. Humber Wetlands Project.

—— & ——1995*b*. Recommendations. *In*: VAN DE NOORT, R. & ELLIS, S. (eds) *Wetland Heritage of Holderness*. Humber Wetlands Project, School of Geography & Earth Resources, University of Hull, Hull, 361–364.

—— & —— (eds) 1997. *Wetland Heritage of the Humberhead Levels*. Humber Wetlands Project.

—— & —— (eds) 1998. *Wetland Heritage of the Ancholme and Lower Trent Valleys*. Humber Wetlands Project. Centre for Wetland Archaeology, University of Hull.

VOS, P. C. & DE WOLF, H. 1993. Diatoms as a tool for reconstructing sedimentary environments in coastal wetlands: methodological aspects. *Hydrobiologia*, **269/270**, 285–296.

ZONG, Y. & HORTON, B. P. 1998. Diatom zones across intertidal flats and coastal saltmarshes in Britain. *Diatom Research*, **13**, 375–394.

—— & ——1999. Diatom-based tidal-level transfer functions as an aid in reconstructing Quaternary history of sea-level movements in Britain. *Journal of Quaternary Science*, **14**, 153–167.

Holocene sediment storage in the Humber Estuary

J. G. REES,[1] J. RIDGWAY,[1] S. ELLIS,[2] R. W. O'B. KNOX,[1]
R. NEWSHAM[1] & A. PARKES[2,3]

[1] *British Geological Survey, Keyworth, Nottingham NG12 5GG, UK*
(e-mail: j.rees@bgs.ac.uk)
[2] *School of Geography and Earth Resources, University of Hull, Hull HU6 7RX, UK*
[3] *Present address: Northsea Software Systems, 18 Newlands House, Newlands Science Park, Inglemire Lane, Hull HU6 7TQ, UK*

Abstract: In order to determine the processes that have governed the accumulation and erosion of sediments in the Humber Estuary (English North Sea coast) through the Holocene, the character, volume and source of sediments were studied. Eight sediment suites were identified on the basis of chemostratigraphy, lithostratigraphy and mineralogy. The locally sourced, freshwater, Basal Suite is overlain by the Newland and Butterwick Suites, deposited between *c.* 8 and 7.4 cal. ka BP in brackish environments behind a morainic barrier at St Andrew's Dock, Hull. These are overlain by the largely marine, saltmarsh sediments of the Garthorpe Suite, which in turn are overlain, with erosional contact, by the channel sandflat and mudflat deposits of the Saltend, Sunk Island and Skeffling Suites. Most of the Saltend Suite is likely to have been deposited since *c.* 4 ka ago, whilst the Sunk Island and Skeffling Suites are likely to have been deposited since Medieval times and from the late eighteenth century onwards, respectively, as indicated by their concentrations of anthropogenic metals. On the coast, the Spurn Suite consists of sediments, associated with a spit system, which are almost entirely marine in origin. The suites show a progressive increase in marine influence; sediments of the oldest suite being entirely from the terrestrial catchment, those of the younger suites from erosion of the North Sea floor and coast. The relationships between suites show that during the last 4 ka the geomorphological evolution of the estuary has been marked by widespread erosion episodes that have led to the partial removal or redistribution of earlier deposits. By modelling the volumes of the suites it can be shown that, of the total volume of the estuarine fill (9.6 km^3), over half is likely to have been deposited during this period.

The growing vulnerability of coasts to both natural and human-induced threats has highlighted the need for a greater understanding of the processes that affect them. The Land–Ocean Interaction Study (LOIS) aimed to quantify the exchange, transformation and storage of materials at the land–ocean boundary and determine how these parameters vary in time and space (NERC 1994; Wilkinson *et al.* 1997). The historical component of LOIS, the Land–Ocean Evolution Perspective Study (LOEPS), focused on the characterization and quantification of materials stored along the English North Sea coast over the last 10 ka. It aimed to increase our understanding of long-term coastal changes, which are difficult to identify by looking solely at modern systems. In particular it focused on the response of the coast to changes in relative sea-level, human influence and climate.

The LOIS studies centred on the Humber Estuary, which receives freshwater runoff from two major river system catchments, the Trent and the Yorkshire Ouse. These have a combined area of 24 000 km^2, or approximately one-fifth of the area of England. The LOEPS projects focused on the storage of materials in the estuary during the Holocene and build on the understanding of the Holocene stratigraphy of the estuary gained by earlier projects such as the Humber Wetlands project (Van de Noort & Davies 1993; Dinnin 1995, Dinnin & Lillie 1995; Van de Noort & Ellis 1995, 1997). These, reviewed by Long *et al.* (1998), were based on the stratigraphy of sediments in natural surface exposures (e.g. Crowther 1987; Dinnin 1997; Didsbury 1988), excavations (e.g. Switsur 1981; Buckland & Sadler 1985), boreholes (e.g. Gaunt & Tooley 1974; Berridge & Pattison 1994), and determined through geophysical methods (Institute of Geological Sciences 1968, 1973). The results of the Humber LOEPS projects are set out in this volume: the methodologies used are

From: SHENNAN, I. & ANDREWS, J. (eds) *Holocene Land–Ocean Interaction and Environmental Change around the North Sea*. Geological Society, London, Special Publications, **166**, 119–143. 1-86239-054-1/00/$15.00

summarized in Ridgway *et al.* (this volume), the storage of carbon and sulphur are described by Andrews *et al.* (this volume); analysis of the Holocene depositional environments is set out by Metcalfe *et al.* (this volume) and the storage of sediments in the estuary during the Holocene in relation to changes in relative sea-level, climatic conditions and human activities is described here.

In addressing the likely effects on the estuary of changing human and climatic influences and rapid changes in sea-level, coastal zone managers need to know how the estuary has responded to similar changes in the past. In particular there is a need to identify the character, volume and source of sediments that have accreted through time and to understand the processes that have governed their accumulation and erosion. This paper describes aspects of the LOEPS projects that were designed to:

(1) Identify distinctive sediment bodies within the fill of the estuary and establish their composition (chemical and mineralogical), textures and volumes.
(2) Compare the characteristics of sediment bodies in the estuary with those of possible original sources (both from the river catchments and offshore), and thus determine the likely source of the estuarine sediments through the Holocene.

Sampling, analysis and additional data sources

Samples for geochemical, mineralogical and particle size analysis were taken from boreholes drilled into the Holocene fill of the estuary and from likely source areas. The geochemistry of the estuarine fill was also compared with existing onshore and offshore geochemical data sets. A summary of the sampling and analytical techniques employed in the study is given by Ridgway *et al.* (1999).

Sampling

The estuarine samples used in the LOEPS projects were obtained from 22 boreholes drilled

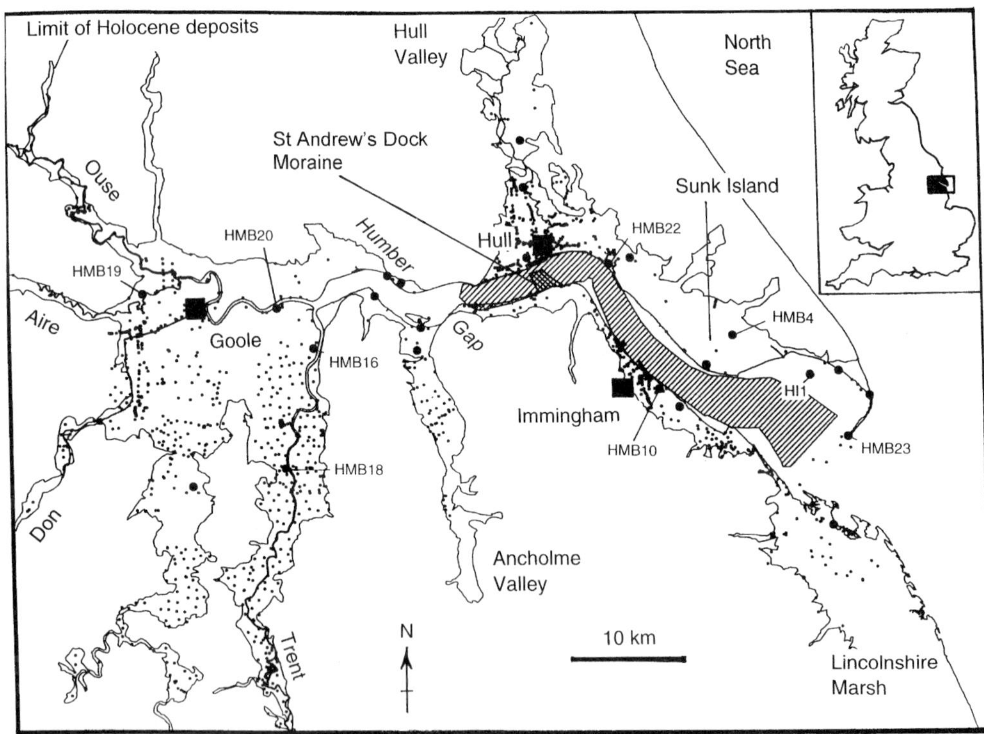

Fig. 1. The distribution of boreholes penetrating the Holocene fill and of seismic data, used in modelling of the Holocene fill of the Humber Estuary. The boreholes (dots) all occur in the National Geosciences Records Centre (NGRC). The boreholes drilled within the LOIS programme are shown separately as filled circles. The seismic (pinger, sparker) data were obtained during Institute of Geological Sciences (1968, 1973) surveys in the hatched area. The inset map shows the position of the estuary on the North Sea coast.

through the Holocene sequence (Ridgway *et al.* 1998*a*) (Fig. 1), of which five (HMB10, HMB16, HMB18, HMB19 and HMB20) were sampled more extensively than the others. Because of uncertainty regarding the degree of chemical variation in the Holocene sequences, a pilot project was undertaken, using a portable X-ray fluorescence (XRF) analyser. This showed that, in 27 cores from throughout the LOEPS study area (Tees–north Norfolk), most distinctive lithological units had consistent chemical signatures (Ridgway *et al.* 1998*a*, and this volume), suggesting that the chemistry of composite samples from cores could provide a basis for the development of a stratigraphy for the Holocene sediments. A programme of sampling for high precision, multi-element, laboratory-based XRF analysis, heavy mineralogy and clay mineralogy was thus undertaken on the Humber cores. Sampling intervals were chosen taking particular account of properties such as the grain size, colour, presence of laminations and organic material (including shells and shell debris) and drilling-related features (cavings, material recovered in bags, etc.). Where units appeared homogeneous material from several core runs was combined into a single composite sample. Some particularly thick stratigraphic units were represented by several such composites. Disturbed surface soils were avoided as far as possible, but the upper units of most cores were sampled over smaller intervals in an attempt to examine possible anthropogenic contamination.

In order to establish the character of potential source materials from the North Sea, 30 samples were taken from a range of tills (by Jon Cox, University of East Anglia) in the Holderness cliffs at Dimlington, Tunstall, Aldbrough, Hornsea and Barmston (Fig. 2). In addition, offshore Pleistocene tills, which may have contributed sediment to the early Holocene fill of the estuary, were sampled in six British Geological Survey (BGS) boreholes (Fig. 2). To characterize the sediments within the Ouse and Trent river systems, samples of river bedload sediments were taken from 19 sites (Fig. 3). The character of inland Pleistocene sediments was examined by taking 15 samples of the glaciolacustrine sediments from the Vale of York ('25 foot drift') and two of tills from south of York. To determine the character of sediments on the south side of the estuary, a sample of the (Anglian) till on the Lincolnshire Wolds and three samples of Jurassic rocks forming the basement to the Holocene succession were taken.

For clay mineralogy studies, samples were taken from HMB10 and HMB22 in the outer Humber and from HMB16 and HMB20 in the

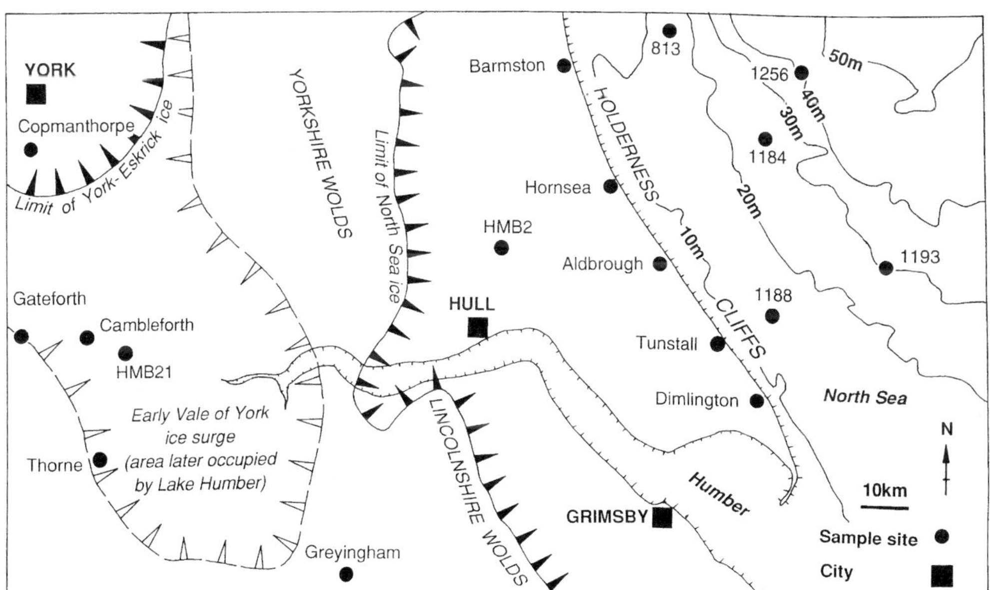

Fig. 2. Sites of samples of Pleistocene sediments taken as part of the provenance studies. The samples from the Holderness cliffs were taken by Jon Cox of the University of East Anglia. Note that no samples were taken for heavy mineral analysis from the sample sites at Hornsea or Tunstall. The provenance studies also used geochemical data from the BGS offshore database.

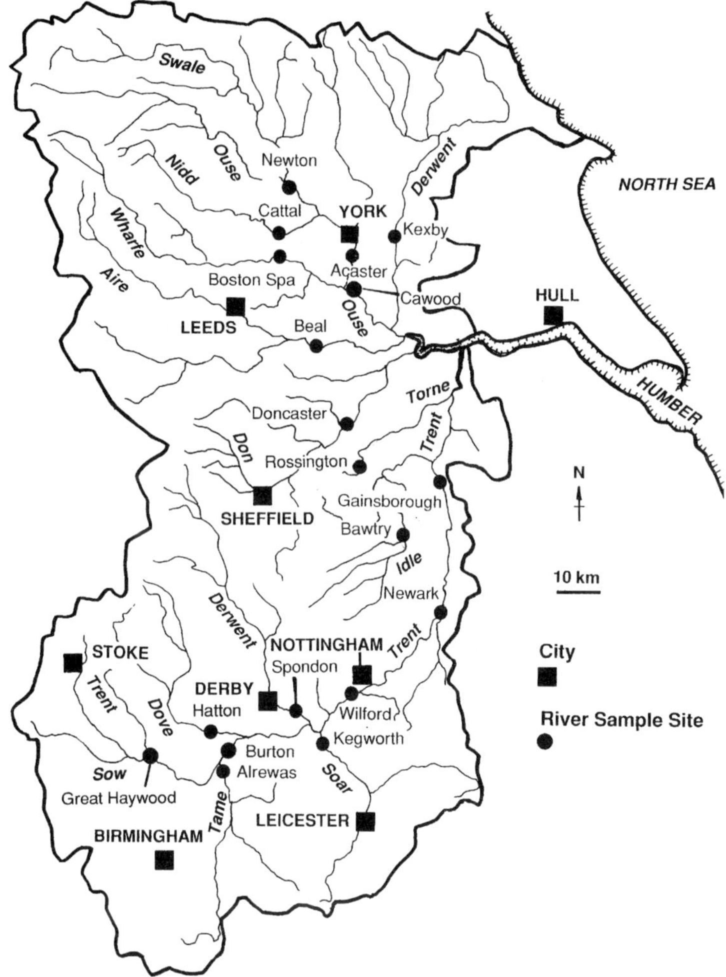

Fig. 3. Sites of river samples taken as part of the provenance studies. All samples were taken of submerged sand and mud mixtures within 2 m of the river bank.

inner Humber. Samples from potential source materials were also taken from BGS reference archives, including North Sea sediments offshore of the Humber and the Skipsea Till from Holderness, and also from floodplain sediments of the catchment areas of three of the main tributaries (Ouse, Don and Trent) of the inner estuary.

Analytical methods

Geochemistry. The use of cores placed a constraint on the amount of sediment obtainable and whole sediment samples were analysed, rather than a specific size fraction, in order to ensure that sufficient material was available for analysis. Analyses (for MgO, Al_2O_3, P_2O_5, K_2O, CaO, TiO_2, MnO, Fe_2O_3, Cr, Co, Ni, Cu, Zn, As, Rb, Sr, Y Zr, Ba, Pb, La and Ce) were performed using the XRF facilities at the BGS. Details of the methodology, along with sampling and analytical quality control procedures, are given in Ingham & Vrebos (1994) and Ridgway *et al.* (1998*a*, this volume). Collection and analysis of 21 replicates of original samples yielded correlation coefficients between original and replicate pairs which were better than: 0.98 for 15 elements; 0.95 for five elements; and 0.92 for one element; all with regression line slopes of approximately 1:1. The remaining element (Cr) showed one aberrant replicate result, which reduced the correlation coefficient to 0.89 (Ridgway *et al.* 1998*a*). These results demonstrate the reliability of the sampling and analytical methods employed.

Heavy mineralogy. Samples for heavy mineral analysis were taken over the same sampling intervals as used for geochemical analysis. The samples were ultrasonically cleaned and wet sieved. Heavy minerals were separated from the 63–125 m grain size fraction by gravity-settling in bromoform (sp. gl. 2.90). Heavy mineral fractions were mounted in Canada Balsam and analyses were done by conventional optical microscopy; this involved a count of 200 non-opaque detrital grains per sample (where grain recovery permitted) to determine the overall composition of the suite.

Clay mineralogy. Quantitative clay mineralogy was carried out at the University of Hull. Air-dried samples were gently crushed placed in an agate mortar. The disaggregated samples were then suspended in distilled water; and a few drops of Dispex were added to prevent clay flocculation. The suspensions were stirred and allowed to stand for 4 h, after which time the top 5 cm was extracted by pipette. The resulting slurry was mounted on slides and dried overnight at room temperature. The air-dried samples were then scanned on a Phillips PW1130/00 X-ray diffractometer from 2 to 20° 2θ using CuK_a radiation, then glycolated and re-run, and heated to 350°C and re-run again. Clay contents were quantified by the measurement of peak areas and the results normalized.

Additional data sources

Humber Estuary. Stratigraphical data from the Humber Holocene sequence were derived from many sources (see review by Long *et al.* 1998), though most come from boreholes. Over 4000 of these, mostly rotary or percussion holes, were drilled in onshore areas of the estuary, but the amount and quality of information supplied varied enormously. Many, drilled as water wells or for shot holes in seismic reflection surveys, have logs with only very basic lithological descriptions. Others were logged in detail and provide textural information, notably the 25 LOEPS boreholes (see above) as well as many drilled by site-investigation companies, engineering companies and universities. The data were collated by several organizations, including the BGS. Of particular note is a BGS database (Ridgway *et al.* 1998*a*) that contains stratigraphic and sedimentological data from about 3500 boreholes, archived at the National Geosciences Records Centre (NGRC) at Keyworth, UK. Several hundred gravity cores, vibrocores and piston cores have been obtained from intertidal and subtidal areas throughout the estuary, many by engineering companies and Associated British Ports. The vast majority of the data were compiled by the BGS, who also collected several hundred subtidal sediment grab samples. Seismic surveys of the modern estuary channel were undertaken by the BGS (Institute of Geological Sciences 1968, 1973); those shot by oil companies have insufficient resolution to be useful in studies of Holocene stratigraphy.

Catchments. Apart from the samples taken from rivers draining into the Humber, further information about the geochemistry of sediments in the Ouse and Trent catchments was available from the data set of the Wolfson Geochemical Atlas. Based on stream sediments, this was produced by the Applied Geochemistry Research Group of Imperial College, London. Sampling started in 1969, with a mean sample density of about one per 2.5 km^2. At each site two 100-g-samples of active stream-bed sediment were taken at least 20 m upstream of any road and sieved at 200 μm. No samples were taken in urban areas. Most of the elements were analysed spectrographically, though four, including As and Zn, were determined by atomic absorption spectrophotometric (AAS) and colorimetric methods. Sampling, analytical and quality control procedures are described in the atlas (Webb *et al.* 1978).

Offshore areas. Geochemical data from the North Sea were obtained from the BGS offshore database. This is based on the analysis of seabed sediments collected by Shipek grab between 1979 and 1982. A sample was collected for each 43 km^2 area. Analysis of the <2 mm fraction of the sediments was made using direct current (d.c.) arc emission spectrometry. Full details of the sampling, analysis and quality control are given by Stevenson *et al.* (1995).

Characterization of the estuarine fill

Previous Holocene stratigraphical investigations (see earlier) focused on widely correlatable peats within the sequence, these being the only obvious, laterally persistent sedimentary units. Using geochemistry and, to a lesser extent, mineralogy it has been possible to identify distinct bodies of sediment, here termed suites, which

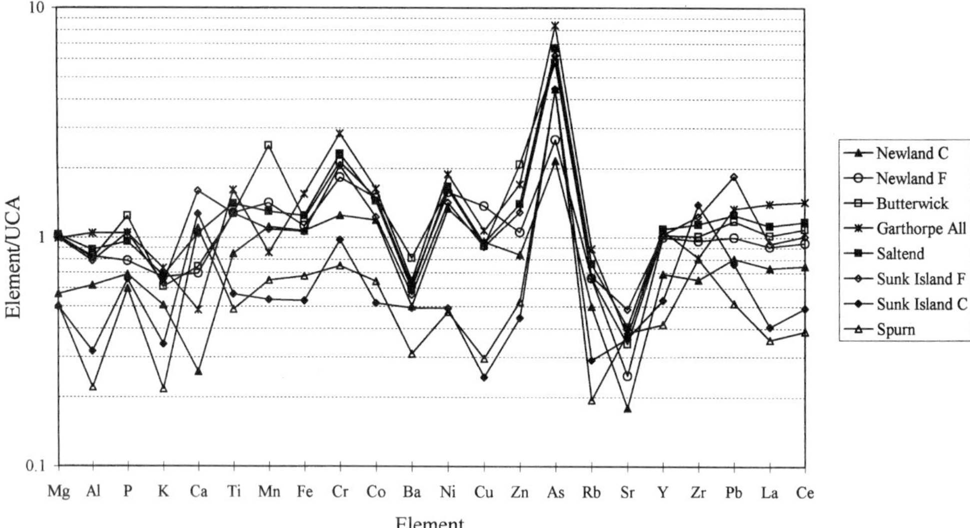

Fig. 4. Spidergram of mean geochemical signatures for the main sediment suites (Skeffling not shown). C and F denote coarse and fine grained units, respectively.

provide a firmer basis for the interpretation of the Holocene succession.

The suites have been identified chiefly through the use of spidergrams, in which element concentrations normalized to the upper crustal average (UCA) values of Wedepohl (1995) are plotted on a logarithmic scale. The geochemical signatures on the spidergrams were then grouped, according to similarity, into chemostratigraphic suites. Spidergrams have the advantage that no assumptions are made and no subjective judgement or interpretation is involved in their preparation. They enable the basic data to be seen at a glance, in a way that is easy to understand, and provide a picture of: (a) the characteristic geochemical pattern of a particular group of samples; (b) the spread of element concentrations within the group; (c) any consistent and systematic differences between groups of samples (Haslam & Plant 1990).

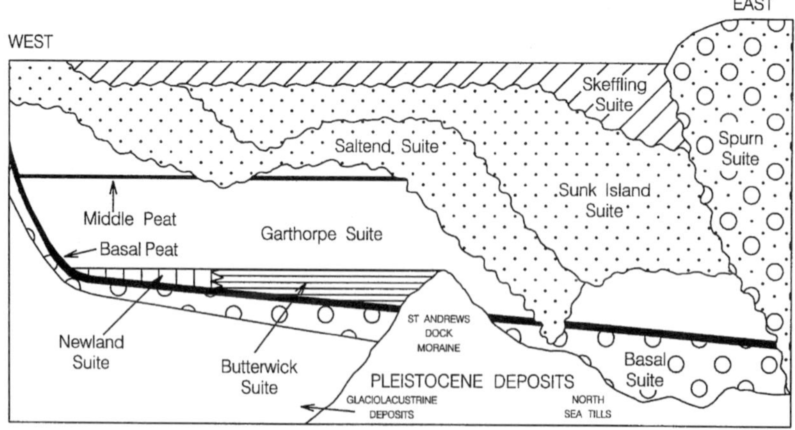

Fig. 5. Schematic longitudinal section of the Humber Estuary. This shows the configuration of the suites discussed in the text. The Basal Peat, shown here between the Basal and Garthorpe Suites, locally occurs near the junction within either of the suites. The Middle Peat occurs towards the top of the Garthorpe Suite.

For several reasons, no attempt has been made to compensate for grain-size variation by the use of an element proxy such as Al (Loring & Rantala 1992). Correlation coefficients show that whilst some elements (e.g. Co, Cr, Cu, Ni) are strongly correlated with Al, others (e.g. Ca, Mg, Mn, Zr, Ba) are not (Ridgway *et al.* 1998*a*; Plater *et al.* 1999). Wholesale normalization to Al would thus have an unpredictable effect on some elements. In addition, iron and manganese oxyhydroxides and organic matter complex trace metals and so must also be considered when attempting normalization (Plater *et al.* 1999). It would be possible to normalize only those elements that correlate strongly with Al, but this also has drawbacks. Normalizing when both metal and normalizer are at low concentrations yields problematical results (Rowlatt & Lovell 1994) and in the LOEPS studies gave abnormally high metal values for sands in both the Tees and Humber regions. In practice, spidergrams work well with sediments of silt to clay size without normalization for grain size, particularly if the shape of the geochemical signature is considered rather than the absolute values. Sands remain problematical with or without grain size normalization, but if direct comparisons are restricted to sediments of similar grain size (i.e. muds with muds, sands with sands) the technique works well.

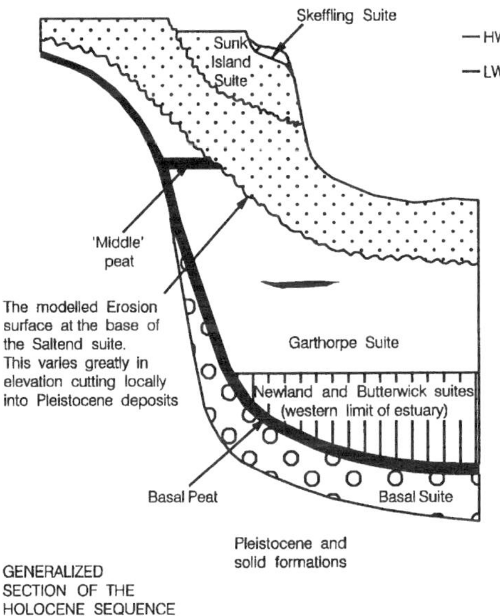

Fig. 6. Schematic cross section of the western part of the Humber Estuary, showing the relative positions of the suites discussed in the text, and the occurrence of the Basal Peat and 'Middle' Peat. Although the Skeffling Suite is only shown here 'perched' on the side of the estuary, sediments of it are likely to be widespread within present tidal limits.

Eight distinct bodies, referred to here as sediment suites, have been identified and named as the Basal, Newland, Butterwick, Garthorpe, Saltend, Sunk Island, Skeffling and Spurn Suites; these are described below. Peat horizons within the suites provided material for ^{14}C dating, from which related ages were deduced. Mean geochemical signatures for the groups are shown in Fig. 4. Differences between suites are sometimes subtle, but are consistent and can be confirmed by scatterplots of element or element ratio pairs (see later discussion under provenance). Fuller details of the geochemical characteristics of the suites are given in Ridgway *et al.* (1998*b*). Schematic sections of the estuary showing the configuration of the suites are shown in Figs 5 and 6. Average grain size distributions of the suites are shown in Fig. 7.

Basal Suite

The Basal Suite unconformably overlies Pleistocene sediments and Mesozoic rocks. Its thickness varies considerably, though in most parts of the estuary (particularly in the east) it is less than 5 m. The top is defined by the base of the overlying Newland, Butterwick or Garthorpe Suites. It is dominated by fine- to medium-grained sands, with subordinate greenish-grey, sandy muds. A peat, commonly split into several levels and referred to as the 'Basal Peat', occurs at the top of the suite in many areas, though locally it is overlain by muds. The suite commonly forms one or more fining-up cycles with muds containing peats, woody horizons and roots, overlying pebbly sands and is interpreted as being wholly of freshwater origin as no marine fauna or flora were found within it. It is likely to be diachronous, possibly being pre-Holocene in age in some areas (for instance below Sunk Island) and considerably younger in others. In the inner part of the estuary, peats within 0.5 m of the top of the suite in boreholes HMB18 (AA23435), HMB16 (AA25585) and HMB20 (AA25589) have mean ages of 8176, 8034 and 7954 cal. a BP, respectively (Table 1), and it is possible that in some areas deposition may have continued until much later. Geochemically, the Basal Suite is characterized by extreme variability.

Newland Suite

The Newland Suite has been proved only in the lower part of the Ouse Valley, between 7.0 and

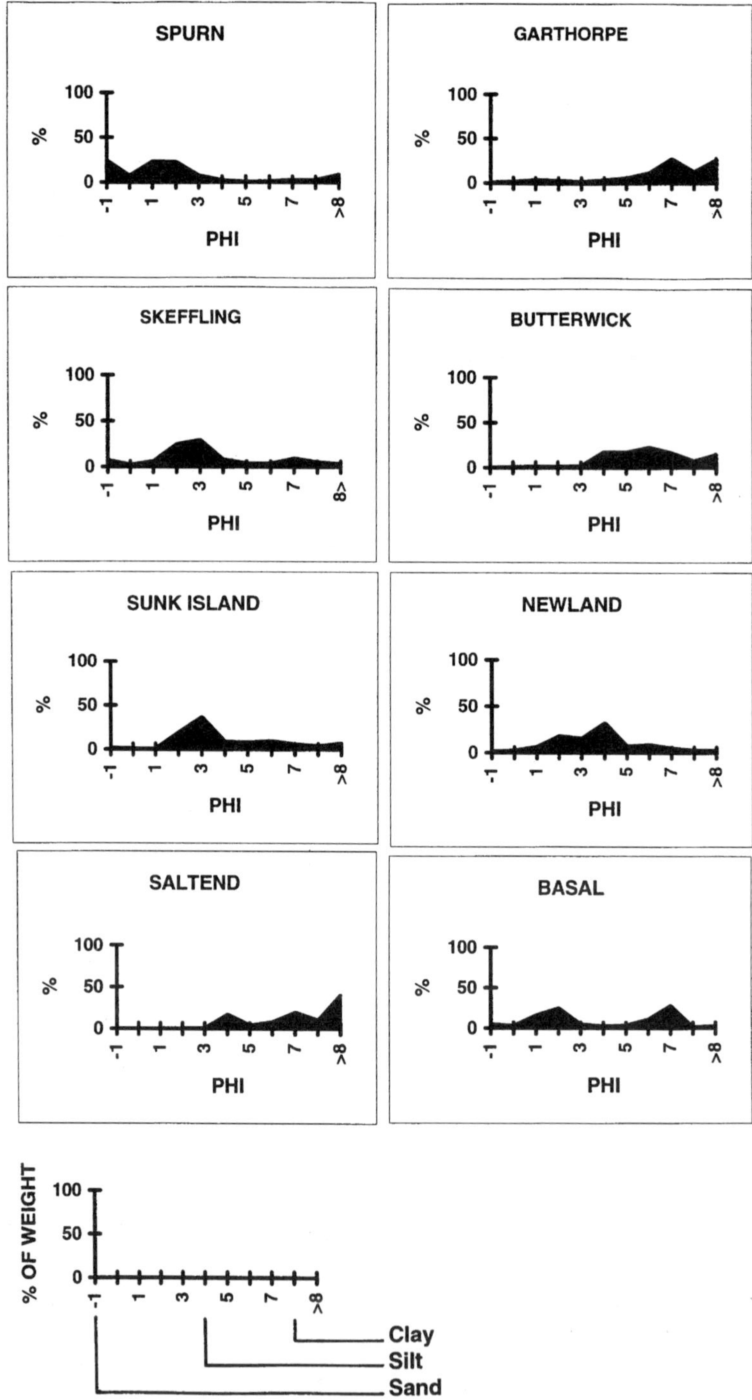

Fig. 7. Average grain-size distributions of the sediment suites. Subdivisions are in ϕ units, where <-1, granules; -1–4, sand; 5–7, silt; >8, clay. Particle size analyses were made by sieving and Sedigraph granulometer (see text). Data for the Skeffling Suite are those supplied to the Environment Agency by McLaren. These are based on analysis using the Malvern Laser Granulometer, and include samples extending beyond the modelled limits of the estuary (Figs 16–20) and may include samples from underlying suites.

Table 1. *Radiocarbon ages and calibration data for specific dates quoted in the text**

Borehole	Material	Laboratory code	^{14}C age ± 1σ (years)	Calibrated age (yrs BP)			Altitude (m)	Alt. Int.†
				Max.	Mean	Min.		
HMB8	Peat	AA25582	8555 ± 65	9646	9491	9397	−12.00	
HMB13	Peat	AA22674	3910 ± 45	4439	4372	4155	−1.88	4401, 4354
HMB16	Peat	AA25585	7265 ± 60	8135	8034	7922	−11.40	8060, 7999
HMB18	Sandy peat	AA23435	7450 ± 75	8370	8176	8007	−8.02	
HMB20	Sandy peat	AA25589	7205 ± 85	8132	7954	7814	−12.10	
HMB20	Silt with roots	AA25588	6550 ± 75	7534	7392	7268	−5.82	

* Ages in this table were calibrated with CALIB 3.0 using the bidecadal atmospheric curve (Stuiver & Reimer 1993 and references therein).
† Alt. Int., alternative interpretations of the mean calibrated age based on different intersections on the curve.

15.7 m in borehole HMB19 [SE 6868 2464], where it is overlain by the Garthorpe Suite. It comprises two upward-fining units: much of the lower unit consists of grey, laminated muds with few burrows or root traces; the upper comprises coarse- to fine-grained sands grading up into muds. The bases of many sands are erosional and contain pebbles, including mud rip-up clasts. Laminae within some beds are clearly disturbed suggesting that the sediments have slumped. The erosional bases of beds, presence of lags and likely bank collapse structures are consistent with the Newland Suite being deposited in a channel. The suite contains microfaunal evidence of limited marine influence (Metcalfe *et al.* this volume) and is considered to be a partially mixed fluvial–marine deposit. Sediments of the suite have not been dated, but on the basis of radiocarbon evidence of the age of the Basal and Garthorpe Suites, the suite is likely to have been deposited between about *c.* 8 and 7.4 cal. ka BP (Table 1).

Butterwick Suite

The Butterwick Suite occurs in the lower part of the Trent Valley, in the Ouse Valley below Goole, probably in the Ancholme Valley and in the Humber as far east as St Andrew's Dock, Hull (see below). It overlies the coarser, less well-laminated, Basal Suite and is overlain by the more organic Garthorpe Suite. Difficulties in distinguishing it on textural grounds from the Newland Suite and thus of identifying it in boreholes where geochemical data are unavailable, prevent calculation of its maximum thickness, though east of Goole it is at least 4 m thick. The suite comprises interbedded grey-brown sands and (commonly) laminated muds. The sands tend to fine upwards and have steeply dipping, erosional bases. The muds contrast with those of the Newland Suite, as they contain many roots and stem fragments. They also contain marine shells, as may be seen between 8.61 and 10.86 m in borehole HMB18 [SE 8288 0744], the reference section. The laminated, fine-grained nature of the suite suggests that it may have been deposited within a tidally influenced environment. On the basis of radiocarbon evidence (see above and below) the suite was probably deposited between about *c.* 8 and 7.4 cal. ka BP (Table 1).

Garthorpe Suite

The Garthorpe Suite overlies the Newland and Butterwick Suites or, where these are absent, the Basal Suite. It is most commonly overlain by the Saltend Suite, as in the reference section between 2.0 and 8.61 m in borehole HMB18 [SE 8288 0744], though locally it is overlain by the Sunk Island Suite. The boundary between these suites and the Garthorpe Suite is commonly erosional. The suite varies considerably in thickness, reaching 13 m in the western part of the estuary. It is dominated by mottled pale-grey, poorly laminated muds, locally containing shell fragments at several levels. Root and stem traces and peaty horizons are common. In the eastern part of the estuary the Garthorpe Suite contains the 'Basal Peat', which more commonly occurs within the Basal Suite. Higher in the suite, a widespread peat horizon, the 'Middle Peat' or 'Bronze-Age Peat', occurs a few metres below Ordnance datum. The presence of these peats, extensive rooted horizons, stems and woody fragments suggest that the suite was largely deposited in saltmarsh environments. In the eastern part of the estuary, seaward of Sunk Island, deposition of the suite started at least 8.5 ka BP (HMB8, AA25582). In the west, where it overlies the Butterwick and Newland Suites, deposition probably commenced c. 7.4 cal. ka BP, as the oldest dated materials within the basal metre of the suite were deposited 7392 cal. a BP (HMB20, AA25588) (Table 1). The age of the youngest sediments of the suite is not known. Some sediments near Whitton Ness (HMB13, AA22674) were deposited 4372 cal. a BP (Table 1), and it is probable that some parts of the suite were deposited much later than this (possibly today, in some areas).

Saltend Suite

The Saltend Suite overlies the Basal, Newland, Butterwick and Garthorpe Suites, commonly with erosional contact, over most parts of the Humber Estuary and is locally erosionally overlain by the Sunk Island and Skeffling Suites. It is thickest towards the axis of the present channel system, reaching 23 m midway between Trent Falls and Goole. The suite comprises grey, mottled grey-brown or brown, laminated muds and fine-grained sands, commonly extensively burrowed, which are readily differentiated from the poorly laminated organic sediments of the Garthorpe Suite. Shell fragments are common, though most are finely comminuted. Distinction of sediments in the suite from those of the overlying Sunk Island and Skeffling Suites is based mainly on subtle differences in geochemistry, as they are compositionally and texturally similar. The reference section of the suite is that between the surface and 6.06 m depth in borehole HMB22 [TA1603 2748]. The sands and muds are interpreted to have been deposited on intertidal sandflats and mudflats, respectively. The suite is probably diachronous. Based on the elevation of the base of the suite and the relative mean sea-level curve for the estuary (Metcalfe *et al.* this volume), parts of it may have been deposited before 6 ka ago (this is, as yet, unproven). Over extensive peripheral areas of the estuary the suite is likely to be less than 4.4 ka old, as proved by dated sediments within the underlying Garthorpe Suite (see above).

Sunk Island Suite

The Sunk Island Suite occurs from the North Sea landwards at least as far as the Goole area. It overlies the Basal, Newland, Butterwick, Garthorpe and Saltend Suites, commonly with erosional contact and is overlain by the Skeffling Suite. The proven thickness of the suite in boreholes (including HMB8 [TA 2572 1753], which contains the reference section above 10.5 m depth) ranges from 3.38 to 12.0 m. The suite comprises sands and muds. The sands are cross-stratified and predominate towards the base, where they are fine- to medium-grained and fine upwards into mottled grey-brown, laminated and burrowed muds. Small shell fragments are scattered throughout; roots only occur near the modern surface. A twofold division of the suite into lower, sand-rich and upper, mud-rich parts is typical and distinguishes the suite from the Saltend Suite, in which sand bodies are smaller and more evenly distributed vertically and horizontally. The laminated muds are interpreted as being deposited on mudflats; the sands on sandflats or subtidal channel bars. Sediments of the suite have not been dated directly, though the high levels of Pb they contain suggest that they post-date the onset of mining in the Pennines and were probably deposited from Medieval times (since c. AD 1250) onwards (Hudson-Edwards *et al.* in press).

Skeffling Suite

The Skeffling Suite has only been proven in the uppermost 1.05 m of Borehole HI1 [TA 3629 1623], the reference section. Here it is texturally indistinguishable from laminated and burrowed sands of the Sunk Island Suite, but contains elevated levels of several anthropogenic metals, which suggests deposition since the late eighteenth century. The suite is likely to occur extensively over most parts of the channel of the estuary and is still being deposited today.

Spurn Suite

The sediments forming the modern spit of Spurn Point are included within the Spurn Suite. The reference section is an upward-fining sequence containing several upward-fining units of yellow-brown, cross-stratified, well-sorted, well-rounded gravels, sandy gravels and medium to very-coarse sands containing shells, found in the 20-m-deep borehole, HMB23, drilled at the head of the spit [TA 3982 1072]. The only other sediments assigned to the suite are 0.34 m of laminated, pebbly, coarse-grained sands containing marine shells, which occur immediately above the Pleistocene till in borehole HMB4 [TA 2480 2050] in Sunk Island, and which are overlain by sands of the Sunk Island Suite. The sediments of the suite are interpreted to be beach sands and gravels, derived largely from tills, or deposits of washovers or channel lags, transported landwards from the beach or spit. There

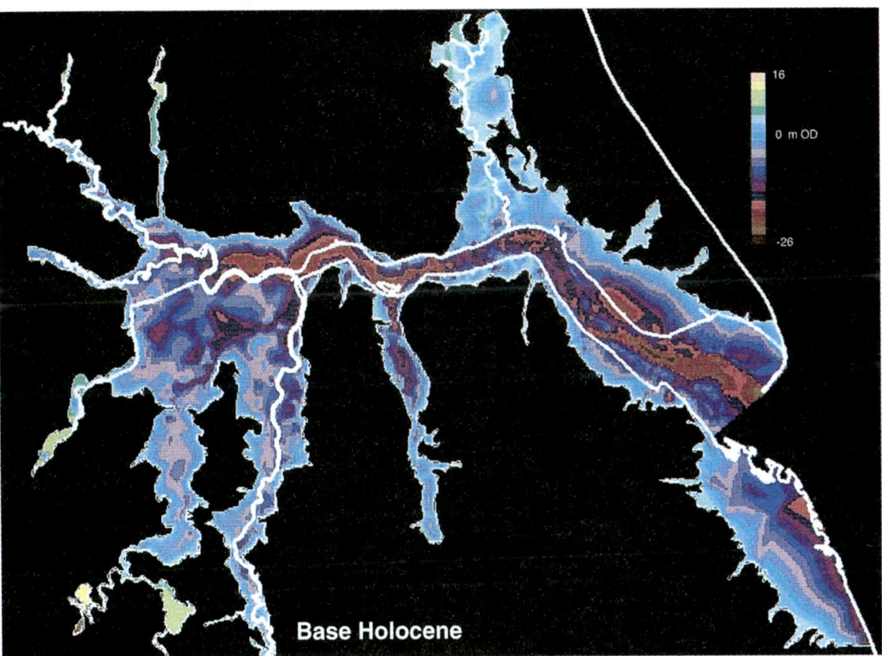

Fig. 8. Base Holocene grid. This is the most tightly constrained grid of those shown. Several features of the grid in the subtidal part of the eastern estuary are likely to be the product of erosion within the Holocene, though are impossible to identify as such.

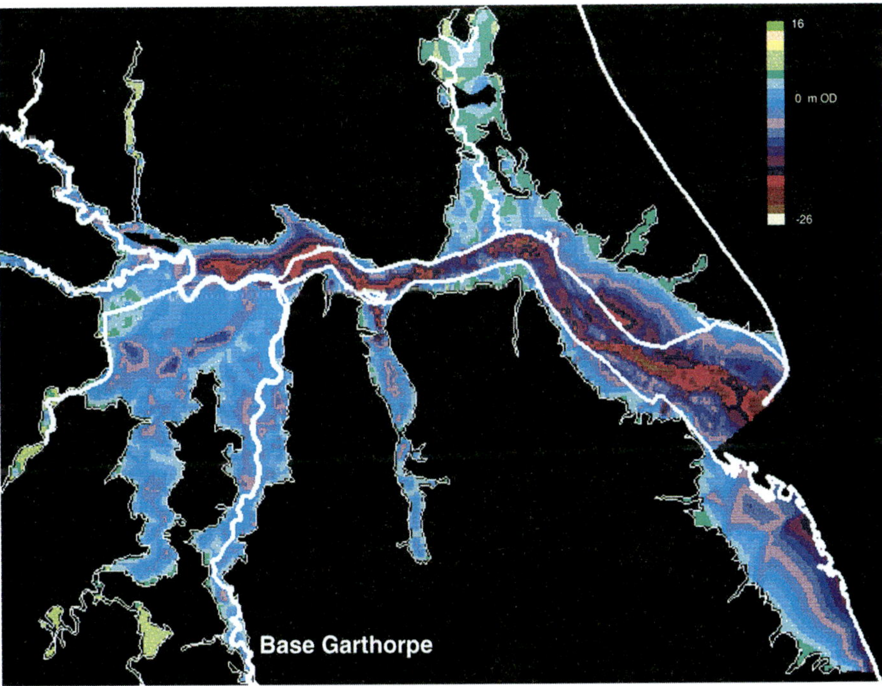

Fig. 9. Basal grid of the Garthorpe Suite. The likely depositional extent of the Newland and Butterwick Suites is the area west of the moraine in the vicinity of St Andrew's Dock (Fig. 1). The grid east of this is the same as that (not illustrated) which represents the top of the Basal Suite.

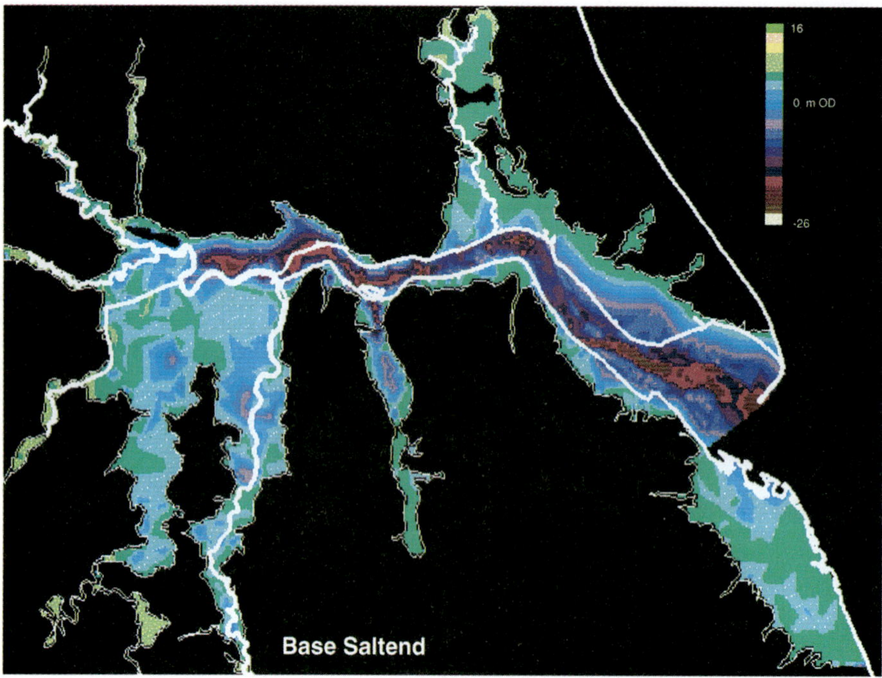

Fig. 10. Basal grid of the Saltend Suite. This model shows the base of the main erosional unit in the estuary.

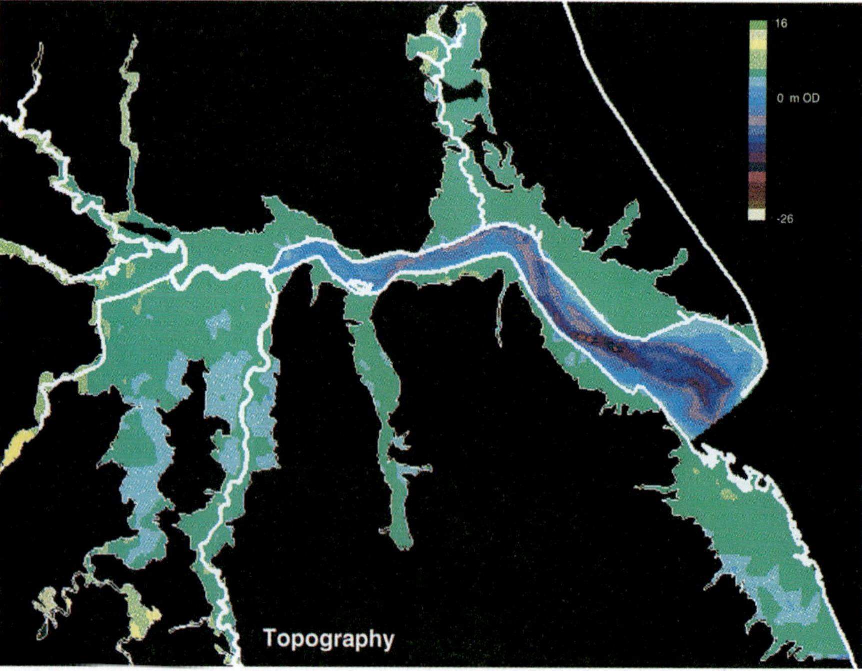

Fig. 11. Present topography and bathymetry. This model was generated in BGS on the basis of elevation data from borehole records in the National Geosciences Data Centre and 1968 bathymetric data.

are only poor constraints on the age of the suite. Pollen from detrital peats between 13.3 and 13.8 m depth in borehole HMB23 suggest that sediments of the suite were being deposited at least 6 ka ago (Metcalfe *et al.* this volume). Sediments of the suite are still being deposited today.

Volumetric modelling of the estuarine fill

To enable three-dimensional models of the sediment suites to be produced, their boundaries in boreholes within the BGS borehole database were identified and coded. Identification of the boundaries in boreholes without geochemical data was based on recognition of textural and colour characteristics of the suites. However, it was impossible to differentiate the Saltend, Sunk Island and Skeffling Suites in this manner and the boundaries between them were not identified. Likewise, it has not been possible to differentiate the sediments of the Newland and Butterwick Suites. (The relative volumes of the Saltend and Sunk Island Suites and of the Newland and Butterwick Suites, given in Table 2 are estimates; the volume of the Skeffling Suite was derived from ABP-NUDCM 1998.) Following coding, the elevations of the boundaries of the stratigraphical divisions were copied into *Intergraph Microstation Terrain Modeler*, in which all the LOEPS three-dimensional modelling was undertaken. The boundaries of some suites were created simply by gridding the elevation of the tops of the suites in boreholes in which they occur, or are interpreted to occur. The base Holocene surface (Fig. 8) was modelled in this manner, though it is also based on data from IGS seismic reflection (sparker, pinger) surveys collected in 1968, 1973 (Institute of Geological Sciences, 1968, 1973); the boundary between the Pleistocene sediments (or bedrock) and the Holocene deposits was interpreted on each of the seismic lines and then modelled.

Some surfaces, containing only limited amounts of data, were constrained by other surfaces that contain more data. The modelled surface of the base of the Newland and Butterwick Suites (not illustrated) was generated in this manner: a grid was made of the thickness of the Basal Suite in all of the boreholes in which it could be confidently identified and this was added to the base Holocene grid to generate the basal grid of the Newland and Butterwick Suites. The grid representing the base of the Garthorpe Suite (Fig. 9) is the same as that of the base of the Newland and Butterwick Suites in the eastern part of the estuary, as these suites do not occur east of a line extending approximately between St Andrew's Dock, Hull and New Holland. Their distribution appears to be controlled by the presence of a till–bedrock sill

Table 2. *Principal characteristics of the Humber suites*

Sediment suite	Present volume[1] (km^3)	Relative mean sea-level rise ($mm\,a^{-1}$)[3,4]	Mean grain size (ϕ)[5]	Skew	Kurtosis	Mean USi index	Depositional environment salinity	Main sediment sources
Spurn	–	1.5	2.6	0.86	1.81	–	Marine	N. Sea
Skeffling	0.2[2]	2.5–3.5[2]	1.67	0.96	1.55	–	Marine	N. Sea/Pennines
Sunk Island	?2.0	0.8	4.51	0.23	1.98	87	Marine	N. Sea/Pennines
Saltend	?2.5	0.9	2.6	−0.48	0.32	81	Estuarine	N. Sea/Vale'York
Garthorpe	2.8	2.3	6.87	−0.24	0.43	75	Estuarine	N. Sea/Vale'York
Butterwick	?0.3	8.3	6.41	−0.13	0.50	57	Brackish	Vale'York/N. Sea
Newland	?0.1	8.3	4.21	0.22	1.29	4	Brackish	Vale'York/N. Sea
Basal	1.7	>8.3	6.06	−0.22	3.96	10/70[6]	Freshwater	Local

[1] The relative volumes of the Saltend and Sunk Island Suites (total 4.5 km^3) are estimates, as are the relative volumes of the Newland and Butterwick Suites (total 0.4 km^3). The volumes of the Garthorpe, Butterwick, Newland and Butterwick Suites reflect their present volumes, prior to erosion associated with the base of the Saltend Suite.
[2] The volume of the Skeffling Suite and estimate of relative sea-level rise is taken from ABP-NUDCM (1998).
[3] The relative mean sea-level rise in millimetres per year is averaged over the period during which the suite was deposited.
[4] Estimates of sea-level rise are based on Shennan *et al.* (this volume).
[5] The grain size data for the Skeffling Suite were provided by Patrick McLaren to the Environment Agency in 1994 (Rees *et al.* 1998) and are based on Malvern Granulometer analyses.
[6] The two values for the Basal Suite in the Mean USi index relate to the suite in the western (10) and eastern (70) parts of the estuary.

extending across the estuary at this point on the eastern side of the Humber Gap. The sill may be a remnant of a moraine which extended across the gap, between the Yorkshire and Lincolnshire Wolds, at the end of the Pleistocene and which probably dammed glacial Lake Humber. The existence of the sill explains the limited marine influence in the western part of the estuary during the deposition of the Newland and Butterwick Suites, indicated by their chemostratigraphy, heavy mineralogy and biofacies (Metcalfe *et al.* this volume).

The main erosional surface in the estuary (which cuts-out parts of the Basal, Newland, Butterwick and Garthorpe Suites) occurs at the base of the Saltend Suite and locally the Sunk Island Suite (Fig. 10). This can be recognized with reasonable confidence across large parts of the estuary, though the age of the erosion is likely to vary widely in different places. It is likely that between Hull and the Ancholme Valley the erosion may be early Holocene (or even latest Pleistocene) in age and related to the catastrophic events associated with the drainage of the glacial Lake Humber, which existed west of the gap during the latest Pleistocene. Over most of the estuary however, the erosional surface is likely to be considerably younger; but unfortunately the sequence above the erosion surface has not been dated because of the paucity of peats within it. The surface certainly cuts out the 'Middle Peat' (which is approximately of Bronze Age) over large parts of the estuary and is probably less than 4 ka old in most places.

The volumes of the suites (Table 2) were calculated by the generation of 'difference grids' between two existing grids. For example, the volume of the Garthorpe Suite was derived by calculating a difference grid between the grid representing the base of the Garthorpe Suite and that of the base of the Saltend Suite, which effectively marks the top of the Garthorpe Suite; the volume of the Garthorpe Suite was then determined by analysis of the difference grid. The volumes of the Basal, Newland and Butterwick (combined), Garthorpe, Saltend and Sunk Island (combined) Suites were calculated in this manner. The volumes of the Saltend and Sunk Island Suites were determined using a digital terrain model (DTM) generated at the BGS of the surface of the estuarine sediments. Because of the impact which the erosion at the base of the Saltend Suite has had on the (preserved) volumes of the sediment suites, the volumes of the Basal Suite and of the Newland and Butterwick Suites (combined) prior to the erosion can only be calculated on the basis of simple assumptions about the estuarine geomorphology at the time of deposition.

It is likely that the net accretion of sediments in the estuary over the last 150 years (ABP-NUDCM 1998) gives a gross estimate of the volume of the sediments of the Skeffling Suite (*c.* 0.2 km^3), as the suite is likely to have formed since the start of the Industrial Revolution, which approximately coincides with this period. The present topography and bathymetry of the estuary is illustrated in Fig. 11.

Provenance of the estuarine fill

Geochemistry

The provenance of sediments of the estuarine fill can also be examined through the use of spidergrams. Figure 12 shows how sediments of the Basal Suite in the eastern part of the estuary (HMB11) are clearly related to a North Sea till source, whilst in the western estuary Basal Suite sediments in HMB16 show a strong similarity to Vale of York drift. A more complex situation is demonstrated in Fig. 13, where the geochemical signature of the Butterwick Suite is compared with those from North Sea tills (mean of 36 tills from Holderness cliffs and offshore), average sediment from the Trent basin (catchment weighted mean of three river sediment samples from Gainsborough on the Trent, Bawtry on the Idle and Rossington on the Torne) and a mean value for stream sediment data from the Trent catchment (data from the Wolfson Geochemical Atlas). The spidergram for the Trent catchment is incomplete because the range of elements determined for the Wolfson Atlas differs from the range used in the LOEPS study. If the effects of anthropogenic contamination (P, Cu, Zn and Pb) in the river and catchment sediment are discounted, the pattern of variation in the Butterwick signature is perhaps more similar to the Trent catchment-derived sediment than to that of the North Sea tills, suggesting a fluviatile (Trent) provenance.

At the eastern end of the estuary, the signatures of sediments from the Spurn Suite and the coarser material of the Sunk Island Suite show similarities to the signature of a mean of over 1000 sea-bed sediment samples from the BGS Offshore Geochemistry Database (Fig. 14). As with the Wolfson data (above), parts of the signature are incomplete because of differences in the range of elements determined. There is a clear indication of a marine provenance for these coarse-grained suites.

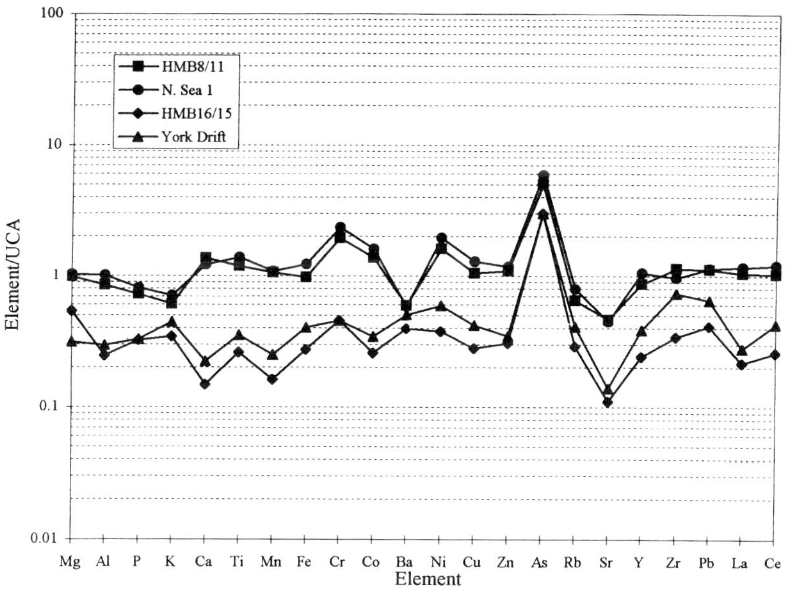

Fig. 12. Basal Suite spidergrams. This figure illustrates well how spidergrams may be grouped. It shows how the Basal Suite in the eastern part of the estuary (here represented by sample HMB8/11) is compositionally similar to one of the North Sea tills. Likewise, the Basal Suite in the western part of the estuary (represented by sample HMB16/15) is compositionally similar to the average composition of the Vale of York drift.

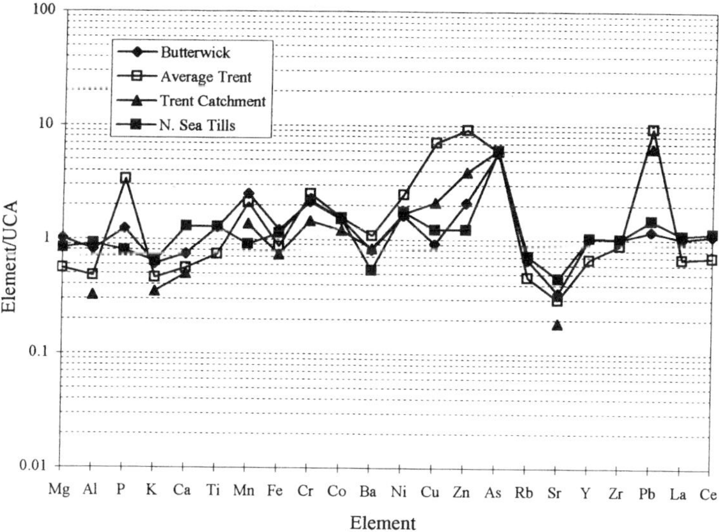

Fig. 13. Butterwick Suite spidergram. The average geochemical signature of the Butterwick Suite is compared with those from North Sea tills (mean of 36 tills from Holderness cliffs and offshore), average sediment from the Trent basin (catchment weighted mean of three river sediment samples from Gainsborough on the Trent, Bawtry on the Idle and Rossington on the Torne) and a mean value for stream sediment data from the Trent catchment (data from the Wolfson Geochemical Atlas supplied by I. Thornton, Imperial College, London. The spidergram is incomplete because the range of elements determined is different).

The situation for the finer Sunk Island Suite sediments and the Saltend and Garthorpe Suites is less clear cut. The signatures of all three (Fig. 15) are similar over much of the range with only relatively subtle differences appearing and all compare relatively well with an average signature for North Sea tills. However, close examination shows the Sunk Island and Saltend sediments to have a slightly closer affinity to the tills, suggesting a stronger marine influence in their provenance. It must be remembered that some of the North Sea tills are exposed in the country behind Hull and the Holderness coast and it is possible that these may be the source of

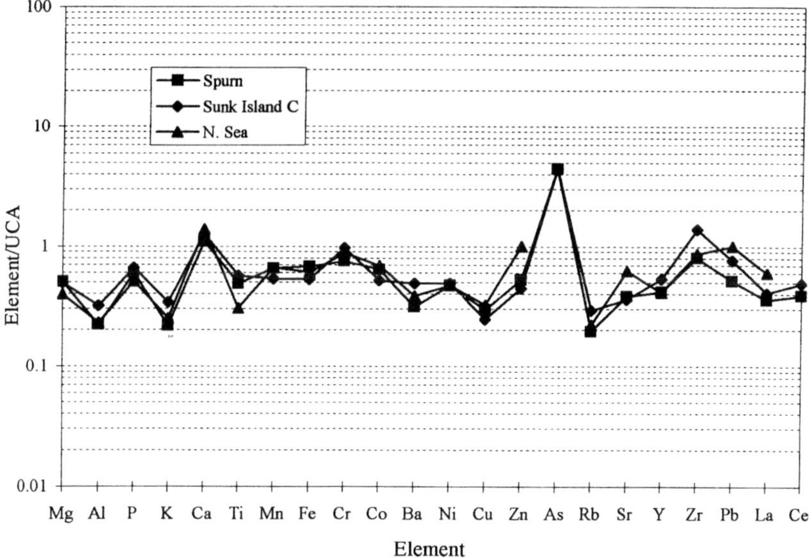

Fig. 14. Spidergram of coarse sediments in the Sunk Island and Spurn Suites. This shows the similarity of the coarse sediments of these suites with those of the mean of over 1000 sea bed sediment samples from the BGS offshore geochemistry database.

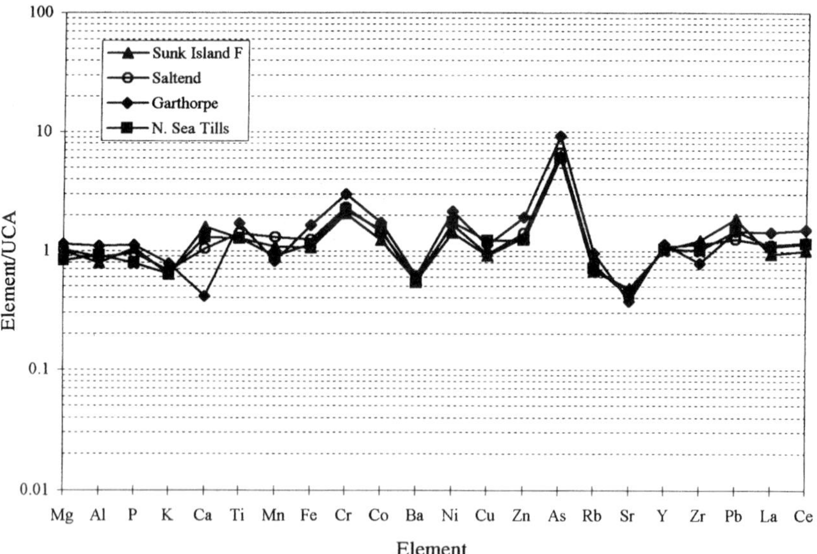

Fig. 15. Spidergram of fine-grained sediments from the Garthorpe, Saltend and Sunk Island Suites. Only subtle differences distinguish the suites. That of the Garthorpe Suite is least similar to that of the mean of 36 North Sea tills, suggesting that this suite contains smaller amounts of marine-derived sediments.

some of the sediments. The sediments would have been transported to the estuary in watercourses, such as the River Hull. The differences in grain size between the coarse sea bed sediments shown in Fig. 14 and finer-grained estuarine sediments makes direct comparison difficult.

The stratigraphic divisions of the estuarine sediments and their relationship with potential sources can be explored further by means of scatterplots (Figs 16–19). Variables for the scatterplots were largely chosen empirically, but generally involve element pairs that have low correlation coefficients for the Humber Holocene sediments, are not directly grain size related (except Rb) and/or that relate to specific environments (e.g. Ca and Sr in marine sediments; Zn and Pb in the central and southern Pennine orefields and offshore sediments derived from the Tyne–Tees and northern Pennine orefield). The suites recognized through comparison of spidergram signatures form distinctive groups on the scatterplots, although the degree of separation and/or affinity varies from diagram to diagram.

On the Rb/Sr plot (Fig. 16), the Spurn and Sunk Island Suites form separate groups, but with a distinctive trend linking them, with the exception of the oldest three Spurn samples, which lie off the trend. The Garthorpe, Butterwick and Newland Suites also form separate groups with some anomalous Garthorpe samples. The Basal Suite shows the variability described in the previous section, whilst the Skeffling Suite samples plot with the Sunk Island group. A wide scatter across the Garthorpe and Sunk Island groupings characterizes the Saltend Suite. Similar relationships characterize other plots (e.g. Ni/CaO; not illustrated).

The Sunk Island and Skeffling Suites are easily distinguished on other plots, such as Zn/Pb, as are the Butterwick and Newland Suites (Fig. 17). High Pb values generally characterize the Sunk Island and Skeffling Suites (cf. Rees *et al.* 1999). It is unclear whether the Pb is derived mainly from mining in the Trent and Ouse catchments, or whether it is derived principally from the Tyne and Tees, via the North Sea (offshore geochemistry maps show large amounts of Pb being transported down the North Sea coast; Stevenson *et al.* 1995). The trend linking the marine Spurn Suite to the Sunk Island sediments suggest that the latter is more

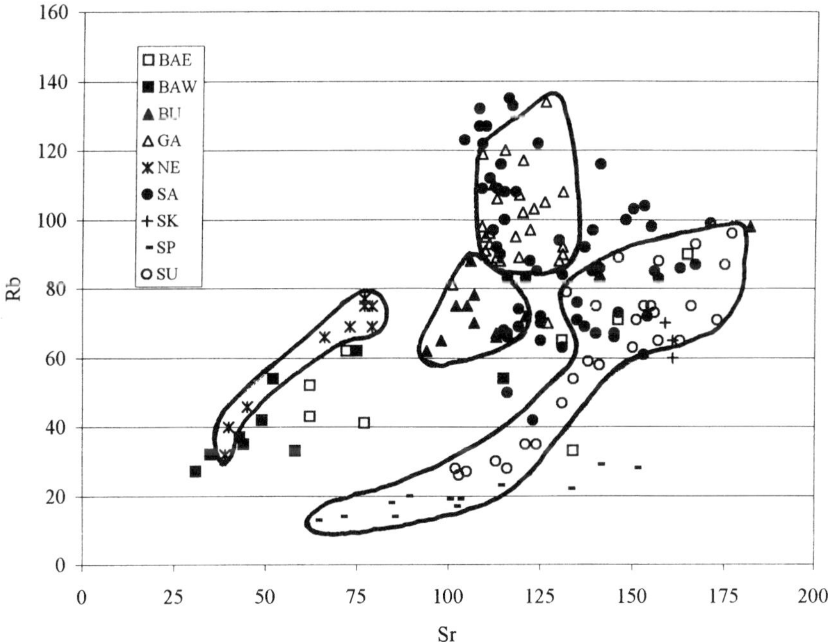

Fig. 16. Rb/Sr scatterplot. This plot is one of several in which the samples from individual suites are closely clustered. Note that the fields of the Newland, Butterwick, Garthorpe and Sunk Island/Spurn Suites (marked) have no overlap. BAE, Basal Suite east; BAW, Basal Suite west; BU, Butterwick Suite; GA, Garthorpe Suite; NE, Newland Suite; SA, Saltend Suite; SK, Skeffling Suite; SP, Spurn Suite; SU, Sunk Island Suite.

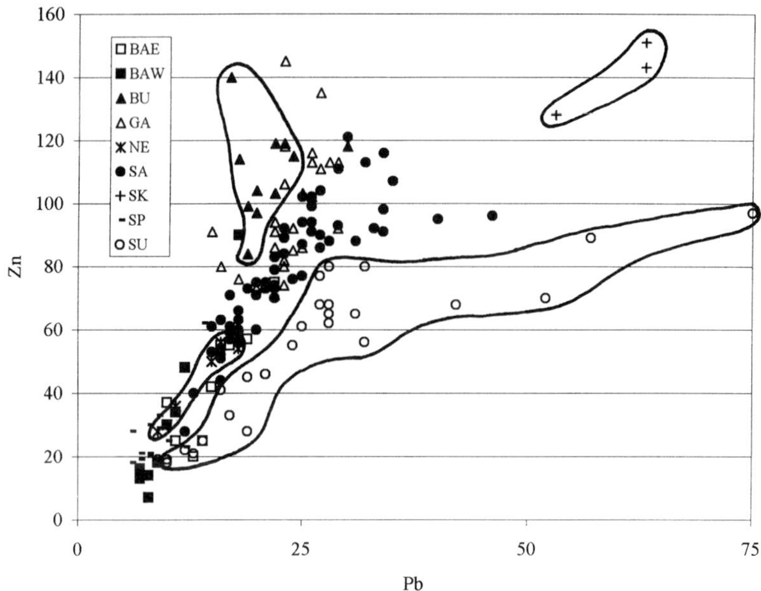

Fig. 17. Zn/Pb scatterplot. This plot is useful in separating the Sunk Island and Skeffling Suites (fields marked). It also shows that the Newland Suite is closer in composition to the Saltend Suite than the Butterwick Suite, which may have more of a Pennine (Trent, riverine) influence (fields marked). BAE, Basal Suite east, BAW, Basal Suite west; BU, Butterwick Suite; GA, Garthorpe Suite; NE, Newland Suite; SA, Saltend Suite; SK, Skeffling Suite; SP, Spurn Suite; SU, Sunk Island Suite.

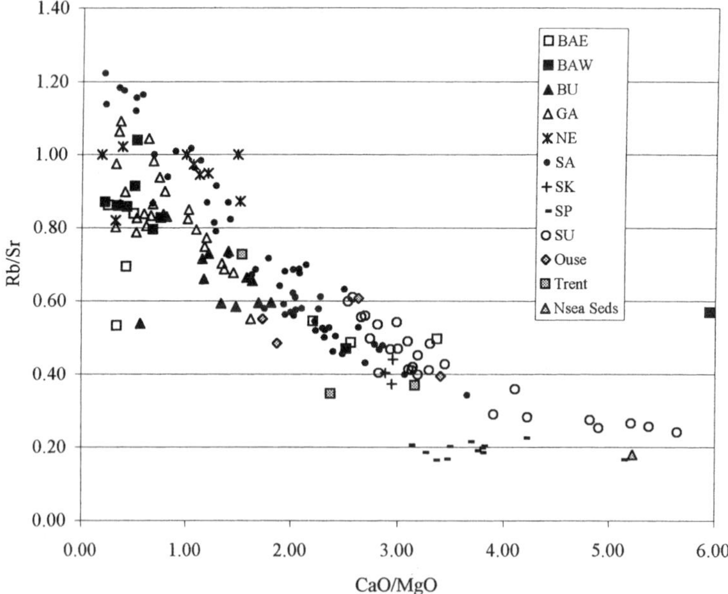

Fig. 18. Rb/Sr versus Ca/Mg scatterplot of Holocene estuarine sediments and selected source sediments. This plot shows a clear separation of the Garthorpe and Saltend Suites not seen on other scatterplots.

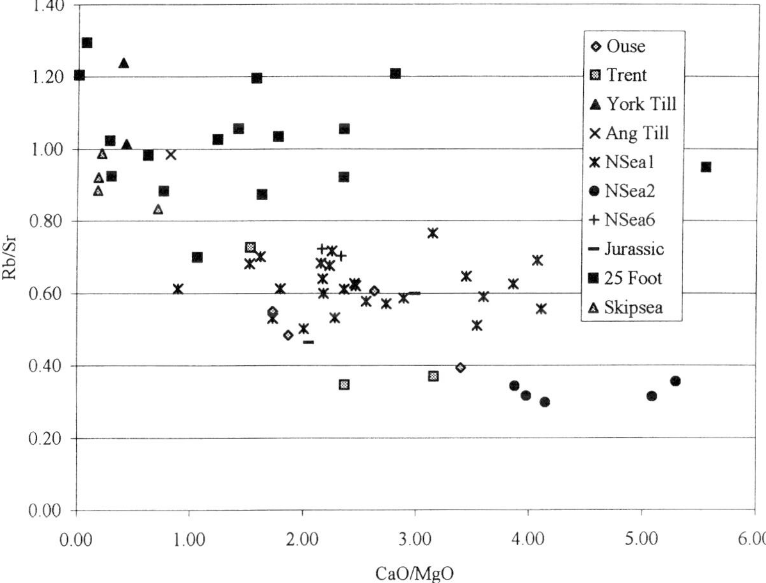

Fig. 19. Rb/Sr versus Ca/Mg scatterplot of river sediments, Pleistocene sediments and Jurassic rocks for comparison with Fig. 18.

probable, although the Zn/Pb ratios are generally higher than those found in mining related Tees estuary sediments (Plater *et al.* 1999).

The Rb/Sr versus CaO/MgO plots (Figs 18 and 19) are useful for comparing the similarity between the various sediment suites and potential sources, although they need to be interpreted with caution. There is a clear affinity between some of the North Sea sediments and the Spurn, Sunk Island and Skeffling Suites, as these plot in a similar area towards the bottom-right area of the diagrams, suggesting that the suites were largely derived from North Sea sources (as noted above). There is some similarity in the distribution of sediments from the Trent river system and the sediments of the Butterwick Suite, again suggesting that some were derived from the Trent system. By way of contrast the sediments of the Newland Suite plot in a different area to those of sediments of the Ouse system. They have a composition which is more similar to the Pleistocene sediments of the Vale of York drift ('25 foot' drift), suggesting that many of the sediments in the suite may have a relatively local provenance. The Saltend Suite is clearly separated from the Garthorpe and Sunk Island Suites on Fig. 18.

The Garthorpe Suite has close affinities with both the Vale of York ('25 foot') drift and the Skipsea Till (Figs 18 and 19), suggesting that many of its sediments were derived from relatively local (Humber lowland) sources, but with a significant input from the till on the Holderness coast or from rivers (east of Hull) draining into the estuary. The samples from the Saltend Suite plot in part of the Vale of York drift ('25 foot' drift) field, but also considerably further towards the bottom right into the field of North Sea tills, suggesting that they contain sediment derived both from river systems and North Sea.

Heavy mineralogy

Heavy mineral suites in sediments reflect the heavy mineral content of the source area and, like geochemistry, can be used to characterize sedimentary units and to provide information on provenance. The relative abundance of unstable to stable minerals was determined as a means of assessing the relative contributions to the assemblages of northerly derived tills and local pre-Quaternary sandstones. In order to achieve this, an unstable : stable mineral index (USi) has been devised:

$$\mathrm{USi} = 100\frac{\mathrm{Ca} + \mathrm{Px} + \mathrm{Ep}}{\mathrm{Ca} + \mathrm{Px} + \mathrm{Ep} + \mathrm{Ap} + \mathrm{Ru} + \mathrm{To} + \mathrm{Zr}}$$

where Ca, calcic amphibole; Px, pyroxene; Ep, epidote; Ap, apatite; Ru, rutile; To, tourmaline; and Zr, zircon.

The mean USi index of each of the sediment suites is indicated in Table 2. This shows that the index ranges from 4 in the Basal Suite in the west of the estuary to 87 in the Sunk Island Suite. The suites with a high USi index, such as the Sunk Island Suite and the Saltend Suite, are characterized by heavy mineral assemblages with a very high proportion of unstable minerals (primarily calcic amphibole, clinopyroxene and epidote). By way of contrast, the Basal Suite in the west of the estuary contains a high proportion of relatively stable minerals, such as zircon, tourmaline, rutile and apatite.

The distinctive contrasts in USi value are considered to represent genuine differences in sand provenance. They cannot be ascribed to selective burial-related dissolution of unstable minerals since there is no indication of increased mineral etching within samples with low USi values.

The unstable mineral assemblages contain a high proportion of metamorphic minerals, indicating derivation from the Scottish Highlands, since no such assemblages occur in potential source rocks of the catchments of the Trent and Ouse river systems. High proportions of unstable minerals in the sediments from the eastern estuary suggests that these assemblages are of first-cycle origin and thus may have reached the Humber Estuary by long-shore drift or, more plausibly, through reworking of Pleistocene tills. The stable mineral assemblages, with their high proportion of apatite and consistent occurrence of monazite, are comparable to assemblages in the Triassic Sherwood Sandstone Group. This indicates derivation by fluvial drainage of the hinterland dominated by Triassic strata.

Clay mineralogy

The proportions of illite, kaolinite and expansible clays found in the Butterwick, Garthorpe and Saltend Suites, in river floodplain sediments, and in North Sea sediments and tills are shown on Fig. 20.

The mineralogical similarity of the riverine and North Sea sources makes it difficult to identify particular provenances within the samples. Mixing of clay suspensions by estuarine transport processes has probably contributed to this difficulty. The mix of clays within the Butterwick Suites is unlike that of river or North Sea materials, suggesting possible derivation from onshore Pleistocene sources. Kaolinite is more common in the Garthorpe Suite than other suites

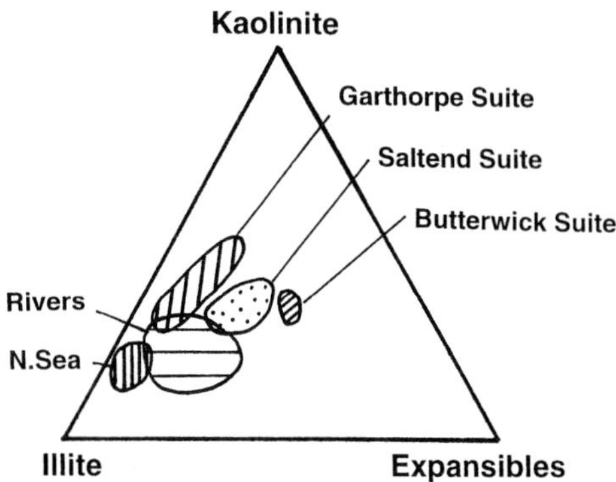

Fig. 20. Triangular diagram showing the clay mineral characteristics of sediment suites and source materials.

or source materials, but the reasons for this are not clear. The mix of clays in the Saltend Suite plot in a field in the left centre of the triangular diagram, between the North Sea and riverine sources and the Garthorpe and Butterwick samples, suggesting that the estuary was well-mixed during deposition of the Saltend Suite.

Summary: sources of the sediment suites

The Basal Suite is considered to be entirely non-marine in origin. Its composition generally reflects local sources. Hence, in the west it is largely derived from the Pleistocene Vale of York drift and in the east from North Sea tills.

The Newland Suite is very distinctive compositionally and is thought to be derived mainly from Pleistocene sediments of the Vale of York drift, as indicated by the low USi heavy mineral index and similar Rb/Sr versus CaO/MgO ratios (Figs 18 and 19). However, the presence of marine microfauna suggests that it was deposited partially under marine influence. The major contribution of the Vale of York drift indicates that sediments were being preferentially eroded in the vale, in comparison with other areas, such as the Pennines. The reason for this is uncertain as it is unlikely that deforestation was taking place at this time (pre-Bronze Age). The relative paucity of marine sediment (compared with overlying suites) is likely to be due to the sill between St Andrew's Dock and New Holland (see above), which prevented much marine influence in areas west of the sill at the time of deposition.

The presence of the sill may have impacted on the Butterwick Suite in a similar manner, although the terrestrial sediments in this case were probably derived from the Trent river system (see Fig. 13). The suite is also likely to have been deposited under a degree of marine influence.

The similarity of the mean spidergrams of the Garthorpe Suite and North Sea tills (Fig. 14) and the closeness of the Garthorpe and some North Sea till fields on the Rb/Sr versus CaO/MgO plots (Figs 18 and 19) suggest that the suite is derived partially from North Sea tills. Whilst most of these occur on the coast and offshore, the Skipsea Till where it is exposed onshore between Hull and the coast may have been a source of sediment. Similarly, Pleistocene sources in the Vale of York are also likely to have made a contribution, as indicated by the similar ratios of these sediments and those of the Garthorpe Suite. The Trent and Ouse river system samples have higher CaO/MgO ratios than most of the Garthorpe samples and it is thus unlikely that the suite has a significant river component in its provenance.

A dominantly marine origin for the Saltend Suite is suggested by the close similarity of the spidergram with those of North Sea tills (Fig. 15). However, the comments made regarding the terrestrial component of the Garthorpe Suite could be applied also to the Saltend sediments, although the river sediments of the Trent and Ouse systems are closer to some of the Saltend Suite samples on Fig. 18. The Saltend Suite differs geochemically from the underlying Garthorpe Suite in its higher Ca concentrations (Fig. 15). This could be caused by the slightly more abundant shell content of the suite and increased marine influence. Although these factors are probably important, it is also possible that the reduced levels of Ca in the Garthorpe Suite reflect leaching. This conclusion is suggested by the identification of subdivisions within the Saltend Suite based on minor variations in element levels through the suite. Although these are not set out here, the subdivision below the present ground surface has markedly decreased levels of Ca (and enhanced Mn) suggestive of leaching and soil development.

The Sunk Island Suite is distinguished from the very similar underlying Saltend Suite by its markedly higher levels of anthropogenic metals, such as Pb (Fig. 17), and higher CaO/MgO ratios. The latter are also similar to those of the Spurn Suite and in some cases to average sea bed sediments from the North Sea off the Humber and tills of the North Sea 2 group (Fig. 19). A close geochemical association between the finer-grained members of the suite and the North Sea tills can be seen also in Fig. 15. The suite thus seems to have a marine provenance, most sediments being derived from the North Sea.

The Skeffling Suite comprises sediments that are almost indistinguishable from most of those of the Sunk Island Suite on textural grounds, but which carry a greater load of metal contaminants (Fig. 17). They appear to have a mainly mixed source, although mostly derived from the North Sea.

The Spurn Suite consists almost entirely of sand- and gravel-grade sediment that has either been transported down the coast via longshore drift, or has been eroded from the shoreface surface during the Holocene transgression. The sediments are chemically very similar to those of equivalent grain size in the North Sea. Many of the sediments of the suite are likely to have been eroded and re-mobilized several times during the transgression. Those on Spurn Point will have migrated with it (De Boer 1964).

Discussion

The relative influence of the factors forcing changes in sedimentation storage patterns has been reviewed by Long *et al.* (1998). Those which control the balance of supply of marine versus on-shore sourced sediment are discussed here.

Supply of marine sediment to the estuary

The progressive rise in sea-level through the Holocene period is clearly the main reason for the increasing storage of marine sediments in the estuary. However, the character of sediments stored in the estuary has not changed gradually over time, but shows marked changes and unconformities, suggesting that the estuary has responded morphodynamically to the rise. This is illustrated by the transition from freshwater and brackish environments and associated sediments (largely derived from the Vale of York and from the Trent river system) in the Newland and Butterwick Suites, to environments with a notable marine influence in the Garthorpe Suite. Although marine waters initially reached the western part of the estuary during deposition of the former suites, the sill between St Andrew's Dock and New Holland (see above) effectively reduced the marine influx. This scenario changed during deposition of the Garthorpe Suite, when marine waters reached all sections of the western part of the estuary.

Overall, during most of the Holocene the rate of relative sea-level rise has declined through time (see Table 2 and Metcalfe *et al.* this volume), although this trend is probably punctuated by fluctuations that have important impacts for sediment storage and erosion (Long *et al.* 1998). Between *c.* 5000 and 100 a BP the average rate was probably less than 1 mm a^{-1}. The effective 'stabilization' of relative sea-level at the beginning of this period is likely to have caused some changes in sediment storage in the estuary. The decreasing rate of sea-level rise would have caused the amount of space available for sediment storage to diminish. As a consequence, assuming an undiminished rate of sediment supply, it is likely that lateral migration and associated erosion of the estuary channel would have been more common than earlier in the Holocene. Factors such as these may (in part) explain the widespread vertical change from saltmarsh deposits of the Garthorpe Suite to the mudflat and sandflat deposits of the Saltend Suite over large parts of the estuary.

The supply of marine sediments to the estuary could also be affected by storm surges. However, there is no good evidence of these having influenced the pattern of sediment storage within, or the form of, the Humber Estuary during its Holocene evolution. The only deposits which may have resulted from a storm surge, are those of the Spurn Suite at the base of HMB4, which are likely to have been transported westwards into the estuary from the open coast at the time.

Supply of sediment from on-shore sources to the estuary

The supply of terrestrial material to the estuary is likely to be controlled both by climatic change and human activities. The relative importance of these in sediment delivery to the estuary is debated by Macklin *et al.* (this volume).

The rate of discharge to the estuary from rivers draining into it will have been influenced by climatic changes, particularly in wetness. Recent palaeoenvironmental studies (Smith 1985; Chiverell 1998) demonstrate a number of wet periods in the Holocene, some of which are likely to have had a significant impact on lowland river incision and alluviation. The first major episode of Holocene river erosion and channel bed incision associated with one of these wet periods commenced in the late Bronze Age. In the piedmont reaches of the Swale catchment this took place sometime between 3850 and 2900 cal. a BP and caused enhanced rates of sedimentation in the lower reaches between 2900 a BP and 2450 cal. a BP (Macklin *et al.* 1999). This wet period coincided with the development of the Saltend Suite in the estuary and may have influenced the amount of channel erosion and migration associated with it.

The history of humankind in the Humber Estuary and its catchment since the last glaciation has been well-documented by the archaeological community (see Van de Noort & Ellis 1997 and references therein). Anthropogenic activities have probably had a substantial impact on sedimentation patterns by *inter alia* increasing sediment flux through agriculture (and associated runoff), removing parts of the intertidal area by reclamation and flood-defence and increasing the size of the channel by dredging.

It is unlikely that humankind had any notable influence on the form of, or storage of sediments within, the estuary before about 3850 cal. a BP. However, pollen records show that from this time onwards, until *c.* 2.1 cal. ka BP, rapid deforestation took place and agriculture began (Tinsley 1975; Atherden 1976), particularly in the Humber lowlands, for instance in the lower part of the Vale of York. Whilst it may be expected that these practices would have caused considerable amounts of soil erosion and increased sediment supply to rivers entering the estuary, the rate of sediment delivery to the estuary during this period is unlikely to have been high (Macklin *et al.* this volume). It is more likely that agricultural practices had a significant impact on sediment delivery to the estuary in the High Middle Ages, with the massive expansion of tillage into areas that had not been cultivated until the Medieval warm period (Lamb 1995; Macklin *et al.* this volume). The effect of this, in combination with climatic deterioration (above), is likely to have influenced the development of the Sunk Island Suite. Chemically, the suite is distinguished from the Saltend Suite mainly by its higher Pb content as a result of anthropogenic contamination (probably at least, in part from the Tyne–Tees, via the North Sea). However, the Sunk Island Suite is also distinct on the Rb/Sr versus CaO/MgO diagram (Fig. 18), suggesting a more fundamental change in provenance, with a greater marine influence in addition to an increase in industrial or mining activities.

Apart from demonstrating the increasing marine character of sediments in the estuary through the Holocene, the provenance studies suggest a change in the source of on-shore material in the estuary. During deposition of the Basal, Butterwick, Newland, Garthorpe and Saltend Suites most onshore sourced materials were derived from the Pleistocene Vale of York drift. In the overlying Saltend and Sunk Island Suites most on-shore sourced sediments were probably derived from Pennine river sources. These sediments are likely to have reached the estuary both directly, through the Ouse and

Trent river systems, as well as via the North Sea from the Tyne and Tees Rivers. A similar switch in sediment source has been identified in sediments stored within floodplains and terraces of rivers in the Ouse system by Macklin *et al.* (this volume). They note that the transition from Pleistocene to 'Carboniferous' Pennine sources took place at different times in the various rivers of the system, though all had switched by *c.* 2 cal. ka BP. Unfortunately the switch in sediment signature cannot be used to date the age of the Saltend Suite, as it would be expected to occur later in the estuary than in the rivers feeding it. The influx of sediments from Pennine sources into the estuary is also likely to have been caused by deforestation and the spread of agricultural practices further into upland areas. Despite the effects described above, it is likely that humankind's greatest impacts were during deposition of the Skeffling Suite, when significant changes were made to the morphology of the estuary, though reclamation of intertidal areas and to a lesser degree dredging of the channel.

Conclusions

This work demonstrates the following.

(1) Distinctive sediment bodies may be recognized within the Humber Estuary fill, which have become progressively more marine in character through the Holocene.
(2) The limited influence of marine conditions in the western parts of the estuary in the early Holocene was largely caused by the presence of a morainic ridge near St Andrew's Dock, Hull.
(3) A major change in sedimentation took place during the Bronze Age, when widespread migration of the estuarine channel took place (forming the Saltend Suite). This may have been caused by increased runoff, caused by climatic change, in conjunction with a decrease in the space available for sedimentation caused by a slowing rate of sea-level rise.
(4) A later episode of channel migration occurred during Medieval times (Sunk Island Suite). This may have been caused by enhanced sediment supply, probably caused by a rapid spread of agriculture during a wet period.

This paper is based on work carried out within the Holocene Evolution of the Humber Estuary project. J. Cox (University of East Anglia) kindly collected the samples from the Holderness Cliffs). We are grateful to I. Thornton, Imperial College, London for the supply of data from the Wolfson Geochemical Atlas. J. G. Rees, J. Ridgway, R. W. O'B Knox and R. Newsham publish by permission of the Director of the British Geological Survey (NERC). LOIS publication no. 579.

References

ABP-NUDCM 1998. *Humber Estuary Shoreline Management Plan: Geomorphology and Processes.* Associated British Ports–Newcastle University Department of Coastal Management.

ANDREWS, J. E., SAMWAYS, G., DENNIS, P. F. & MAHER, B. A. 2000. Origin, abundance and storage of organic carbon and sulphur in the Holocene Humber Estuary – emphasising human impact on storage changes. *This volume.*

ATHERDEN, M. A. 1976. The impact of late prehistoric cultures on the vegetation of the North York Moors. *Transactions of the Institute of British Geographers, New Series*, **1**, 284–300.

BERRIDGE, N. G. & PATTISON, J. 1994. *Geology of the Country around Grimsby and Patrington.* Memoirs of the British Geological Survey, Sheets 90, **91**, 81 and 82 (England and Wales). Stationery Office, London.

BUCKLAND, P. C. & SADLER, J. 1985. The nature of the Flandrian alluviation in the Humberhead levels. *East Midlands Geographer*, **8**, 239–251.

CHIVERELL, R. C. 1998. *Moorland Vegetation History and Climate Change on the North York Moors during the last 2000 years.* PhD thesis, University of Leeds.

CROWTHER, D. R. 1987. Sediments and archaeology of the Humber foreshore. *In*: ELLIS, S. (ed.) *East Yorkshire Field Guide.* Quaternary Research Association, Cambridge, 99–105.

DE BOER, G. 1964. Spurn Head: its history and evolution. *Transactions of the Institute of British Geographers*, **34**, 71–89.

DIDSBURY, P. 1988. Evidence for Romano-British settlement in Hull and the lower Hull Valley. *In*: PRICE, J. & WOLFSON, P. R. (eds) *Recent Research in Roman Yorkshire, Oxford.* British Archaeological Reports, British Series, **193**, 21–35.

DINNIN, M. 1995. Introduction to the palaeoenvironmental survey. *In*: VAN DE NOORT, R. & ELLIS, S. (eds) *Wetland Heritage of Holderness: an Archaeological Survey.* Humber Wetlands Project, University of Hull, 27–48.

——1997. The palaeoenvironmental survey of the Rivers Idle, Torne and Old River Don. *In*: VAN DE NOORT, R. & ELLIS, S. (eds) *Wetland Heritage of Holderness: an Archaeological Survey.* Humber Wetlands Project, University of Hull, 81–155.

—— & LILLIE, M. 1995. The palaeoenvironmental survey of southern Holderness and evidence for sea-level change. *In*: VAN DE NOORT, R. & ELLIS, S. (eds) *Wetland Heritage of Holderness: an Archaeological Survey.* Humber Wetlands Project, University of Hull, 87–120.

GAUNT, G. D. & TOOLEY, M. J. 1974. Evidence for Flandrian sea-level changes in the Humber Estuary and adjacent areas. *Bulletin of the Institute of Geological Sciences*, **48**, 25–41.

HASLAM, H. W. & PLANT, J. A. 1990. *Rock Geochemistry: Guidelines for the Acquisition and Interpretation of Lithogeochemical Data. Part 1: General. Part 2: Sedimentary Rocks*. British Geological Survey Technical Report, WP/90/1.

HUDSON-EDWARDS, K. A., MACKLIN, M. G., FINLAYSON, R. & PASSMORE, D. G. 1999. Medieval Lead Pollution in the River Ouse at York, England. *Journal of Archaeological Science*, **26**, 809–819.

INGHAM, M. N. & VREBOS, B. A. R. 1994. High productivity geochemical XRF analysis. *Advances in X-ray Analysis*, **37**, 717–724.

INSTITUTE OF GEOLOGICAL SCIENCES. 1968. *Humber Investigations*. Marine Geophysics Unit Report.

——1973. *Humber Investigations*. Marine Geophysics Unit Report.

LAMB, H. H.. 1995. *Climate History and the Modern World*. Routledge: London.

LONG, A. J., INNES, J. B., KIRBY, J. R., LLOYD, J. M., RUTHERFORD, M. M., SHENNAN, I. & TOOLEY, M. J. 1998. Holocene sea-level change and coastal evolution in the Humber estuary, eastern England: an assessment of rapid coastal change. *The Holocene*, **8**, 229–247.

LORING, D. H. & RANTALA, R. T. T. 1992. Manual for the geochemical analysis of marine sediments and suspended particulate matter. *Earth Science Reviews*, **32**, 235–283.

MACKLIN, M. G., TAYLOR, M. P., HUDSON-EDWARDS, K. A. & HOWARD, J. A. 2000. Holocene environmental change in the Yorkshire Ouse Basin and its influence on river dynamics and sediment fluxes to the coastal zone. *This volume*.

METCALFE, S. E., ELLIS, S., HORTON, B., INNES, J. B., MCARTHUR, J. J., MITLEHNER, A., PARKES, A., PETHICK, J. S., REES, J. G., RIDGWAY, J., RUTHERFORD, M. M., SHENNAN, I. & TOOLEY, M. J. 2000. The Holocene evolution of the Humber estuary: reconstructing change in a dynamic environment. *This volume*.

NERC 1994. *Land-Ocean Interaction Study (LOIS) Implementation Plan*. Natural Environment Research Council, Swindon.

PLATER, A. J., RIDGWAY, J., APPLEBY, P. G., BERRY, A. & WRIGHT, M. R. 1999. Historical contaminant fluxes in the Tees estuary: geochemical, magnetic and radionuclide evidence. *Marine Pollution Bulletin*, **37**, 343–360.

REES, J. G., RIDGWAY, J., NEWSHAM, R., KNOX, R. W. O'B., OUTHWAITE, K. & BALSON, P. S. 1998. *Post-glacial Sediment Storage in the Humber Estuary: An Overview for Estuarine Management*. British Geological Survey Technical Report, WB/98/35C.

——, ——, KNOX, R. W. O'B., WIGGANS, G. & BREWARD, N. 1999. Sediment-borne contaminants in rivers discharging into the North Sea through the Humber Estuary. *Marine Pollution Bulletin*, **37**, 316–329.

RIDGWAY, J., ANDREWS, J. E., ELLIS, S., HORTON, B. P., INNES, J. B., KNOX, R. W. O'B., MCARTHUR, J. J., MAHER, B. A., METCALFE, S. E., MITLEHNER, A., PARKES, A., REES, J. G., SAMWAYS, G. M. & SHENNAN, I. 2000. Analysis and interpretation of Holocene sedimentary sequences in the Humber Estuary. *This volume*.

——, REES, J. G., GOWING, C. J. B., INGHAM, M. N., COOK, J. M., KNOX, R. W. O'B., BELL, P. D., ALLEN, M. A. & MOLINEAUX, P. J. 1998*a*. *Land–Ocean Evolution Perspective Study (LOEPS) Core Programme, Geochemical Studies, 1: Methodology*. British Geological Survey, Technical Report, WB/98/55.

——, —— & KNOX, R. W. O'B. 1998*b*. *Land–Ocean Evolution Perspective Study (LOEPS) Core Programme, Geochemical Studies, 2: The Humber Estuary*. British Geological Survey, Technical Report, WB/98/54.

ROWLATT, S. M. & LOVELL, D. R. 1994. *Methods for the normalisation of metal concentrations in sediments*. International Council for the Exploration of the Sea, Marine Quality Committee, *E:20*.

SHENNAN, I., HORTON, B. P., INNES, J. B., LLOYD, J. L., MCARTHUR, J. J. & RUTHERFORD, M. M. 2000. Holocene crustal movements and relative sea-level changes on the east coast of England. *This volume*.

SMITH, B. M. 1985. *A palaeoecological Study of Raised Mires in the Humberhead Levels*. PhD thesis, University of Wales, Cardiff.

STEVENSON, A. G, TAIT, B. A. R., RICHARDSON, A. E., SMITH, R. T., NICHOLSON, R. A. & STEWART, H. R. 1995. *The Geochemistry of Sea-bed Sediments of the United Kingdom Continental Shelf; the North Sea, Hebrides and West Shetland Shelves and the Malin–Hebrides Sea Area*. British Geological Survey Technical Report WB/95/28C.

STUIVER, M. & REIMER, P. J. 1993. Extended ^{14}C data base and revised CALIB 3.0 ^{14}C age calibration program. *Radiocarbon*, **35**, 215–230.

SWITSUR, V. R. 1981. Radiocarbon dating. *In*: MCGRAIL, S. (ed.) *The Brigg 'Raft' and her Pre-historic Environment*. British Archaeological Reports, British Series, Oxford, **89**, 117–121

TINSLEY, H. M. 1975. The former woodland of the Nidderdale Moors (Yorkshire) and the role of early man in its decline. *Journal of Ecology*, **63**, 1–26.

VAN DE NOORT, R. & DAVIES, P. 1993. *Wetland Heritage: an Archaeological Assessment of the Humber Wetlands*. Humber Wetlands Project, University of Hull.

—— & ELLIS, S. 1995. *Wetland Heritage of Holderness: an Archaeological Survey*. Humber Wetlands Project, University of Hull.

—— & ——1997. *Wetland Heritage of Humberhead Levels: an Archaeological Survey*. Humber Wetlands Project, University of Hull.

WEBB, J. S., THORNTON, I., HOWARTH, R. J., THOMPSON, M. & LOWENSTEIN, P. 1978. *The*

Wolfson Geochemical Atlas of England and Wales. Clarendon Press, Oxford.

WEDEPOHL, K. H. 1995. The composition of the continental crust. *Geochimica et Cosmochimica Acta*, **59**, 1217–1232.

WILKINSON, W. B., LEEKS, G. J. L., MORRIS, A. & WALLING, D. E. 1997. Rivers and coastal research in the land–ocean interaction study. *The Science of the Total Environment*, **194–195**, 5–14.

Origin, abundance and storage of organic carbon and sulphur in the Holocene Humber Estuary: emphasizing human impact on storage changes

J. E. ANDREWS,[1] G. SAMWAYS,[1,2] P. F. DENNIS & B. A. MAHER[1]

[1] *School of Environmental Sciences, University of East Anglia, Norwich NR4 7TJ, UK (e-mail: j.andrews@uea.ac.uk)*

[2] *Present address: Badley Ashton and Associates, Winceby House, Winceby, Horncastle, Lincolnshire LN9 6PB, UK*

Abstract: An organic carbon (C_{org}) and sulphur (S) storage inventory for Holocene sediments in the Humber Estuary is established; sources of organic matter and their variation over time are identified, and with chronological control, the importance of estuarine sediments as C_{org} and S stores is demonstrated. Humber Holocene sediments are grouped into seven widespread environmental facies with statistically significant geochemical data sets: (1) *oak–hazel fenwood* (OHF); (2) *alder carr* (AC), appearing as peats in core; (3) *river channel muds or sands* (Rcm/s); (4) *high saltmarsh* (HSM); (5) *low saltmarsh* (LSM); (6) *intertidal mudflat* (ITMF); and (7) a *sandy* facies (S). Carbon, nitrogen and sulphur (CNS) abundances show that these facies have diagnostic geochemical signatures and $\delta^{13}C$ values for bulk organic matter exhibit a range of average values: −28‰ (terrestrial peats), −27‰ (HSM), and −24.5‰ (ITMF) reflecting the up core transition from terrestrial peats through saltmarshes to more open marine mudflat environments as regional sea-level rose. Chronology and average sedimentation rates are partly constrained by radiocarbon dates; palaeomagnetic techniques helped define discrete sediment packages and discontinuities (time gaps). Although the Humber Holocene sediment record is not continuous, long-term sedimentation rates (about $1\,mm\,a^{-1}$) show that sediment accretion kept pace with regional sea-level rise between 6 and 2 cal. ka BP. This sedimentation rate, combined with core evidence to allow a geographic reconstruction of the palaeo-Humber (3–2 cal. ka BP), is used to calculate storage values for C_{org} and S in the various environments of the palaeo-Humber. Comparison of the C_{org} and S sedimentation and storage terms for the palaeo-Humber with modern values highlights the impacts of reclamation and commercial/urban development in the estuary in the last 300 years. OHF and AC peats, which were the largest C_{org} and S stores in the palaeo-estuary, are now absent (reclaimed), while saltmarshes are no longer widespread. Conservative calculations show a net decrease in C_{org} deposition from about 3.2×10^5 tonne in the palaeo-estuary to no more than 2.5×10^3 tonne today, a >99% reduction in potential C_{org} storage capacity. The total modern yearly S deposition is approximately 2% of its value 2 ka ago. Removal of saltmarsh and associated brackish–freshwater wetland suggests that suspended sediment and associated C_{org} and S are currently bypassing former (Holocene) storage areas and may be impacting North Sea biogeochemical cycling.

The coastal zone is a potentially important site of deposition and regeneration of organic carbon (see e.g. Berner 1982; 1989), but little is known about the specific role of this environment in the global carbon balance of the past, present and future (Mackenzie *et al.* 1998). Coastal zone net organic metabolism, i.e. the difference between primary production and respiration of organic matter, is probably a significant term in the global oceanic carbon budget (Smith & Hollibaugh 1993). The coastal environment is one of the four major interactive domains on the Earth's surface (Ver *et al.* 1994; Mackenzie *et al.* 1998) and in this domain organic matter metabolism is profoundly impacted by human activities (Smith & Hollibaugh 1993). It is therefore clear that a good understanding of the material balance of organic carbon in the coastal zone is necessary, especially in estuaries, where data pertinent to constraining the relationship between primary production and net metabolism are sparse (Smith & Hollibaugh 1993). The Land–Ocean Interaction Study (LOIS) was conceived in part to provide this new data, and to give

From: SHENNAN, I. & ANDREWS, J. (eds) *Holocene Land–Ocean Interaction and Environmental Change around the North Sea*. Geological Society, London, Special Publications, **166**, 145–170. 1-86239-054-1/00/$15.00

detailed, regional-scale, long-term perspectives of material fluxes in the coastal zone on a range of time-scales including the last 200 years, the last two millennia and the Holocene period as a whole (LOIS Science Plan 1992).

The study described here focuses on the Humber Estuary in the central part of the river–atmosphere–coast study (RACS) site. The aim is to demonstrate how effective estuaries are at storing organic matter on 10 ka (Holocene) time-scales, with particular reference to the way the storage term has changed as relative sea level has risen (Shennan *et al.* this volume). Most of the environmental changes impacting the estuary over the last 6–2 ka have been natural (ignoring human-induced deforestation for the time being). This contrasts strongly with the last 300 years, when anthropogenic effects such as reclamation, industrial, commercial and agricultural development and pollution have been great. We are able to highlight this contrast between the natural and human-impacted estuary and suggest how future management of temperate estuaries in developed countries might impact organic matter storage and hence the carbon metabolism of the estuarine and wider coastal zone.

The Humber Estuary

The Humber Estuary on the east coast of England (Fig. 1) is one of the largest UK estuaries with a tidal length of 120 km and maximum width of 15 km (Pethick 1988), comparable with the Thames and Severn Estuaries. The Humber receives drainage principally from the Rivers Trent and Ouse, catchments that drain almost one-fifth of England (Neale 1988). The Humber Estuary is typically defined (Fig. 1) as the area from the confluence of the Ouse and Trent at Trent Falls [SE 86 23] to the sea between Spurn Head [TA 39 10] and Cleethorpes [TA 31 09]. This area can be conveniently divided at the Humber Gap, where the bridge crossing is located (Fig. 1) into the upper or inner Humber area, between Humber Gap and Trent Falls (Fig. 1), and the lower or outer Humber area, east of Humber Gap to the open sea.

At the end of the Devensian glaciation, *c.* 13 cal. ka BP, the melting ice exposed a till plain in the Holderness area; the retreating ice front probably forced the Humber drainage to flow southeastwards from the present position of Hull towards the coast (at this time many kilometres east of its present position) accounting for the southeast–northwest orientation of the present outer estuary (Pethick 1988). As sea-level rose in the early part of the Holocene (see maps in Lambeck 1995) the English North Sea coastline began to approach its present position in this area about 8–7 cal. ka BP. As sea-level rose and the 6 cal. ka BP outer estuary formed, the outline was probably irregular (see Pethick 1988) defined by the topography of the Holderness

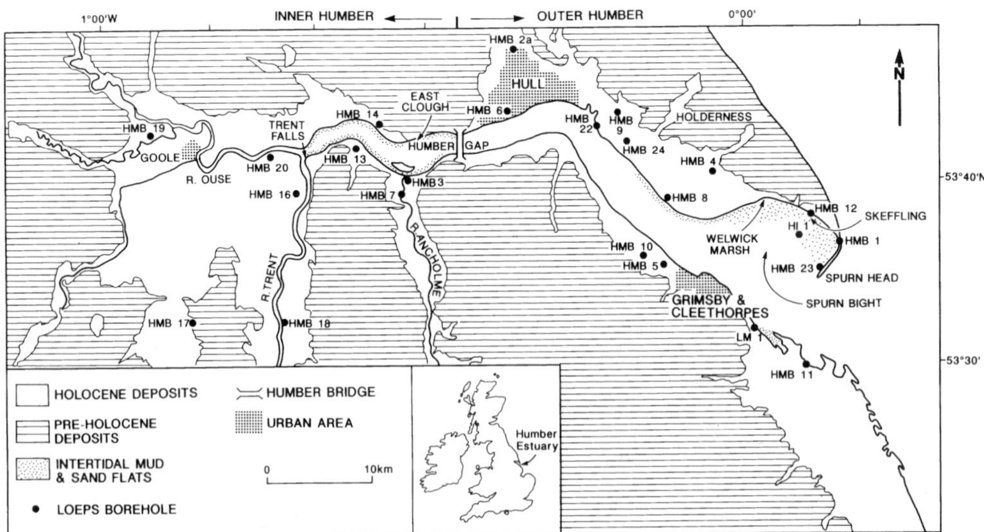

Fig. 1. Location map of the study area showing borehole sites and numbers (HMB13, etc.) and geographical features.

tills. The smoother profile of the modern estuary presumably developed as marine-derived sands and muds were deposited by tidal flows.

A detailed knowledge of the evolution of Holocene sedimentary facies in the Humber was not well known before the onset of LOIS; although general information, pre-1950s references and some simplified borehole logs and radiocarbon dates on peat deposits can be found in Smith (1958), Gaunt & Tooley (1974), Gaunt *et al.* (1992), Gaunt 1994; Berridge & Pattison (1994) and, recently, Long *et al.* (1998). More detailed information on the facies evolution of the Holocene Humber has recently been collected as part of LOIS (Metcalfe *et al.* this volume), while the Humber Wetlands Project has provided useful new data on Holocene environmental changes in parts of the estuary and its tributaries (Van de Noort & Ellis 1995, 1997, 1998).

From medieval times onward the low-lying areas in the Humber Estuary and in adjoining rivers suffered a number of drainage improvements and reclamation. The history of reclamation, most effective over the last 300 years and locally known as 'warping', is reasonably well-documented (e.g. de Boer 1988; Berridge & Pattison 1994; Lillie & Weir 1997; 1998). Reclamation was accompanied by various phases of port development and industrialization, culminating in the current developed configuration of the estuary banks.

The modern Humber Estuary is a well-mixed macrotidal system and the water is typically turbid. It has been estimated that most of the sediment entering the Humber comes from the flooding marine tide, *c.* $2.2 \times 10^6\,m^3\,a^{-1}$ according to O'Connor (1987), compared with just $0.3 \times 10^6\,m^3\,a^{-1}$ from the rivers. Most of the 'marine' sediment probably comes from erosion of the Holderness cliffs (Al-Bakri 1986; Fig. 1) with very little muddy sediment supplied by offshore erosion (McCave 1987). While the marginal mudflat and saltmarsh deposits are predominantly composed of muds and silts, sands and gravels are common in the axial parts of the channel, on the south shore southeast of Grimsby (Fig. 1), and in some areas of tidal flats (Pethick 1988).

Tidal flats fringe large parts of the outer estuary, but may be narrow (<100 m wide) or absent where seawalls or commercial development are extensive (e.g. areas around Grimsby and Hull in Fig. 1). Tidal flats are typically <100 m wide in the inner estuary. The widest area of tidal flats (>4 km wide) is in Spurn Bight [TA 4013 3117] immediately northwest of Spurn Head (Fig. 1). In parts of the outer estuary, mudflats give way landward to saltmarsh environments, typically in front of clay embanked sea walls. The mudflat to saltmarsh transition is typically marked by the presence of *Spartina anglica*, which makes a broad lower saltmarsh zone (Armstrong 1988) in areas like Welwick Marsh in the northwest corner of Spurn Bight (Fig. 1). Landward of the lower saltmarsh, higher and drier marshes are colonized by *Puccinellia* and *Halimione*.

The natural succession of marine to terrestrial environments usually ends with mature saltmarsh, the succession having been truncated by the construction of seawalls and by reclamation; however, at a few locations in the inner Humber area (e.g. at Broomfleet [SE 905 268] and Blacktoft Sand [SE 850 235]) managed *Phragmites* marshes stand at the landward edge of saltmarshes. It is important to note that this truncated succession from saltmarsh to reclaimed agricultural land is wholly artificial, resulting from 300 years of human activity. Before extensive human involvement the natural succession was probably from much wider tracts of saltmarsh, to progressively less saline fen and carr environments. The word carr is a common element of place names in low-lying areas surrounding the inner Humber area and lower Ancholme Valley (de Boer 1988), while coastal reedswamp, alder carr and oak fenwood environments have been identified by Metcalfe *et al.* (this volume) in many of the inner Humber cores collected during the LOIS. These types of marginal marine–terrestrial marshy environments are no longer present in the Humber system, however, close comparisons can be made to parts of Broadland in Norfolk (see e.g. George 1992). The conversion of these types of transitional marsh environments to agricultural land potentially has important effects on long term carbon storage in the Humber.

Methods

This study is centred on 24 cored boreholes (locations shown on Fig. 1) provided as part of the Land–Ocean Evolution Perspective Study (LOEPS) programme (see coring, storage and logging protocols in Ridgway *et al.* (this volume). Sedimentological, geochemical and lithofacies studies were based on these cores, augmented by information from hand gouge augers (typically <4 m penetration) and <1 m deep push cores in recent estuarine and saltmarsh sediments. Grain-size analyses of selected samples (Ridgway *et al.* this volume) were made using a Malvern laser particle sizer (University

Table 1. *Radiocarbon ages and calibration data for specific dates quoted in the text*

Site	Material	Laboratory code	^{14}C age ($\pm 1\sigma$)	Calibrated age (a BP)‡			Altitude (m OD)
				Max.	Mean	Min.	
South Farm, HMB8	Decomposed peat	AA25581	7145 ± 60	8063	7923	7806	−11.67
South Farm, HMB8	Peat (undifferentiated)	AA25582	8555 ± 65	9646	9491	9397	−11.95
South Marsh, HMB5	Peat (undifferentiated)	AA23433	6520 ± 75	7526	7387	7233	−6.77
Immingham, HMB10	Peat (undifferentiated)	AA23431	6050 ± 70	7153	6887	6736	−6.48
Newlands, HMB19	Peat (undifferentiated)	AA23440	5075 ± 55	5930	5821	5670	−3.25
Whitton Ness, HMB13	Decomposed peat	AA22674	3910 ± 45	4439	4376	4155	−1.88
Newlands, HMB19	Bulk organic matter*	AA27141	3395 ± 75	3833	3631	3464	−0.76
Newlands, HMB19	Woody root (*in situ*)†	AA29905	2810 ± 50	3062	2876	2779	−1.87
Newlands, HMB19	Peat (undifferentiated)	AA23439	4800 ± 55	5646	5527	5330	−2.84
Newlands, HMB19	Bulk organic matter*	AA23440	5075 ± 55	5930	5821	5670	−3.25
Garthorpe, HMB16	Peat (undifferentiated)	AA25585	7265 ± 60	8135	8034	7922	−11.43
Garthorpe, HMB16	Bulk organic matter*	AA25565	6540 ± 60	7524	7391	7281	−9.91
Garthorpe, HMB16	Saltmarsh peat	AA23437	3425 ± 65	3836	3664	3475	−2.56
Garthorpe, HMB16	Roots (*in situ*)	AA25564	2160 ± 60	2327	2138	1985	−0.56
Lockham, HMB12	Amorphous peat (*in situ*)	AA23890	4040 ± 65	4816	4479	4354	−1.16
Lockham, HMB12	Roots (*in situ*)	AA25559	3060 ± 50	3370	3261	3082	−0.29

* Date may be unreliable (allochtonous organic matter).
† Sampled on duplicate core taken for palaeomagnetic work.
‡ The calibrated ages shown are the age ranges which contain 95.4% of the area under the probability curve. All ages in this table were calibrated with CALIB 3.0 using the bidecadal atmospheric curve (Stuiver & Reimer 1993; and references therein).

of Hull) and a Coulter laser particle sizer (University of East Anglia).

Total carbon, nitrogen and sulphur concentrations were measured with a Carlo Erba EA 1108 elemental analyser (see method in Ridgway *et al.* this volume). Replicate analysis of laboratory standards with compositions close to the samples (3.50 wt% total C, 2.0 wt% C_{org}, 0.15 wt% total N and 0.50 wt% total S) gave 1σ precisions of $\pm$0.04 wt% total C, $\pm$0.12 wt% organic C, $\pm$0.006 wt% total N and $\pm$0.06 wt% total S. The total S values should be conserved, even though reduced reactive phases such as iron monosulphides will have oxidized, probably to elemental sulphur (see Canfield *et al.* 1986). Pore waters were not removed before analysis and we assumed that dissolved sulphate contributions to total S were trivial in these low porosity sediments. Organic carbon $\delta^{13}C$ was measured on selected samples (see method in Ridgway *et al.* (this volume); results are expressed relative to the Vienna Pee Dee belemnite (VPDB) scale and replicate analyses of an L-alanine laboratory standard gave a 2σ precision of $\pm$0.1‰ for organic carbon $\delta^{13}C$.

Samples for sulphur speciation analysis were fixed in the field with 20% ZnAc solution (Duan *et al.* 1997), frozen immediately (either in dry ice or liquid nitrogen) and stored frozen until analysis. The laboratory method was based on that of Canfield *et al.* (1986), adapted to a multi-step distillation (Fossing & Jorgensen 1989). The technique yielded sulphur values for:

(a) acid volatile sulphur (AVS), mainly iron monosulphides and probably greigite;
(b) cold chromous chloride reduced sulphur (CCrS), taken to represent well-crystallized greigite and newly formed pyrite (Duan *et al.* 1997);
(c) hot chromous chloride reduced sulphur (HCrS) representing mature pyrite and elemental sulphur (Duan *et al.* 1997).

As this sulphur speciation method has minimal effect on organically bound S (Bottrell *et al.* 1994), any residual sulphur detected by elemental analysis was assumed to be organic sulphur.

A sub-set of Holocene sediment samples was also analysed for total sulphide S (analysed as HCrS) to compare with total S values from elemental analysis. Reproducibility of all HCrS values was <3% of the concentration value (at HCrS concentrations around 0.6 wt%) and comparison of data from these methods suggests that in the range 1–2.5 wt% S the HCrS was underestimated by *c.* 15% (because recovery can be less than 100%).

Accelerator mass-spectrometry (AMS) radiocarbon dates from 27 samples including peats, bulk disseminated organic matter, *in situ* roots at sedimentological surfaces and $CaCO_3$ shells were provided for this study as part of the LOEPS (core programme) allocation at the National Environmental Research Council (NERC) facility (East Kilbride). All samples were scrutinized and approved by the LOEPS radiocarbon committee and dates were calibrated to calendar years before present (BP) using the CALIB 3 program (Stuiver & Reimer 1993). The bidecadal atmospheric curve was used for calibration because $\delta^{13}C_{org}$ values were typically lower than −25‰ PDB, indicating that most of the organic matter was of terrestrial origin. Freshwater gastropod shell carbonate in core HMB16 had $\delta^{13}C$ values around −13‰: at depositional temperatures, inferred at about 10°C, this carbonate was in equilibrium with dissolved CO_2 with a $\delta^{13}C$ of about −24 to −25‰, implying that the water body had not equilibrated with atmospheric CO_2. Under these circumstances it is appropriate to calibrate dates from the freshwater shell carbonate with the bidecadal atmospheric curve (Stuiver & Reimer 1993). These samples are part of a substantial database of 103 radiocarbon dates taken from peats and other material within the Holocene sequences (LOEPS core programme) that contributed information to a number of allied studies (Shennan *et al.*; Metcalfe *et al.* this volume). In this paper, calibrated radiocarbon ages are given as the central value of the 2σ range and the full data for dates quoted are given in Table 1.

Results and their interpretation

Lithofacies and sedimentology

The cored Holocene sediments have been classified into 16 environmental facies (Metcalfe *et al.* this volume) based on litho- and biofacies data (see Table 2). In the present geochemical study, statistically significant data sets on common and widespread environmental facies are necessary. Seven environmental facies are common (Table 2):

(1) *oak–hazel fenwood* (OHF);
(2) *alder carr* (AC), facies 1 and 2 both appearing as peaty deposits in core;
(3) *river channel muds or sands* (Rcm/s);
(4) *high saltmarsh* (HSM), with a sub-category of *high saltmarsh peats* (HSM(P));
(5) *low saltmarsh* (LSM);
(6) *intertidal mudflat* (ITMF);

Table 2. *Humber depositional environments, sedimentology and distribution in the Holocene and modern estuary*

Abbreviation*	Depositional environment	Sedimentological characteristics	Distribution	
			Palaeo-Humber (with examples, see also Figs 9 and 12)	Modern Humber
RB	Raised bog		Localized: upland areas of the inner basin, most notably Hatfield Moor and Thorne Moor e.g. HMB17 (Hatfield Moor, +0.37 to −1.77 m OD)	Extensively mined as peat workings, some areas now restored to peat forming environments, but extensively degraded by drainage
FWM	Fresh/brackish marsh (with standing water)	Mud dominated sequence with disseminated freshwater gastropods	Localized: base of HMB16 (Garthorpe, −9.76 to −11.44 m OD), inner basin, most likely a standing body of water on the alluvial floodplain marsh	Virtually absent due to drainage and reclamation
OHF	Oak/hazel fenwood	Woody peat	Widespread: typically occurs at the base of inner basin sequences, e.g. HMB20 (Ousefleet, −13.51 to −13.83 m OD). Locally at the base of marginal outer estuary sequences (e.g. HMB12, Lockham, −3.30 to −3.46 m OD)	Virtually absent due to drainage and reclamation
AC	Alder carr	Predominantly minerogenic, typically mud/silt sized, with locally abundant, well preserved woody material	Widespread: typically succeeds OHF or RCm/RCs in the inner basin (e.g. Ousefleet, HMB20, −11.58 to −13.51 m OD). Locally influenced by washed-in marine faunas (e.g. Newlands, HMB19, −0.58 to −1.50 m OD)	Virtually absent due to drainage and reclamation

SF	Sedge fen	Fibrous peat	Localized: succession to OHF in the inner basin (Butterwick, HMB18, −8.02 to −8.16 m OD)	Virtually absent due to drainage and reclamation
CRS	Coastal reedswamp	Fibrous peat	Localized: succession to OHF in HMB12 (Lockham, −3.30 to −3.40 m OD). and marking the first permanent marine influence in HMB16 (Butterwick, −2.56 to −2.60 m OD)	Restricted to local areas at the head of the estuary due to reclamation
RCm/RCs	River (beyond tidal limit)	Mud and sand dominated sequences with abundant scattered woody fragments (predominant in the muddy sequences)	Widespread: dominating the lower parts of the inner basin sequences (e.g. Whitton Ness, HMB13, −2.48 to −2.65 m OD)	Less common: due to the constriction of the main fluvial channels by reclamation
HSM/HSM(P)	High saltmarsh/high saltmarsh peats	Bimodal distribution: Muds (HSM) often with oxidized root traces, etc., or peats (HSM(P)) with high mud content, locally transitional into overlying low saltmarsh deposits	Widespread: base of most outer estuary sequences, also occurs towards the base of some axial sequences in the inner basin (e.g. Winterton Carrs, HMB7; South Ferriby, HMB3)	Virtually absent due to reclamation and the construction of sea defences
LSM	Low saltmarsh	Muddy sequences characterized by *in situ* root traces and locally faint lamination	Widespread succession to HSM in the outer estuary, and axial sequences in the inner estuary	Rare: saltmarshes are generally undergoing erosion in the modern estuary, except in isolated areas such as Spurn Bight
ITMF	Intertidal mudflat	Mixed silt–mud sequences, commonly organized in bundled laminae, evidence of bioturbation and disseminated shell fragments	Widespread succession to LSM in the outer estuary and axial sequences in the inner estuary	Significant in Spurn Bight, but short-lived elsewhere in the estuary
ITSF	Intertidal sandflat	Typically fine to medium grained sands, commonly laminated	Widespread succession to LSM in the outer estuary and axial sequences in the inner estuary	Significant in Spurn Bight, but short-lived elsewhere in the estuary

(*continued*)

Table 2. (*continued*)

Abbreviation*	Depositional environment	Sedimentological characteristics	Distribution	
			Palaeo-Humber (with examples, see also Figs 9 and 12)	Modern Humber
ICm/ICs	Intercreek areas (flooded by high spring tides)	Generally muddy silt–sand sequences with little evidence of rooting, bundled laminae locally well-developed	Rare: succession to LSM in HMB1 (Kilnsea), and virtually the entire sequence in Thorngumbald (HMB24)	Rare: due to the paucity of creeks
ITCm/ITCs	Creeks and creek levees (intertidal channels)	Generally muddy silt–sand sequences with little evidence of rooting, bundled laminae locally well-developed	Common, particularly towards the top of inner basin sequences	Less common: due to the restriction of the main channel by reclamation
STCm/STCs	Subtidal channel	Bioturbated, locally shelly muddy silts and sands. Bundled laminae locally well-preserved	Common: sporadically present throughout the outer estuary sequences, interbedded with ITC	Less common: due to the restriction of the main channel by reclamation
BS	Beach sands and gravels (including dunes)	Ranging from fine to coarse grained, locally coarse grained horizons, shell fragments, some faint lamination preserved (mainly lost during coring induced dewatering)	Cored only at Kilnsea (HMB1) and Spurn Point (HMB23)	Occur only at Spurn Head and south shore beaches east of Grimsby. However, active dune systems are restricted by sea defences
GFm/GFs/TILL/B	Pre-Holocene (glaciofluvial, Pleistocene till, bedrock)	A wide variety of sands and gravels and boulder clay containing diagnostic chalk fragments. Commonly rootletted by the overlying Holocene colonizers	Sequence bases	Not applicable

* m, mud; s, sand.

(7) intertidal and subtidal sandy facies (ITSF, ICs, STCs and BS) are geochemically similar and have been amalgamated into a *sandy* facies (S).

The distribution of these facies is outlined in Table 2 and discussed further by Metcalfe *et al.* (this volume).

Carbon, nitrogen and sulphur geochemistry

Comparison of weight per cent total C and weight per cent C_{org} (Fig. 2) demonstrates that carbon in the OHF and AC peaty facies, saltmarsh and freshwater marsh lithofacies is almost entirely organic (total C = organic C). However, in the intertidal mudflat and sandy lithofacies, total carbon is greater than organic carbon, indicating the presence of significant inorganic carbon, mainly calcite shelly material, and in some localities, authigenic ferroan dolomite. Overall weight per cent total C can be used as an indicator of weight per cent C_{org} except in the intertidal and sandy lithofacies.

In these Holocene core samples, weight per cent total N correlates positively with weight per cent C_{org} (Fig. 3) suggesting that total N principally reflects weight per cent organic N (Hedges *et al.* 1986; Andrews *et al.* 1998). This interpretation is supported by data from modern sediments on Welwick Marsh. Here, exchangeable N (i.e. that bound to clay minerals and particle surfaces) is present only in the upper 25–30 cm of short cores. Lower in the cores organic N accounts for the entire N present.

Total carbon, nitrogen and sulphur (CNS) abundances show that the environmental facies identified have broad but distinctive geochemical signatures. In general, the peaty facies (OHF, AC, SF and CRS; Table 2) have the highest organic carbon content, up to 38 wt% C_{org} (Fig. 4), which covaries with total N and S (Figs 3 and 5), suggesting that N and S are associated principally with organic matter. Intertidal mudflat deposits, typically with around 1–2 wt%

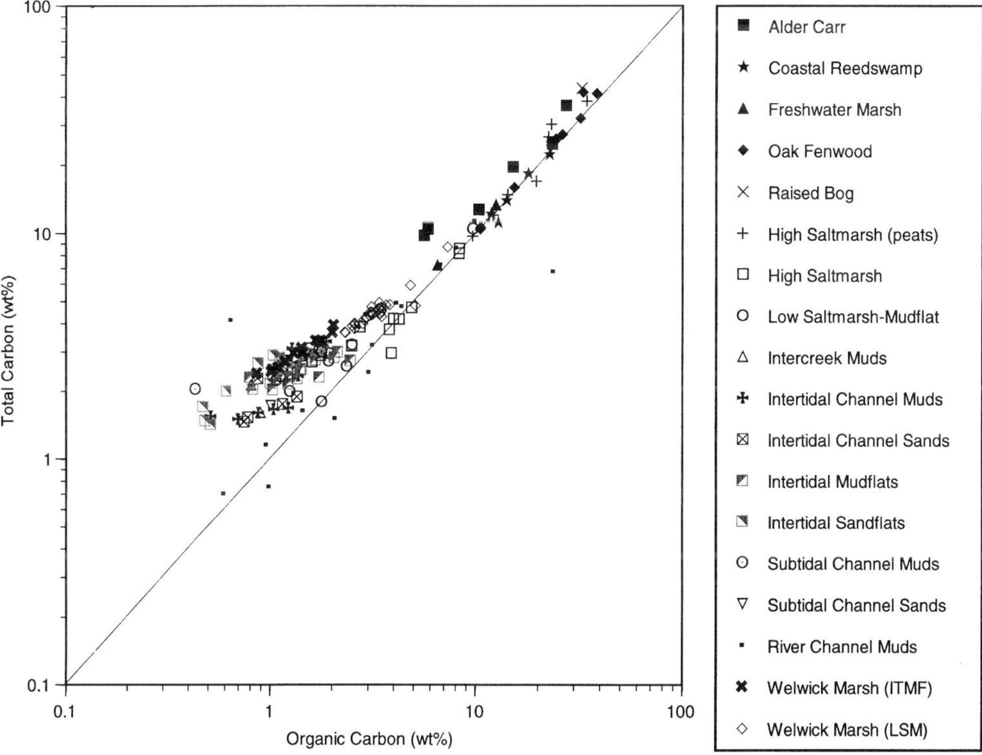

Fig. 2. Weight per cent organic carbon plotted against weight per cent total carbon for Holocene and modern sediments. Total carbon values are a reasonably good indicator of organic carbon content except for intertidal and some lower saltmarsh sediments, which both contain a significant amount of shelly $CaCO_3$.

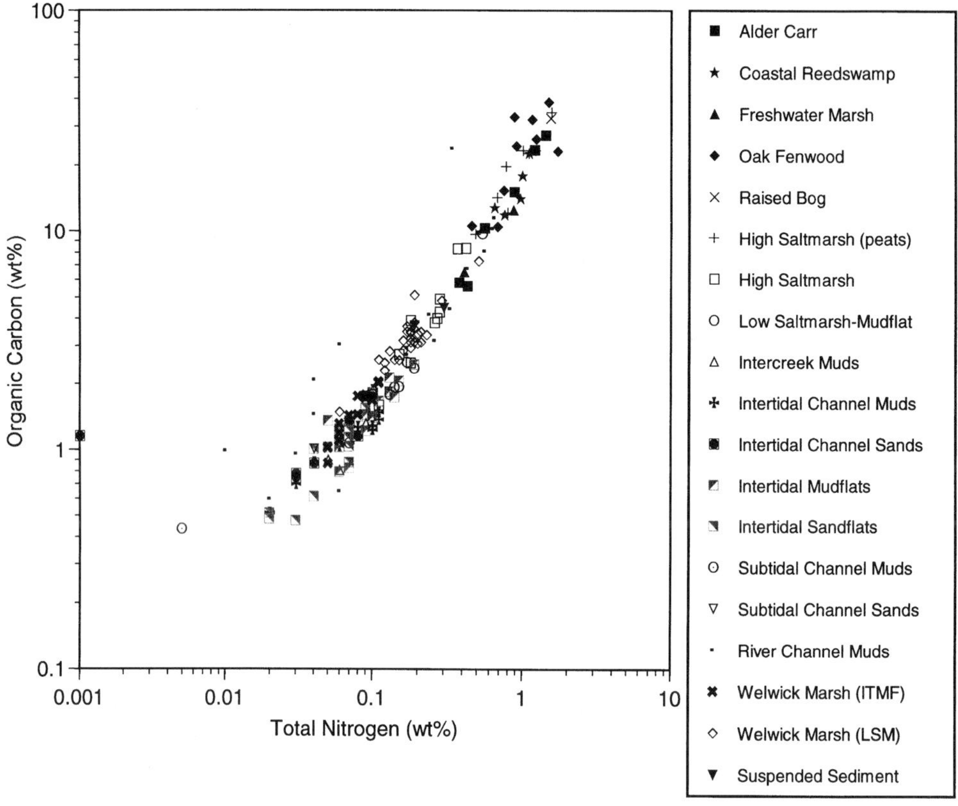

Fig. 3. Weight per cent total nitrogen plotted against weight per cent organic carbon for Holocene and modern sediments. These parameters are positively correlated, suggesting that total N is mainly composed of organic N.

total C have the lowest total N and S abundances (Figs 3 and 5), whereas saltmarsh sediments have intermediate C abundances (2–20 wt% total C).

As would be expected from the above discussion, the peaty, intertidal mud and saltmarsh lithofacies exhibit clear variations in C_{org}/total N ratio (Fig. 3). Organic matter from OHF, AC and HSM(P), i.e. 'peats', have mean C/N ratios of 23, 17 and 20 (Table 3 and Fig. 3), ratios fairly typical of degraded terrestrial organic matter (Bordovsky 1965). The lower value for AC suggests that a larger percentage of less woody material was present in these environments. Higher saltmarsh to lower saltmarsh facies have C/N ratios between 18:1 and 17:1 (Fig. 3), indicating a mixture of terrestrial plant and more marine organic matter. Intertidal mudflat facies have values around 16.5:1, fairly typical of near-shore marine sediments with a mixture of marine (e.g. N-rich marine phytoplankton material has a C/N range of 6–9; (Bordovsky 1965; Coffin *et al.* 1989) and terrestrial organic matter.

In general, C_{org}/total S ratios for intertidal mudflat, sand and some lower saltmarsh lithofacies (all with <5 wt% C_{org}) plot within the lower part of the normal marine field (Fig. 5) of Berner & Raiswell (1984); the relationship between weight per cent C_{org} and weight per cent total S is thus similar to that of marine sediments (Berner & Raiswell 1984). Bacterial sulphate reduction of labile organic matter produces iron monosulphides and eventually pyrite (Berner 1984); however, subaerial oxidation of sulphides in the Humber tidal flat and saltmarsh environments accounts for the lower S values especially in siltier and sandy sediments (see also Fig. 5). C_{org} rich (>10 wt% C_{org}) samples from upper saltmarsh, freshwater peat (OHF, AC facies) and freshwater marsh samples also have high total S values (up to 8–9 wt% total S; Fig. 5). In most organic-rich facies, e.g. high saltmarsh, coastal reedswamp and alder carr peats, increasing C_{org} is matched by increasing total S (Fig. 5); some oak fenwood facies

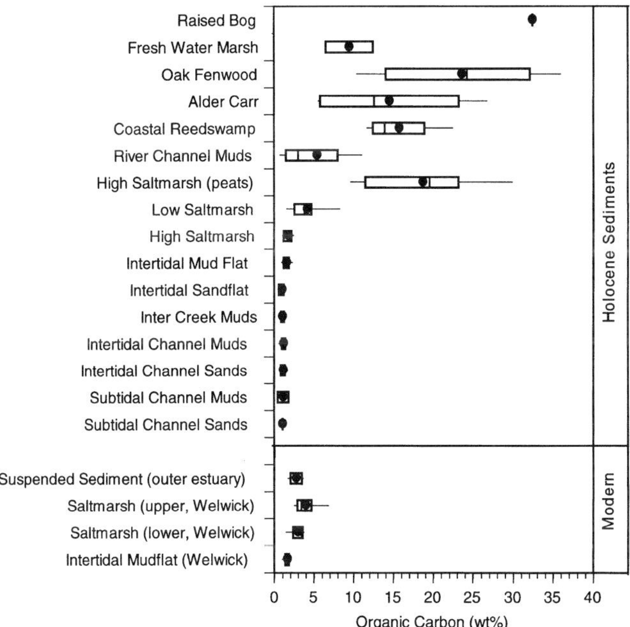

Fig. 4. Box and whisker plot of weight per cent organic carbon plotted by environmental facies for Holocene and modern sediments. In each plot the dot represents the mean value, the vertical bar represents the median value, the ends of the horizontal box represent the 25th and 75th percentiles and the ends of the horizontal line represent the tenth and 90th percentiles. The mean values for the seven common environmental facies are given in Table 3.

have very high C_{org} but intermediate total S values suggesting incorporation of some woody material. Comparison of total S values from elemental analysis versus total reduced (sulphide) S (HCrS method) from a range of Holocene facies (Fig. 6) confirms that a number of the organic-rich samples with high total S values have a large component of non-sulphide S: this is interpreted as organically bound sulphur (see also Bottrell *et al.* 1998), and consequently there is no clear relationship between marine conditions and high total S in the organic-rich samples.

Mobilization of S above and below organic-rich horizons is clear from elemental S staining of adjacent low S facies, e.g. pre-Holocene till underlying basal peats. This probably occurred by oxidation of organic matter (liberating organic S) and AVS/greigite, followed by upward and downward diffusion of sulphate S and subsequent formation of bacterially mediated diagenetic AVS/greigite in low permeability sediments. It is not certain that S diffusion occurred before coring; it is possible that diffusion occurred after coring when the confining pressure was released. It is, however, clear that the sulphide oxidized to form elemental S after coring (because elemental S was not visible when the cores were first recovered). Mobilization of S might explain a number of lower saltmarsh samples with intermediate C_{org} values (3–9 wt%) but correspondingly high total S (2–5 wt%, Fig. 5). These samples came from LSM facies overlying saltmarsh or coastal reedswamp peats, and it is likely that their S values have been enhanced by post-depositional migration.

Holocene samples collected specifically for sulphur speciation studies from a gouge augured core on the mudflats at East Clough [SE 973 247] (Fig. 1), showed that in SM facies (<*c.* 4 cal. ka BP; based on peat date in Long *et al.* 1998), sulphide is mainly in the CCrS phase (0.26–0.5 wt% S), suggesting a dominant mineralogy of greigite and poorly crystalline pyrite. AVS and HCrS values are typically low (<0.07 wt% S) suggesting that: (a) oxidation of the most metastable sulphides has occurred; and (b) that pyrite has not formed a stable crystalline phase despite having had 2–3 ka to do so. A 4000-year-old peat sample (based on the date in Long *et al.* 1998) from the bottom of the same core has a high total reduced S (1.68 wt% S), mostly as CCrS and an additional 4.82 wt% organic S.

These interpretations are supported by data from modern ITMF, LSM and HSM samples

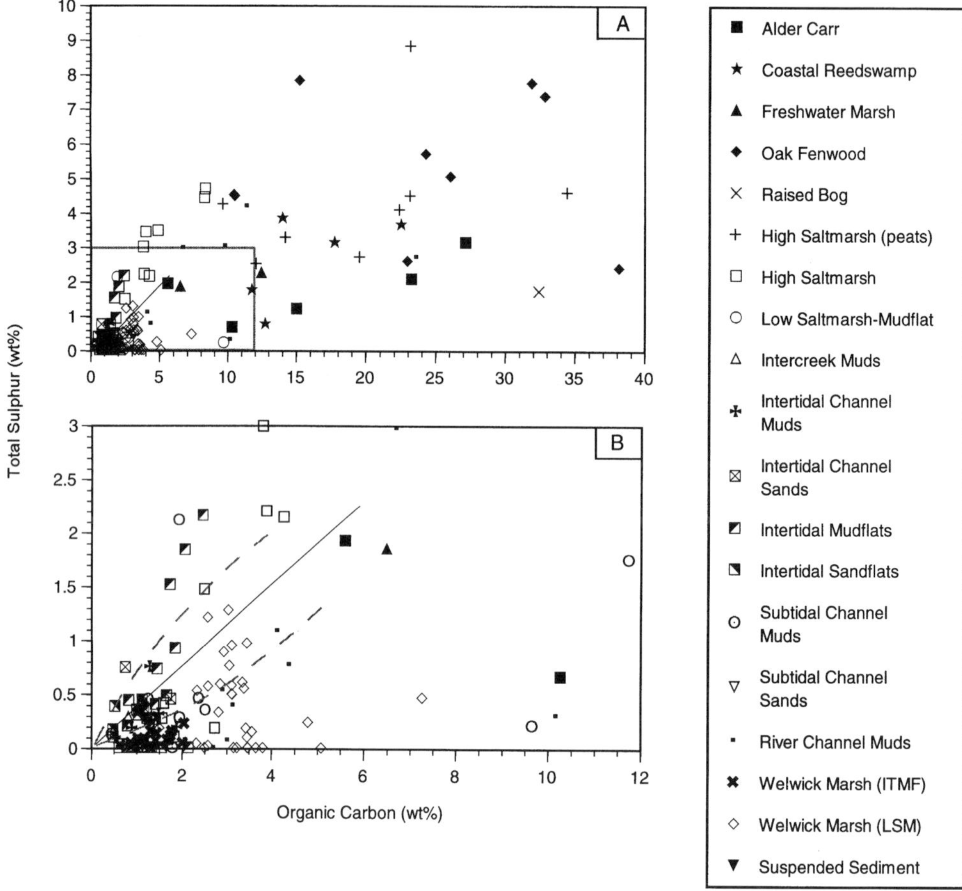

Fig. 5. Weight per cent organic carbon plotted against weight per cent total sulphur for Holocene and modern sediments. Plot A shows the full data range and plot B shows the inset data from A. The envelope (and mean line) of normal marine sediments from Berner & Raiswell (1984) is shown on B.

from Skeffling intertidal flats and Welwick saltmarsh (assumed to be representative of Holocene ITM and SM facies). Here, sulphur speciation showed that total reduced S is low (0.12–0.57 wt%), consistent with partial oxidation in saltmarsh facies, and mainly composed of AVS and CCrS. This implies that most of the reduced S is present as iron monosulphides, with greigite likely to be the main phase in CCrS. HCrS is a small component of the total reduced S suggesting that conversion to pyrite is either slow, or inhibited (especially in the HSM facies) by reoxidation.

The mean values for total S for each of these environments (Fig. 7) show clearly that Holocene saltmarsh and freshwater peaty facies account for quite high sulphur storage (typically 2–8 wt% total S). These environments are no longer present in the modern Humber Estuary, while the modern saltmarshes appear to be storing less S (typically <0.5 wt% total S). This latter effect may be because the mean grain size of the modern marshes is coarser than in the earlier Holocene (see also Rees *et al.* this volume, table 2), promoting higher rates of oxidation; however, we do not yet have enough data to answer this definitively.

Stable carbon isotopes in bulk organic matter

$\delta^{13}C_{org}$ values (Fig. 8) for mean bulk Holocene organic matter range from −25.0 (intertidal mudflats), −26 to −27 (saltmarsh facies) and

Table 3. *Mean geochemical values and data ranges for the seven common Holocene Humber depositional environments*

Depositional environment*	Parameter	Total C (wt%)	C_{org} (wt%)	$\delta^{13}C_{org}$ (‰VPDB)	Total N (wt%)	C_{org}/total N	Total S (wt%)
OHF	**Mean**	**24.4**	**23.5**	**−28.3**	**0.92**	**23.3**	**6.4**
	Min.	10.3	10.4	−29.1	0.46	13.3	2.4
	Max.	44.1	38.1	−27.4	1.72	37.2	14.9
AC	**Mean**	**15.4**	**14.5**	**−27.9**	**0.82**	**17.0**	**2.3**
	Min.	0.8	5.6	−29.0	0.02	12.9	0.0
	Max.	44.4	27.1	−26.3	2.12	19.3	7.8
RCm	**Mean**	**5.9**	**5.4**	**−27.1**	**0.32**	**29.8**	**1.6**
	Min.	0.6	0.6	−28.8	0.01	10.7	0.0
	Max.	13.1	23.6	−23.6	0.73	98.0	4.8
HSM(P)	**Mean**	**21.7**	**18.7**	**−27.6**	**1.02**	**20.2**	**4.2**
	Min.	9.6	9.6	−27.8	0.49	15.1	0.2
	Max.	37.7	34.4	−27.3	1.57	25.0	8.8
HSM	**Mean**	**3.9**	**1.7**	**−24.8**	**0.2**	**15.3**	**1.45**
	Min.	0.6	1.1	−26.9	0.0	12.5	0.0
	Max.	35.1	2.5	−23.4	1.6	22.4	5.2
HSM mean†	**Mean**	**8.8**	**10.7**	**−25.7**	**0.43**	**17.9**	**2.3**
	Min.	0.6	1.1	−27.8	0.0	12.5	0.0
	Max.	37.7	34.4	−23.4	1.58	25.0	8.8
LSM	**Mean**	**4.5**	**4.1**	**−26.2**	**0.25**	**17.0**	**2.3**
	Min.	1.6	1.5	−28.3	0.0	13.7	0.0
	Max.	14.7	8.3	−24.0	0.98	22.2	11.2
ITMF	**Mean**	**2.5**	**1.4**	**−25.2**	**0.10**	**16.5**	**0.8**
	Min.	1.2	0.2	−27.2	0.0	12.1	0.0
	Max.	3.7	2.4	−24.2	0.25	27.3	4.5
Sands (ITSF, ITCs, STCs)	**Mean**	**2.1**	**1.0**	**−24.6**	**0.08**	**19.0**	**0.5**
	Min.	0.2	0.5	−26.3	0.0	12.4	0.0
	Max.	4.7	1.7	−21.1	0.86	26.0	4.2

* See Table 2 for descriptions of depositional environments. Total CNS data based on between 19–77 analyses per depositional environment. Organic C data based on between 6–20 analyses per depositional environment. Stable isotope data based on between 7–14 analyses per depositional environment.
† The mean values for HSM combine the data for both HSM and HSM(P) as described in Table 2. It should be noted that these are mean values of a bimodal data set. This is acceptable for scaling exercises, as we have no means of assessing the actual distribution of these Holocene HSM sub-facies.

−27.5 to −29‰ (OHF, AC and FWM facies). The intertidal mudflat samples have the most marine (i.e. least negative) signature (compare for example with data in Chmura & Aharon 1995), whereas the peats have a clear terrestrial C_3 organic matter signature (Deines 1980). The saltmarsh sediments exhibit quite a wide range of values, representing a continuum from marine to terrestrial inputs, although dominated by the latter.

The range of $\delta^{13}C_{org}$ values in the Holocene sediments is comparable with data from modern saltmarsh, suspended sediments and intertidal samples from the Humber (Fig. 8). Modern intertidal surface muds have a $\delta^{13}C_{org}$ range from −19‰ at the mouth of the estuary to −26‰ at the riverine end of the estuary. This range suggests that marine organic matter in the estuary has a $\delta^{13}C_{org} \geq -19$‰ (a conclusion also reached by Fichez *et al.* 1993, working in the Wash, 60 km south of our study site), while terrestrial organic matter inputs are ≤ -26‰, i.e. C_3 organic matter.

Overall the modern ITM values are *c.* 2‰ heavier than those from comparable Holocene facies, and a similar offset is seen between modern (Welwick Marsh) and Holocene saltmarsh $\delta^{13}C_{org}$ values. This offset is probably due to some input of carbon in modern sediments from *Spartina anglica*, an isotopically heavy C_4 lower saltmarsh plant (mean $\delta^{13}C_{org}$ of −13.6‰; Jackson *et al.* 1986, and −13.2‰, Nithart 1995) that first arose in the UK as a natural hybrid from crossing alien and native species some 120 years ago (Armstrong 1988), and first artificially planted in the Humber Estuary in 1926 in the Skeffling area of Spurn Bight (Wilson 1938; Pethick 1987). It is also possible that the offset is

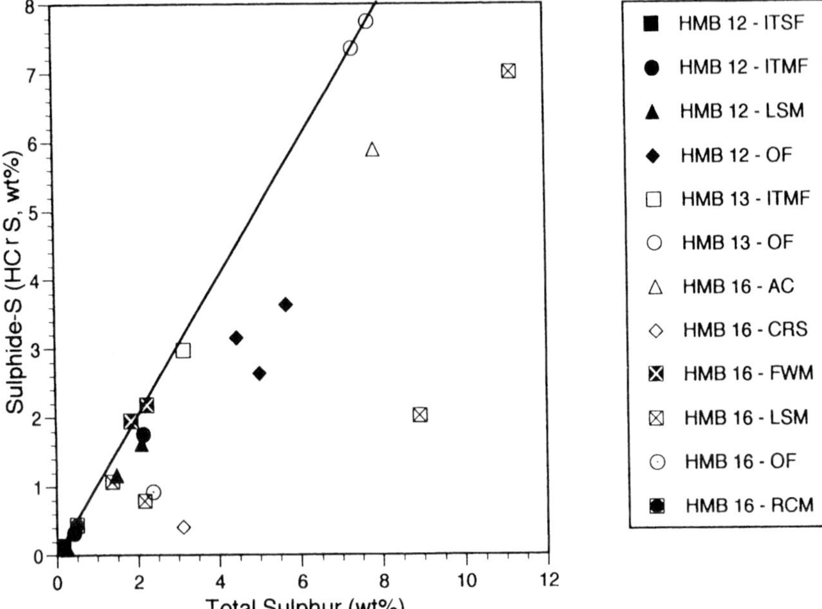

Fig. 6. Weight per cent total S by elemental analysis plotted against weight per cent total sulphide S by hot chromous chloride reduced sulphur (HCrS) for Holocene sediments. The plot shows that total S values by elemental analysis are reasonably representative of sulphide S except for some oak fenwood, alder carr, coastal reedswamp and lower saltmarsh samples that have a large component of non-sulphide, organically bound sulphur.

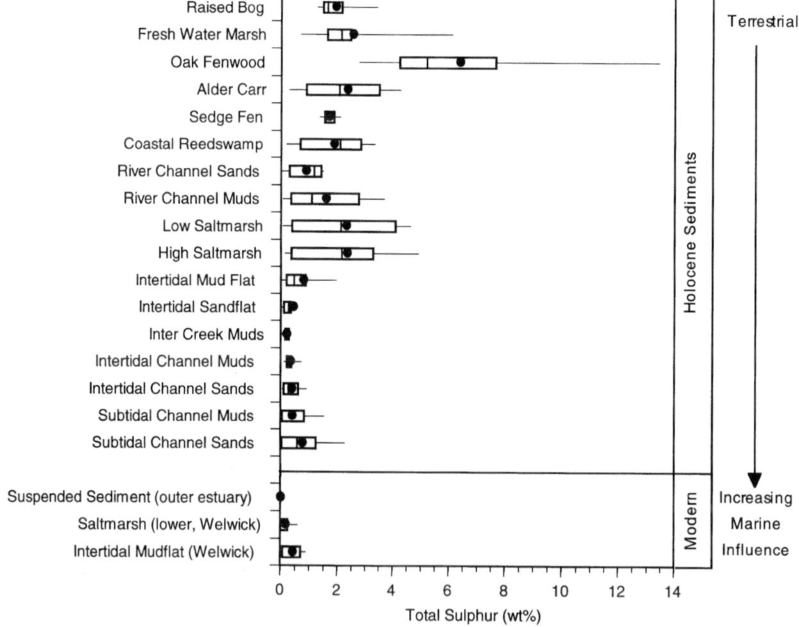

Fig. 7. Box and whisker plot of weight per cent total sulphur plotted by environmental facies for Holocene and modern sediments. In each plot the dot represents the mean value, the vertical bar represents the median value, the ends of the horizontal box represent the 25th and 75th percentiles and the ends of the horizontal line represent the tenth and 90th percentiles. The mean values for the seven common environmental facies are given in Table 3.

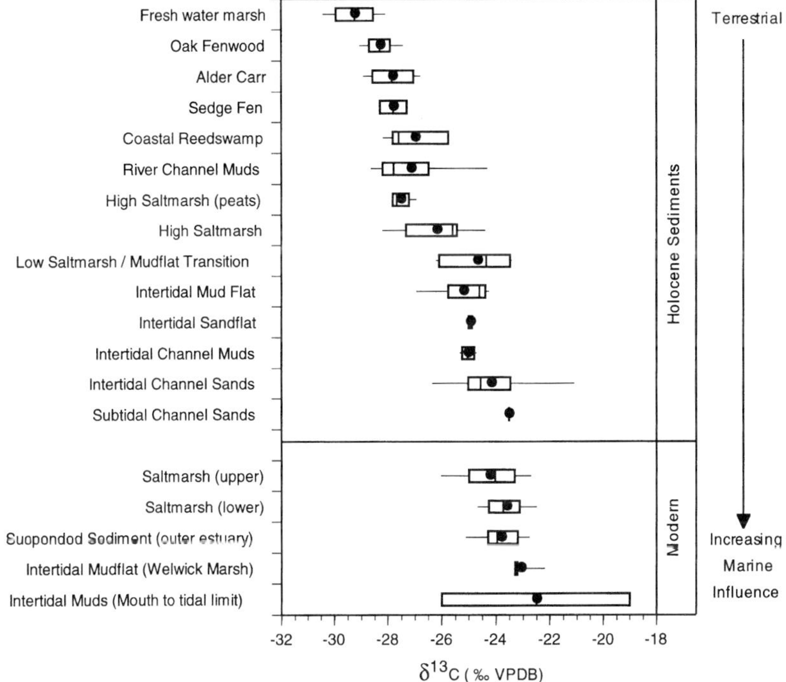

Fig. 8. Box and whisker plot of $\delta^{13}C_{org}$ values plotted by environmental facies for Holocene and modern sediments. In each plot the dot represents the mean value, the vertical bar represents the median value, the ends of the horizontal box represent the 25th and 75th percentiles and the ends of the horizontal line represent the tenth and 90th percentiles. The mean values for the seven common environmental facies are given in Table 3.

in part caused by: (a) increased input of isotopically heavy marine organic matter and/or benthic cyanobacterial material in the modern estuary, however, this material tends to be labile and oxidized before incorporation into sediment (Chmura & Aharon 1995); (b) decreased input of isotopically light C_3 terrestrial organic matter from oak fenwood and alder carr environments over the last 300 years due to reclamation of these environments for agricultural and urban land.

It is unclear whether some of the isotopic offset could be due to the $\delta^{13}C_{org}$ values of the older Holocene sediments being altered by diagenetic processes. Some authors suggest that during microbially mediated respiration of organic matter $\delta^{13}C$ values are either not much affected, or increase slightly as the nitrogenous, isotopically light organic matter is consumed (Hayes 1983; Macko & Estep 1984). Others suggest that the refractory residual lignin is isotopically lighter than the bulk organic matter (Benner *et al.* 1987). Our observed shift, like that discussed by Chumra & Aharon (1995), is towards isotopically lighter values in the older sediments but there is no related offset in C/N ratios between modern and Holocene sediments from similar facies (Fig. 3).

It is clear from the above that each Holocene lithofacies has distinctive carbon, nitrogen and sulphur geochemistry. These data, particularly the combined $\delta^{13}C_{org}$ and C/N ratios allow reasonably confident characterization of organic matter types based on data means and ranges (summarized in Table 3); abundance values of C N and S for different palaeoenvironments can therefore be readily calculated (see below).

Establishing a chronology and sedimentation rates

Establishing a chronology in these types of Holocene sequences is problematic. Radiocarbon dates on peats are generally reliable, but most of these occur at the bases of sequences (Fig. 9) and therefore, even where well-dated, do not aid chronology in the upper parts of cores. Mid-section peats are uncommon, but present in some cores (e.g. HMB12, HMB13, HMB16 and HMB19). Although a number of bulk

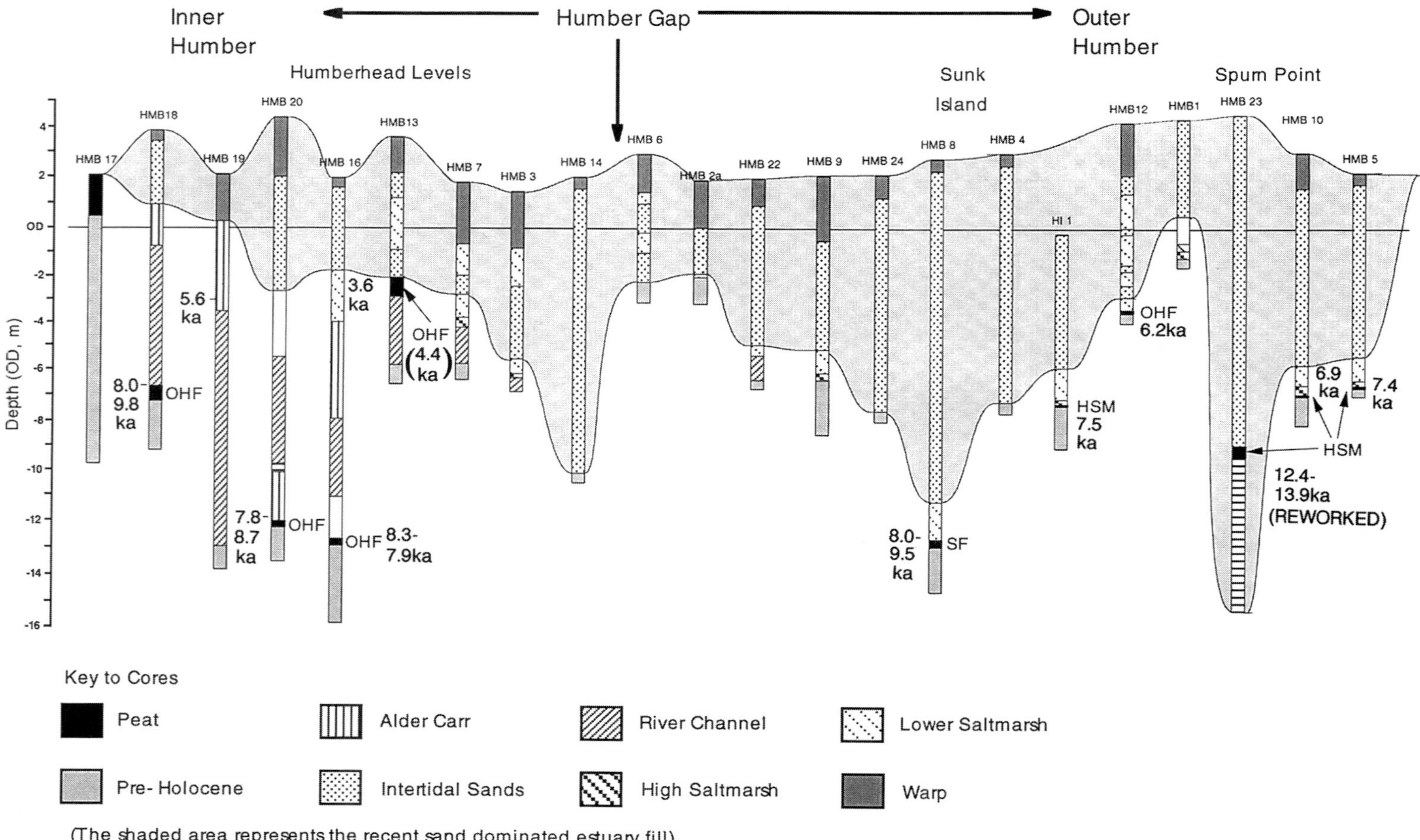

Fig. 9. Summary diagram of general facies variation in HMB cores relative to Ordnance datum (OD). The locations of the boreholes are shown in Fig. 1. Note that in most cores Holocene sedimentation starts with a basal peat (dates are given as 4.4 ka, etc.). In the outer Humber these peats formed in saltmarsh environments and pass upward into lower saltmarsh sediments. In the inner Humber, riverine sediments are common at the bases of cores, giving way to carr environments and eventually more marine sediments. Detailed chronologies for HMB10, HMB12 and HMB16 are given in Ridgway *et al.* (this volume).

organic matter samples were taken from various levels in some cores, these generally gave what appear to be spurious ages, too old for their depth, indicating that old allochthonous organic matter has been reworked in the estuary during the Holocene.

The more reliable radiocarbon dates on *in situ* peats etc. suggest that in the outer Humber area, sedge fen peat was forming between 9491 and 7923 cal. a BP in core HMB8 (Fig. 9), while high saltmarsh peat formation was initiated in the Spurn Bight area between 6887 and 7387 cal. a BP (HMB5 and HMB10). In most outer Humber cores (e.g. HMB1 HMB5, HMB8 and HMB10) these peats are followed by an upward change to muddy saltmarsh deposits, typically after *c.* 7 cal. ka BP, eventually followed by an up core change (undated) to more open-marine mudflat and eventually mud-sandflat environments with time (Fig. 9). This transgressive facies change is reflected in a rapid up core change in $\delta^{13}C_{org}$ from terrestrial (−28 to −30‰; Fig. 10) to more marine values (−25‰ Fig. 10). In HMB10, this change occurred after 7170 cal. a BP, and after 6140 cal. a BP in HMB12.

Cores from the inner Humber area have a varied lithofacies transition depending on their proximity to riverine or more axial estuarine position. Basal oak fen peats are typically older than 8 cal. ka BP, while mid-section oak fen peats in HMB19 and HMB13 began forming at about 5.9 cal. ka BP and were locally present until at least 4376 cal. a BP (HMB13). In general, fully marine lithofacies did not occur until after *c.* 3.5–4 cal. ka BP in this area, although biofacies data show marine influence in alder carr environments before this time. This gradual increase in marine influence is clear in the up core $\delta^{13}C_{org}$ profile of HMB16 near Trent Falls (Fig. 10), where the values become least negative (most marine) above −2.00 m OD (between 2 and 3 cal. ka BP based on radiocarbon and palaeomagnetic techniques, see below). Core HMB18 is located upriver of HMB16 in the Trent Valley (Fig. 1) and is dominated by river channel muds and sands and alder carr organic sediments. The up-core $\delta^{13}C_{org}$ profile (Fig. 10) shows a progressive change towards more marine values, reflecting a slowly increasing marine influence, though never resulting in open saltmarsh or mudflat facies.

The lack of reliable radiocarbon chronology in the upper parts of cores prompted us to explore the possibility of using palaeomagnetic secular

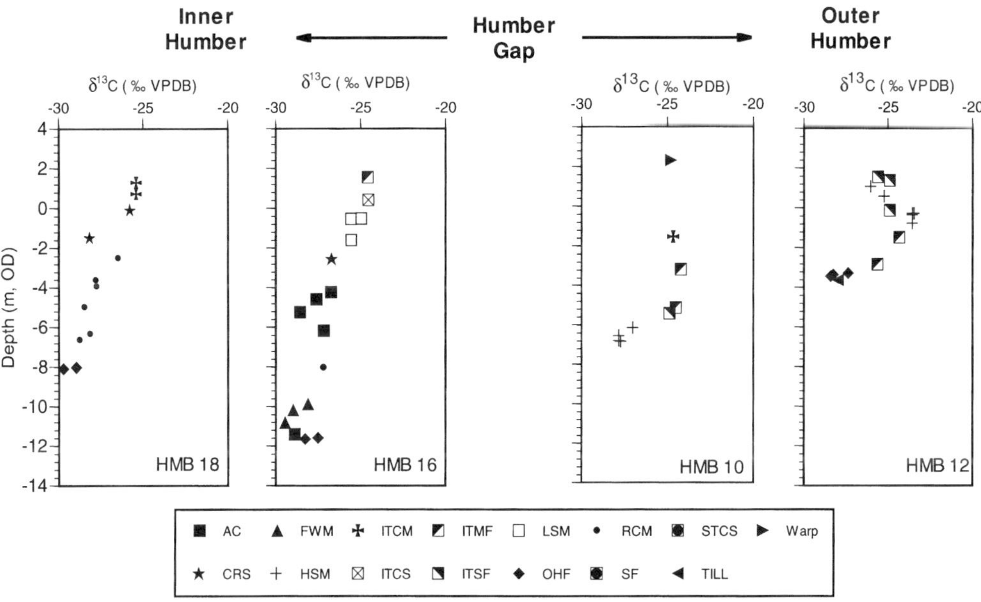

Fig. 10. Variation in $\delta^{13}C_{org}$ values with depth in four HMB cores (located in Fig. 1). In all cases, basal values are most negative (down to −30‰), giving way to less negative values up core (typically around −25‰). In the outer Humber cores this change occurs <1 m above the basal peat, reflecting early transgression in this area. In the inner Humber area the change in values is more gradual, reflecting gradual marine flooding of carr and reedswamp environments.

variation (PSV) records (Turner & Thompson 1981, 1982) to constrain the chronology (see method details in Ridgway *et al.* (this volume). Where possible we tried to verify this record with radiocarbon dates from carefully selected *in situ* rootlets at sedimentological surfaces, autochthonous organic matter in lake and freshwater marsh facies and freshwater gastropod shell carbonate. The palaeomagnetic data are probably most informative in indicating discrete sedimentary packages and hiatuses (Ridgway *et al.* 1999); however, preliminary attempts to use the data for absolute dating are also discussed in (Ridgway *et al.* (this volume).

In the outer Humber area, PSV work was done on duplicates of cores HMB12 and HMB14. In HMB12 the PSV record is reliable between −2.83 and −6.00 m OD and the data, while not diagnostic, can be interpreted as consistent with radiocarbon dates in the same depth interval (details in Ridgway *et al.* (this volume). Similarly, the PSV record in duplicate HMB14 can be interpreted as consistent with the suggestion that the distinctive lead content of these sediments, part of the Sunk Island Suite of Rees *et al.* (this volume), implies a Roman or post-Roman age (i.e. <2 ka old).

In the inner Humber area, PSV work was done on duplicates of cores HMB16 and HMB19. A full discussion of the PSV data in HMB16 from the lower Trent Valley is given in Ridgway *et al.* (this volume). Overall the interpretation of results in the upper 5 m of the core are consistent with associated radiocarbon dates and suggest a number of significant time breaks in the sequence.

HMB19 is located upriver of Goole in the Ouse Valley (Fig. 1), where the upper 4.5 m of core show marine influence in coastal reedswamp facies. A palaeomagnetic record was obtained below 0.00 m OD, above which level the core is of oxidized reclaimed sediment. Between 0.00 and −1.66 m OD the inclination record, when compared to Turner & Thompson (1981, 1982), suggests ages between 2–4.3 cal. ka BP. A date of 2876 cal. a BP on possible *in situ* woody rooted material at −1.87 m OD and a maximum age of 3631 cal. a BP on allochthonous organic silt at −0.76 m OD gives tentative support to the suggested PSV age. Between −1.66 and −3.27 m OD the magnetic intensity was too low to measure reliably due to the organic-rich composition of the sediment. However, the peat at −2.84 to −3.16 m OD has top and base ages of 5527 and 5821 cal. a BP, respectively. Below this peat the PSV data are ambiguous and are not interpreted further.

Long-term sedimentation rates

In the few core sections where the chronology appears reliable it is possible to calculate notional minimum rates of sedimentation for various Holocene lithofacies aged between 8 and 2 cal. ka BP (Table 4). It is not possible to be very confident about the validity of these rates because of uncertainties surrounding the degree of compaction in muddy, silty and organic rich sediments, the possibility of time breaks, and the accuracy of some dates. However, the results suggest that minimum long-term sedimentation

Table 4. *Minimum Holocene long-term sedimentation rates based on best available chronologies in apparently continuous sedimentation intervals*

Core	Depositional environment*	Depth (mOD)/age (cal. a BP) base of interval	Depth (m OD)/age (cal. a BP) top of interval	Interval thickness (mm)	Time (a)	Sedimentation rate (mm a^{-1})
HMB19	CRS	−2.84/5527	−0.76/631†	2080	1891	1.10
HMB16	FWM	−11.44/8034	−9.91/7391	1530	643	2.38
HMB16	LSM	−2.56/3664	−0.56/2138	2000	1526	1.31
HMB12	LSM	−1.16/4479	−0.29/3261	870	1218	0.71
					Mean	**1.37**
				6000–2000 cal. a data only	**Mean**	**1.04**
HMB14‡	ITCm	−6.00/*c.* 2000	0.00/modern	8000	*c.* 2000	*c.* 4.0

* See Table 2 for descriptions of depositional environments.
† Date on organic silt probably too old (contaminated by allochthonous carbon).
‡ Based on SU geochemical suite of Rees *et al.* (1999) being post-Roman age, thus sedimentation rate is very approximate.

rates in freshwater marsh, coastal reedswamp and lower saltmarsh facies varied between 0.7 and 2.4 mm a^{-1} (mean = 1.4 mm a^{-1}). The three rates in the 6–2 cal. ka BP range have a mean long-term sedimentation rate of 1.04 mm a^{-1}, comparable to the proposed rate of sea-level rise of 1.2 mm a^{-1}, based on Humber sea-level index points between 6 and 2 cal. ka BP (Shennan *et al.* this volume, Fig. 7), suggesting that sedimentation was roughly in equilibrium with the accommodation space created by sea-level rise in the palaeo-Humber Estuary.

These long-term sedimentation rates are an order of magnitude smaller than linear accretion rates based on ^{137}Cs chronology (G. Samways & G. Shimmield pers. comm. 1999) in a modern outer Humber saltmarsh. At Welwick, intertidal mudflats are accreting at *c.* 11 mm a^{-1}, the marsh front is accreting at *c.* 17–18 mm a^{-1}, while the marsh top is accreting at 13–15 mm a^{-1}. It is possible that the modern Humber accretion rates are significantly higher (i.e. affected by anthropogenic activities) than those in the palaeo-Humber, although compaction in the older sediments will certainly cause some of the apparent difference.

Storage estimates

The volume of sediment deposited in a year and its weight per cent of C_{org}, N or S can be computed knowing:

(a) the depositional area of the modern or palaeo-Humber;
(b) the mean C_{org}, total N and total S weight per cent values for these depositional environments (Table 3);
(c) that about 1 mm of sediment was deposited in a year (consistent with the long-term sea-level record and sedimentation records as discussed above) to fill available accommodation space.

In addition, by assuming a dry bulk density appropriate for the deposited sediments (2000 kg m^3, except peats where 100 kg m^3 was used), the number of tonnes of sediment deposited and its percentage weight of C_{org}, total N and total S can be calculated.

Area of the modern Humber Estuary

The area of the modern estuary can be calculated from Admiralty charts while the types of sediments have been mapped by the British Geological Survey (1:250 000 Sea Bed Sediments; Sheet 53N 00 Spurn, 1990). The total area of the Humber Estuary from Trent Falls to the Humber mouth (here defined as a line between Spurn Head and Cleethorpes [NGR TA 320075]) is 265 km^2, of which 111 km^2 is intertidal area and 154 km^2 is subtidal area (Table 5). To estimate the maximum possible depositional area we assumed that sediment is deposited and stored on this whole area.

To provide a more realistic estimate one might argue that only the muddy (i.e. unwinnowed) areas are a long-term sediment store. From the Admiralty charts and sediment map it is possible to estimate the areas of:

(a) subtidal mud banks, which are at least temporary sediment stores (*c.* 74 km^2; Table 5);
(b) channel areas, which are probably non-depositional (*c.* 80 km^2);
(c) mud covered intertidal areas, which are probably accretionary (*c.* 76 km^2; Table 5).

Estuary wide, the intertidal areas are 68% mud (<63 μm sediment) covered and 32% sand covered, while the subtidal area is 48% mud covered and 52% sand covered.

Finally, to give a minimum estimate of sediment deposition one might argue that only Spurn Bight has available accommodation space to allow long term sediment accumulation today. Elsewhere in the estuary erosional

Table 5. *Area calculations* for the modern Humber Estuary*

Σ area	Σ subtidal area	Σ intertidal area	Σ mud covered† sub-tidal area	Σ mud covered† intertidal area	Mud covered† intertidal area (Spurn Bight)
265	154	111	74	76	24

* All units are in kilometres square and were measured from Admiralty Chart and British Geological Survey (1:250 000 Sea Bed Sediments; Sheet 53N 00 Spurn, 1990). Intertidal and subtidal areas were measured between Trent Falls [SE 8623] and at the Humber Mouth a line between Spurn Head [TA 3910] and Cleethorpes [TA 320075].

† Where mud (i.e. <63 μm material) covered is assumed to represent a net depositional area.

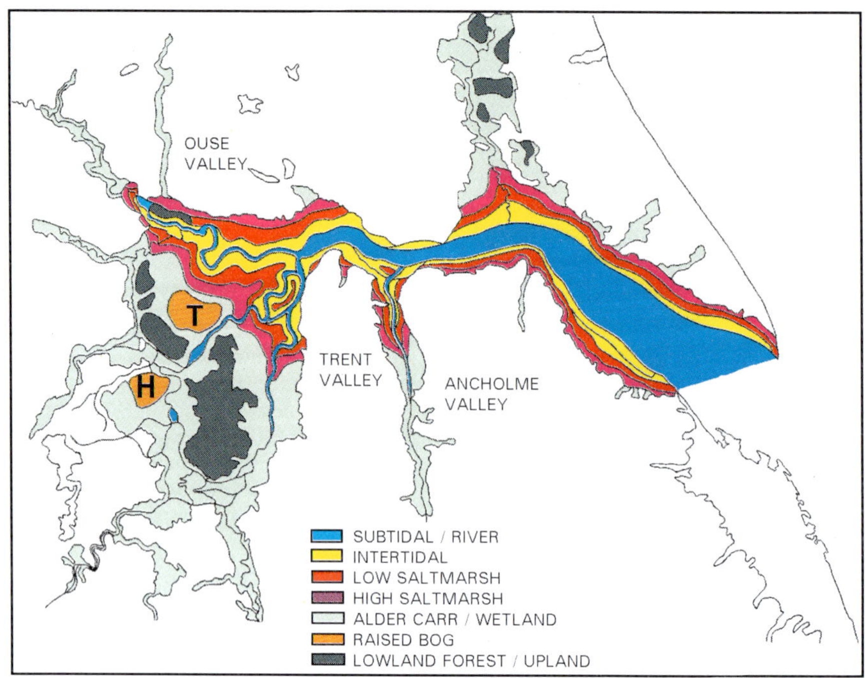

Fig. 11. Palaeogeography of the Humber area (3–2 cal. ka BP). Areas of the various environmental facies in Tables 6 and 7 were calculated from this reconstruction. Drainage in the Trent and Ouse area is based on data in Dinnin (1997*b*, fig. 4.1). T, Thorne Moor; H, Hatfield Moor.

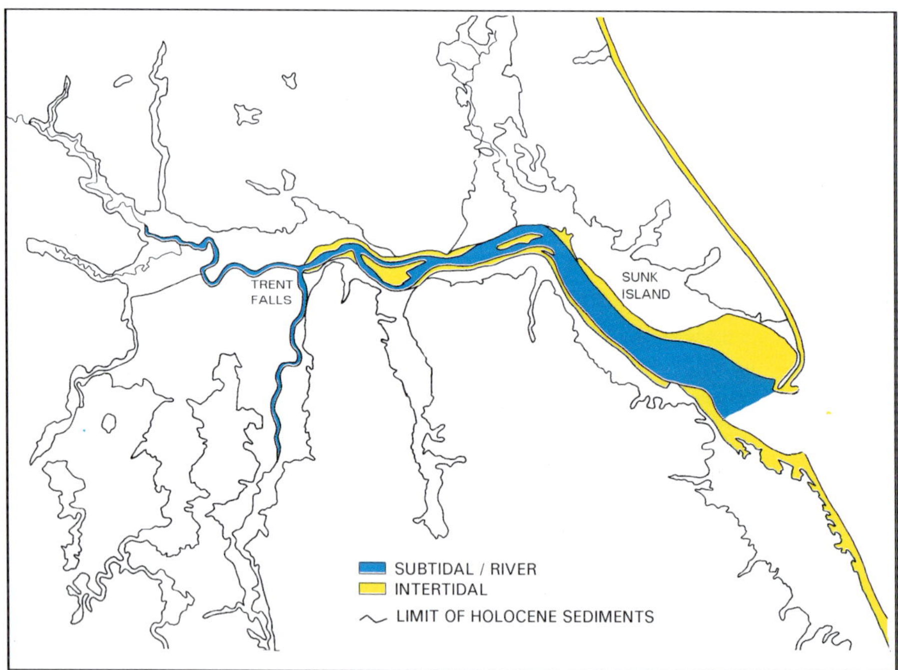

Fig. 12. Geography of the modern Humber. Although both lower and mature saltmarsh are present in parts of the landward fringes of the intertidal zone they are too small to be detailed at this scale. The black line marks the landward boundary of Holocene deposits, now reclaimed as agricultural or urban land. Note that this reclamation has totally removed the freshwater sedge fen, alder carr and oak fenwood environments, and partially removed saltmarsh and intertidal sediments (compare with Fig. 11).

features are common on intertidal and saltmarsh areas (see discussion in Pethick 1990), suggesting that over decadal time-scales these areas are not sediment stores. The mud covered area of Spurn Bight today is *c.* 24 km^2 (Table 5).

Area of the palaeo-Humber Estuary

The depositional area of the palaeo-Humber can be estimated by integrating all of the available geological evidence. For example, the palaeogeographic reconstruction for 3 to 2 cal. ka BP (Fig. 11; see also Metcalfe *et al.* this volume) has been drafted to satisfy the geological evidence available from borehole and published information, principally the LOEPS boreholes and information in Smith (1958); Gaunt & Tooley (1974); Gaunt *et al.* (1992); Gaunt 1994; Berridge & Pattison (1994); Van de Noort & Ellis (1995; 1997; 1998) and Long *et al.* (1998). It is important to realize that much of this palaeogeography is conjectural, based on the height relative to OD that modern environments form in the tidal frame and extrapolations of these, and the known succession of environments, to the ancient land surface. In many areas these extrapolations are not testable due to subsequent urbanization and reclamation. None the less, this palaeogeography is a reasonable representation of the expected areal extent of environmental facies in the study area before major influence by human activity; excepting deforestation and its potential effects on riverine sediment yield (Long *et al.* 1998). There is not enough geological information to identify environmental changes between 3 and 2 cal. ka BP, so this palaeogeography represents both the environmental distribution at the time of maximum extent of marine inundation (see discussion in Long *et al.* 1998; Metcalfe *et al.* 1999) and the situation before reclamation and land-use changes.

At this time the estuary was flanked by wide lower saltmarshes and mudflats, (Fig. 11), and fringed on the landward side by up to 1 km of upper saltmarsh. Tidal flats probably existed in the Ouse and Aire valleys around Goole (Gaunt 1994; Dinnin 1997*a*) and into the lower parts of the Trent and Ancholme Valleys (Lillie & Neuman 1998). Landward of the saltmarshes, in the low-lying river valleys, were extensive tracts of freshwater swamp and alder carr (Fig. 11), forming where drainage was impeded by the relatively high sea-level. In the Humber Head lowlands, tracts of raised bog had been forming from *c.* 5 cal. ka BP onwards (Dinnin 1997*a*; Dinnin *et al.* 1997), building the peat lowlands of Hatfield and Thorne Moors.

Changes in storage potential

Comparison of the estimated sedimentation area and potential storage of C_{org}, and total S in the modern and palaeo-estuary demonstrate, firstly, how the storage potential has changed over time, and secondly, the implications this has for managing the estuary in the future. Storage of N and other nutrients will be discussed in detail elsewhere and are not considered further here.

Organic carbon

Alder carr environments were the most import sinks of C_{org} 3–2 cal. ka BP (Fig. 11), accounting for about 276 000 tonne C_{org} a^{-1} (Table 6); while saltmarshes were also important, sedimenting some 34 000 tonne C_{org} a^{-1} (Table 6). The net yearly potential storage of C_{org} in the Humber (including lowland raised bog peats, which accounted for about 656 tonne C_{org} a^{-1}) was thus about 316 000 tonne (Table 6).

This representation of the estuary 3–2 ka ago can be compared to the present-day environmental facies distribution in the same area (Fig. 12), to demonstrate the net effect of human activity on the storage potential of the estuary. The modern estuary is constrained by sea-walls in the commercial docklands of Kingston-upon-Hull and much of the south bank from the Humber Bridge to Grimsby (Fig. 1). Elsewhere, notably in the Sunk Island area [TA 2619] (Fig. 12), extensive reclamation has replaced former saltmarsh, mudflat and in places channel environments, with agricultural land. Similarly, in the area around Trent Falls [SE 8623] (Fig. 12) the extensive mudflats and saltmarshes of 3–2 cal. ka BP have been reclaimed for agricultural land. The estuary is now flanked by mudflats and thin strips of saltmarsh; many of which are subject to decadal-scale erosion and redeposition cycles related to the position of the main channels in the estuary. The lowland freshwater swamps and alder carr environments have been completely removed from the modern Humber system by reclamation for agriculture (e.g. see Gaunt 1987; Lillie & Weir 1997; 1998), while the lowland raised bog peats (Hatfield and Thorne Moors) have been partially drained and stripped of peat for horticultural and domestic use (Dinnin 1997*a*; Dinnin *et al.* 1997). Data on peat removal in Eversham (1991) suggests that about 2×10^6 tonne of C_{org} were mined from Thorne Moor alone between 1750 and the late 1980s.

Table 6. *Comparison of area, C_{org} content, yearly sediment and C_{org} deposition in the palaeo (3–2 cal. ka BP) Humber and modern Humber Estuaries*

Environment	Area (km^2)	Average C_{org} content (wt%)	Sed deposited in one year (tonne)	C_{org} deposited in one year
3–2 cal. ka BP estuary				
Raised bog*	21	30	2 187	656
AC	918	15	1 837 080	275 562
HSM	117	10	233 280	23 328
LSM	129	4	262 440	10 498
Intertidal flats	305	1	612 360	6 124
	1490		**2 947 347**	**316 167**
Modern estuary				
Raised Bog	neg	30	0	0
AC	neg	15	0	0
HSM	neg	10	0	0
LSM	5	4	10 000	400
Intertidal flats	106	1	212 000	2 120
	111		**222 000**	**2 520**

* The value for raised bog is poorly constrained, based on an assumed bulk density of 100 kg m^3, yielding a yearly tonnage consistent with estimates of peat removal from Thorne Moor (Eversham 1991)

Reclamation and conversion of former mudflat or saltmarsh to agricultural land has a severe effect on potential C_{org} storage. The first step of reclamation amounts to instantaneous filling (and overfilling) of accommodation space, and the weight percentage of C_{org}, total N and total S reflects the source of the warped material (usually suspended sediment in the Humber water (de Boer 1988)). This stage results in a net short term increase in storage. However, effectively permanent removal of accommodation space means that the year-by-year millimetre-scale filling of accommodation created by rising sea-level (i.e. by the sedimentation of relatively organic-rich sediments in saltmarsh environments) is lost, and the slow build up of organic-rich sediment does not occur. For example, reclamation of 320 km^2 of high saltmarsh 300 years ago, by warping to a height of 1 m with intertidal muddy sediment causes instantaneous storage of *c.* 12.5×10^6 tonne (assuming ITM warp has 1.5 wt% C_{org}). The equivalent natural HSM sedimentation would have achieved this storage in 150 years, showing that the net effect of reclamation is a 50% decrease in storage over 300 years. If the reclaimed land is used for agriculture, ploughing efficiently oxidizes organic matter in the upper 0.3 m of the land, leading to release of C_{org} to the atmosphere as CO_2. Artificial input of C_{org} as manure to agricultural land is thus not stored, but recycled in cropping and further lost by oxidation, such that agricultural use of reclaimed land further decreases its former natural storage potential. Creation of made- ground for urban or commercial purposes caused instant cessation of storage as space was typically filled with inert industrial waste and in places, local drift deposits (Berridge & Pattison 1994).

The net effect of human-induced changes to the Humber estuary has been a decrease in depositional area from about 1490 km^2 *c.* 3–2 ka ago (Table 6), to somewhere between 265 (max.) to 24 (min) km^2 (Table 5), resulting in a drastic decrease in yearly sediment deposition from about 2.9×10^6 to about 2.2×10^5 tonne (assuming an overestimate of depositional area to include the whole modern intertidal area of 111 km^2, Table 6). Even this conservative comparison results in a net decrease in C_{org} deposition from about 3.2×10^5 tonne in the palaeo-estuary to no more than 2.5×10^3 tonnes today, a $>99\%$ reduction in potential C_{org} storage capacity.

The total effect of human activity on the Humber Estuary through reclamation over the last 300 years, assuming a linear decrease in depositional area over that time, has been a loss from potential storage of *c.* 0.5×10^9 tonne of C_{org}. We also note that the modern estuary waters contain about 3.5×10^6 tonne of suspended sediment (Pethick 1990), much of it resuspended from the bed and mudflats on tidal cycles because there is very little accommodation for longer-term storage. This tonnage is similar to the yearly sedimentation potential of the 3–2 ka old estuary (Table 6), suggesting that the 3–2 ka old estuary, with significantly greater accommodation space, was much less turbid than today.

Table 7. *Comparison of area, total S content, yearly sediment and total S deposition in the palaeo (3–2 cal. a BP) and modern Humber Estuaries*

Environment	Area (km^2)	Average S content (wt%)	Sed deposited in one year (tonne)	Total S deposited in one year
3–2 cal. ka BP estuary				
Raised bog	21	4.0	2187	87
AC	918	2.3	1837080	42253
HSM	117	2.3	233280	5365
LSM	129	2.3	262440	6036
Intertidal flats	305	0.5	612360	3062
	1490		**2947347**	**56803**
Modern estuary				
Raised bog	neg	4.0	0	0
AC	neg	2.3	0	0
HSM	neg	2.3	0	0
LSM	5	2.3	10000	230
Intertidal flats	106	0.5	212000	1060
	111		**222000**	**1290**

Sulphur
The complete reclamation of lowland freshwater swamp and alder carr environments from the modern Humber system has had a severe effect on S-storage (Table 7). Some 3–2 ka ago these environments sequestered some 42 000 tonne of S per year (much of it as organic S), while saltmarsh environments accounted for a further 11 000 tonne of yearly S deposition. Today, almost all Humber S deposition is occurring in the much reduced intertidal zone as metastable sulphide, and the total yearly S deposition is approximately 2% of its value 2 ka ago (Table 7).

Implications for future management of the Humber Estuary

The discussions above show not only the very large effect that past human management of the Humber Estuary has had on the C_{org} and total S deposition budgets, but also set the context for the potential effects of future management.

In attempts to combat tidal flooding of lowland coastal areas it is now common in shoreline management plans (SMPs) to see proposals for 'managed retreat' of sea defences. This typically involves letting existing defences fail to allow the coastline to recede to a new line of defence, while at the same time encouraging rebuilding of natural saltmarshes, often on areas of formerly reclaimed saltmarsh (see e.g. Wakeford *et al.* 1992; Clayton & O'Riordan 1995). While new saltmarsh creation by managed retreat is designed principally to dissipate the energy of the flooding sea, the recreation of marsh on formerly reclaimed land will also have an impact on the deposition and sequestering of C_{org}, N and S. In effect, 'managed retreat' should reverse some of the effects of reclamation.

The data in Tables 6 and 7 can be used to predict the impact of 'managed retreat' on C_{org} and S budgets (the effects on nutrients, including N will be discussed elsewhere); two points stand out. Firstly, managed retreat will principally recreate new mud and sandflats and lower saltmarsh environments; mature upper saltmarsh may form given time and space. Creation of large areas of mud and sandflat has only a moderate effect in increasing C_{org} and S storage, as these environments sediment the lowest amounts of C_{org} and S. More efficient C_{org} and S storage will occur if saltmarsh, and particularly mature saltmarsh forms. The second point is that even with extensive managed retreat, the estuary will never approach its former capacity to sediment C_{org} and S because it is most unlikely that alder carr and lowland freshwater swamp will be re-created to any great extent. It is these environments that had the greatest capacity for both C_{org} and S (especially organic S) storage and it is these environments that have been permanently removed, not just in the Humber, but in temperate estuarine settings globally.

J. Rees (British Geological Survey) did the initial core logging and a number of other BGS staff did the levelling. P. King and T. Hardman (UEA) assisted with CHNS elemental analyses and grain size analysis. Technical support from L. Rix, R. Lachowycz, G. Lee,

S. Bennett, I. Marshall, R. Bryant, A. Etchells, S. Davies and P. Judge (all UEA) is very much appreciated. This research was supported by NERC LOIS Special Topic allocation GST/02/736 and the paper is LOIS publication number 584.

References

AL-BAKRI, D. 1986. Provenance of the sediments in the Humber Estuary and the adjacent coasts, Eastern England. *Marine Geology*, **72**, 171–186.

ANDREWS, J. E., GREENAWAY, A. M. & DENNIS, P. F. 1998. Combined carbon isotope and C/N ratios as indicators of source and fate of organic matter in a poorly flushed, tropical estuary: Hunts Bay, Kingston Harbour, Jamaica. *Estuarine, Coastal and Shelf Science*, **46**, 743–756.

ARMSTRONG, W. 1988. Life in the Humber (A): Salt-marshes. *In*: JONES, N. V. (ed.) *A Dynamic Estuary: Man, Nature and the Humber*. Hull University Press, Hull, 46–57.

BENNER, R., FOGEL, M. L., SPRAGUE, E. K. & HUDSON, R. E. 1987. Depletion of ^{13}C in lignin and its implications for stable carbon isotope studies. *Nature*, **329**, 708–710.

BERNER, R. A. 1982. Burial of organic carbon and pyrite sulphur in the modern ocean: its geochemical significance. *American Journal of Science*, **282**, 451–473.

——1984. Sedimentary pyrite formation: an update. *Geochimica et Cosmochimica Acta*, **48**, 605–615.

——1989. Biogeochemical cycles of carbon and sulphur and their effect on atmospheric oxygen over Phanerozoic time. *Palaeogeography, Palaeoclimatology, Palaeoecology*, **75**, 97–122.

—— & RAISWELL, R. 1984. C/S method for distinguishing freshwater from marine sedimentary rocks. *Geology*, **12**, 365–368.

BERRIDGE, N. J. & PATTISON, J. 1994. *Geology of the Country Around Grimsby and Patrington*. Memoirs for 1:50 000 geological sheets 90 and 91 and 81 and 82 (England and Wales), HMSO, London.

BORDOVSKY, O. K. 1965. Sources of organic matter in marine basins. *Marine Geology*, **3**, 5–31.

BOTTRELL, S. H., HANNAM, J. A., ANDREWS, J. E. & MAHER, B. A. 1998. Diagenesis and remobilization of carbon and sulphur in mid-Pleistocene organic-rich freshwater sediment. *Journal of Sedimentary Research*, **68**, 37–42.

——, LOUIE, P. K. K., TIMPE, R. C. & HAWTHORNE, S. B. 1994. The use of stable sulfur isotope ratio analysis to assess selectivity of chemical analyses and extractions of forms of sulfur in coal. *Fuel*, **73**, 1578–1582.

CANFIELD, D. E., RAISWELL, R., WESTRICH, J. T., REAVES, C. M. & BERNER, R. A. 1986. The use of chromium reduction in the analysis of reduced inorganic sulphur in sediments and shales. *Chemical Geology*, **54**, 149–155.

CHMURA, G. L. & AHARON, P. 1995. Stable carbon isotope signatures of sedimentary carbon in coastal wetlands as indicators of salinity regime. *Journal of Coastal Research*, **11**, 124–135.

CLAYTON, K. & O'RIORDAN, T. 1995. Coastal processes and management. *In*: O'RIORDAN, T. (ed.) *Environmental Science for Environmental Management*. Longman, Harlow, 151–164.

COFFIN, R. B., FRY, B., PETERSON, B. J. & WRIGHT, R. T. 1989. Carbon isotope composition of estuarine bacteria. *Limnology and Oceanography*, **34**, 1305–1310.

DE BOER, G. 1988. History of the Humber coastline. *In*: JONES, N. V. (ed.) *A Dynamic Estuary: Man, Nature and the Humber*. Hull University Press, Hull, 16–30.

DEINES, P. 1980. The isotopic composition of reduced organic carbon. *In*: FRITZ, P. & FONTES, J. C. (eds) *Handbook of Environmental Isotope Geochemistry*. **1**, Elsevier, Amsterdam, 329–406.

DINNIN, M. 1997*a*. Introduction to the palaeoenvironmental survey. *In*: VAN DE NOORT, R. & ELLIS, S. (eds) *Wetland Heritage of the Humberhead Levels an Archaeological Survey*. The Humber Wetlands Project, University of Hull, Hull, 31–45.

——1997*b*. The drainage history of the Humberhead Levels. *In*: VAN DE NOORT, R. & ELLIS, S. (eds) *Wetland Heritage of the Humberhead Levels an Archaeological Survey*. The Humber Wetlands Project, University of Hull, Hull, 19–30.

——, ELLIS, S. & WEIR, D. 1997. The palaeoenvironmental survey of West, Thorne and Hatfield Moors. *In*: VAN DE NOORT, R. & ELLIS, S. (eds) *Wetland Heritage of the Humberhead Levels an Archaeological Survey*. The Humber Wetlands Project, University of Hull, Hull, 157–189.

DUAN, W. M., COLEMAN, M. L. & PYE, K. 1997. Determination of reduced sulphur species in sediments – an evaluation and modified technique. *Chemical Geology*, **141**, 185–194.

EVERSHAM, B. 1991. Thorne and Hatfield Moors: implications of land use change for nature conservation. *Thorne and Hatfield Moors Papers*, **2**, 3–18.

FICHEZ, R., DENNIS, P., FONTAINE, M. F. & JICKELLS, T. D. 1993. Isotopic and biochemical composition of particulate organic material in a shallow water estuary (Great Ouse, North Sea, England). *Marine Chemistry*, **43**, 263–276.

FOSSING, H. & JORGENSEN, B. B. 1989. Measurement of bacterial sulphate reduction in sediment: evaluation of a single-step chromium reduction method. *Biogeochemistry*, **8**, 205–222.

GAUNT, G. 1987. The geology and landscape development of the region around Thorne Moors. *Thorne Moors Papers*, **1**, 5–28.

——1994. *Geology of the Country around Goole, Doncaster and the Isle of Axholme*. Memoirs for the one-inch sheets 79 and 88 (England and Wales). HMSO, London.

——, FLETCHER, T. P. & WOOD, C. J. 1992. *Geology of the Country around Kingston upon Hull and Brigg*. Memoirs for the 1:50 000 Geological Sheets 80 and 89 (England and Wales). HMSO, London.

—— & TOOLEY, M. J. 1974. Evidence for Flandrian sea-level changes in the Humber estuary and adjacent areas. *Bulletin of the Geological Survey of Great Britain*, **48**, 25–41.

GEORGE, M. 1992. *The Land Use, Ecology and Conservation of Broadland.* Packard Publishing Limited, Chichester.

HAYES, J. M. 1983. Practice and principles of isotopic measurements in organic geochemistry. *In*: MEINSCHEIN, W. G. (ed.) *Organic Geochemistry of Contemporaneous and Ancient Sediments.* Society of Economic Paleontologists and Mineralogists.

HEDGES, J., CLARK, W., QUAY, P., RICHEY, J., DEVOL, A. & SANTOS, U. D. M. 1986. Compositions and fluxes of particulate organic material in the Amazon River. *Limnology and Oceanography*, **31**, 717–738.

JACKSON, D., HARKNESS, D. D., MASON, C. F. & LONG, S. P. 1986. *Spartina anglica* as a carbon source for salt-marsh invertebrates: a study using ^{13}C values. *Oikos*, **46**, 163–170.

LAMBECK, K. 1995. Late Devensian and Holocene shorelines of the British Isles and North Sea from models of glacio-hydro-isostatic rebound. *Journal of the Geological Society of London*, **152**, 437–448.

LILLIE, M. & NEUMANN, H. 1998. Introduction to the palaeoenvironmental survey. *In*: VAN DE NOORT, R. & ELLIS, S. (eds) *Wetland Heritage of the Ancholme and Lower Trent Valleys an Archaeological Survey.* The Humber Wetlands Project, University of Hull, Hull, 21–31.

—— & WEIR, D. 1997. Alluvium and warping in the Humberhead Levels: the identification of factors obscuring palaeoland surfaces and the archaeological record. *In*: VAN DE NOORT, R. & ELLIS, S. (eds) *Wetland Heritage of the Humberhead Levels an Archaeological Survey.* The Humber Wetlands Project, University of Hull, Hull, 191–218.

—— & ——1998. Alluvium and warping in the lower Trent valley. *In*: VAN DE NOORT, R. & ELLIS, S. (eds) *Wetland Heritage of the Ancholme and Lower Trent Valleys an Archaeological Survey.* The Humber Wetlands Project, University of Hull, Hull, 103–122.

LOIS 1992. *Science Plan for a Community Research Project.* Natural Environment Research Council, Swindon.

LONG, A. J., INNES, J. B., KIRBY, J. R., LLOYD, J. M., RUTHERFORD, M. M., SHENNAN, I. & TOOLEY, M. J. 1998. Holocene sea-level change and coastal evolution in the Humber Estuary, eastern England: an assessment of rapid coastal change. *The Holocene*, **8**, 229–247.

MACKENZIE, F. T., LERMAN, A. & VER, L. M. B. 1998. Role of the continental margin in the global carbon balance during the past three centuries. *Geology*, **26**, 423- 426.

MACKO, S. E. & ESTEP, M. L. F. 1984. Microbial alteration of stable nitrogen and carbon isotopic compositions of organic matter. *Organic Geochemistry*, **6**, 787–790.

MCCAVE, I. N. 1987. Fine sediment sources and sinks around the East Anglian coast (UK). *Journal of the Geological Society of London*, **144**, 149–152.

METCALFE, S. E., ELLIS, S., HORTON, B. P., INNES, J. B., MCARTHUR, J. J., MITLEHNER, A., PARKS, A., PETHICK, J. S., REES, J., RIDGWAY, J., RUTHERFORD, M. M., SHENNAN, I., TOOLEY, M. J. 2000. The Holocene evolution of the Humber Estuary: reconstructing change in a dynamic environment. *This volume.*

NEALE, J. W. 1988. The geology of the Humber area. *In*: JONES, N. V. (ed.) *A Dynamic Estuary: Man, Nature and the Humber.* Hull University Press, Hull, 1–15.

NITHART, M. 1995. *Role of Two Polychaete Species* Nereis diversicolour, Scelopholos assinger *in Processing Organic Matter and Nutrient Cycling in a North Norfolk saltmarsh, UK.* PhD thesis, University of East Anglia.

O'CONNOR, B. A. 1987. Short and long term changes in estuary capacity. *Journal of the Geological Society of London*, **144**, 187–195.

PETHICK, J. 1987. *The Distribution and Status of* Spartina anglica *in the Humber Estuary.* Report to the Nature Conservancy Council, Hull.

——1988. The physical characteristics of the Humber. *In*: JONES, N. V. (ed.) *A Dynamic Estuary: Man, Nature and the Humber.* Hull University Press, Hull, 31–45.

——1990. The Humber Estuary. *In*: ELLIS, S. & CROWTHER, D. R. (eds) *Humber Perspectives: A Region Through The Ages.* Hull University Press, Hull, 54–67.

REES, J. G., RIDGWAY, J., ELLIS, S., KNOX, R. W. O'B., NEWSHAM, R. & PARKES, A. 2000. Holocene sediment storage in the Humber Estuary. *This volume.*

RIDGWAY, J., ANDREWS, J. E., ELLIS, S., HORTON, B. P., INNES, J. B., KNOX, R. W. O'B., MCARTHUR, J. J., MAHER, B. A, METCALFE, S. E., MITLEHNER, A., PARKES, A., REES, J. G., SAMWAYS, G. & SHENNAN, I. 2000. Analysis and interpretation of Holocene sedimentary sequences: techniques applied in the Humber Estuary. *This volume.*

SHENNAN, I., LAMBECK, K., HORTON, B., INNES, J., LLOYD, J., MCARTHUR, J. & RUTHERFORD, M. 1999. Holocene isostacy and relative sea-level changes on the east coast of England. *This volume.*

SMITH, A. G. 1958. Post-glacial deposits in south Yorkshire and north Lincolnshire. *New Phytologist*, **57**, 19–49.

SMITH, S. V. & HOLLIBAUGH, J. T. 1993. Coastal metabolism and the oceanic organic carbon balance. *Reviews of Geophysics*, **31**, 75–89.

STUIVER, M. & REIMER, P. J. 1993. Extended ^{14}C data base and revised CALIB 3.0 ^{14}C age calibration program. *Radiocarbon*, **35**, 215–230.

TURNER, G. M & THOMPSON, R. 1981. Lake sediment record of the geomagnetic secular variation in Britain during Holocene times. *Geophysical* Journal of the Royal Astronomical Society, **65**, 703–725.

—— & ——1982. Detransformation of the British geomagnetic secular variation record for Holocene times. *Geophysical Journal of the Royal Astronomical Society*, **70**, 789–792.

VAN DE NOORT, R. & ELLIS, S. 1995. *Wetland Heritage of Holderness an Archaeological Survey.* The Humber Wetlands Project, University of Hull, Hull.

—— & ——1997. *Wetland Heritage of the Humberhead Levels an Archaeological Survey*. The Humber Wetlands Project, University of Hull, Hull.

—— & ——1998. *Wetland Heritage of the Ancholme and Lower Trent Valleys an Archaeological Survey*. The Humber Wetlands Project, University of Hull, Hull.

VER, L. M. B., MACKENZIE, F. T. & LERMAN, A. 1994. Modeling pre-industrial C–N–P–S biogeochemical cycling in the land–coastal margin. *Chemosphere*, **29**, 855–887.

WAKEFORD, R. G., HATHAWAY, R. A. & BOLLINGTON, H. R. 1992. *Development and Flood Risk*. Joint circular, DoE, MAFF and Welsh Office, London & Cardiff.

WILSON, A. K. 1938. *The Adventive Flora of the East Riding of Yorkshire*. Hull Scientific and Field Naturalists' Club, Hull.

Sediment provenance and flux in the Tees Estuary: the record from the Late Devensian to the present

A. J. PLATER,[1] J. RIDGWAY,[2] B. RAYNER,[1] I. SHENNAN,[3]
B. P. HORTON,[3] E. Y. HAWORTH,[4] M. R. WRIGHT,[3]
M. M. RUTHERFORD[3] & A. G. WINTLE[5]

[1] *Department of Geography, University of Liverpool, P.O. Box 147, Liverpool L69 3BX, UK (e-mail: gg07@liv.ac.uk)*

[2] *Coastal and Engineering Geology Group, British Geological Survey, Kingsley Dunham Centre, Keyworth, Nottingham NG12 5GG, UK*

[3] *Environmental Research Centre, Department of Geography, University of Durham, Science Laboratories, South Road, Durham DH1 3LE, UK*

[4] *Institute of Freshwater Ecology, Windermere Laboratory, The Ferry House, Far Sawrey, Ambleside, Cumbria LA22 0LP, UK*

[5] *Luminescence Laboratory, Institute of Geography and Earth Sciences, University of Wales, Aberystwyth SY23 3DB, UK*

Abstract: The influences of sea-level, climate, human activity and coastal morphology on post-glacial sediment flux and deposition in the Tees Estuary were considered in a multidisciplinary investigation of the Late Pleistocene and Holocene sedimentary record. The following tripartite division was identified using a combination of lithostratigraphic and geochemical data: a Late Glacial laminated clay providing evidence of a former proglacial lake and a proxy record of climate change; an early–mid-Holocene intercalated sequence of tidal silts and clays and peats; and a late Holocene succession characterized by increasing evidence of human activity and metal contamination. Sea-level change has been identified as the main control on sedimentation *via* decelerating sea-level rise and changing tidal dynamics between *c.* 8 and 3 ka BP. Climate controlled the sequence of rhythmite thickness in the Late Glacial clays, whilst increased wetness after *c.* 3 ka BP may have encouraged terrestrial sediment influx. Enhanced sediment supply to the coastal zone can also be attributed to increasing human activity in the catchment from the Bronze Age onward, first as a consequence of clearance, and subsequently as a result of mining and industrial expansion. Fine-grained sediment flux has almost exclusively been from the Tees catchment to the coast, extending offshore during rebound-induced collapse and erosion of the Late Glacial lake basin. The only notable onshore sediment flux has been the deposition of marine sands in the outer estuary between *c.* 6.5 and 3.5 ka BP.

As part of the National Environmental Research Council (NERC) funded Land–Ocean Interaction Study (LOIS) special topic (NERC 1994), research into the post-glacial evolution of the Tees Estuary was undertaken as a contribution to the Land–Ocean Evolution Perspective Study (LOEPS). The rationale for much of the work to be completed on the Tees was to link processes operating over time-scales of the order of the Holocene to those which characterize the present estuary and intertidal environments, thus providing a bridge between long-term perspectives and the study of contemporary processes in the coastal zone. In terms of specific objectives, the first was to determine the history of sediment flux and storage in the Tees Estuary during the Holocene. This was achieved by establishing the nature and morphology of the successive depositional environments as well as sediment sources. As a consequence, it was possible to assess the influence, and relative importance, of sea-level change, coastal morphology and catchment land-use on sediment flux with a view to developing models for predicting the response of the Tees Estuary to future climate and environmental change.

A multidisciplinary study was completed on deep (*c.* 10–30 m), intermediate (*c.* 5 m) and

From: SHENNAN, I. & ANDREWS, J. (eds) *Holocene Land–Ocean Interaction and Environmental Change around the North Sea*. Geological Society, London, Special Publications, **166**, 171–195. 1-86239-054-1/00/$15.00

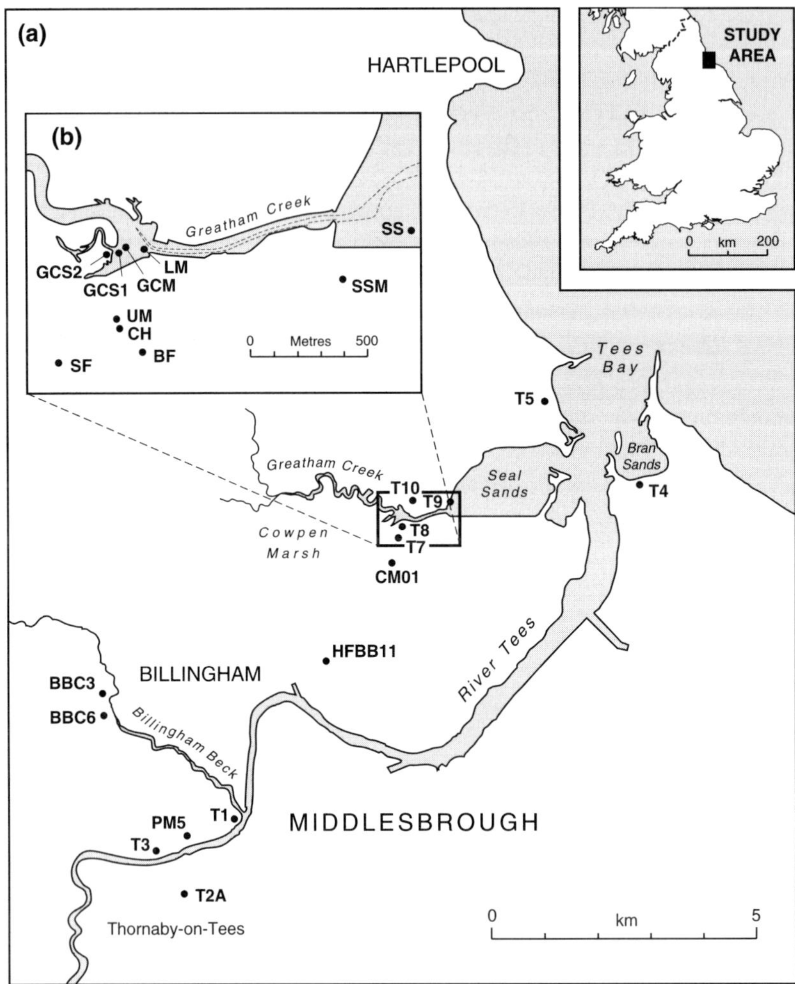

Fig. 1. Location of LOIS boreholes in the Tees coastal lowlands and shallow cores in the present and reclaimed intertidal zone.

shallow cores (*c.* 0.5 m) from the coastal lowlands and present inter- and supratidal zones of the Tees Estuary (Fig. 1). Palaeoenvironmental reconstruction was based primarily on stratigraphic investigations in combination with diatom, pollen and grain size analyses. Chronological data on the deep and intermediate cores were obtained from radiocarbon and luminescence dating of organic and minerogenic horizons, whilst a combined radionuclide (^{210}Pb and ^{137}Cs) and environmental pollution (environmental magnetism and X-ray fluorescence (XRF) geochemistry) approach was used to obtain sediment accumulation rates for shallow cores from mudflat, saltmarsh and 'reclaimed' marsh environments. Sediment provenance was determined by environmental magnetism and XRF geochemical analyses, with complementary investigations being undertaken on potential source materials (Tees catchment and North Sea bed). This paper reviews the results of this research programme and presents a summary of the post-glacial evolution of the Tees Estuary, particularly in the context of sediment provenance, flux and storage.

Interpretative framework and sediment geochemistry

The deep boreholes (T1–T10) drilled in the coastal lowlands of the Tees Estuary during the course of the LOIS project by the British Geological Survey (Fig. 1) provide the overall

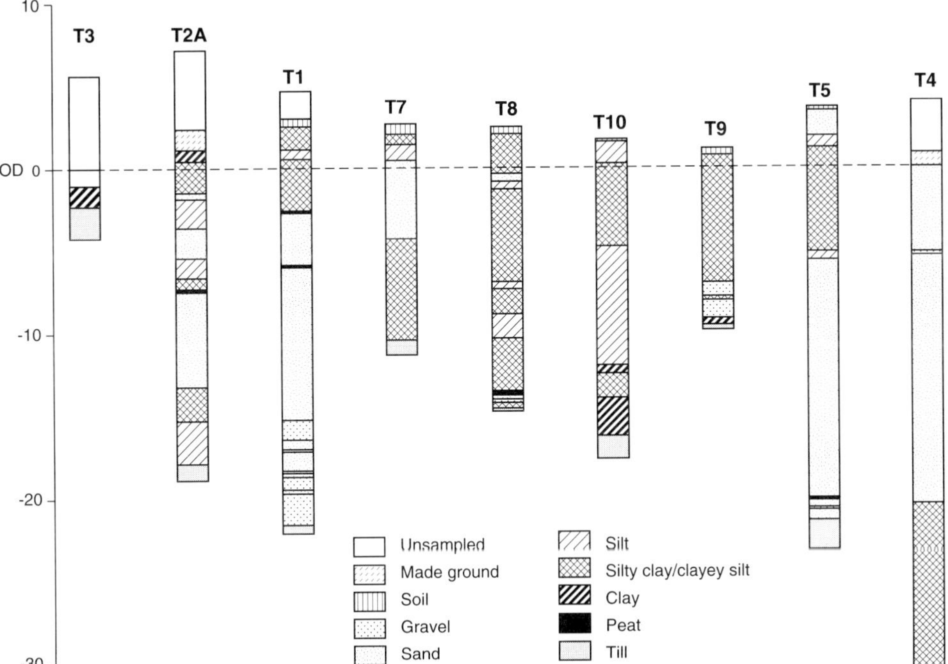

Fig. 2. Stratigraphic cross-section of the Tees coastal lowlands from the deep borehole record.

stratigraphic framework for the interpretation of the post-glacial history of sedimentation and sediment flux. It is clear from Fig. 2 that the lithostratigraphy is rather complicated, and that correlation between different units is extremely problematic on the basis of physical characteristics, component parts, grain size or, indeed, altitude. Perhaps two sandy lithofacies can be identified, one located in the inner (T2A, T1 and T7) and one in the outer estuary (T5 and T4), with a more fine-grained succession of sediments in mid estuary. Although chronological data in the form of ^{14}C and luminescence ages (Table 1) provide additional information to aid the interpretation of temporal and spatial trends in the nature of sedimentation, it is XRF geochemistry that provides the key.

The geochemical study of the Tees boreholes followed the same lines as similar studies in the Humber (Rees *et al.* this volume; Ridgway *et al.* this volume). Because the scale of geochemical variation in the Holocene successions was unknown, a pilot project was undertaken using a portable XRF analyser. This pilot project, described in more detail in Ridgway *et al.* (1998; this volume), showed that recognizable stratigraphic units in the cores had distinctive multi-element geochemical signatures. A total of 125 samples were then taken from cores T1–T10 and analysed for 22 major and trace elements (MgO, Al_2O_3, P_2O_5, K_2O, CaO, TiO_2, MnO, Fe_2O_3, Cr, Co, Ba, Ni, Cu, Zn, As, Rb, Sr, Y Zr, Pb, La and Ce) using high-precision laboratory-based XRF facilities at the British Geological Survey, in order to characterize the general chemistry and contamination histories of the major stratigraphic units. The sampling interval varied according to significant changes in lithostratigraphy. The use of cores places constraints on the amount of sample material available and consequently whole sediment samples were analysed, rather than a specific size fraction, in order to ensure that sufficient sediment was available for analysis. Details of the analytical methodology along with sampling and analytical quality control procedures are given in Ridgway *et al.* (1998; this volume). Correlation coefficients for replicate sampling and analysis of 21 pairs of samples were better than: 0.98 for 15 elements, 0.95 for five elements, and 0.92 for one element. The remaining element (Cr) showed one aberrant replicate result, which reduced the correlation coefficient to 0.89 (Ridgway *et al.* 1998). These results demonstrate the reliability of the methods employed.

Interpretation of the data relied heavily on the use of spidergrams, in which element concentrations normalized to the upper crustal average

Table 1. *Radiocarbon and luminescence ages obtained for the post-glacial stratigraphic record in the Tees coastal lowlands*

Core No.		Depth (m)	Radiocarbon age (cal. a BP)
	Material		
T2A	Wood	13.65	7646–7937
T2A	Sandy peat with wood	13.65–13.78	9437–9843
T2A	Wood	18.40	7681–8128
T1	Peat with silt and sand	6.80	6786–7200
T1	Peat with sand and wood	9.66	10 141–10 923
T8	Peat with clay and wood	14.95	8562–9183
T9	Peat with silt and clay	2.98	7173–7431
T4	Peat with sand and wood	20.70	25 580 ± 360
BBC3	Clayey peat	4.99	6864–7207
BBC3	Peat with trace clay	5.25	7337–7549
BBC3	Peat	6.49	7656–7945
BBC3	Peat	6.59	7995–8334
BBC6	Phragmites peat with clay	5.63	7017–7392
PM5	Silty peat	2.62	2733–2955
PM5	Phragmites peat	4.35	6212–6852
PM5	Peat with silt	4.54	6864–7207
HFBB11	Peat with silt and wood	1.55	2800–3253
HFBB11	Peat with silt	2.26	5739–6169
HFBB5	Peat with clay	5.17	8210–9361
HFBB5	Peat	5.28	7833–8166
	*Dating approach**		
T1	Fine grain	5.40	3200 ± 781
T1	Coarse grain	7.50	1952 ± 433
T2A	Fine grain	8.00	6299 ± 2020
T2A	Fine grain	8.30	10 611 ± 3242
T2A	Fine grain	14.10	7945 ± 2917
T2A	Fine grain	18.20	43 594 ± 7804
T4	Coarse grain	10.10	3543 ± 281
T4	Coarse grain	22.20	6242 ± 1381
T5	Coarse grain	1.50	424 ± 40
T5	Coarse grain	2.10	273 ± 36
T5	Coarse grain	8.70	4871 ± 397
T5	Coarse grain	20.20	5969 ± 508
T9	Fine grain	3.90	18 365 ± 10 015

* For the dating approach, the luminescence age (years BP $\pm 1\sigma$) is given.

(UCA) values of Wedepohl (1995) are plotted on a logarithmic scale. The geochemical signatures from the spidergrams were then grouped according to similarity into chemostratigraphic suites (Fig. 3), for which typical concentration data are listed in Table 2. This methodology, albeit somewhat subjective, was chosen in preference to multi-element statistical procedures because it allows the user to take account of chemical variation brought about by effects such as leaching of Ca at a weathering surface, the gradual onset of contamination, grain size differences or mis-sampling across a genuine geochemical boundary, all of which would normally adversely affect statistical grouping. It is robust in that strong contaminant signatures, involving only one or two elements, can be recognized in sediment groups which are chemically similar in other respects.

No attempt has been made to compensate for grain size variation by the use of an element proxy such as Al. There are several reasons for this. Correlation coefficients for the Tees samples show that whilst many elements (e.g. Co, Cr, Cu, Ni) are strongly correlated with Al, others (e.g. Ca, Mg, Mn, Zr, Ba) are not (Ridgway *et al.* 1998; Plater *et al.* 1999). Wholesale normalization to Al would thus have an unpredictable effect on some elements. In addition, iron and manganese oxyhydroxides and organic matter complex trace metals and must also be considered when attempting normalization (Plater *et al.* 1999). It would be possible to normalize only those elements that correlate strongly with Al, but this also has drawbacks. Normalizing when both metal and normalizer are at low concentrations yields problematical results (Rowlatt & Lovell 1994) and in this study

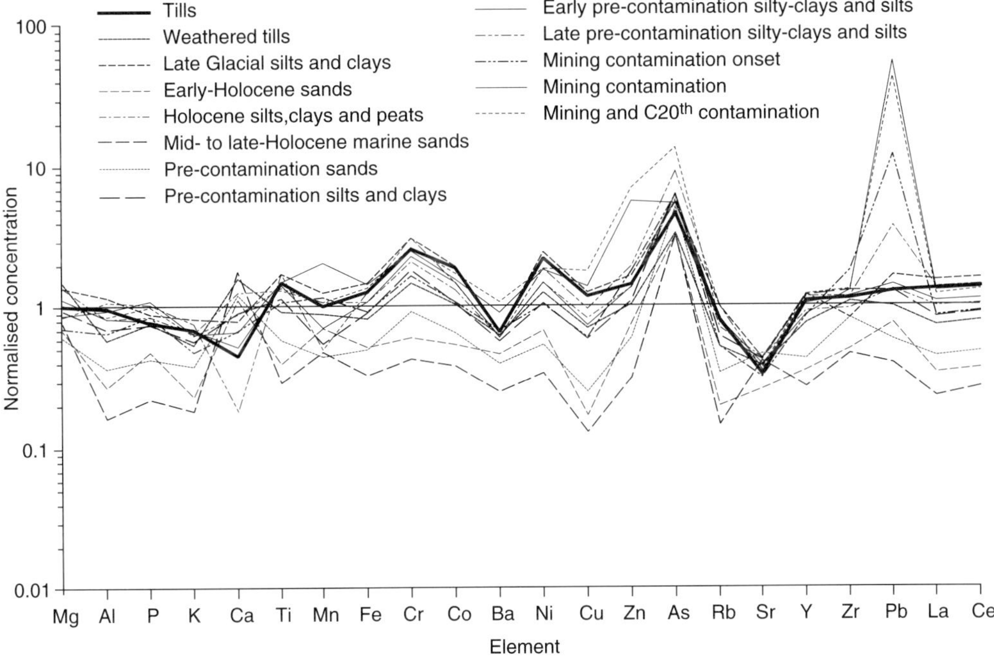

Fig. 3. Multi-element spidergrams of chemostratigraphic suites defined on the basis of element concentration patterns, showing concentration data normalized to upper crustal average (UCA) values.

gave abnormally high metal values for sands. In practice, spidergrams work well with non-normalized data for sediments of silt–clay size, particularly if the shape of the geochemical signature is considered rather than the absolute values. Sands remain problematical with or without grain size normalization.

Together with the chronological data, a combined litho-chemostratigraphy can be established for the post-glacial evolution of the Tees. This shows a basal suite of tills together with weathered tills and fluvio-glacial sandy-silts and gravels (Fig. 4), which slope from west to east, thus forming the pre-Holocene Tees basin. However, a third litho-chemostratigraphic succession directly overlies the basal sediments in the majority of cores. This succession is predominantly clay and silt with occasional sands (Fig. 4), and is considered to date from the Late Glacial period as a result of its stratigraphic position and from previous research (Radge 1939; Agar 1954). Whilst this suite of sediments is difficult to define lithologically in the transition from the Devensian to the Holocene over much of the basin, a thick sequence of laminated clays and silts can be identified in core T9. From additional borehole records in the BGS archive, the laminated sediments appear to form a morphological high in the outer part of the estuary. This feature has had a marked influence on the Holocene evolution of the Tees, particularly in response to early–mid-Holocene sea-level rise, and may have been reflected in recent times by the branching of the main tidal channels in the outer estuary until nineteenth and twentieth-century land claim. The main body of the Holocene stratigraphy is made up of two sandy lithofacies associated with a widespread succession of intercalated silts, clays and peats (Fig. 4). The more landward of these sandy facies appears to date from the early Holocene (see bracketing radiocarbon ages given in Fig. 4), whilst the seaward sands and gravels are more mid–late Holocene in age (see coarse grain luminescence ages). The Holocene silts and clays pass upward into a litho-chemostratigraphic record of contamination from the late Holocene to the present-day (Fig. 4). This record shows some variability due to grain size effects, but it is possible to identify geochemical signatures related to mining activities in the Tees catchment during archaeological and historical times and, more recently, industrial development in the coastal lowlands of the Tees Estuary (Fig. 3).

The litho-chemostratigraphic record described above forms the basis for the following investigation of sediment provenance, flux and deposition in the Tees, in which particular focus is

Table 2. *Mean geochemical data for chemostratigraphic suites defined from multi-element spidergrams*

Grouping	*n*	MgO (%)	Al_2O_3 (%)	P_2O_5 (%)	K_2O (%)	CaO (%)	TiO_2 (%)	MnO (%)	Fe_2O_3 (%)	Cr (ppm)	Co (ppm)	Ba (ppm)
Tills	5	2.20	14.26	0.12	2.32	1.86	0.75	0.07	5.53	88	22	431
Weathered till	11	3.25	8.65	0.11	1.86	6.45	0.48	0.06	3.71	51	12	413
Late Glacial	14	3.06	17.40	0.14	2.82	3.33	0.90	0.09	6.47	107	22	432
Early Holocene sands	12	1.55	3.89	0.08	0.81	4.68	0.21	0.05	2.26	21	6	311
Holocene silt & clay	22	3.10	11.72	0.13	2.21	5.29	0.66	0.07	4.67	72	15	409
Marine sand	15	1.58	1.88	0.03	0.53	7.31	0.13	0.03	1.32	12	4	150
Pre-contamination sand	2	1.63	3.27	0.05	0.84	7.13	0.18	0.03	1.53	21	5	209
Pre-contamination silt & clay	5	1.95	9.30	0.11	1.88	3.87	0.56	0.04	3.68	59	12	356
Early Pre-contamination	5	2.56	12.20	0.12	2.31	2.11	0.73	0.05	4.68	80	17	401
Late Pre-contamination	5	1.93	16.09	0.16	2.62	0.75	0.90	0.03	5.50	107	17	410
Mining contamination onset	4	1.56	9.46	0.13	1.64	2.85	0.55	0.08	4.07	58	13	448
Mining contamination	10	1.91	13.75	0.16	2.17	2.68	0.77	0.13	6.32	85	18	603
Mining + twentieth century contamination	2	2.30	12.45	0.16	2.13	3.67	0.72	0.07	6.08	92	20	713

Grouping	Ni (ppm)	Cu (ppm)	Zn (ppm)	As (ppm)	Rb (ppm)	Sr (ppm)	Y (ppm)	Zr (ppm)	Pb (ppm)	La (ppm)	Ce (ppm)
Tills	40	16	74	9	86	104	23	266	22	42	90
Weathered till	23	10	55	7	56	121	16	256	17	24	52
Late Glacial	45	18	87	11	110	133	25	194	28	50	103
Early Holocene sands	13	3	39	6	22	83	7	120	13	11	24
Holocene silt & clay	27	12	70	11	75	133	19	231	22	31	67
Marine sand	5	1	15	7	13	138	5	101	6	7	16
Pre-contamination sand	8	2	20	5	22	116	6	120	7	8	20
Pre-contamination silt & clay	19	8	53	12	62	132	17	287	20	25	56
Early Pre-contamination	28	10	71	12	84	102	21	275	24	34	74
Late Pre-contamination	34	14	99	18	111	111	24	272	64	45	91
Mining contamination onset	20	9	93	12	56	100	18	432	205	28	60
Mining contamination	33	20	286	10	88	121	25	299	931	41	88
Mining + twentieth century contamination	34	26	360	27	84	117	25	317	732	39	86

placed on the period of laminated silt and clay deposition during the Late Glacial; the influence of relative sea-level rise and catchment vegetation and land-use history on Holocene sedimentation; and the flux of both sediment and contaminants into Tees Estuary during the late Holocene and the historical past.

Late Glacial laminated silts and clays

The results of the XRF geochemistry enable the definition of an extensive litho-chemostratigraphic succession of variable thickness which occupies the temporal transition from the Devensian to the early Holocene (Fig. 4). Of particular significance in the Late Glacial evolution of the Tees Estuary is the sequence of laminated silts and clays, with occasional sandy partings, which is clearly represented in core T9. This laminated clay was considered in detail by Agar (1954), although the laminated sediments of the Teesside region were first noted by Wollacott (1921), Raistrick (1934) and Radge (1939). Indeed, Radge (1939) recorded two distinct laminated clay units on the basis of altitude, i.e. a high level and a low level blue clay, both of which were considered to have been deposited in an ice-dammed Lower Tees lake basin. According to Agar (1954), the Late Glacial laminated clay was deposited in association with a 'marginal sand', the latter forming at the shore of the proglacial lake. The average thickness of the laminated clay was approximately 7.5 m, with the thickest part of the sequence located in the central part of the lake basin. Towards the periphery of the former

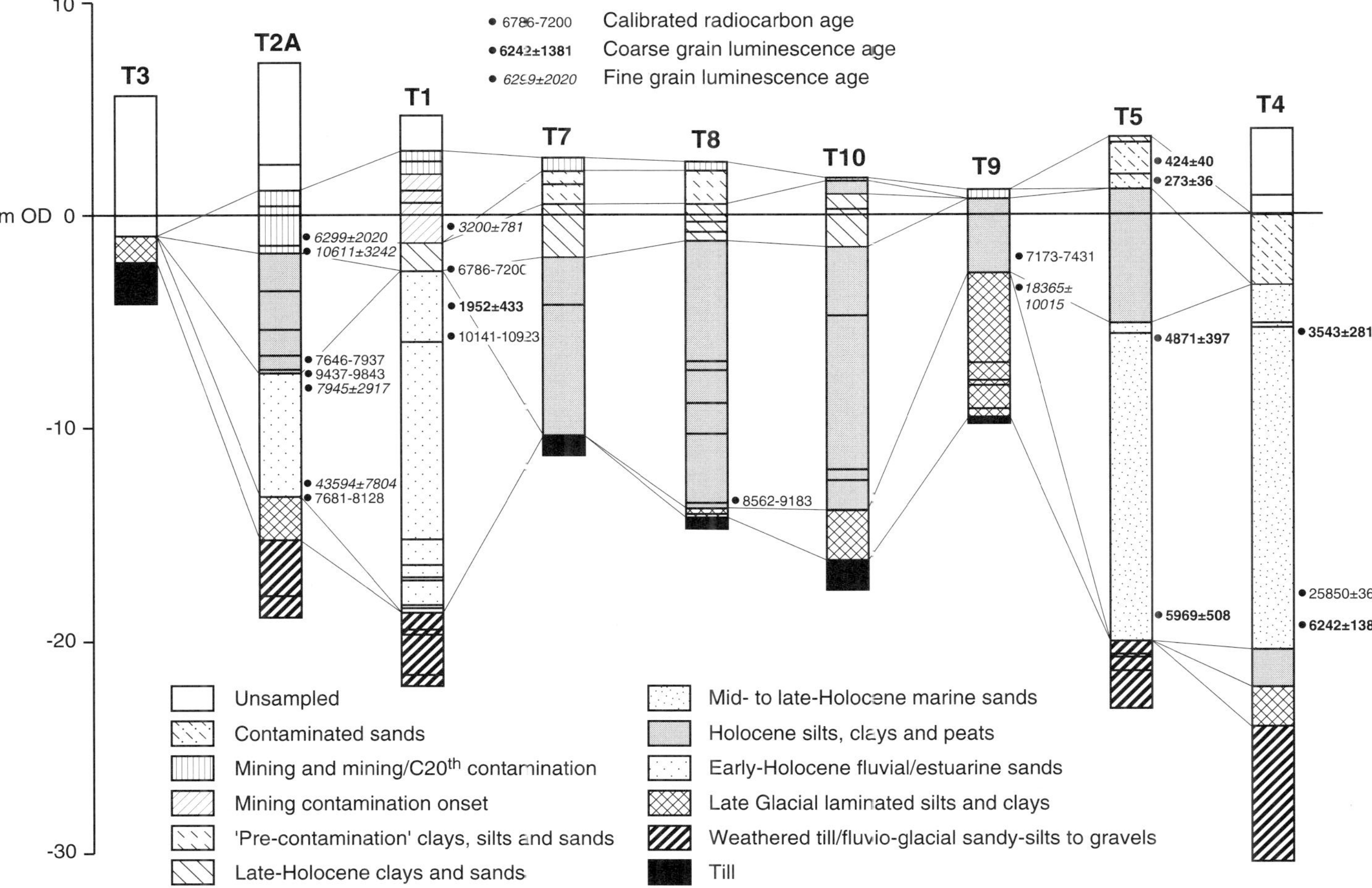

Fig. 4. Stratigraphic cross-section showing litho-chemostratigraphic groupings defined on the basis of their lithology and elemental concentration patterns.

lake basin, a higher sand content was noted with only thin clay laminae present.

Palaeontological, granulometric and magnetic analyses were undertaken in an attempt to establish the environment of deposition for this laminated sedimentary record, thought previously to have been a proglacial lake (Radge 1939), and to determine whether the sequence of lamination thickness records Late Quaternary climate change, as observed in varved lake sediments (de Geer 1912; Jopling 1975; Renberg & Sergerström 1981; Strömberg 1994; Anderson 1996; Björk *et al.* 1996; Kemp 1996; Wohlfarth 1996). Initial measurements of lamina thickness between depths of 3.03 and 5.86 m in core T9 reveal that each potential varve, and therefore total varve thickness, is made up of a light brown silty-sand, overlain by a reddish-brown silty-clay and a black to dark grey clay (units 3, 2 and 1, respectively). The base of each varve was defined as the transition from the dark grey clay to the succeeding light brown silty-sand, this being considered to represent the change from winter to spring/summer sedimentation within the proposed lake. Below 5.86 m, the laminae are generally distorted and are interbedded with glacial till.

Palaeoecological evidence concerning the environment in which the laminated silts and clays were deposited proved hard to come by. No diatoms are present in the laminated sediments, suggesting that diatom preservation was either extremely poor or the environment of deposition was such that diatoms were not able to thrive in the turbid, quiet-water conditions. Diatoms are present in the uppermost few centimetres of the laminated silts and clays, where the laminae are either disturbed or indistinct. According to the interpretation scheme of Vos and de Wolf (1993), the diatom assemblage between depths of 3.10 and 2.20 m is characteristic of a saltmarsh becoming replaced by a tidal flat and channel system (Rayner pers. comm. 1997). Pollen data obtained from an isolated peat horizon at a depth of 2.98 m also reveal a saltmarsh environment with sedges and grasses. Subsequent radiocarbon dating returned an age of 7173–7431 cal. a BP for this detrital peat horizon (possibly an allochthonous peat ball). Consequently, although the environment of deposition cannot be confirmed, the laminated deposits are established as being at least pre- to mid-Holocene in age.

The environmental magnetic properties of the three separate 'varve' units were determined on approximately 0.5 ml of sample using a vibrating sample magnetometer (VSM). The resulting hysteresis loops for units 1 (dark grey clay) and 2 (red-brown silty-clay) display single-domain characteristics, whilst unit 3 (silty-sand) is multi-domain (Rayner pers. comm. 1997). Unit 3 exhibits a lower coercivity than units 1 and 2 but unit 1 has a lower susceptibility than units 2 and 3 which are statistically indistinguishable (although the silty-sand does exhibit higher magnetic susceptibility values). Hence, the silty-sand (unit 3) is probably derived from the same source as the red-brown clay (unit 2), although the coarser-grained component is multi- rather than single-domain and has a lower contribution from higher coercivity minerals. The detrital minerogenic component of the dark grey clay is of the same provenance as the red-brown clay, but the lower susceptibility is an obvious consequence of a lower proportion of minerogenic material and, hence, a lower concentration of magnetic minerals. These results agree well with those of Björck *et al.* (1996), who concluded that the clayey winter laminae in Lake Sännen display a lower magnetic susceptibility than the more silty summer laminae. Furthermore, the fact that the red-brown clay and silty-sand appear to be derived from the same source, albeit with some disparity due to differing grain size, is indicative of detrital inwash during spring and summer. The silty-sand, therefore, represents higher energy inwashing, and the occasionally thicker sandy units may reflect either more extreme runoff (warmer springs and/or wetter summers) or deposition in closer proximity to the incoming jet.

The particle size distributions of the upper and lower parts of the three units were determined in order to establish whether they exhibited normal grading, i.e. the heavier, coarser grains settling out of suspension first from the inwash. The silty-sand proves to be coarse silt, whilst the red-brown and dark grey units (units 2 and 1, respectively) are very fine silt to clay (Rayner pers. comm. 1997). Although the dark grey silty-clay units do not exhibit any notable grading, the red-brown clay and the silty-sand units both have a coarser-grained lower part. This is in good agreement with the provenance information obtained from the magnetic properties in that the two inwash units display evidence of grading due to particle settling from the incoming terrestrial jet.

Although the palaeoenvironmental evidence for the nature of laminated silty-clay and silty-sand deposition is equivocal, the standing water body is most likely to have been a lake in which sediment supply was controlled by seasonal variations in meltwater and runoff input. The cold climate conditions necessary for this degree of seasonal variation, combined with the need

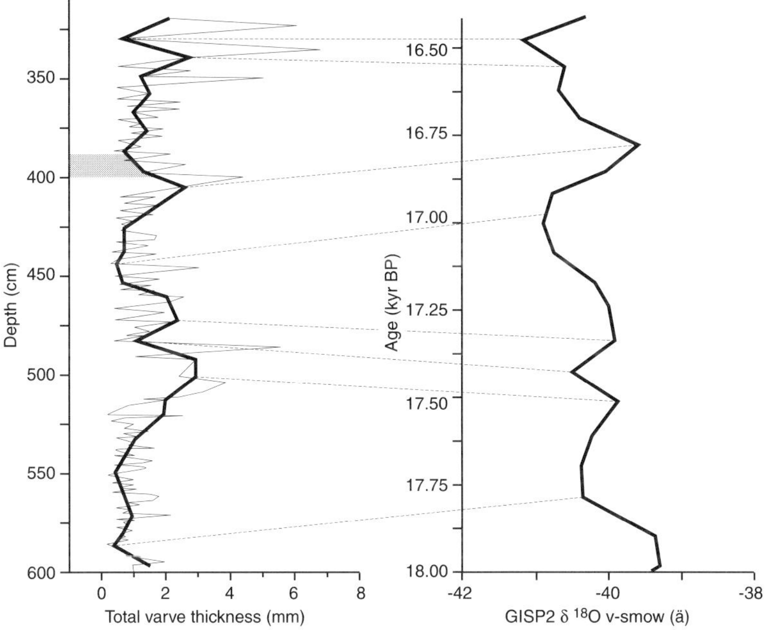

Fig. 5. Comparison between total varve thickness in core T9 (light line, individual measurements; thick line, 30-point running mean) and the GISP $\delta^{18}O$ record between 18 and 16.5 ka BP (after Rayner, pers. comm. 1997). The shaded part of the varve record is that dated by fine grain luminescence dating to 18 365 ± 10 015 a BP.

for the Lower Tees to have been impounded, are indicative of a proglacial lake environment between the North Yorks Moors and an ice mass to the north and northeast. This varved lake sediment record may, therefore, reflect climate change in the form of rhythmite thickness. However, the thickness of the silty-sand unit suggests that deposition may have been influenced periodically by proximity to the lake shore. Consequently, the rhythmite index of Hart (1992) was used to confirm whether the laminated sequence was indeed a varve sequence *sensu stricto* or a series of more marginal deltaic units, potentially turbidites, deposited from the waning inwash. The results of this analysis, using the ratio of the dark grey clay thickness (C_N) to the clay/silt thickness $(C/Z)_N$ (determined by combining units 1 and 2), reveal that much of the laminated sediment has a turbiditic or waning flow origin (Rayner pers. comm. 1997). Indeed, the rhythmites would appear to be category IV/P and V according to the classification scheme of Smith (1978), i.e. annual clastic varves deposited closest to the inlet river (IV/P) and turbidites (V).

Given the fact that the laminated silts and clays in T9 are a combination of varved lake sediments and more deltaic units deposited nearer the lake margin, it is unlikely that the sequence of rhythmite thickness provides a true proxy record of climate as a consequence of temporal changes in the extent of annual melting. However, the marginal silty-sand units merely act to accentuate varve thickness during warmer periods when spring meltwater and summer inwash is enhanced. Comparison with other proxy records of climate change for the period of laminated clay deposition reveals the extent to which the laminated sedimentary record is 'contaminated' by its proximity to the former lake shore. For this purpose, Rayner (pers. comm. 1997) proposed a comparison between the sequence of varve thickness and the Greenland Ice Sheet Project 2 (GISP2) oxygen isotope record (Fig. 5). Initial inspection reveals a reasonable agreement between a 30-point running mean of varve thickness and the GISP2 $\delta^{18}O$ record (Grootes *et al.* 1993). Although an independent correlation based on Fourier series was attempted using CORPAC, some interpretation was required to link peaks in the two data sets. Hence, the period of 18–16 ka BP was chosen for correlation, as the laminated lake sediment record must have been deposited during a period of ice impoundment prior to glacial retreat from the region. If significant retreat had taken place before approximately 16 ka BP, relative sea-level would have been of sufficient altitude to inundate the former lake basin (Lambeck 1991). However, as illustrated previously, marine inundation of the upper part of the laminated silts and clays did not take place until the mid-Holocene. Although absolute dating of the laminated sediments in T9 is problematic, they would appear to be post-Devensian maximum and the radiocarbon chronology places their deposition prior to 7173–7431 cal. a BP. In addition, fine grain luminescence dating of a series of particularly thick silty-sand laminae between depths of 3.80 and 3.99 m gives an age of 18 365 ± 10 015 a BP. The large uncertainty associated with this luminescence age is due primarily to water content and water content history reconstruction, and to sample variability in sensitivity to radiation dose. In the absence of any additional chronological data, the correlation between varve thickness and $\delta^{18}O$ (V-SMOW) must be considered as tentative. However, a climate control on sediment supply can at least be inferred for the period of laminated silt and clay deposition.

The morphology and elevation of the Late Glacial laminated sediments, particularly in relation to younger coastal and marine sediments preserved at lower altitudes within the Tees stratigraphic record, is indicative of a resistant clay island at least in the outer part of the Tees Estuary during the early–mid-Holocene. This island was eventually inundated by the sea at *c.* 7.3 ka BP. Consequently, an extensive hiatus is envisaged between the deposition of the laminated clays and subsequent submergence by the sea. This is considered to be due to post-glacial sea-level fall as glacio-isostatic crustal rebound exceeded the rate of eustatic sea-level rise, as predicted by Lambeck (1991) for the Tees region between 16 and 12 ka BP.

Holocene sediments and sediment flux

The transition from the Late Glacial to the Holocene in the Tees Estuary is difficult to define stratigraphically and geochemically. Comparison of the typical multi-element spidergrams for the Late Glacial and Holocene silts and clays (Fig. 3) reveals that the latter can only really be distinguished on the basis of lower Cr concentrations. Hence, the Late Glacial laminated clays are part of a sedimentary succession that evolved from the initial weathering and reworking of the pre-Holocene till surface of the Lower Tees basin to the deposition of sands, silts and clays in an early Holocene estuarine setting. In cores T2A and T1, the Late Glacial to Holocene transition is represented by a sequence of sands and gravels (Fig. 4). A sandy lithofacies in core T2A is bracketed by ^{14}C dates on wood at a depth of 18.40 m and a sandy peat with wood at 13.65–13.78 m. Whilst the sandy peat places deposition of the underlying sand prior to 9437–9843 cal. a BP, the woody detritus towards the base of the sequence returns a somewhat contradictory age of 7681–8128 cal. a BP. Further equivocal chronological data are provided by fine-grain luminescence ages of 43 594 ± 7804 and 7945 ± 2917 a BP on sandy units at depths of 18.20 and 14.10 m, respectively. Although the ^{14}C and fine-grain luminescence ages provide a complicated chronology, the deposition of this sandy facies can be dated to approximately 10–8 ka BP in core T2A. The same facies must have been deposited prior to 6786–7200 cal. a BP in core T1, from a ^{14}C date on a thin peat horizon at a depth of 6.80 m, with the lower part being laid down at least by 10 141–10 923 cal. a BP. However, this latter age is on a sandy peat with wood, which shows limited evidence of being *in situ*. Although there is limited evidence in the way of sedimentary structures, it is likely that these sands and gravels were deposited in the newly formed Tees Estuary under the influence of early Holocene sea-level rise, perhaps in a channel or as a bay-head delta.

The provenance of the sandy facies is also uncertain. In terms of CaO/MgO versus Rb/Sr ratios, these early Holocene sands and gravels lie at an interface between two groups of sediments (Fig. 6), which can be identified as 'marine' and 'terrestrial' on the basis of their palaeogeography and the similarity of their geochemical signatures to the present-day sediments from the fluvial and marine end-members of the Tees system (data from BGS geochemical databases; Stevenson *et al.* 1995; British Geological Survey 1996). Furthermore, the early Holocene sands are geochemically similar to both the average composition of 548 North Sea bed samples (data from Stevenson *et al.* 1995) and the weathered till and fluvio-glacial sands in core T5. It is tempting to conclude that the early Holocene transgression swept sands and gravels onshore from the floor of the North Sea. Alternatively, both the early Holocene sands and gravels and the North Sea sediments may have their origins in the weathering of the Devensian tills. Although an underlying grain-size control on the apparent similarities cannot be ruled out, it is likely that the early Holocene sands and gravels reflect both terrestrial and marine influences on sediment supply in the newly established and expanding Tees Estuary.

The main body of the Holocene stratigraphic record is made up of a sequence of tidal silts and

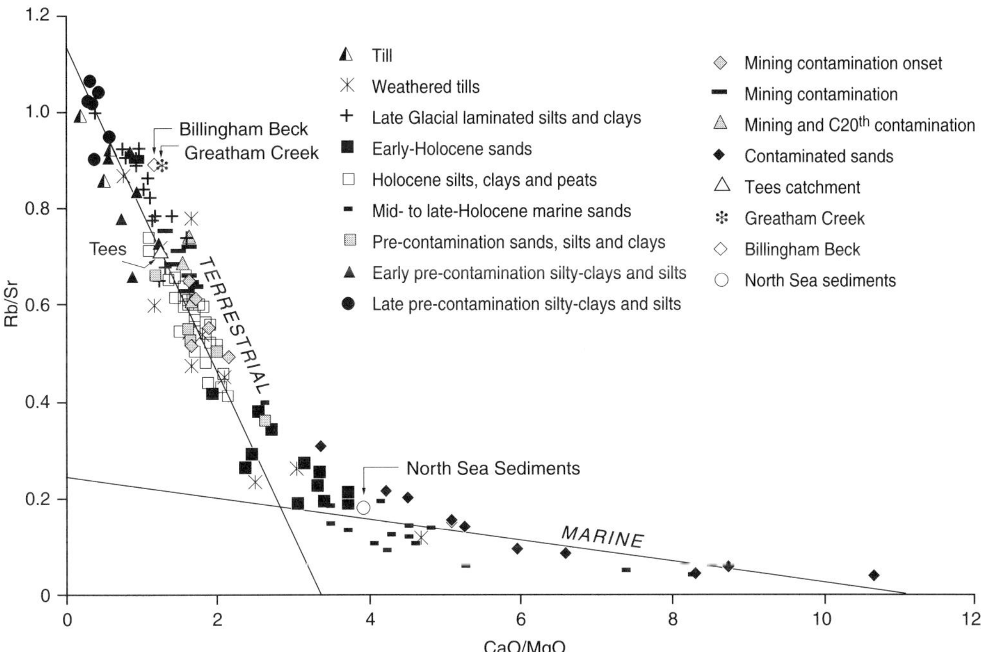

Fig. 6. Scatterplot of CaO/MgO ratio against Rb/Sr ratio for sediment samples analysed for their geochemical composition. The different symbols reflect groupings as defined from the spidergrams. The 'terrestrial' and 'marine' groupings are identified according to palaeogeography and similarity to present-day source materials.

clays, terrestrial peats and some lagoonal sediments of variable composition and grain size (Plater & Poolton 1992; Wright *et al.* 1996, 1997). The earliest evidence for tidal mud deposition dates from approximately 10 960 a BP (Shennan 1983) where marine sediments overlie a thin (*c.* 5 cm) peat bed at −10.61 m OD in the region of Thornaby-on-Tees. Radiocarbon dates on peat deposits at the base of the Holocene silts and clays in cores T2A, T1 and T8 support an early Holocene marine influence on deposition in the Tees Estuary (Fig. 4), although the altitudes of these chronological data suggest that sea-level change was not alone in controlling the spatial and vertical extents of tidal deposition at this time. Whilst spatial variations in post-depositional compaction of the underlying sediments may account for the rather low altitude for the peat with clay and wood in core T8, this unit possesses detrital characteristics, i.e. sharp stratigraphic contacts and limited evidence of 'succession' from pollen and diatom analyses. The sandy peat with wood at the base of the Holocene silts and clays in core T2A, dated to 9437–9843 cal. a BP, also appears to be detrital, although its altitude is more in accordance with the possible early Holocene sea-level trend inferred from the altitude of the sea-level index point from Thornaby, i.e. −10.61 m OD at 10 420–11 007 cal. a BP. Hence, these organic units provide equivocal limiting ages for the subsequent period of marine inundation. In addition, sedimentation in the inner estuary at this time will clearly have been influenced by the morphological high of the Late Glacial laminated clays in the outer estuary, which may have had to have been overtopped or at least by-passed before marine inundation of the inner estuary took place. If this is so, tidal sedimentation prior to *c.* 7.3 ka BP will have been limited by the morphological high, and may then have taken place through rather sudden inundation after this time. Hence, erosion and reworking of the early Holocene sedimentary record in the inner estuary is likely.

Coastal sedimentation from approximately 8 to 2.5 ka BP was clearly influenced by relative sea-level rise, the record of which shows a decreasing trend through the mid-Holocene (Fig. 7). Whilst the morphological high in the outer estuary became submerged during this time, the deceleration in the rate of relative sea-level rise during the mid-Holocene led to the encroachment of perimarine and terrestrial peats over much of the Tees Estuary, and to the formation of fresh to brackish water lagoons at the very limit of tidal

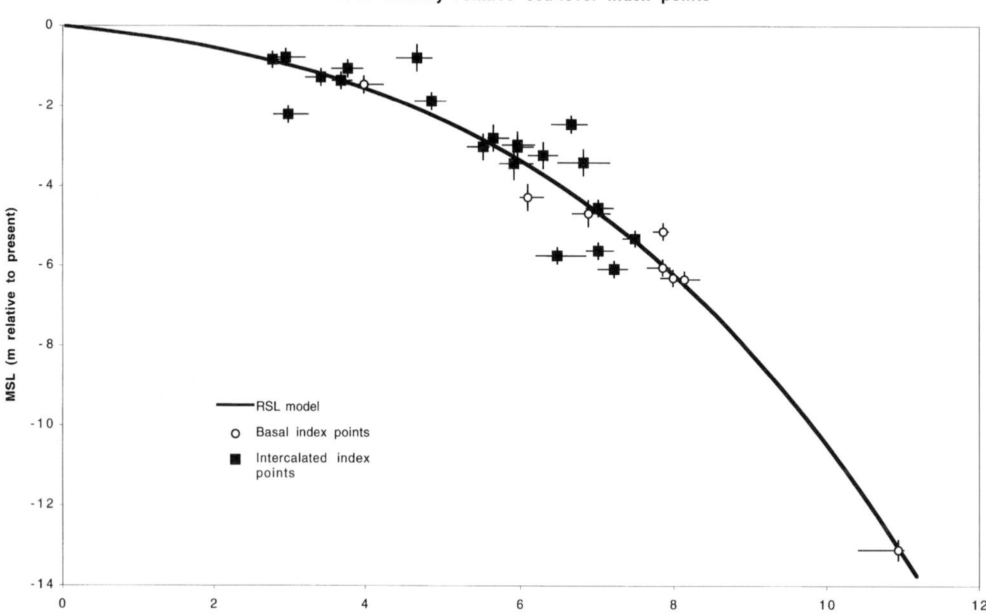

Fig. 7. Relative sea-level curve for the Tees constructed from ^{14}C-dated sea-level index points according to Shennan (1986). The trend line from the relative sea-level (RSL) model is from Shennan *et al.* (this volume *a*).

penetration, and beyond, in Billingham Beck (Wright *et al.* 1997) The sequences of environmental change in both the mid–outer and inner estuary environments are preserved in the diatom and pollen records obtained from cores CM01, HFBB11, PM5, BBC6 and BBC3, and can be summarized with reference to the diatom data for the two end-members of this transect, i.e. CM01 (mid–outer estuary) and BBC3 (inner estuary) (Figs 8 and 9, respectively).

The stratigraphic record from core CM01 has been described previously in Plater and Poolton (1992), but the corresponding palaeoenvironmental record was based on a limited number of levels for which sufficient diatom valves were counted. The earliest evidence of Holocene sedimentation is the deposition of a thin peat bed, for which a ^{14}C age of 7737–7929 cal. a BP was determined. This overlies a reworked pre-Holocene surface comprised of weathered till and contorted laminated clay. The thin peat passes upward into a browny-grey clay which, from the additional diatom data presented in this study (Fig. 8), was deposited on an intertidal to lower supratidal mudflat (Fig. 8; diatom assemblage zone 2). This environment of deposition then appears to have undergone some degree of shallowing, with sedimentation taking place at the mudflat–saltmarsh interface in the upper 10 cm or so of the browny-grey clay. Between depths of 4.55 and 3.25 m, the clay facies becomes mottled and contains dispersed organic material. The diatom data from this overlying mottled, organic clay are indicative of deposition in a tidal embayment on an intertidal to lower supratidal mudflat (diatom assemblage zone 3), suggesting a deepening and an increase in the marine influence on sedimentation. Diatoms are largely absent from the upper part of this mottled unit and the overlying grey clay, but the transition from the grey clay to the overlying peat bed preserves a few diatom valves, which are indicative of saltmarsh–mudflat deposition (diatom assemblage zone 4). The continuing reduction in the marine influence, probably as a consequence of a decelerating rise of relative sea-level (Fig. 7), resulted in the initiation of peat accumulation at 5918–6174 cal. a BP, which continued until 3579–3831 cal. a BP. The period of peat accumulation was brought to a gradual end by the encroachment of saltmarsh and mudflat environments within a tidal inlet (diatom assemblage zones 5 and 6), which become further developed in the lower 15 cm of the overlying mottled grey clay with organic laminations (diatom assemblage zone 7).

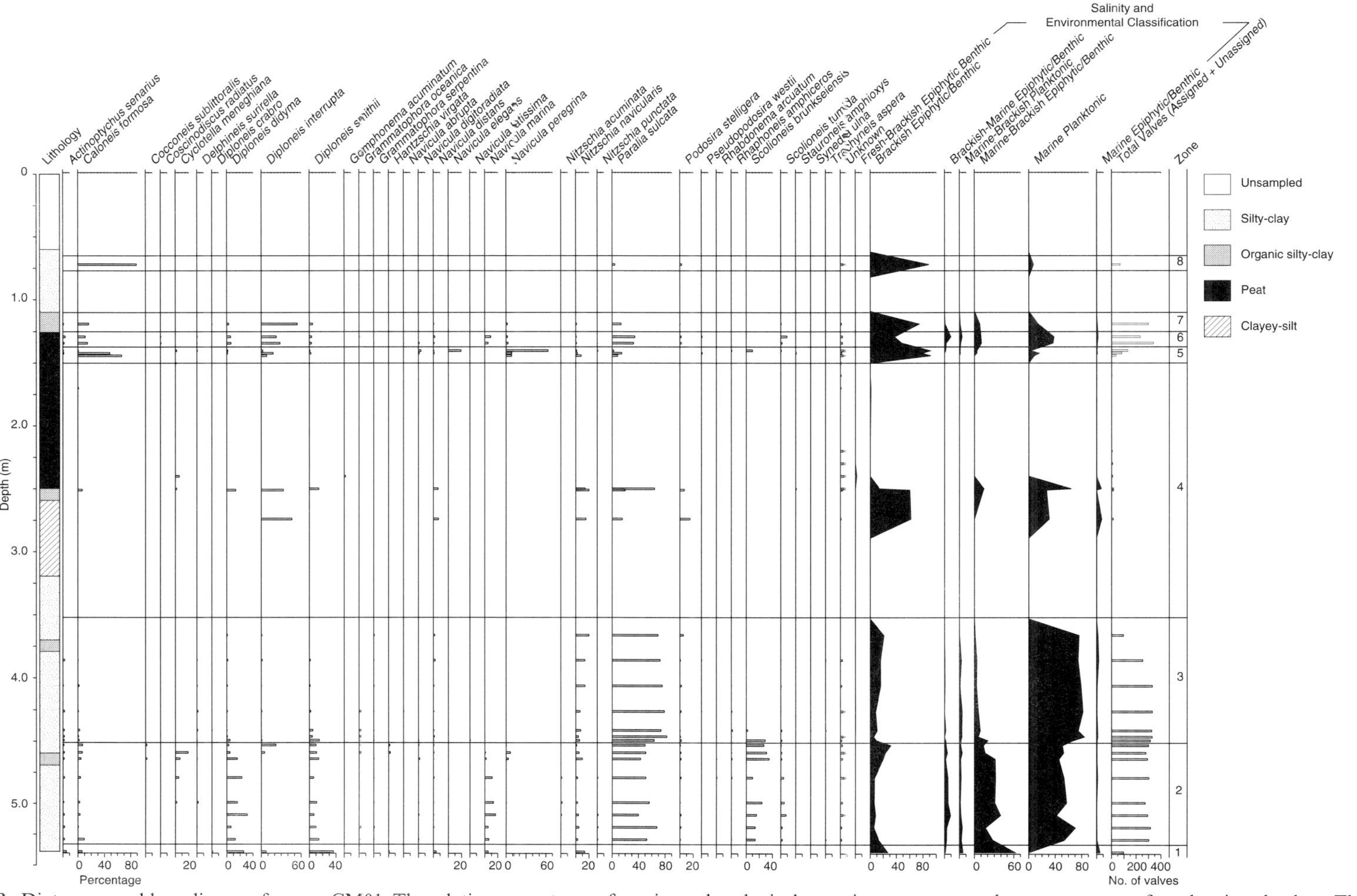

Fig. 8. Diatom assemblage diagram for core CM01. The relative percentages of species and ecological groupings are expressed as a percentage of total assigned valves. The diatom assemblage zones denote parts of the record where the environment of deposition does not change significantly.

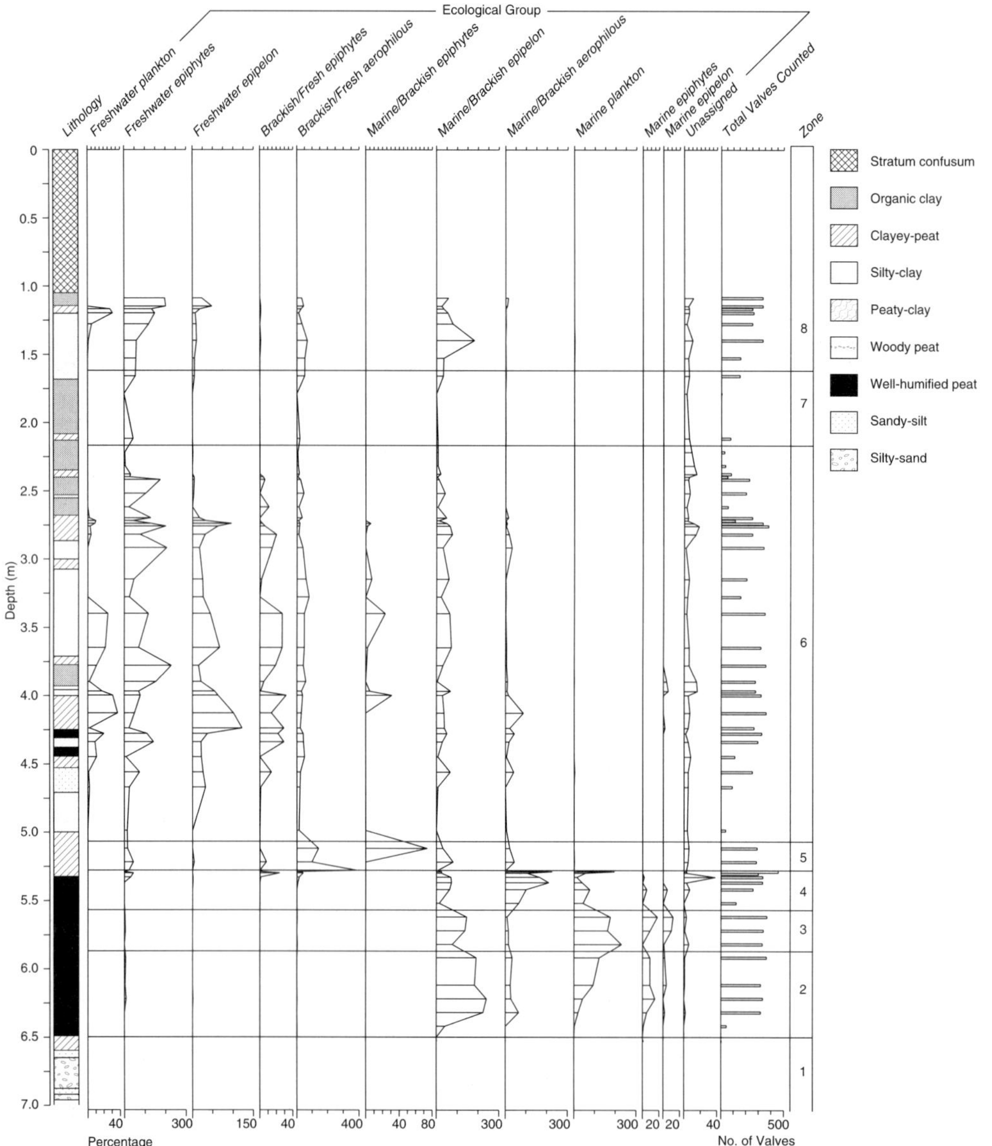

Fig. 9. Diatom assemblage (environmental classification) diagram for core BBC3.

In contrast to core CM01, the diatom record from Billingham Beck illustrates a more permanent change in the environment of deposition during the mid-Holocene. A 7 m core was recovered from the region of Billingham Beck Country Park. The stratigraphic sequence in core BBC3 proved to be extremely variable. Peat beds, particularly those at depths of 6.59, 6.49, 5.25 and 4.99 m, enabled a reasonable chronological framework to be established for the preserved sequence of environmental change. The earliest environment of deposition was that of a tidal flat (Fig. 9 diatom assemblage zone 2), which may have resulted from marine inundation of Billingham Beck after *c.* 8.3–7.6 cal. ka BP (Table 1). From the diatom assemblage data, this gave way to a subtidal environment and tidal channel (diatom assemblage zone 3) under increasing marine influence. The reversal of this trend resulted in a return to shallow intetidal conditions (diatom assemblage zone 4) and, eventually, a supratidal setting (diatom assemblage zone 5) by 7656–7549 cal. a BP. By 6864–7207 cal. a BP, a lagoonal environment had developed, which then characterized sedimentation in this region of Billingham Beck through the mid–late Holocene, with some degree of temporal variation in both the fresh and brackish water influences. Hence, Billingham Beck appears to have been at the limit of the tide since approximately 7.5 ka BP. This transition from a marine to a perimarine sedimentary environment corresponds to the deceleration in the rate of relative sea-level rise at this time (Fig. 7). Further evidence of a sea-level control on the change from tidal flat to lagoonal environments is provided by a neighbouring core (BBC6) in which the transition takes place at a slightly later date (7017–7392 cal. a BP), suggesting a gradual, but not extensive, retreat of the marine influence from Billingham Beck.

In addition to the widespread deposition of the intercalated peat and tidal mud facies in the outer estuary during the mid–late Holocene, a sequence of sands, up to 19 m in thickness, was deposited seaward of the Late Glacial laminated clay morphological high. The coarse grain luminescence dating of these sands in cores T5 and T4 places their deposition between *c.* 6.3 and 3.5 ka BP, with the ages falling in correct chronological order, and three of the four exhibiting a reduced degree of scatter and, therefore, more efficient zeroing at the time of deposition. Furthermore, the sands in T4 overlie a suite of silty-clays and clayey-silts, the top of which are geochemically similar to the main body of Holocene silts and clays (Fig. 3). The radiocarbon date of $25\,850 \pm 360$ ^{14}C a BP on a thin peat with sand and wood at a depth of 20.70 m in core T4 is, therefore, out of sequence and is probably derived from allochthonous organic material. It would seem that fine-grained tidal deposition in the outermost part of the Tees Estuary gave way to an extensive phase of marine sand deposition during the early part of the mid-Holocene. This continued until the late Holocene, although the intercalated Holocene tidal muds had advanced seaward of the Late Glacial morphological high by approximately 5 ka BP.

This second phase of sand deposition, the first being located in the inner part of the Tees Estuary, appears to be derived from the input of material from the North Sea rather than the Tees catchment. Although these marine sands cannot be distinguished from the early Holocene fluvial–estuarine sands on the basis of their typical spidergrams (Fig. 3), the former have higher CaO/MgO and lower Rb/Sr ratios (Fig. 6) and are geochemically similar to typical North Sea sediments (Stevenson *et al.* 1995). Hence, a period of onshore sand transport accompanied the decelerating rise of relative sea-level during the mid-Holocene.

As with many palaeoenvironmental studies, the Holocene record of sediment flux and accretion in the Tees Estuary is of limited temporal resolution due to the relatively small number of radiocarbon and luminescence dates distributed throughout the stratigraphic record. As a consequence, rates of sedimentation can only be averaged over thick sedimentary sequences for which ages are available on organic units at their upper and lower limits. In previous studies on coastal lowland sequences, the age information is provided by ^{14}C dates on intercalated peat beds (Tooley 1978; Shennan 1986) and, hence, the temporal resolution on sedimentary history of minerogenic sequences is very poor. In this study, a diatom-based transfer function approach (Horton *et al.* this volume) has been used in combination with an established relationship between present-day sediment accumulation rates and altitude in the tidal frame (see Late Holocene sediment supply and contamination) to provide more detailed information on Holocene sediment flux and storage in the Tees Estuary.

From the preceding discussion on the history of sedimentation in the Tees, the succession of sedimentary environments can be linked with temporal changes in the marine influence during the Holocene, primarily as a consequence of changes in the rate of relative sea-level rise. Assuming that diatoms have not changed their ecological responses to certain environmental variables through time, it is possible to reconstruct not only changes in the marine influence but, more significantly, changes in the altitude of the site of deposition relative to various reference tidal levels (Horton 1997; Zong & Horton 1999). Indeed, research on the present distribution of diatom species on the saltmarshes and tidal flats of Greatham Creek has shown that their distribution is related to ground altitude as a function of tidal inundation frequency and duration, sediment grain size and organic content (Horton 1997; Zong & Horton 1999). Furthermore, as illustrated in Fig. 10, sediment accumulation rate is a decreasing linear function of altitude in the vertical range between the mudflat and upper saltmarsh environments of Greatham Creek. Thus, a two stage analysis enables, firstly, the reconstruction of a former altitude or tidal level from the diatom assemblage data and, secondly, a sediment accumulation rate from the reconstructed altitude.

The results of this reconstruction for cores CM01, HFBB11, PM5 and BBC3 are given in Fig. 11a–d. In addition, where sediment accumulation rates can be calculated for intercalated

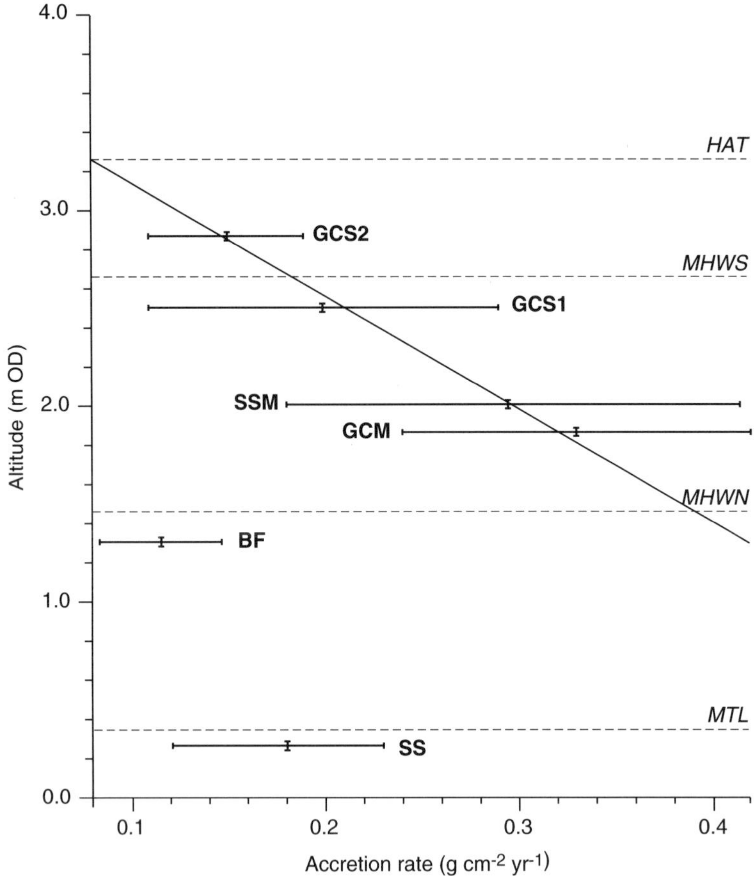

Fig. 10. Relationship between sediment accumulation (grams per square centimetre per annum) and altitude in the tidal frame as determined from radionuclide dating of short cores from the present inter- and supratidal zones. (HAT, highest astronomical tide; MHWS, mean high water of spring tides; MHWN, mean high water of neap tides; MTL, mean tidal level.)

sequences where ^{14}C and luminescence dates are available, 'actual' accumulation rates are displayed as a comparison with those 'predicted' by the transfer function. It is clear that the predicted accumulation rates largely reflect changes in the marine influence, and that the lower parts of the cores preserve faster sediment accumulation rates than the upper parts. This is probably a consequence of decreasing relative sea-level rise through the Holocene (as discussed earlier). However, cores CM01, HFBB11 and PM5 also exhibit an increase in sediment accumulation in their upper parts, which does not appear to be linked to any significant acceleration in the rate of relative sea-level rise (Fig. 7). In terms of age, the faster sediment accumulation rates below approximately −1.5 m OD date from approximately 7.8 to 7 ka BP, whilst those between *c.* +0.3 and +1.1 m OD are post 3 ka BP.

Comparison of the predicted and actual sediment accumulation rates reveals that the transfer function approach overestimates sediment accumulation by an average factor of three, although there is a closer match between the two data sets during the period from approximately 7.8–6 ka BP. This disparity is unlikely to be related to autocompaction, particularly as dry mass accumulation rates (grams per square centimetre per annum) are used in preference to linear sedimentation rates (millimetres per annum). Hence, the contemporary tidal sediment regime would appear to be an inappropriate analogue for the majority of the Holocene. If, however, attention is focused on the periods of time during the Holocene where the match between the predicted and actual sediment accumulation rates is closer, information on enhanced sediment flux may be

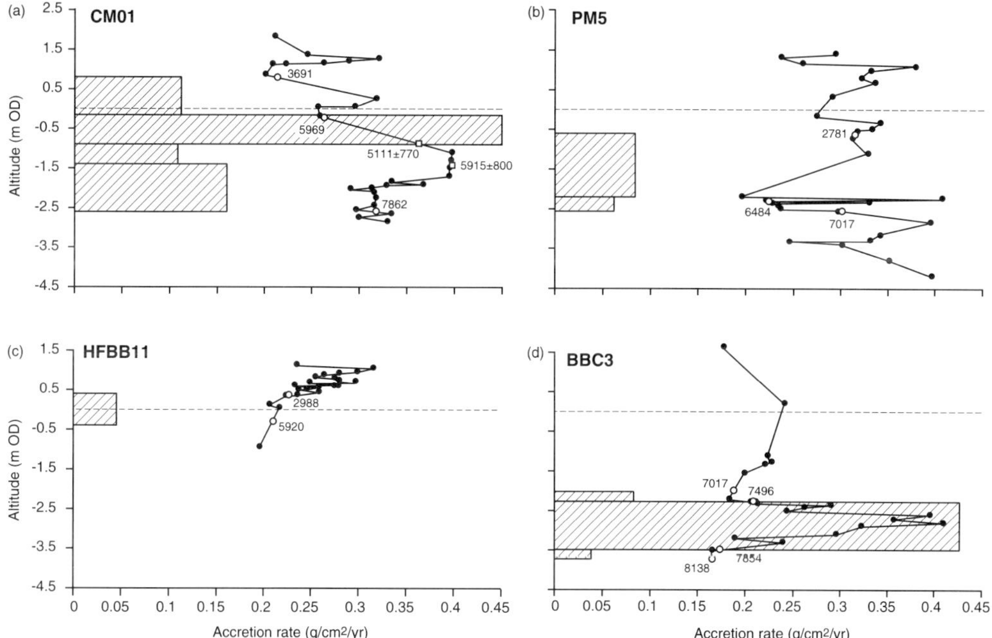

Fig. 11. Rates of sediment accumulation reconstructed from the diatom-based transfer function approach for cores CM01 (**a**), PM5 (**b**), HFBB11 (**c**) and BBC3 (**d**). The predicted rates of accumulation (line and solid circles) are compared with actual rates (shaded bars) obtained from sedimentary units with confining radiocarbon (open circle, mean calibrated age) and luminescence ages (open box, mean age).

obtained. The first of these periods, between 7.8 and 6 ka BP, is probably due to relative sea-level rise and changes in the tidal regime. Although climate change and Mesolithic clearance activity cannot be ruled out as a contributory factor (Smith 1970; Tallis & Switsur 1973; Jacobi *et al.* 1976; Jones 1976; Simmons & Innes 1985; Innes & Simmons 1988; Simmons 1993), the rate of sediment accumulation in the Tees Estuary will have increased in response to changing tidal dynamics during the early–mid-Holocene. McArthur *et al.* (1998) and Shennan *et al.* (this volume *b*) propose that *c.* 8 ka BP, spring tidal range in the Tees region was approximately 63% of its present magnitude. Tidal range is considered to have increased to approximately 90% of the present range by *c.* 6 ka BP, after which it continued to increase roughly linearly to the present day. The relatively rapid increase in tidal range between approximately 8 and 6 ka BP is likely to have been accompanied by an increase in tidal asymmetry due to increased flow velocity of the tidal wave and reduced bottom friction in deeper water, as observed for the Dutch coast during the early–mid-Holocene (Franken 1987; Hulsen 1994; Vos & van Heeringen 1997). Increasing tidal asymmetry not only accounts for increasing tidal sediment accumulation between 7.8 and 6 ka BP, but is also a crucial reason why the present-day tidal sediment regime is not typical of much of the Holocene.

The increase in sediment accumulation after approximately 3 ka BP is more difficult to explain by relative sea-level rise, mainly because there is no evidence of any acceleration in the rate of rise at this time (see Fig. 7). Although crustal flexure due to residual forebulge collapse may have encouraged renewed relative sea-level rise, many studies of post-glacial rebound propose that forebulge migration should have been complete well before this time (Lambeck 1991, 1995). Similarly, an acceleration in the rate of sea-level rise at *c.* 3 ka BP cannot be entirely discounted, particularly as several other sites in the UK preserve a record of an increase in the marine influence at this time (e.g. Liverpool Bay, Northumberland, the Fenland and Romney Marsh). Alternatively, an enhanced influx of sediment from the terrestrial system may have brought about the transition from organic to minerogenic sedimentation and an acceleration in the rate of tidal sediment accumulation. For example, the first significant opening of the tree canopy is thought to date from the Bronze Age, *c.* 3.3 ka BP (Turner *et al.* 1973; Chambers 1978), with a more extensive sequence of clearance from the Iron Age, *c.* 2.2 ka BP (Chambers 1978; Turner 1979, 1981; Roberts 1989; Simmons *et al.* 1993), to the Roman occupation of northern Britain, *c.* 1.8 ka BP (Barber *et al.* 1993; Simmons *et al.* 1993). In their study of lakes in the English Lake District, Pennington & Lishman (1984) note a marked increase in terrestrial sediment influx at approximately 2.5–2.3 ka BP, whilst Macklin & Lewin (1993) have linked alluviation in the river valleys of northern England from *c.* 2 ka BP onward to the release of soil. Although soil erosion as a result of human activity is no doubt an important factor in catchment sediment yield during the late Holocene, perhaps climate change, i.e. increased runoff as a consequence of increased wetness, may be responsible for these large-scale changes in sediment flux (Macklin *et al.* this volume). Indeed, significant alluviation in the Ouse catchment between 900 and 430 BC can be related to a wet-shift (Dinnin 1997), with the climate in Northern Britain from *c.* 850 to 760 BC being characterized by a change from warm and continental to oceanic conditions (Van Geel *et al.* 1996; Chiverrell 1998). Hence, records of peat bog surface wetness (e.g. Haslam 1987; Barber *et al.* 1993) may hold the key to understanding enhanced sediment flux in the Tees Estuary after *c.* 3 ka BP.

Whilst the diatom-based transfer function approach to the reconstruction of sediment accumulation may be flawed due to temporal changes in tidal dynamics and sediment supply, the former can be considered as being relatively insignificant after *c.* 6 ka BP and, hence, the results are of some significance in the study of mid–late Holocene sediment flux where more conventional data may be unobtainable.

Late Holocene sediment supply and contamination

Dating from the Roman occupation of northern Britain to the present day, the industrial activity of humans in the Tees catchment and estuary has had an increasing influence on the coastal lowland sedimentary record. As reported in Berry & Plater (1998) and Plater *et al.* (1999), both the deep boreholes and shallow (*c.* 0.5 m) cores contain a record of far- and near-field contaminant sources. However, further XRF geochemical analyses have revealed that the depth and extent of this contamination, particularly that related to mining activity in the Tees catchment, is greater than first assumed.

According to Plater *et al.* (1999) the evidence for the onset of mining contamination in the Tees Estuary is found in core T1, in which increasing concentrations of Zn, Ba and Pb upcore record a progressive increase in metal contamination (Fig. 12). Unlike its nearest neighbour, core T2A exhibits a more sudden increase in metal contamination, which has been interpreted as a depositional hiatus in a former southern meander loop of the Tees channel. Closer inspection of the XRF spidergrams reveals evidence of earlier, less marked metal concentrations in cores T7, T8, T10 and T5. For example, the typical spidergrams (Fig. 3) and mean concentration data (Table 2) both illustrate an almost progressive increase in Pb and Zn concentration from the early- and late-stage 'pre-contamination' silty-clays, through the 'mining contamination onset', to the 'mining contamination' and 'mining/twentieth-century

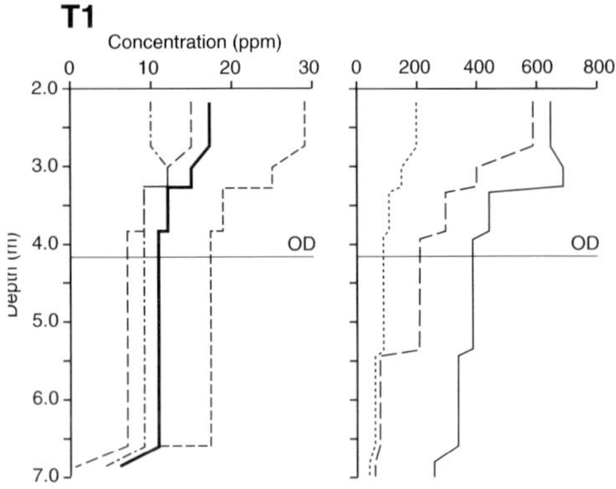

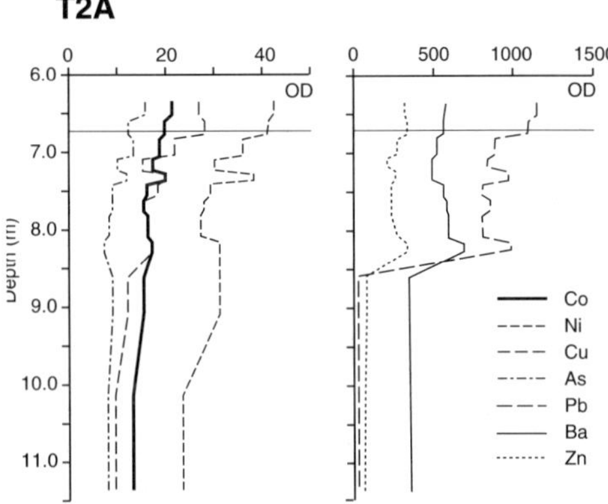

Fig. 12. Down-core plots of metal concentration for cores T1 and T2A illustrating a gradual onset of contamination in T1 and a sudden increase in metal contamination at approximately 8.20 m depth in T2A (after Plater *et al.* 1999).

contamination' litho-chemostratigraphic facies. Hence, the transition from late Holocene to 'contaminated' sediments can be seen in a series of sands, silts and clays, which, in the cases of cores T7 and T8, pass upward into a succession of early- and late-stage 'pre-contamination' sediments (Fig. 4), the latter of which show the first significant signs of enhanced Pb concentrations (Fig. 3 and Table 2). The chronology of contamination then continues upward into the 'mining contamination onset' sediments noted above for cores T1 (Fig. 4), to be completed by a widespread phase of sedimentation associated with mining and subsequent twentieth-century industrial activity, which is characterized by significant heavy metal contamination (Fig. 3 and Table 2), in particular Cu, Ni, Zn, Co, Pb and As (Berry & Plater 1998; Plater *et al.* 1999).

Chronological data for the earliest sedimentary records of enhanced Pb concentration are very difficult to establish. A fine-grain luminescence age on the lowermost 'contaminated' sediments at a depth of 5.40 m in core T1 dates the onset of contamination to 3200 ± 781 a BP; whilst a sandy-silt in core T5, exhibiting only limited evidence of enhanced metal concentrations, has been dated using a coarse-grain luminescence approach to 273 ± 36 and 424 ± 40 a BP at depths of 2.10 and 1.50 m, respectively. It proved particularly difficult to obtain an equivalent dose for the fine-grain luminescence age in core T1, although a second determination was more successful. Hence, the age of *c.* 3.2 ka BP may be older than the true age of deposition. The coarse-grain luminescence ages also exhibited some degree of scatter during the determination of equivalent dose, which is likely to be a consequence of poor bleaching on deposition. Although the earliest luminescence age agrees well with the proposed onset of enhanced terrestrial sediment flux from the Bronze Age onward, it is difficult to account for the marked increase in metal concentration merely as a consequence of land clearance. Alternatively, the onset of contamination may be linked to the earliest unequivocal record of metal contamination in the Northern Pennines Orefield which, following the discovery of two pigs of smelted lead dated to AD 81 (Raistrick & Jennings 1965), dates from the Roman occupation of northern Britain (Dunham 1990). It seems most likely that the sequence of metal contamination preserved in core T1 illustrates increasing Zn and Pb production in the Tees catchment between the seventeenth and nineteenth centuries, the Pb mining peaking from AD 1815 to 1880 and the Zn from AD 1880 to 1920 (Dunham 1944). Indeed, as illustrated in Plater *et al.* (1999), the Zn/Pb ratios of the contaminated sediments in cores T1 and T2A are within the 0.3–0.4 range observed in overbank sediments in the Tees catchment, which reflect contamination from mining activity (after Hudson-Edwards *et al.* 1997). Clearly some redistribution of this sedimentary record in the coastal lowlands and the River Tees floodplain (Macklin *et al.* 1997; Hudson-Edwards *et al.* 1997) will have influenced the recent record of contamination, but enhanced concentrations of As in association with elevated Zn/Pb ratios and magnetic concentrations can be attributed to a combination of limited far-field zinc mining and near-field fertilizer production after the 1940s (Plater *et al.* 1999).

The record of twentieth-century metal contamination has been used in combination with a combined radionuclide chronology (^{210}Pb and ^{137}Cs) to establish chemozones (Allen & Rae 1986) and rates of tidal sediment accumulation for the present intertidal and supratidal environments of the Tees (Plater *et al.* 1999). Due to the human disturbance of the main Tees channel, primarily through dredging, this study has focused on Greatham Creek, which is probably the least disturbed of the tidal channels in the Tees Estuary. The radionuclide data reveal that peak metal and magnetic concentrations in cores GCS2, GCS1 and GCM (upper saltmarsh, lower saltmarsh and mudflat, see Fig. 1) date from the mid-1950s. Whilst the onset of enhanced metal and magnetic concentrations dates from approximately AD 1925 in the upper saltmarsh core, the lower saltmarsh and mudflat cores preserve a reduction in contaminant levels after the early 1980s (Plater *et al.* 1999).

As discussed previously, a broadly linear relationship is observed between altitude in the tidal frame and dry mass accumulation rate from mudflat to upper saltmarsh environments (Fig. 10). Although, the contamination record in the sandflat environment of Seal Sands (SS; Fig. 1) was originally interpreted as being very recent (Berry & Plater 1998), i.e. post-1950, the radionuclide chronology reveals a much slower rate of sediment accumulation. In addition, whilst the 'reclaimed' marshland (BF; Fig. 1) exhibits a low rate of sedimentation for its present altitude, this environment was effectively taken out of the tidal sedimentary system at the turn of the century. Hence, very little sediment has been delivered to this area since land claim, and site BF (the ICI Brine Fields) may well have been a saltmarsh at the time of embankment. From Table 3 it is encouraging that the rate of sediment accumulation (grams per square centimetre per annum) appears to be a direct function of tidal inundation frequency and duration.

Table 3. *Mean sedimentation rates for short cores in the present inter- and supratidal zones from Greatham Creek*

	Altitude (m OD)	Linear rate ($mm\,a^{-1}$)	Mass accumulation rate ($g\,cm^{-2}\,a^{-1}$)
GCS2	+2.86	3.16 ± 0.75	0.15 ± 0.04
GCS1	+2.50	3.74 ± 2.65	0.20 ± 0.09
SSM	+2.00	2.78 ± 1.03	0.30 ± 0.12
GCM	+1.84	3.29 ± 0.87	0.33 ± 0.09
BF	+1.31	1.22 ± 0.53	0.11 ± 0.03
SS	+0.26	2.50 ± 0.65	0.18 ± 0.05

However, there is no unequivocal increase in the rate of sedimentation in the pioneer marsh, which would be expected where the vegetation interrupts the tidal flow and, hence, encourages turbulence and sediment deposition. The model of tidal sedimentation is, therefore, perhaps oversimplified in that the rate of accumulation cannot merely be a linear function of the amount of time available for sediment to settle out of suspension. It is also of some significance that rates of sediment accumulation expressed as linear sedimentation rates (millimetres per annum) exhibit no clear relationship with altitude. This is perhaps a function of the balance between minerogenic and organic sedimentation on the marsh surface (Allen 1990). Sedimentation on the upper and lower saltmarshes may be more organic and, therefore, have a lower bulk density. Temporal variation in the rate of sediment accumulation should also be considered. The sedimentation rates given in Table 3 are average values calculated for the entire period of accretion preserved in each core. This is a little misleading since Plater *et al.* (1997) noted an acceleration in the rate of sediment accumulation for the mudflat and lower saltmarsh sediments of Greatham Creek from the mid-1970s to the mid-1980s, perhaps as a consequence of land claim and engineering works during the latter part of the twentieth century. Hence, the rates of sediment accumulation for the more active parts of the intertidal zone (mudflat and lower saltmarsh) may be higher than those expected if the tidal dynamics had not been disturbed during the period of record.

From the above discussion on recent sediment accumulation in Greatham Creek, it is clear that the response of the intertidal zone to sea-level rise during the twentieth century has not been limited by sediment supply. Whilst sea level has risen by, on average, 1–2 $mm\,a^{-1}$ over the last 100–150 years (Woodworth 1987; Douglas 1991), the Earth's crust has remained relatively stable in the Tees region as a result of it being located on a glacio-isostatic rebound hinge-line (Shennan 1989). Thus, it would appear that the saltmarshes and tidal flats of Greatham Creek have actually prograded during the latter part of the twentieth century, with the sediments accreting faster than the rise in tidal levels. Although this may seem encouraging in terms of the future response of the estuarine sedimentary environments to accelerated sea-level rise, the sediment budget during the recent past has been significantly enhanced by both industrial and agricultural activity, and has been disturbed aperiodically by land claim.

Controls of sediment flux and sedimentation

From the preceding sections, it is clear that climate change, sea-level trends, human activity and, to a more limited extent, coastal morphology have had an important bearing on the evolution of the Tees Estuary since the last glacial maximum, particularly in the context of temporal variations in sediment provenance, flux and storage.

The earliest evidence of a climate control on sediment supply dates from the Late Glacial, when influx to a proglacial lake in the Lower Tees basin appears to have been a function of mean annual temperature, the extent of spring meltwater and the degree of runoff generation during the summer months. Hence, the Late Glacial laminated silts and clays may provide a proxy record of climate between *c.* 18 and 16.5 ka BP (subject to more rigorous chronological control), even though they are perhaps a sequence of turbiditic or waning flow marginal deposits rather than true varves.

After a significant hiatus, during which the lake system collapsed following glacial retreat and a fall in relative sea-level due to post-glacial glacio-isostatic rebound of the crust, an early Holocene transgressive event led to marine inundation of the relict lake basin and weathered till surface. Although a morphological high, made up of Late Glacial silts and clays, remained in the outer part of the estuary, perhaps resulting in a branching of the main tidal channels, which then persisted until the late-nineteenth century, early Holocene sedimentation in the inner estuary was dominated by sand deposition and by the formation of variably intercalated tidal muds and terrestrial peats, initially in the mid estuary, but eventually extending over almost the entire estuary during the mid–late Holocene.

From approximately 8 to 2.5 ka BP, sedimentation over much of the Tees Estuary took place under the influence of a decreasing rate of relative sea-level rise and, during the earlier part of this period, increasing tidal range and asymmetry. Throughout this period, the fine-grained sediment appears to have been derived from the Tees catchment, although a simultaneous phase of marine deposition in the outer-estuary, i.e. seaward of the morphological high, can be linked to an influx of relatively coarse material to the Tees from the North Sea. It is of some significance that the majority of fine-grained sediment input to the estuary was derived from the Tees catchment, with perhaps any incoming coarse material being limited to the inner estuary and fringing brackish water lagoons. This is in accordance with the findings of Emery & Milliman (1978) who proposed that rivers provide more than 90% of the sediment input to the coastal zone. Indeed, even under a regime of enhanced (but decreasing) sea-level rise relative to the present day, the only evidence of a significant marine sediment input during the mid–late Holocene is that of the sandy facies deposited in the outer estuary. This is unlikely to be a function of available energy, but rather available sediment, in that cores from immediately offshore reveal the adjacent floor of the North Sea to be either an exposed bedrock ramp or glacial till armoured with a shelly lag or tidal sand deposits (Harland & Long 1996; Brew *et al.* 1999).

The late Holocene appears to have been a period of renewed sediment influx from the Tees catchment as a consequence of clearance and industrial activity by humans from the Bronze Age onward. Although the earlier part of this sediment input may also have been driven by relatively minor changes in climate, i.e. enhanced runoff generation of soil loss due to increased wetness, the later stages are characterized by increasing metal contamination due to Zn and Pb mining along the River Tees, perhaps from the Roman period but more likely from the seventeenth to the nineteenth centuries. Metal contamination in the estuarine sedimentary record peaked during the mid-twentieth century due to a combination of decreasing catchment mining activity, sediment reworking, and industrial and urban expansion in the Tees coastal lowlands. Indeed, near-field industrial development has left its mark in the sedimentary record in the form of elevated Zn/Pb ratios and enhanced As concentrations, although a notable decrease in metal contamination from the early 1980s onwards can be observed in the most recent record of tidal sedimentation.

In the context of temporal changes in the relative significance of factors influencing sediment flux and the post-glacial evolution of the Tees, sea-level appears to have been a fundamental control from deglaciation of the region until the late Holocene. Sea-level fall as a consequence of post-glacial rebound resulted in pre-Holocene erosion of the former Late Glacial lake basin and, hence, a net transport of material to the offshore zone. As glacio-isostatic rebound attenuated and eustatic sea-level rise became dominant, marine inundation of the Lower Tees basin resulted in an extensive period of sand and mud deposition in the inner- and mid-estuary environments, although the direction of sediment movement was primarily from the Tees catchment to the coastal zone. However, the outer part of the estuary records a landward flux of material from offshore, this being controlled by decelerating sea-level rise during the mid–late Holocene coupled with the grain size of the available sediment source. Sea level appears to have had a more limited influence on sedimentation in the Tees Estuary during the late Holocene, giving way to climate change and, in particular, human activity. Again, the net direction of sediment flux during this time was from the land to the coastal zone.

The most recent phase of sedimentation in Greatham Creek has been controlled by sea-level rise and human activity in the form of far-field mining and agriculture and near-field industrialization, urbanization and land claim. Indeed, at no stage in the recent past has the response of the saltmarshes and tidal flats to changes in sea-level been restricted by insufficient sediment supply. It may seem appropriate to consider the early–mid-Holocene as an analogue for the future response of the Tees Estuary to any greenhouse-induced acceleration in the rate of sea-level rise, but enhanced tidal sedimentation at this time was perhaps encouraged by changes in tidal dynamics. Future sea-level rise may, indeed, be accompanied by increasing tidal range and asymmetry, but the early-stage impacts are likely to be enhanced landward penetration of the tidal influence and wave energy, thus resulting in saltmarsh retreat and erosion of the upper profile (Pethick 1993). The flatter profile which results will introduce a negative feedback, encouraging energy dissipation *via* enhanced bottom friction, thus reducing the incoming wave energy and leading to deposition and a steepening of the profile. This self-regulation will only take place if the sedimentary environments are given sufficient space to retreat in the first instance, and providing that sufficient sediment is available to keep pace with the rising tidal levels. There is no reason to doubt such a response from the preceding discussion of sediment flux during various stages of the Holocene, particularly the late Holocene.

Conclusion

The post-glacial evolution of the Tees Estuary can be subdivided into three distinct periods of sediment accretion. The earliest of these is a period of laminated silt and clay deposition during the Late Glacial in which a proxy record of climate change is preserved in the sequence of rhythmite thickness. Following a period of erosion, consequent upon deglaciation and relative sea-level fall, the early–mid-Holocene witnessed renewed sea-level rise and marine inundation of the former lake basin, leading to deposition of an inner-estuary sand facies overlain by a sequence of intercalated tidal muds and perimarine to terrestrial peat deposits. This stratigraphic record was laid down under the influence of a decelerating rise in relative sea-level accompanied by increasing tidal range and asymmetry between approximately 8 and 6 ka BP. After this time, a significant reduction in the marine influence resulted in the formation of fresh–brackish water lagoonal environments at the limit of tidal penetration and the encroachment of peat deposits over the mid–outer estuary. At the same time, a marine sand facies was deposited in the outer estuary against a morphological remnant of the Late Glacial silts and clays. Whilst a renewed increase in the marine influence after approximately 3 ka BP resulted in a third period of sedimentation, this late Holocene phase is characterized by increasing evidence of human activity and metal contamination, which culminated during the early 1980s.

In considering the main controls on sediment provenance and flux since the last glacial maximum, sea-level change has been the underlying control from the Late Glacial to the late Holocene, initially in the context of relative sea-level fall due to post-glacial rebound, but mainly in the form of a decelerating rate of relative sea-level rise and changing tidal dynamics during the early–mid-Holocene. Climate appears to have had an influence on sediment flux during the Late Glacial and perhaps from *c.* 3 ka BP onward, but this later period has also been characterized by increasing human activity via land clearance, mining, agriculture and, more recently, industrial and urban development in the coastal lowlands. Coastal morphology has also had some control on the evolution of the Tees since the Late Glacial in the form of a morphological high in the outer estuary, preventing marine sands from entering the mid estuary during the mid–late Holocene. Most recently, tidal sedimentation in Greatham Creek during the twentieth century has been controlled by the frequency and duration of tidal inundation (a function of relative sea-level trends) and human disturbance of the inter- and supratidal zones.

In the context of temporal variations in sediment flux, deglaciation and post-glacial rebound led to the collapse of the Late Glacial lake basin and, hence, the transport of eroded lake sediments offshore. Marine inundation of the Tees Estuary during the early Holocene resulted in the deposition of sands in the inner estuary and served to limit the extent of offshore sediment movement, eventually leading to the deposition of sedimentary material derived from the Tees catchment in the coastal zone during much of the Holocene. Indeed, the only evidence of any onshore flux of sediment during the mid–late Holocene is that of marine sand deposition, derived from the North Sea, in the outer estuary. The combined influences of climate change and human activity during the last 3 ka or so have enhanced the flux of sediment to the coastal zone from the catchment of the River Tees and, hence, it is likely that the mudflats and saltmarshes of the estuary will respond favourably to any enhanced sea-level rise during the next century.

The authors should like to thank all those associated with the research presented here, especially P. Appleby, A. Berry, S. Björck, D. Brew, R. Chiverrell, J. Dearing, C. Evans, H. Glaves, V. Holliday, K. Hudson-Edwards, J. Innes, A. Long, M. Macklin, F. Musson, S. Nolan, A. Pemberton, E. Santa Maria, N. Shenton, K. Swales, P. Vos, M. Weldon and Yongqiang Zong who have all contributed to the final reporting stage but are not included in the author list. Thanks are also extended to S. Mather for the final production of the figures to a very tight schedule. J. Ridgway publishes by permission of the Director, British Geological Survey (NERC). This research was undertaken as part of the NERC-funded LOIS special topic (grant no. LOEPS 33, GST/02/0738). This is LOIS publication number 588.

References

AGAR, R. 1954. Glacial and post-glacial geology of Middlesbrough and the Tees Estuary. *Proceedings of the Yorkshire Geological Society*, **29**, 237–253.

ALLEN, J. R. L. 1990. Salt-marsh growth and stratification: a numerical model with special reference to the Severn Estuary, Southwest Britain. *Marine Geology*, **95**, 77–96.

ALLEN, J. R. L. & RAE, J. E. 1986. Time-sequence of metal pollution. Severn Estuary, southwestern UK. *Marine Pollution Bulletin*, **17**, 427–431.

ANDERSON, R. Y. 1996. Seasonal sedimentation: a framework for reconstructing climatic and environmental change. *In*: KEMP, A. E. S. (ed.) *Palaeoclimatology and Palaeoceanography from Laminated Sediments*. Geological Society, Special Publications, **116**, 1–15.

BARBER, K. E., DUNMAYNE, L. & STONEMAN, R. 1993. Climatic change and human impact during the late Holocene in northern Britain. *In*: CHAMBERS, F. M. (ed.) *Climate Change and Human Impact on the Landscape: Studies in Palaeoecology and Environmental Archaeology*. Chapman & Hall, London, 225–236.

BERRY, A. & PLATER, A. J. 1998. Rates of tidal sedimentation from records of industrial pollution and environmental magnetism: the Tees Estuary, north-east England. *Water, Air and Soil Pollution*, **106**, 463–479.

BJÖRCK, S., KROMER, B., JOHNSON, S., BENNIKE, O., HAMMARLUND, D., LEMDAHL, G., POSSNERT, G., RASMUSSEN, T. L., WOHLFARTH, B., HAMMER, C. U. & SPURK, M. 1996. Synchronized terrestrial–atmospheric deglacial records around the North Atlantic. *Science*, **274**, 1155–1160.

BREW, D. S., HARLAND, R., RIDGWAY, J., LONG, D. & WRIGHT, M. R. 1999. Changing oceanographic conditions in the post-glacial North Sea – the story preserved in the sediments. *Ocean Challenge*, **8**(3), 26–31.

BRITISH GEOLOGICAL SURVEY 1996. *Regional Geochemistry of North-east England*. British Geological Survey, Keyworth, Nottingham.

CHAMBERS, C. 1978. A radiocarbon-dated pollen diagram from Valley Bog, on the Moor House National Nature Reserve. *New Phytologist*, **80**, 273–280.

CHIVERRELL, R. 1998. *Moorland Vegetation History and Climatic Change on the North Yorks Moors During the Last 2000 Years*. PhD thesis, University College of Ripon and York St John.

DE GEER, G. 1912. A geochronology of the last 12 000 years. *Compte rendu, 11th International Geological Congress (1910)*. Stockholm, **1**, 241–253.

DINNIN, M. 1997. Introduction to the palaeoenvironmental survey. *In*: VAN DE NOORT, R. & ELLIS, S. (eds) *Wetland Heritage of the Humberhead Levels*. Humber Wetlands Project, English Nature, 31–45.

DOUGLAS, B. C. 1991. Global sea level rise. *Journal of Geophysical Research*, **96**(C4), 6981–6992.

DUNHAM, K. C. 1944. The production of galena and associated minerals in the northern Pennines with comparative statistics for Great Britain. *Transactions of the Institution of Mineralogists and Metallurgists*, **53**, 181–252.

——1990. *Geology of the Northern Pennines Orefield*, **1**, *Tyne to Stainmore*. 2nd edition. Economic Memoirs of the British Geological Survey, Sheets 19 and 25 and 13, 24, 26, 31, 32 (England and Wales). HMSO, London.

EMERY, K. O. & MILLIMAN, J. D. 1978. Suspended matter in surface waters: influence on river discharge and upwelling. *Sedimentology*, **25**, 125–140.

FRANKEN, A. F. 1987. *Rekonstruktie van het paleogetijklimaat in de Noordzee*. Rapport Waterloopkundig Laboratorium Delft, X 0029-00.

GROOTES, P. M., STUIVER, M., WHITE, J. W. C., JOHNSEN, S. & JOUZEL, J. 1993. Comparison of oxygen isotope records from GISP2 and GRIP Greenland ice cores. *Nature*, **366**, 552–554.

HARLAND, R. & LONG, D. 1996. A Holocene dinoflagellate cyst record from offshore north-east England. *Proceedings of the Yorkshire Geological Society*, **51**(1), 65–74

HART, J. A. 1992. Sedimentary environments associated with glacial Lake Trimmingham, Norfolk, UK. *Boreas*, **21**, 119–136.

HASLAM, C. J. 1987. *Late Holocene Peat Stratigraphy and Climatic Change – A Macrofossil Investigation from the Raised Mires of North Western Europe*. PhD thesis, University of Southampton.

HORTON, B. P. 1997. *Quantification of the Indicative Meaning of a Range of Holocene Sea-level Index Points from the Western North Sea*. PhD thesis, University of Durham.

——, EDWARDS, R. J. & LLOYD, J. M. 2000. Applications of a microfossil-based transfer function in Holocene sea-level studies. *This volume*.

HUDSON-EDWARDS, K., MACKLIN, M. & TAYLOR, M. 1997. Historical metal mining inputs to Tees river sediment. *Science of the Total Environment*, **194/195**, 437–445.

HULSEN, L. J. M. 1994. *Modellering paleo-getij in de Noordzee*. Werkdocument Waterloopkundig Laboratorium, Arch. nr. vr885.94, Delft.

INNES, J. B. & SIMMONS, I. G. 1988. Disturbance and diversity: floristic changes associated with pre-elm decline woodland recession in North East Yorkshire. *In*: JONES, M. (ed.) *Archaeology and the Flora of the British Isles*. University Committee for Archaeology, Oxford, 7–20.

JACOBI, R. M., TALLIS, J. H. & MELLARS, P. A. 1976. The southern Pennine Mesolithic and ecological record. *Journal of Archaeological Science*, **3**, 307–320.

JONES, R. L. 1976. The activities of Mesolithic man: further palaeobotanical evidence from north-east Yorkshire. *In*: DAVIDSON, D. A. & SHACKLEY, M. L. (eds) *Geoarchaeology: Earth Science and the Past*. Duckworth, London, 355–367.

JOPLING, A. V. 1975. Early studies on stratified drift. *In*: JOPLING, A. V. & MCDONALD, B. C. (eds) *Glaciofluvial and Glaciolacustrine Sedimentation*. Society of Economic Palaeontologists and Mineralogists, Special Publications, **23**, 4–21.

KEMP, A. E. S. 1996. *Palaeoclimatology and Palaeoceanography from Laminated Sediments*. Geological Society, Special Publications, **116**.

LAMBECK, K. 1991. Glacial rebound and sea-level change in the British Isles. *Terra Nova*, **3**(4), 379–389.

——1995. Late Devensian and Holocene shorelines of the British Isles and North Sea from models of glacio-hydro-isostatic rebound. *Journal of the Geological Society, London*, **152**, 437–448.

McArthur, J. J., Shennan, I., Horton, B. P. & Innes, J. B. 1998. Modelling the Holocene palaeogeographies and tidal regimes of the Western North Sea (abstract). *In*: Barrett, R. L. (ed.) *LOIS – Final Annual Meeting, University of East Anglia*, NERC, Swindon, 148.

Macklin, M. & Lewin, J. 1993. Holocene river alluviation in Britain. *Zeitschrift für Geomorphologie, Neue Folge – Supplementband*, **88**, 109–122.

——, Hudson-Edwards, K. & Taylor, M. 1997. The significance of pollution from historical metal mining in the Pennine orefields on river sediment-associated metal fluxes to the North Sea. *Science of the Total Environment*, **194/195**, 391–397.

——, Taylor, M. P., Hudson-Edwards, K. & Howard, A. J. 1999. Holocene environmental change in the Yorkshire Ouse Basin and its influence on river dynamics and sediment fluxes to the coastal zone. *This volume*.

Natural Environment Research Council 1994. *Land–Ocean Interaction Study (LOIS) – Implementation Plan*. NERC, Swindon, 61p.

Pennington, W. & Lishman, J. P. 1984. The post-glacial sediments of Blelham Tarn: geochemistry and palaeoecology. *Archiv fur Hydrobiologie, Suppl.-Bd*, **69**(1), 1–54.

Pethick, J. 1993. Shoreline adjustments and coastal management: physical and biological processes under accelerated sea-level rise. *The Geographical Journal*, **159**(2), 162–168.

Plater, A. J. & Poolton, N. R. J. 1992. Interpretation of Holocene sea-level tendency and intertidal sedimentation in the Tees estuary using sediment luminescence techniques: a viability study. *Sedimentology*, **39**, 1–15.

——, Berry, A., Wright, M. R. & Appleby, P. G. 1997. Rates of tidal sedimentation from industrial pollution and radionuclide chronologies (abstract). *In*: Barrett, R. L. (ed.) *LOIS – second annual meeting, Hull*. NERC, Swindon, 27–29.

——, Ridgway, J., Appleby, P. G., Berry, A. & Wright, M. R. 1999. Historical contaminant fluxes in the Tees Estuary: geochemical, magnetic and radionuclide evidence. *Marine Pollution Bulletin.*, **37**, 343–360.

Radge, G. W. 1939. The glaciation of North Cleveland. *Proceedings of the Yorkshire Geology Society*, **24**, 180–205.

Raistrick, A. 1934. The correlation of glacial retreat stages across the Pennines. *Proceedings of the Yorkshire Geological Society*, **22**, 119.

—— & Jennings, B. 1965. *A History of Lead Mining in the Pennines*. Longman, London, 347p.

Rees, J. G., Ridgway, J., Knox, R. W. O'B., Ellis, S., Newsham, R. & Parkes, A. 1999. Holocene sediment storage in the Humber Estuary. *This volume*.

Renberg, I. & Segerström, U. 1981. Application of varved lake sediments in palaeoenvironmental studies. *Wahlenbergia*, **7**, 125–133.

Ridgway, J., Rees, J. G., Gowing, C. J. B., Ingham, M. N., Cook, J. M., Knox, R. W. O'B., Bell, P. D., Allen, M. A. & Molineaux, P. J. 1998. *Land–Ocean Evolution Perspective Study (LOEPS) Core Programme, Geochemical Studies 1: Methodology*. British Geological Survey, Technical Report, WB/98/55.

——, Andrews, J. E., Ellis, S., Horton, B. P., Innes, J. B., Knox, R. W. O'B., McArthur, J. J., Maher, B. A., Metcalfe, S. E., Mitlehner, A., Parkes, A., Rees, J. G., Samways, G. M. & Shennan, I. 2000. Analysis and interpretation of Holocene sedimentary sequences in The Humber Estuary. *This volume*.

Roberts, N. 1989. *The Holocene: an Environmental History*. Blackwell, Oxford, 227p.

Rowlatt, S. M. & Lovell, D. R. 1994. Methods for the normalisation of metal concentrations in sediments. International Council for the Exploration of the Sea, Marine Quality Committee, *E:20*

Shennan, I. 1983. Flandrian and Late Devensian sea-level changes and crustal movements in England and Wales. *In*: Smith, D. E. & Dawson, A. G. (eds) *Shorelines and Isostasy*. Institute of British Geographers, Special Publication, **16**, Academic Press, London, 255–283.

——1986. Flandrian sea-level changes in the Fenland II: Tendencies of sea-level movement, altitudinal changes, and local and regional factors. *Journal of Quaternary Science*, **1(2)**, 155–179.

——1989. Holocene crustal movements and sea-level changes in Great Britain. *Journal of Quaternary Science*, **4**, 77–89.

——, Lambeck, K., Horton, B. P., Innes, J. B., Lloyd, J., McArthur, J. & Rutherford, M. M. 2000*a*. Holocene crustal movements and relative sea-level changes on the East Coast of England. *This volume*.

——, ——, Flather, R., Horton, B., McArthur, J., Innes, J., Lloyd, J., Rutherford, M. & Wingfield, R., 2000*b*. Modelling western North Sea palaeogeographies and tidal changes during the Holocene. *This volume*.

Simmons, I. 1993. Vegetation change during the Mesolithic in the British Isles: some amplifications. *In*: Chambers, F. M. (ed.) *Climate Change and Human Impact of the Landscape: Studies in Palaeoecology and Environmental Archaeology*. Chapman & Hall, London, 109–118.

—— & Innes, J. B. 1985. Late Mesolithic land-use and its impact in the English uplands. *In*: Smith, R. T. (ed.) *The Biogeographical Impact of Land-Use Change: Collected Essays*. Biogeographical Monographs 2 Biogeography Study Group, University of Leeds, Leeds, 7–17.

——, Atherden, M. A., Cloutman, E. W., Cundill, P. R., Innes, J. B. & Jones, R. L. 1993. Prehistoric environments. *In*: Spratt, D. A. (ed.) *Prehistoric and Roman Archaeology of North-East Yorkshire*. CBA Research Report 87, BAR British Series 104, Council for British Archaeology, London, 15–50.

Smith, A. G. 1970. The influence of Mesolithic and Neolithic man of British vegetation: a discussion.

In: WALKER, D. & WEST, R. G. (eds) *Studies in the Vegetational History of the British Isles.* Cambridge University Press, Cambridge, 81–96.

SMITH, N. D. 1978. Sedimentation processes and patterns in a glacier fed lake with a low sediment input. *Canadian Journal of Earth Science*, **15**, 741–756.

STEVENSON, A. G., TAIT, B. A. R., RICHARDSON, A. E., SMITH, R. T., NICHOLSON, R. A. & STEWART, H. R. 1995. *The Geochemistry of Sea-bed Sediments of the United Kingdom Continental Shelf; the North Sea, Hebrides and West Shetland Shelves, and the Malin-Hebrides Sea Area.* British Geological Survey, Technical Report, WB/95/28C, British Geological Survey, Keyworth, Nottingham.

STRÖMBERG, B. 1994. Younger Dryas deglaciation at Mt Billingen, and clay varve dating of the Younger Dryas/Preboreal transition. *Boreas*, **23**, 177–193.

TALLIS, J. H. & SWITSUR, V. R. 1973. Forest and moorland in the south Pennine uplands in the mid-Flandrian period. I. Macrofossil evidence of the former forest cover. *Journal of Ecology*, **71**, 585–600.

TOOLEY, M. J. 1978. Interpretation of Holocene sea-level changes. *Geologiska I Stockholm Forhandlingar*, **100**, 203–212.

TURNER, J. 1979. The environment of northeast England during Roman times as shown by pollen analysis. *Journal of Archaeological Science*, **6**, 285–290.

——1981. The Iron Age. *In*: SIMMONS, I. G. & TOOLEY, M. J. (eds) *The Environment and British Prehistory*. Duckworth, London, 250–281.

——, HEWETSON, V. P., HIBBERT, F. A., LOWRY, K. H. & CHAMBERS, C. 1973. The history of the vegetation and flora of Widdybank Fell and the Cow Green reservoir basin, Upper Teesdale. *Philosophical Transactions of the Royal Society*, **B265**, 327–408.

VAN GEEL, B., BUURMAN, J. & WATERBOLK, H. T. 1996. Archaeological and palaeoecological indications of an abrupt climate change in The Netherlands, and evidence for climatological teleconnections around 2650 BP. *Journal of Quaternary Science*, **11**(6), 451–460.

VOS, P. C. & DE WOLF, H. 1993. Diatoms as a tool for reconstructing sedimentary environments in coastal wetlands: methodological aspects. *Hydrobiologia*, **269/270**, 285–296.

—— & VAN HEERINGEN, R. M. 1997. Holocene geology and occupation history of the Province of Zeeland. *In*: FISCHER, M. M. (ed.) *Holocene Evolution of Zeeland (SW Netherlands)*. Netherlands Institute of Applied Geoscience, TNO, Nr59, Haarlem, 5–110.

WEDEPOHL, K. H. 1995. The composition of the continental crust. *Geochimica et Cosmochimica Acta*, **59**, 1217–1232.

WOHLFARTH, B. 1996. The chronology of the last termination: a review of radiocarbon-dated, high-resolution terrestrial stratigraphies. *Quaternary Science Reviews*, **15**, 267–284.

WOLLACOTT, D. 1921. The interglacial problem and the glacial and post-glacial sequence in Northumberland and Durham. *Geological Magazine*, **21**, 60.

WOODWORTH, P. L. 1987. Trends in UK mean sea level. *Marine Geodesy*, **11**, 57–87.

WRIGHT, M. R., PLATER, A. J. & HAWORTH, E. Y. 1996. Palaeoenvironments of the Tees estuary (abstract). *In*: BARRETT, R. L. (ed.) *LOIS – First Annual Meeting, Plymouth*, NERC, Swindon, 211–213.

——, ——, RUTHERFORD, M. M. & HAWORTH, E. Y. 1997. Holocene palaeoenvironmental reconstruction of Billingham Beck, Teesside (abstract). *In*: BARRETT, R. L. (ed.) *LOIS – Second Annual Meeting, Hull*, NERC, Swindon 199–201.

ZONG, Y. & HORTON, B. P. 1999. Diatom-based tidal-level transfer functions as an aid in reconstructing Quaternary history of sea-level movements in Britain. *Journal of Quaternary Science*, **14**(2), 153–167.

Holocene coastal dune initiation in Northumberland and Norfolk, eastern UK: climate and sea-level changes as possible forcing agents for dune initiation

J. D. ORFORD,[1] P. WILSON,[2] A. G. WINTLE,[3] J. KNIGHT[2] & S. BRALEY[2]

[1] *School of Geography, Queen's University, Belfast BT7 1NN, UK (e-mail: j.orford@qub.ac.uk)*

[2] *School of Environmental Studies, University of Ulster, Coleraine BT52 1SA, UK*

[3] *Institute of Geography and Earth Sciences, University of Wales in Aberystwyth, Aberystwyth SY23 3DB, UK*

Abstract: Relative sea-level (RSL) control on dune initiation during the Holocene is examined in the context of chronostratigraphies established from 43 vibracores through dunes and into sub-dune sediments taken from the Northumberland and Norfolk (UK) coasts. The chronology is based on 23 accelerated mass spectroscopy and conventional ^{14}C dates, and 37 infra-red-stimulated luminescence dates. The oldest dunes in Northumberland are *c.* 4 cal. ka BP with phases of dune development at 2.8 and 1.5–1 ka BP. Most dune deposition is of last millennium age, with a concentration, especially in Norfolk, around 500–200 a BP. The initiation and survival of coastal dune sequences relate to macroscale RSL changes over the last 4 ka. Northumberland dunes reflect a gradient of RSL change from a northern RSL fall (forced regression) through to a southern RSL rise (normal regression through sediment supply). The north Norfolk coast has been dominated by a rising RSL through the Holocene, though associated with a sediment supply sufficient to offset the transgressive tendency and allow normal regressive deposition at numerous positions along the coast over the last 1 ka. It is suggested that the development of Little Ice Age (LIA) dunes in both Norfolk and Northumberland identifies the onset of specific conditions in which intertidal sediment sources were exposed (falling sea-level) to onshore winds (LIA circulation changes), which reflect a brief west North Sea period in dune initiation and deposition rates. A comparison of this consolidated dune chronology with statements of RSL elevation and climate conditions in the last 2.5 ka leads to some recognition of RSL fall preceding major dune building in two phases post 1.5 ka BP and post 0.6 ka BP.

This paper reports on the chronology of coastal sand dune initiation along the Northumberland and Norfolk coastal sections of the North Sea, and the possible relationship between both climate and relative sea-level changes (RSL) and dune initiation. The study funded by the Natural Environmental Research Council (UK), investigated the litho- and chronostratigraphy of the dunes as part of the wider Holocene Land–Ocean Interaction Study (LOIS). It is important at the outset to appreciate the potential dynamic difference between European west coast dune systems, such as in western Ireland (e.g. Carter & Wilson 1993), western Britain (e.g. Pye & Neal 1993), Holland (e.g. Klijn 1990) and Denmark (e.g. Christiansen *et al.* 1990) with near-consistent prevailing westerly *onshore* winds during the Holocene, compared with east coast sites with near-consistent prevailing *offshore* winds during the Holocene. This major difference has to be emphasized in that regardless of developing a chronology of dunes, there has to be an associated statement as to the change in environmental conditions by which coastal dunes can both be initiated and persist on the east coast of England.

Coastal dunes can be viewed as an environmentally sensitive index of coastal state. Although there is an equivocal connection between dune and beach, the variability in morphodynamics of these two components can help specify both the changing status and changing controls of coastal deposition systems as a whole (Psuty 1992; Sherman & Bauer 1993). In more utilitarian terms, coastal dunes are recognized as a major element of natural coastal protection for marginal coastal lowlands. Understanding dune dynamics is essential in the formulation of a sustainable policy for strategic coastal zone management.

From: SHENNAN, I. & ANDREWS, J. (eds) *Holocene Land–Ocean Interaction and Environmental Change around the North Sea*. Geological Society, London, Special Publications, **166**, 197–217. 1-86239-054-1/00/$15.00

Coastal dune initiation and development

Two key issues related to coastal dunes are the controls on their initiation and their development. Carter (1988) identifies a simple scenario for dune formation: a plentiful supply of beach sediment; and an expanse of dry beach (tidally exposed) over which a dominant onshore wind of entrainment capability can move sediment into an accommodation space, which in the long-term (see below) is immune from wave attack. Although Carter adds a biological element, in the role of vegetation as a dune-form fixer, of more central interest to this study are the geomorphic controls on dune development in the accommodation space.

Dunes, in a gross budgetary sense, reflect the landward output of a coastal conveyor whose sediment discharge is modulated by the speed of the conveyor (wind activity as a function of direction and entrainment velocities) and the input of sediment. In the last decade while there has been a re-invigoration of investigations into aeolian entrainment and transport dynamics (Sherman *et al.* 1998), there has been little investigation into what controls sediment supply to the conveyor apart from general statements that it is related to sea-level change (Pye 1984; Christiansen *et al.* 1990; Clemmensen *et al.* 1996). It is this latter problem that lies at the centre of this paper. Pye's synthesis showed that dunes responded to both falling and rising sea-levels, as well as being dependent on the rate at which sea-levels changed. In the context of a rising sea-level, new sediment was mobilized into the beach system by shoreline erosion hence supplying the 'excess' sediment for dune formation. This type of scenario implies a longshore component of change, in that the source of sediment is unlikely to be in the same spatial position as the dune sink. It also implies that even in a macro-transgressive shoreline context, one can find normal regressive sites (Posamentier *et al.* 1992) dependent on excess sediment supply. The alternative model involves a general regressive shoreline (forced regression: Posamentier *et al.* 1992) in which aeolian reworking of exposed beach and nearshore sediments, due to a lower sea-level position, allows dune accretion at positions immediately landward of the source area as well as down drift of it.

In assessing controls on dune building, there is a need to differentiate between: (a) dune initiation *per se* as a response to external fluctuations in sediment supply triggered by sea-level change, (b) the continuation/diminution of existing dunes as a function of internal geomorphic changes. Although the latter is a logical extension of the former phenomenon, it is worth recognizing the distinction between initial dune appearance as a function of external environmental change, compared with existing dunes showing a history of deposition variability. In essence, the question is: 'Why does the beach–dune conveyor start at all?', rather than 'Why is the conveyor speed variable?'. Although there are numerous studies that identify sequential variation in the chronology and/or morphological development of coastal dunes (e.g. Pye 1983; Carter & Wilson 1990; Klijn 1990; Wilson 1990; Wilson & McKenna 1996), there are few studies concerned with the initiation of coastal dunes *per se*. This is a statement that needs qualification related to the time-scale of persistence of the phenomena. While studies have identified the development of dunes in terms of foredune initiation and accretion (e.g. Hesp 1988), such foredunes are often a function of short time-domains and may reflect a development sequence that is a function of a seasonal wave control on beach and hence foredune development. Such apparent 'seasonality' can be induced by extreme storm-event spacing, which obliterates the presence of foredunes on a punctuated basis (Orford *et al.* 1999). In this paper we consider the origin of dunes that show a temporal persistence measurable over the mesoscale and into the macroscale (using the coastal time-scales specified by Terwindt & Kroon 1993, where microscale is $<10^0$ a, mesoscale is 10^0–10^2 a and macroscale is $>10^2$ a). Such dunes may be initiated as foredunes but are transformed in a context of consistent or high-rate deposition from ephemeral units on an inter-annual time-scale, into phenomena that have temporal persistence over 10^2 a.

Clearly, a knowledge of chronology is a major requirement for this type of dune persistence study. In previous studies of coastal dune chronology, ^{14}C dating of organic remnants within the dunes was the principal means of developing a chronology. An organic-enriched horizon in dunes is thought to represent a relative hiatus in sediment deposition, implying dune stability sufficient for organic carbon concentration from decaying *in situ* natural vegetation. A problem arises as to the scale of enrichment that identifies clear vegetation fixing as opposed to derived wind-blown organic inclusions. Given the fine-sized nature of most organic-rich deposits, there is a high probability of wind-blown (derived) material being incorporated into coastal dunes. Organic-enriched horizons associated with pedogenic processes suggest a higher probability of *in situ* status (Wilson 1996). Thus ^{14}C dating of organic inclusions has to be recognized as an opportunistic approach dependent on differentiation between *in situ* and derived material: a distinction that is not always easily made. In the last decade, direct age-determination of active non-organic sedimentation in dunes has become possible with the advent of infra-red-stimulated luminescence (IRSL: Huntley *et al.* 1985; Wintle 1993; Ollerhead *et al.* 1994; Pye *et al.* 1995). This method, though prone to disturbance from post-depositional environmental changes (i.e. water content), is a substantial improvement in the means of developing a reliable chronostratigraphy for coastal dunes.

Aims

The purposes of this paper are:

(a) to present the results of coastal dune initiation age determination in Northumberland and Norfolk;
(b) to establish if regional phases of dune building occurred in space or time;
(c) to examine similarities and contrasts in such regional phases between the two areas;
(d) to examine the possible relationship between both climate and sea-level change and identified dune-building phases.

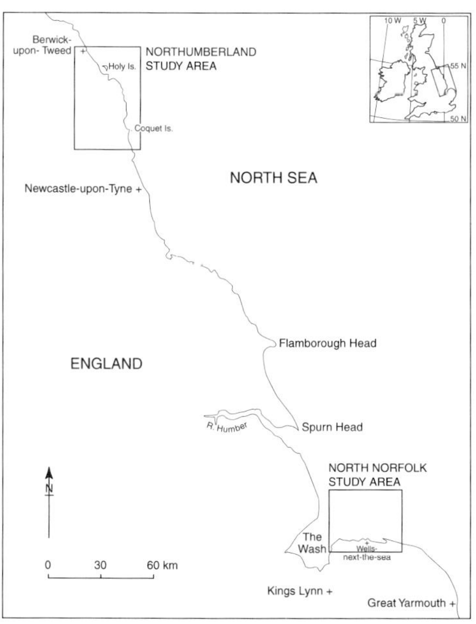

Fig. 1. Location of the two eastern UK coastal areas used in the Holocene dune initiation study: Northumberland and Norfolk.

Areas of study

Figure 1 shows the regional location of dunefields on the eastern coast of the UK in this study: Northumberland and north Norfolk. Both regions show narrow and constrained coastal dunefields with predominant fixed well-vegetated dunes contrasting with the wide and, at times (past and present), mobile dunefields of European west-coast sites.

Northumberland

Dunefields occupy about one-half of the coast between Berwick-upon-Tweed and Druridge Bay (Fig. 2). The crenellate coastline is rock controlled, with a succession of Carboniferous sedimentary units cropping out along this 90 km section (Taylor *et al.* 1971). The limestone and sandstone units of the Carboniferous Limestone dominate the northern three-quarters of this coast forming substantial cliffs, augmented by the resistant igneous intrusive suite comprising the Whin Sill (Bamburgh) and an *en echelon* igneous dyke series cropping out in Holy Island.

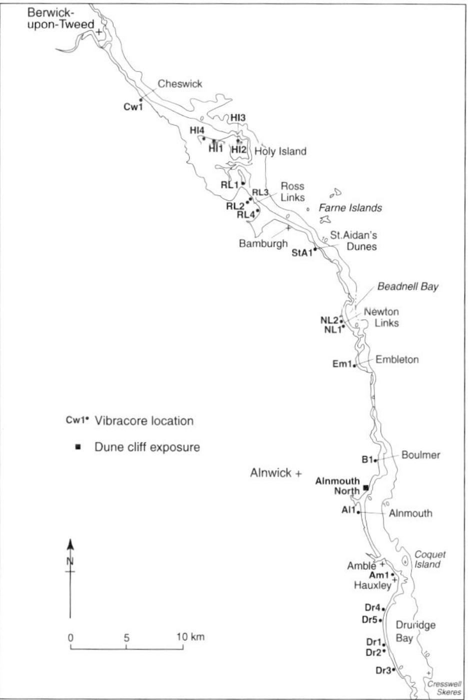

Fig. 2. Vibracore sites (and site code designation) in coastal dunes along the Northumberland coast.

To the south of Boulmer lies a narrow outcrop of resistant Millstone Grit, followed by the Lower Coal Measures cropping out at Alnmouth and Middle Coal Measures cropping out at Amble and underlying Druridge Bay. This variably resistant solid geology defines a series of headlands and bays, the latter providing opportunistic protected accommodation space for dunefields. Most of the solid geology has a cover of Devensian glacigenic materials that extends offshore and whose erosion by the Holocene transgressive shoreline has generally provided much of the sediment now located in the beaches and dunes.

For the purposes of this study, the coast is divided into three zones: Cheswick–Ross Links in the north; St Aidan's–Alnmouth in the centre; and Amble–Druridge Bay in the south (Fig. 2). The tidal range is *c.* 4 m (mesotidal) at springs for the whole coast. These zones show contrasting types of coastal setting for dune accretion. The coastal accretion can be divided into two types related to the above sections.

In the south and central areas a number of bays are defined by resistant rock headlands. These bays show swash-aligned beach and barrier deposition, upon which dunefields of kilometre-scale (longshore) have accreted (Newton Links, Embleton, Alnmouth and Druridge Bay). Some of these bays are estuary exits for small fluvial catchments, and the beach and dune units offer protection to low-energy muds and organic accumulation (Newton Links and Alnmouth). The dune systems range in shoreline length from 9 km at Druridge Bay to 2 km at Embleton and are usually narrow in landward width (<350 m). Dunes at Druridge Bay consist of discontinuous shore-parallel ridges (Fig. 3a) of which the seaward-most (outer) ridge is always the highest (10–16 m Ordnance datum; OD). Lower inland ridges, some of which are wider than the outer ridge, occur along much of the bay. Several deep blowouts, open to the east, occur in the outer ridge; their lack of vegetation and presence of sparsely vegetated blowover plumes indicate that sand continues to be reworked. Elsewhere the dunes are fixed by a continuous vegetation cover. At Newton Links (Beadnell Bay) the dune system extends alongshore for 3 km and is divided by the estuary of Brunton Burn (Fig. 3b). North of the estuary the system is dominated by a single, shore-parallel dune ridge that rises 10–15 m above the beach and with marked scarping on its seaward face. Several blowouts (both active and inactive), open to the east, breach the ridge towards its northern end. Landward of the ridge the dunes are low (<3 m amplitude), hummocky, and lack clearly defined linear elements. South of the estuary for a distance of *c.* 750 m several dune ridges are aligned parallel or sub-parallel to the shoreline. The dunes reach their maximum width (*c.* 350 m) in this area and consist of a high (14 m OD) landward ridge fronted by a series of ridges not exceeding 10 m OD; these lower ridges diverge southwards from a common point. In contrast to the landward ridge the lower ridges are dominated by *Ammophilia arenaria* (marram grass), with few herbaceous species present. Taken together these characteristics indicate that the lower ridges result from recent and marked dune progradation. Further south the dune system narrows and the number of ridges present declines; from thereon the remainder of the system is characterized by a single high ridge at the shoreline backed by low and hummocky dunes. Associated beaches show a range of inner-reflective, outer-dissipative beach forms (Short & Hesp 1982) indicative of a lack of sediment availability for contemporary dune building on the scale of past activity. A second type of dunefield development in the central and southern areas, is found overlying rock-bound headlands; here spatially constrained linear dunefields occur (St Aidan's, Boulmer and Amble). Again these systems are narrow (<300 m), well-vegetated, and dominated by a single ridge 5–15 m above the beach backed by lower hummocky dunes. The only previous work on dune chronology for this coast is that of Frank (1982) and Innes & Frank (1988), who showed that dunes encroached freshwater sites around 2.8–2.5 ka BP at Hauxley (north end of Druridge Bay).

The northern section of the Northumberland coast shows evidence of substantial low-energy peri-marine deposition landward of the contemporary shoreline (Plater & Shennan 1992). The dune systems from Cheswick to Ross Links display marked differences in setting and morphology. The Cheswick dunefield, developed south of a rock headland, extends alongshore for 4 km but does not exceed 400 m in width. Two ridges of similar height (10–20 m OD) dominate a system fixed by vegetation, and are currently immobile. Ross Links is an example of a prograded dunefield, of which most is below 10 m OD. In the western half of the sandy area shown in Fig. 3c the ground surface undulates gently and sand thicknesses are usually less than 2 m; podzolized glacifluvial sediments underlie the sands and are exposed in shallow blowouts. Prominent dune ridges are restricted to a zone *c.* 200 m wide along the outer coast with the highest dunes (14–16 m OD) marking the

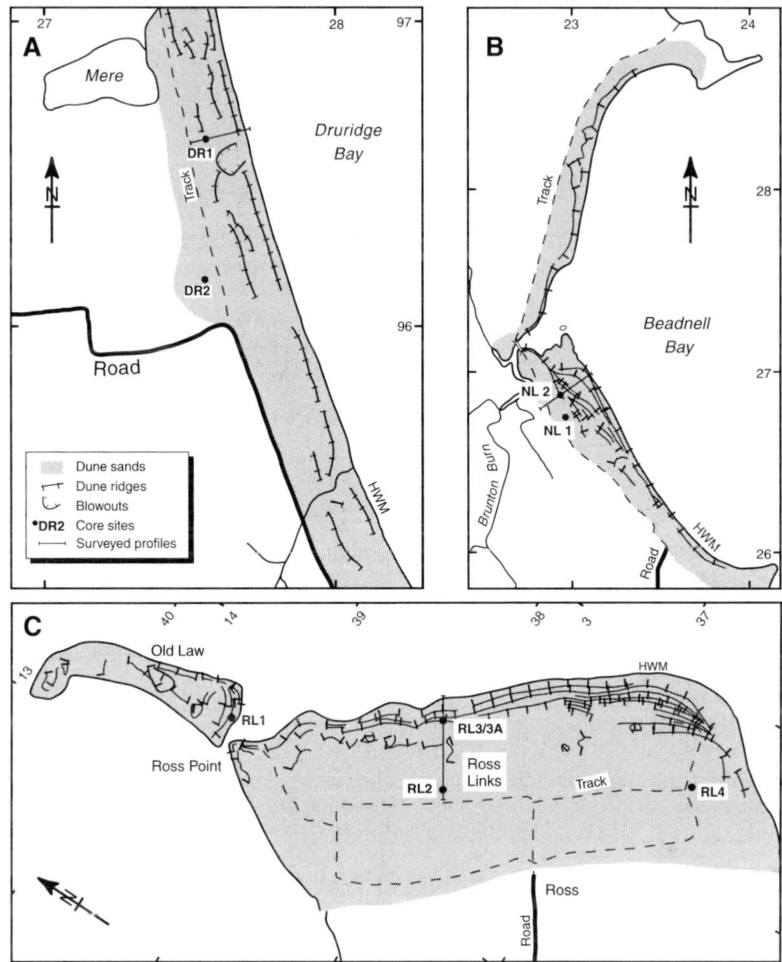

Fig. 3. Examples of dune system setting and morphology from Northumberland: (**a**) south-central Druridge Bay, (**b**) Newton Links (Beadnell Bay), and (**c**) Ross Links. The three sites are examples of the coastal settings gradient from transgressive in Druridge Bay to regressive at Ross Links. Vibracore locations are indicated on Fig. 2. Marginal grid lines are at 1 km intervals.

landward limit of this zone. To seaward, ridges possess more open vegetation dominated by *Ammophilia arenaria*, do not rise above 8 m OD and have amplitudes of 1–4 m. In the southern part of the system dune ridges are both more numerous and continuous within a 300-m-wide zone. The parallelism of these low foredunes suggests the possibility that they are built on prograded beach ridges controlled by swash refraction. Several ridges diverge northwards from a common point near the southeastern curve of the shoreline; ridge-crest elevations are generally within the range 6–9 m OD. A World War II structure amongst these ridges indicates some progradation within the last 50–60 years. Beach faces tend to be broad, low-angle and dissipative.

Shennan *et al.* (1999) have identified the differences in the Holocene sea-level signatures appropriate for northern, central and south Northumberland. The variation between these signatures reflects the north–south gradational change in sea-level over the late and middle Holocene due to variable crustal response. Such signatures act as a first-order control on coastal depositional settings over the last 3000 years. Figure 4 shows the original sea-level index point data of Shennan *et al.* (1999), where the various symbols identify spatially grouped index positions. The north–south continuum of sites has

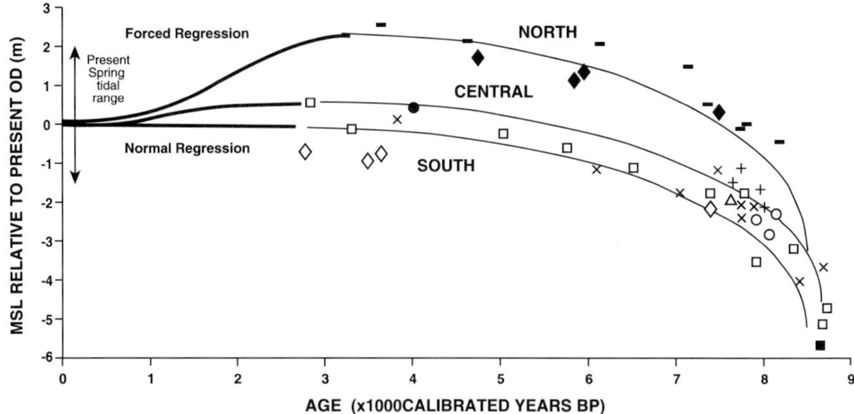

Fig. 4. Relative sea-level index points during the Holocene for Northumberland (after Shennan *et al.* 1999). Symbols identify different geographical locations of cores from which sea-level indicators are derived (as used by Shennan *et al* .1999). The set of curves (fitted by eye) schematically identify the north–south gradient of variable sea-level response along the Northumberland coast and are only approximations of Holocene relative sea-level position up to *c.* 3–4 ka BP. The solid lines representing the response of relative sea-level changes (fall in the north and marginal rise in the south) over the last 3 ka, define the overall control status on coastal deposition: forced regression in the north and normal regression due to sediment supply in the south.

been split into three sets for the purposes of this paper, which are associated with three schematic sea-level trend curves. These curves have not been mathematically calculated given the elevation variance of different indicators. The curves indicate the general elevation of relative sea-level between 4 and 3 ka BP and identify the general sense of overall relative sea-level change post 3 ka BP in order to arrive at present-day mean sea-level. The northern section identifies a forced regressive setting in which a falling sea-level allows the progradation of organic and inorganic sediments diagnostic of the highest levels of the tidal range, as well as beach and dune material. The southern section has a transgressive status, which would allow normal regressive deposition if sediment supply could be spatially constrained, as is the situation in rock-controlled bays. The implication of a transgressive shoreline is that southern shorelines are likely to have moved onshore during the last 3 ka, though the distance is unknown. The central Northumberland section, being close to the hinge-point between regressive and transgressive sea-level positions has shorelines that have been essentially stationary for the last few thousand years and a sedimentation pattern that is likely to be tending towards normal regressive.

North Norfolk

Figure 5 shows the study area of north Norfolk, comprising 35 km of the coast between the entrance to the Wash at Gore Point (site 1) and Blakeney gravel ridge (site 12). The contemporary tidal range shows a fall from macrotidal (6 m) at Gore Point to microtidal (2 m) at Blakeney. The coast is dominated by a broad suite of Holocene coastal depositional environments, which appear to have little direct relationship *per se* with the underlying solid Cretaceous chalk. The chalk is masked by a sequence of glacigenic materials which offshore has been extensively reworked by the Holocene transgressive shoreline (Fig. 6, Shennan *et al.* 1999). The whole north Norfolk coast can be defined as transgressive with potential for substantial normal regressive deposition due to excess longshore sediment supply. The existing coastline is comprised of a number of sand to sandy-gravel barriers separated by tidal inlets connecting extensive offshore ebb-tide orientated deltas to low-energy back-barrier saltmarshes, which grade upwards into limited freshwater fen and carr (Steers 1989; Pearson *et al.* 1989). In some areas (Stiffkey) the barriers are missing, yet the low energy status of the extremely dissipative upper tidal areas is still reflected by the dominance of saltmarshes. The barriers appear to have been principally formed from east to west, moving onshore with rising sea-level: witness the truncated recurves on Scolt Head Island (Fig. 5, sites 13–18) (Bristow *et al.* 1993) and on Blakeney gravel ridge. The Holocene stratigraphy of the barriers and the back-barrier environments are explored by Andrews *et al.* (1999)

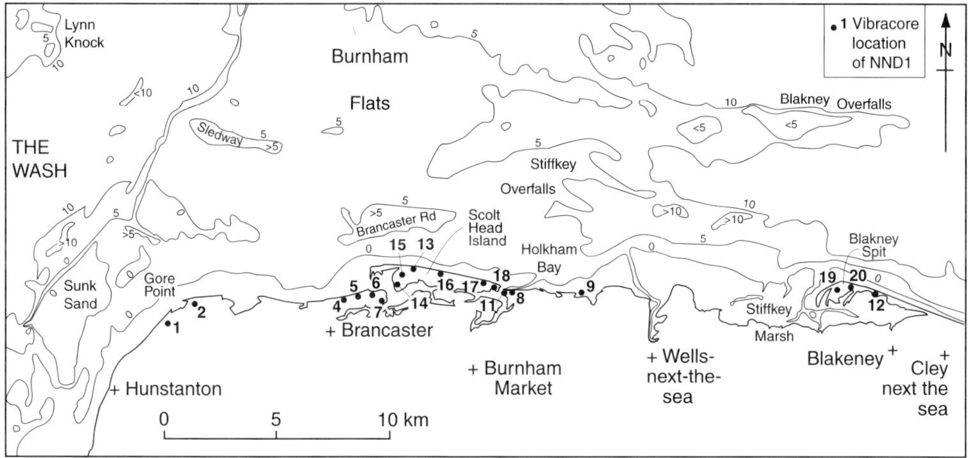

Fig. 5. Vibracore sites (and site code designation) in coastal dunes along the Norfolk coast used in this study. Bathymetry is set to depths (metres) below spring low tide position.

complementing and extending the stratigraphy described by Funnell & Pearson (1989).

The Norfolk dunes can be split into a number of longshore units, which are dominated by their relationship to coastal barrier structures (both sand and gravel based), between Hunstanton in the west and Blakeney to the east. These five sections are defined in terms of both sand and gravel transport corridors and sinks, identifying a wave–sediment cell approach as a basis for discussion of dune evolution: (a) Holme and Gore Point (Fig. 5, Hunstanton to mid-way between sites 2 and 4); (b) Brancaster (Fig. 5, sites 4–7); (c) Scolt Head Island (Fig. 5, sites 11, 13–18); (d) Holkham Bay; and (e) Blakeney Spit.

The central coastal sections (b–d) are dominated by the presence of Scolt Head, such that most of the recent sand accumulations probably relate in some way to the presence of an original

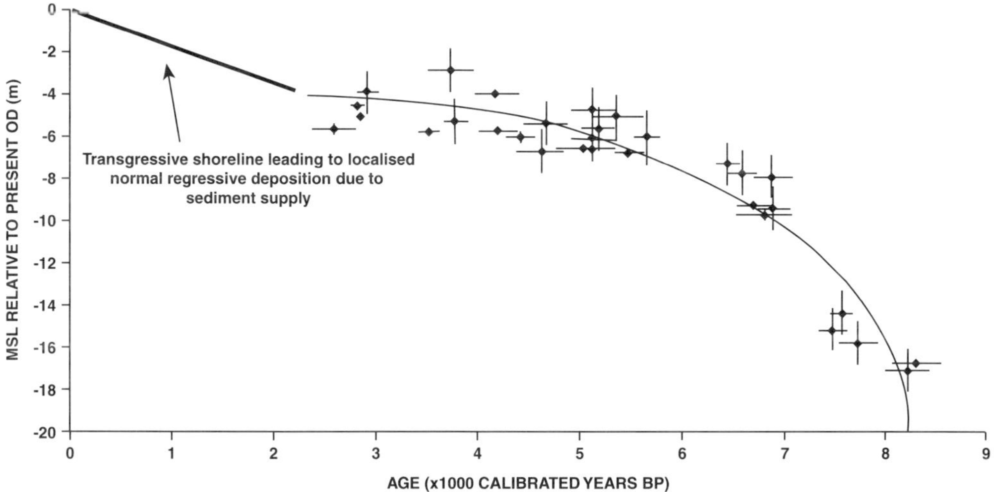

Fig. 6. Relative sea-level index points during the Holocene for Norfolk (after Shennan *et al.* 1999). The vertical and horizontal ranges on indicators indicate the error terms. The relative sea-level trend has been fitted by eye and is only an approximation to the relative sea-level position through the Holocene to 2.5 ka BP. The straight line identifies the sense of sea-level change over the last 2.5 ka, indicating a general transgressive control on coastal deposition along the Norfolk coast.

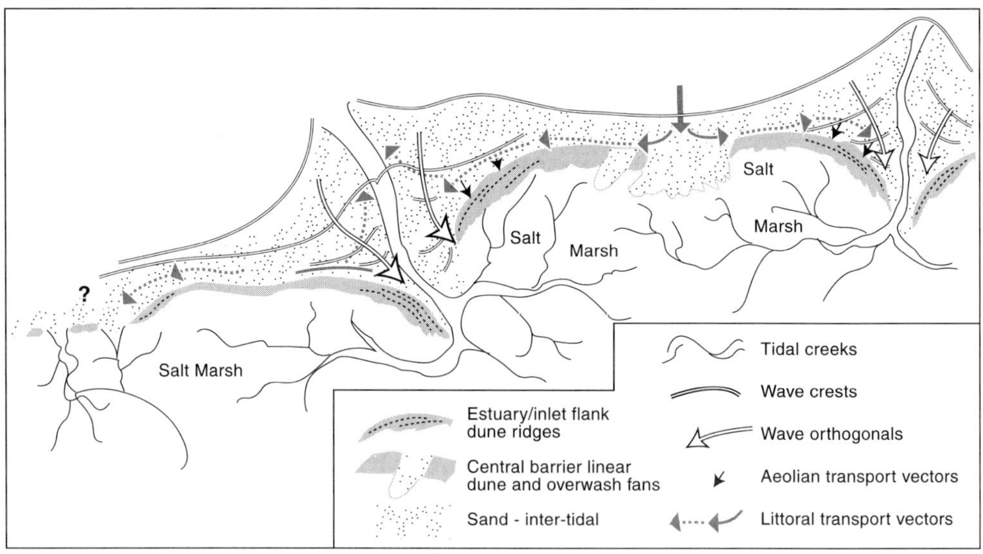

Fig. 7. A schematic geomorphic model of dune development along the north Norfolk coast based on net sediment accretion within tidal inlets. Overall length of coastal section shown would be *c*. 10–15 km. Two sets of dunes are identified: (A) an initial, linear dune line found in the centre of the barriers, and which is now failing due to longshore sediment losses; and (B) a subsequent set of multiple dune ridges tied to the flanks of an estuary/inlet mouth. The latter set reflects dune decoration on top of a series of wave-refracted beach ridges. Cores for dune initiation dating were taken from the linear dune ridge at the barrier centre, or a dune ridge on a recurve, or the landward ridge of an estuary mouth set.

proto-barrier transgressing over the Holocene coastal marsh sequences (Jones 1996). The problem of sediment source for each section has yet to be understood, though the presence of Scolt Head and its antecedent barrier is likely to be an initial source of the present dune accumulations in the central sections. Although there is general support for a westerly transport of littoral sediment, there is a case for localized reversals, i.e. sand at Blakeney from the west (Hardy 1964). There is also growing evidence for onshore feeding of the littoral zone (Chang & Evans 1992).

There is a readily apparent longshore spacing to these wave–sediment cells of *c*. 5 km, which is identified by the spacing of tidal inlets/outlets, which in some cases have accumulated substantial sediment volumes in ebb-delta structures and which act as boundaries to the coastal accumulation systems. The deltas are important both as potential sediment traps and as features perturbing the incident wave field as they control the nearshore and breaking wave through refraction. The spacing of the inlets may be accidental, but the possibility of some shoreface control on their position through trapped wave resonance has been suggested for a potential analogue system in the Friesian barrier islands (Flemming & Antia 1990). The inlets connect to substantial areas of back-barrier marshes and saltings which have been extensively reclaimed since the eighteenth century (e.g. Holkham in 1719), as well as dyked for flood defence purposes.

The Norfolk dunes can be split into two distinct morphological sets, which have a spatial relationship with the corridor–sink structure of the cells. A schematic view of their development is shown in Fig. 7 in the context of two main dune sets: (A) a single linear dune line facing the open coast and (B) multiple dune ridge systems associated with estuary mouths. The linear dune ridge of set A parallels the coastal barrier alignment (generally east–west). In some areas multiple ridges can be observed (e.g. west Holkham). These dune lines have in several places been strengthened and re-vegetated to form a basic defence to sea-flooding, e.g. west Brancaster and Holkham. This morphology is prominent near to the middle sections of the barriers, e.g. Scolt Head Island. The need for defensive strengthening indicates that in several areas these ridges are now in net-erosion status due to reduction in sediment supply. Scolt Head Island shows a good example of mid-island breaching and removal of the dune ridge with a sequence of barrier overwash and intermittent dune construction in a decorative position on washover fans. These linear dunes appear to have a wave-lain core, which has been incremented by aeolian activity into the present dune ridge. There is usually a lack of major onshore aeolian sand movement beyond the ridge, suggesting operation of one or more of the following: (a) fast vegetation stabilization control on sediment influx; (b) limited sediment availability; (c) lack of dominant onshore wind.

Multiple linear dunes of set B prograding at down-drift sink positions, are best observed in the prograded dune series found at recurve positions on both sides of tidal inlets. Contemporary net sediment vectors move material from central barrier erosion, into the tidal inlets where it is either transferred onto the ebb-deltas or is added by refracted waves onto the inlet-mouth recurves (Fig. 7). Comparison of UK Ordnance Survey maps from the early nineteenth century shows that most consistent dune development in the last two centuries is concentrated on these inlet-mouth side-bars, e.g. Brancaster (Fig. 5, sites 4–7). In most cases the dunes show distinct longshore linear forms reflecting wave-lain beach ridges acting as the core to later aeolian deposition as the upper beach prograded. The multiple sub-parallel ridges suggest high transport rates and fast ridge building. Such rapid activity would require a wide upper beach zone as the source area for aeolian winnowing. This sediment-rich beach state is a result of longshore transport of beach material into the inlet mouths. Breaking wave orthogonals are strongly inlet-directed by the shallow bathymetry of the ebb-delta. Brancaster is a good example of such aeolian development on a wave-lain platform to the side of an ebb-delta dominated tidal inlet of Brancaster Harbour lying in the lee of the western end of Scolt Head Island. The back-barrier areas have been extensively 'reclaimed' since the seventeenth century by sea-wall erection for both coastal defence and agricultural enhancement. This has led to reductions in both back-barrier tidal prisms and active cross-sections at inlet mouths, with side-bar build-up as the inevitable response (Gerritsen & Dunsbergen 1998). In the case of Holkham (Fig. 5, site 9), the entire tidal inlet was barred through reclamation, and the old tidal inlet is now the site of major supra-tidal deposition and the emergence of a new offshore barrier island.

Methods

Chronostratigraphy

The study is based on 23 vibracores from the Northumberland dunefields and 20 vibracores from the Norfolk dunes. The cores were taken by the British Geological Survey using a method of coring in 1-m-long sections (run) from an encased depth, extended to depth as coring proceeded. Percentage core sediment extraction varied from run to run. Losses in any run were augered out to the depth of the last core base. Even though cores were through dry sand, retention rates were high, with an average of *c.* 65% core retention. Core penetration was effective even through clast-supported gravels, the limitation being dependent on the gravel size relative to the 60-mm-core diameter. Limited core penetration through silts and clays, into glacigenic sediments as well as compressed peat bands allowed recovery of basal sediments to the dune sequences.

In a number of dunefields, time and cost only allowed us extraction of one overall core. In large fields, a number of spatially separated cores were taken to establish any longshore changes in dune stratigraphy. Core-sites in Northumberland were mainly from the top of the last distinct ridge on the landward side of the dunefield. The dominance of longshore activity in the Norfolk dunes in the form of recurves meant that some cores were taken from within the recurves as well as at the rear of the recurved ridge set. Core run length varied relative to the height of the dune form at the core site. All core ground level heights were reduced to Ordnance datum (UK) using total station and GPS. Characteristic dimensions of cores for the two areas are shown in Table 1; note the longer average length of core obtained from the Norfolk dunes.

Cores were obtained in light-impenetrable plastic liners and split vertically under red-light conditions with one-half being wrapped in black plastic and stored in darkness. One-half of the core was stratigraphically logged and then sampled for sediment size using a Galai CIS laser granulometer to produce grain size spectra between 10–1200 μm based on a 4-μm-interval size. Data were grouped into 0.25ϕ intervals for presentation and analysis using a numerical version of inclusive graphic statistics. The logs formed the basis for a subsequent facies analysis.

Organic material was sampled and submitted for ^{14}C assay. The very small percentages of organic material retained in the cores necessitated accelerator mass spectrometry (AMS) rather than conventional dating for most samples.

Table 1. *Vibracore details*

Variable	Northumberland	Norfolk
Longest core, m	9.29	11.35
Shortest core, m	1.36	2.31
Mean ground elevation, m OD	8.18	8.67
Ground elevation (standard deviation), m	2.48	2.31
Mean core length, m	4.72	6.69
Core length (standard deviation), m	2.16	2.33

The lack of distinct organic horizons *per se* and the low percentage of organic material registered over 5 cm of core run meant that ^{14}C determinations had to be treated with caution. AMS determinations were from Arizona, based on gas reduction carried out at NERC's East Kilbride facility. Also sent for ^{14}C assay were a number of samples from distinct *in situ* peats at the base, or interbedded at the base, of the aeolian deposits. In all, 29 AMS (AA Lab) and five conventional (SURRC and Beta Labs) ^{14}C determinations were undertaken, of which 23 are reported in this study. Age estimates were calibrated using the CALIB 3.0.3A program of Stuiver & Reimer (1993) in order to allow direct comparison with IRSL ages. Results of ^{14}C determinations are given in Table 2. A few dates are anomalous: AA-23493 (>30 ka) is clearly too old, and AA-23487 (2585 ± 45 years), AA-23490 (5030 ± 55 years) and AA-23499 (1070 ± 45 years) are also suspected of being too old because of their unconformable relationship with IRSL ages from adjacent (i.e. underlying and/or overlying) sand samples. The probability is that these

Table 2. *AMS radiocarbon dates from coastal dunes in Northumberland*

Site/core code	Sample elevation (m OD)	Material/ stratigraphic position*	Laboratory code	^{14}C age (a BP ± 1s)	$\delta^{13}C$ (‰)	Calibrated age (a BP ± 2s)		
Holy Island								
NBD2/HI2	7.71–7.60	som/within	AA-23486	Modern	−27.6		–	
NBD2/HI2	7.20–7.10	som/within	AA-23487	2585 ± 45	−26.3	2780	(2735)	2519
NBD2/HI2	6.66–6.59	sp/within	AA-23488	335 ± 45	−28.1	493	(441, 404, 346)	299
NBD2/HI2	6.42–6.40	sp/interbed	AA-23489	990 ± 45	−27.5	971	(927)	776
Ross Links								
NBD6/RL1	2.97–2.93	som/within	AA-23490	5030 ± 55	−21.5	5910	(5750)	5640
NBD7/RL2	3.66–3.62	som/within	AA-23491	Modern	−25.8		–	
NBD9A/RL4A	3.13–3.09	som/within	AA-23492	1030 ± 45	−23.4	1010	(944)	886
NBD9A/RL4A	2.80–2.79	som/within	AA-23493	>30 000	−23.1		–	
St Aidan's Dunes								
NBD10A/STA1A	4.50–4.49	sp/base	AA-23494	3290 ± 45	−27.9	3615	(3495)	3402
NBD10A/STA1A	4.42–4.41	sp/base	AA-23495	3635 ± 45	−27.9	4054	(3924)	3839
NBD10A/STA1A	4.38–4.37	sp/base	AA-23496	3705 ± 50	−28.2	4181	(4028)	3886
NBD10A/STA1A	4.28–4.27	sp/base	AA-23497	4440 ± 50	−28.4	5275	(5015)	4871
Newton Links								
NBD11/NL1	2.84–2.83	sp/base	AA-23498	3690 ± 60	−28.6	4190	(3990)	3860
Amble Dunes								
NBD16/AB1	6.60–6.51	som/within	AA-23499	1070 ± 45	−25.5	1051	(964)	922
NBD16/AB1	4.09–4.06	sp/interbed	AA-23500	980 ± 40	−28.4	959	(922)	776
NBD16/AB1	3.78–3.74	sp/interbed	AA-23501	1170 ± 40	−28.9	1163	(1063)	976
NBD16/AB1	3.71–3.69	sp/interbed	AA-23503	1290 ± 40	−29.1	1288	(1238)	1107
NBD16/AB1	3.52–3.47	sp/base	AA-23502	1590 ± 45	−28.3	1560	(1482)	1365
Druridge Bay								
NBD18/DR2	3.96–3.90	som/within	AA-23504	540 ± 40	−27.9	593	(533)	507
NBD20/DR4	9.47–9.46	sp/base	AA-23505	2420 ± 60	−29.0	2720	(2390)	2330
NBD21/DR5	9.19–9.18	sp/interbed	AA-26346	785 ± 60	−28.6	780	(690)	650
NBD21/DR5	9.08–9.07	sp/interbed	AA-26347	1045 ± 60	−28.8	1060	(950)	810
NBD21/DR5	8.79–8.74	sp/base	AA-26353	1485 ± 60	−28.4	1500	(1355)	1290

* Som, soil organic matter; sp, sandy peat; within, from within the aeolian sequence; interbed, from thin sandy peat layers close to base of aeolian sequence; base, at base of the aeolian sequence.

^{14}C-dated samples contained wind-blown fragments of older carbon derived from offshore peats or coal measures.

Thirty-two samples for IRSL determination (reported in this study) were taken from the dark-stored half of the core, again in red-light conditions. Given the 60-mm-core diameter, there was need to take a minimum of 100 mm of core length to provide the volume of sediment for extraction of K-feldspar grains, which were then stimulated under infra-red light. All IRSL determinations were undertaken at the Aberystwyth Luminescence Laboratory following the method outlined by Wintle *et al.* (1998). *In situ* determination of sample water content was not available during vibracoring. Given the subsequent core splitting, storage and handling history, a water content of 5% was used in IRSL determinations of aeolian sediments. Sample age determination is reported here on a comparable basis to ^{14}C by reducing age to years before present (BP) plus/minus the error term. The error term of IRSL is not a statistical standard error term and therefore is not doubled as a statement of 95% confidence range as in ^{14}C reporting.

Dune initiation age estimates

It is not the purpose of this paper to discuss the facies basis *per se* of the dune stratigraphy recorded in the cores. Knight *et al.* (1998) indicate the basis and structure of the facies analysis for the Norfolk cores, while a similar analysis for the Northumberland cores is to be reported elsewhere (Wilson and coworkers pers. comm.). However, a limited statement as to the facies nature of the cores is made in support of the dune chronology. Facies are identified primarily on a sediment-texture basis. To indicate some of the core variation, Fig. 8 shows three contrasting vibracore logs selected to exemplify the contrasting types of core stratigraphy.

Cores are dominated by massive structureless medium to coarse sand units of aeolian origin. They contain a few units that show structure (cross-stratification and parallel stratification). The dune sediments show a virtual absence of major organic concentrations that could be indicative of vegetated hiatuses, (though the top metre of dunes may show the remnants of limited recent intermittent aeolian reworking). Likewise there is an absence of other than modern roots within the aeolian deposits, reflecting rapid deposition without evidence of biogenic fixing. A few units show some organic accumulation, but such inclusions are difficult to justify as of *in situ* occurrence given the very low percentage of organic content measured.

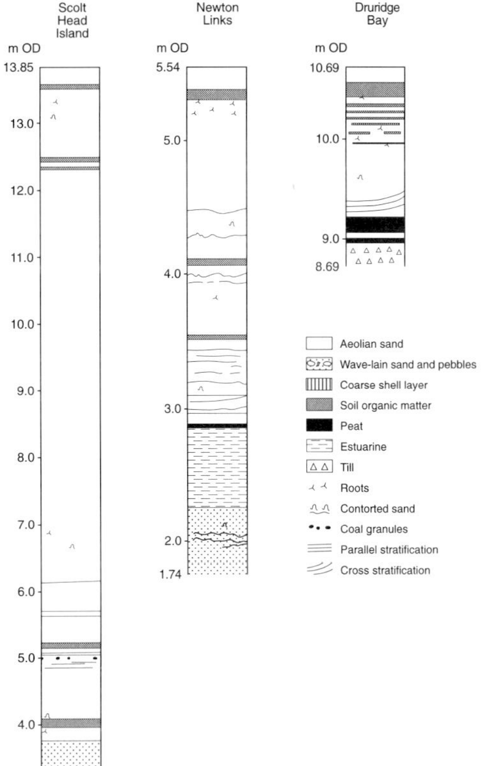

Fig. 8. Lithostratigraphy of selected cores from Norfolk (Scolt Head Island) and Northumberland (Newton Links and Druridge Bay). The stratigraphic signatures displayed by the cores form the basis for the schematic cores/logs in Fig. 9.

Sub-aeolian sediments exhibit a variety of facies. In Norfolk, they generally reflect poorly sorted coarse to medium sand with parallel bedding and frequent shell hash inclusions. Sub-aeolian sediments also show open-worked flat-lying pebbles as well as matrix-supported sandy-gravel deposition. In Northumberland, other facies are identified beneath the dune cover. In southern Northumberland, *in situ* freshwater peats are interdigitated with an overlying aeolian cover, while the peat is superimposed on glacial till. In central Northumberland, the dunes rest on top of laminated estuarine silty-sand. In one location an upper tidal peat lies between the estuarine and aeolian facies. In northern Northumberland dune sediments rest directly on wave-lain beach face sediment analogous to the basal facies identified in the Norfolk cores.

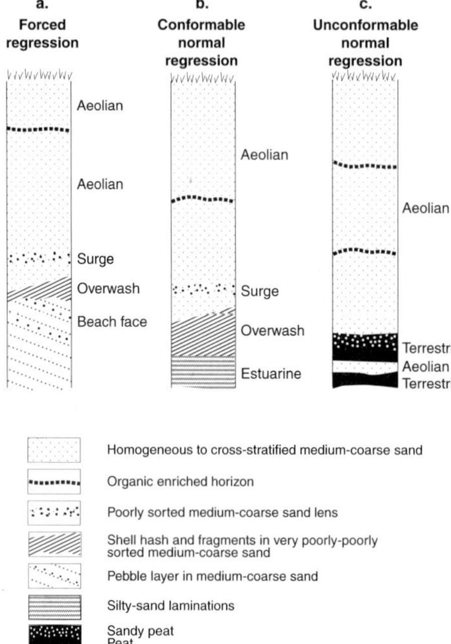

Fig. 9. Models of dune initiation observed from facies analysis of core sequences. (**a**) progradation due to sediment-dominated normal regression (Scolt Head, Norfolk: beach ⇒ overwash ⇒ dune) and associated with relative sea-level rise. The same sequence can be seen with a forced regression associated with relative sea-level fall (Ross Links). (**b**) conformable normal regression (Newton Links: estuary ⇒ back beach ⇒ dune). (**c**) unconformable normal regression (Druridge Bay: till ⇒ freshwater peats ⇒ dunes). The last two sequences are associated with relative sea-level rise.

There are three basic scenarios by which dune initiation can be identified from the facies analysis of the cores (Fig. 9a–c). These scenarios depend on the coastal setting of the cores in the context of overall depositional control, but are not necessarily related solely to sea-level tendency. The issue of sufficient sediment supply for upper beach storage and hence aeolian source area development is also important in this context of dune initiation.

Figure 9a shows a progradational deposition sequence in which wave-lain beach facies pass conformably upwards into aeolian facies with the possibility of overwash facies interdigitated with the lower aeolian unit. In both study areas, gravel is a constituent of beach and overwash facies. The boundary of aeolian–wave-lain sediments (AWB) is taken to be at the grading or junction between gravel, shell hash and very poorly to poorly sorted coarse–medium sand associated with overwash and parallel cross-stratified or massive coarse–fine sand related to aeolian activity. This sequence type dominates the Norfolk cores (Knight *et al.* 1998) and reflects the strong normal regression status of this coastline. This sequence is also typical of cores from the northern section of the Northumberland coast (Cheswick and Ross Links). IRSL material was generally selected from the aeolian component above the AWB boundary which is visually identifiable in the core.

Figure 9b and 9c identifies normal-regressive deposition sequences formed on transgressive shorelines which are sub-divided on the basis of facies conformability within the individual sequence. Conformable normal-regressive deposition is usually associated with overwash and aeolian components overlapping an estuarine sequence (laminated silty-sand plus clay drapes rising sometimes into saltmarsh peats). This sequence appears to be the signature of a coastal barrier being pushed onshore, though not far, given the lack of clear beach face deposition in the sequence. This sequence, or variants of it, is associated with the bay-head dune systems of central Northumberland (Newton Links, Embleton and Alnmouth). Unconformable normal regressive deposition shows an aeolian component overlying or initially interbedded with a terrestrial sequence, which predominantly comprises an *in situ* freshwater peat resting on till. This latter sequence is observed in headland dunes of central and southern Northumberland. In both normal regressive sequences, IRSL samples were taken from the base of the aeolian unit. Samples for ^{14}C dating were also taken from *in situ* peats to identify maximum age for an aeolian invasion. Examination of the ratio of peat and sand content around the peat–sand contact helps to distinguish between gradual, slow inundation and the sharp contacts of fast inundation.

Stratigraphic sequences identify the start of aeolian development at the core-site *per se*, but it is not necessarily the initiation of the dunefield *per se*. In the case of forced regressive sequences, it is assumed that the landward-most dune ridge formed a boundary to dune development and a start position for dune activity. In the case of the conformable normal regressive deposition sequences, the overwash wedge and lack of beach face sediment identifies a near back-beach–beach-crest position, where dunes were initiated as dune decoration before expanding over the older washover sedimentation. However, in the case of the unconformable normal-regressive sequence, dates reflect the timing of aeolian invasion not dune initiation *per se*.

In both cases, material for IRSL determinations were taken from close to the base of the aeolian unit.

Cost limitations prevented sampling of all basal aeolian units. Samples were also taken from higher positions in the aeolian unit of selected cores for the purpose of aeolian budgetary estimates. These dates are included in the chronology to give further temporal control on dune activity.

Results and discussion

Figure 10 shows characterization of the grain size distributions for all samples taken from four core sites reflecting different coastal settings for dune deposition. The range of median sediment size from +1.5 to *c.* $+3\phi$ shows that many of the dune samples would be regarded as coarser than traditional statements of mean dune size for coastal dunes (e.g. Pye & Tsoar 1990), although the median–skewness relationship of the sediment size distribution is as observed elsewhere for dunes (Pye 1982). The clear separation between sediment domains for the four core sites reflects local site control on initial dune-source sediment population. It is noticeable that the finest sediment (Newton Links) occurs in a site where there is an estuary sediment contribution to the littoral zone, whereas the other sites are dominated by onshore and/or longshore sediment supply. The lack of sediment overlap between sites (particularly in Northumberland) suggests that sediment supply is probably spatially constrained to a local area, which may reflect wave focusing along the crenellate coastline.

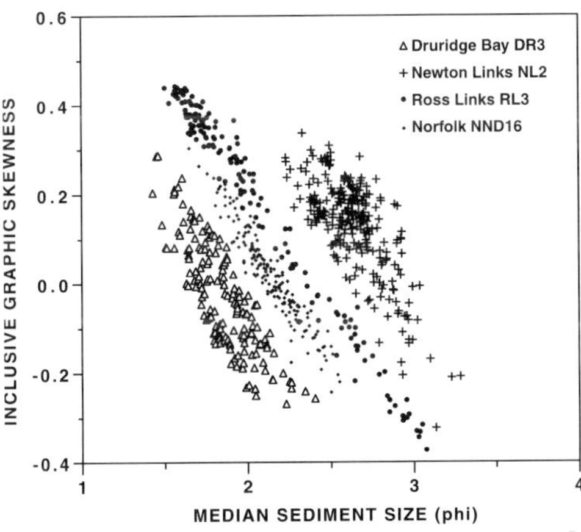

Fig. 10. Median sediment size and skewness for core samples taken from four dune sites in Northumberland and Norfolk to reflect the four coastal settings discussed in the text.

The dune chronologies for the north, central and southern section of the Northumberland coast by core site are shown in Fig. 11a–c. The date type and elevation in the core are shown, as is the transition type of the dune sediments relative to the basal unit. Dates related to the emplacement of the aeolian unit are identified. The Northumberland data have been restructured (Fig. 12) to identify the relative phases of dune activity in terms of the north–south coastal gradient, which reflects coastal setting control imposed by Holocene sea-level change. The dune chronology of Innes & Frank (1988) has also been incorporated (Hauxley dune phase) in this figure.

As there is a consistency of the Norfolk cores in showing a progradational setting to the dunes (Knight *et al.* 1998), only the dates of the lowermost aeolian unit are presented here (Fig. 13).

Northumberland

There are a number of ^{14}C determinations that are unsafe given the low percentage of organic remnants within the 'horizon' and the likelihood of contamination by older organic material (e.g. cores RL4 and AB1). There are also several instances where IRSL and ^{14}C dates do not show conformity of age in the vertical succession. We consider that where adjacent IRSL samples show overlapping error terms these indicate contemporaneous deposition.

In Northumberland, dune-building phases tend to be site specific, though there is considerable site chronology overlap in the last 1 ka. Earlier phases of dune building (>1 ka BP) show limited overlap in time between sites, which makes for difficulty in establishing a regional perspective to dune chronology. Figure 12 identifies the spread of dates relative to the north–south zonation of the coast imposed by relative sea-level tendency over the last 4 ka. The oldest extant dune deposition is seen at St Aidan's (3.9–3.3 ka BP), which is coincident with the deceleration of RSL after the peak of the Holocene transgression in the northern section. The lack of earlier dated dunes may reflect a sampling failure, in that older dunes present in the northern sector would be well landward of the current shoreline, degraded and morphologically indistinct and hence not included in the sampling programme. Alternatively, the rate of Holocene transgression at all Northumberland sites prior to 4 ka BP was likely to be inimical to

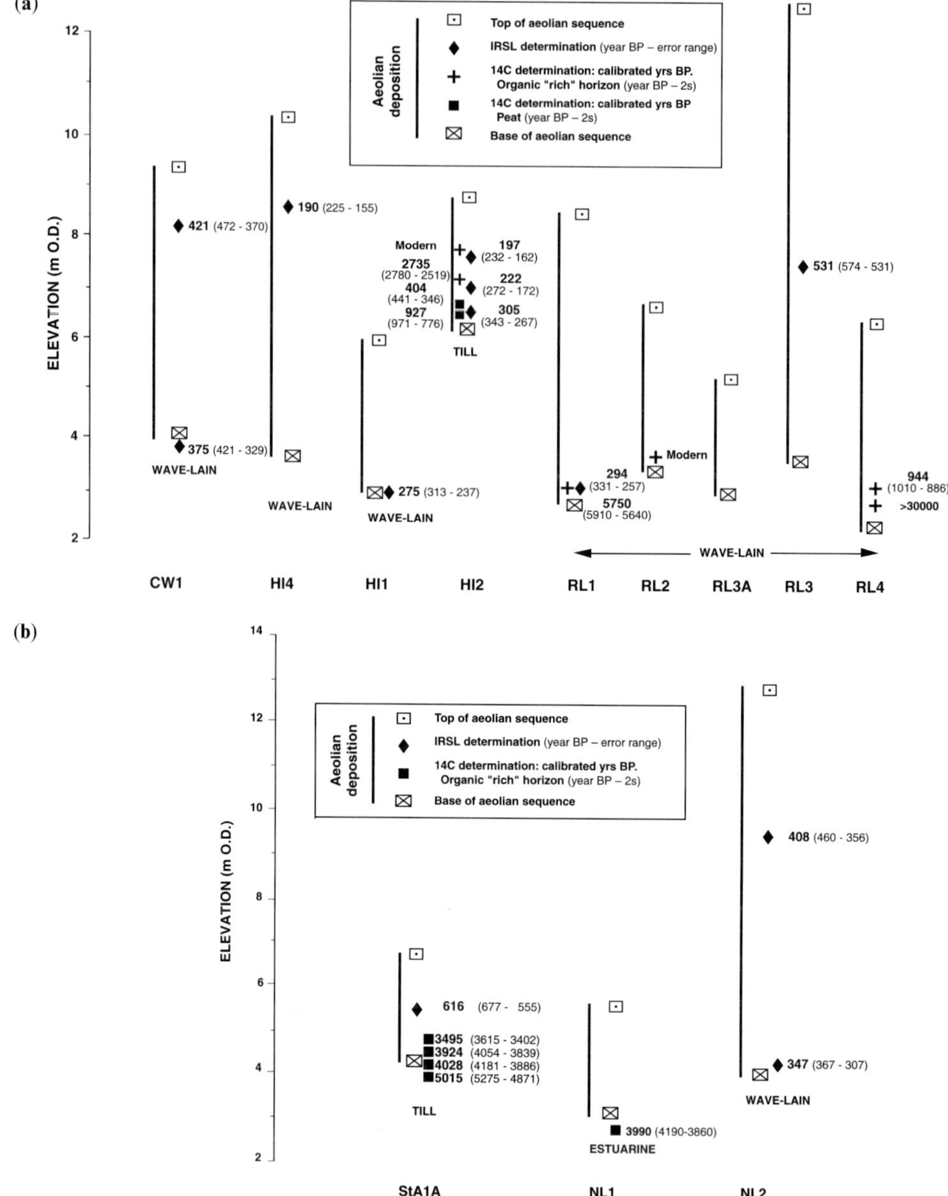

Fig. 11. Northumberland coastal dune chronology: (**a**) north Northumberland, (**b**) central Northumberland, and (**c**) southern Northumberland. Site codes as shown on Fig. 2.

the development and retention of barrier and dune morphology (cf. Jennings *et al.* 1998).

It is clear that there is a pattern of dune presence and absence, which in the first instance maps against the north–south post 4 ka BP sea-level tendency. Northern Northumberland may have landward dune sequences related to the forced regression control, but dated dune initiation shows concentration in the last millennium, probably reflecting the prograding status of this area.

The southern section identifies a number of phases of dune development over the last 3 ka, with the oldest dunes found on protected headlands while the younger dunes are present on the more active leading edges of the transgressive

(c)

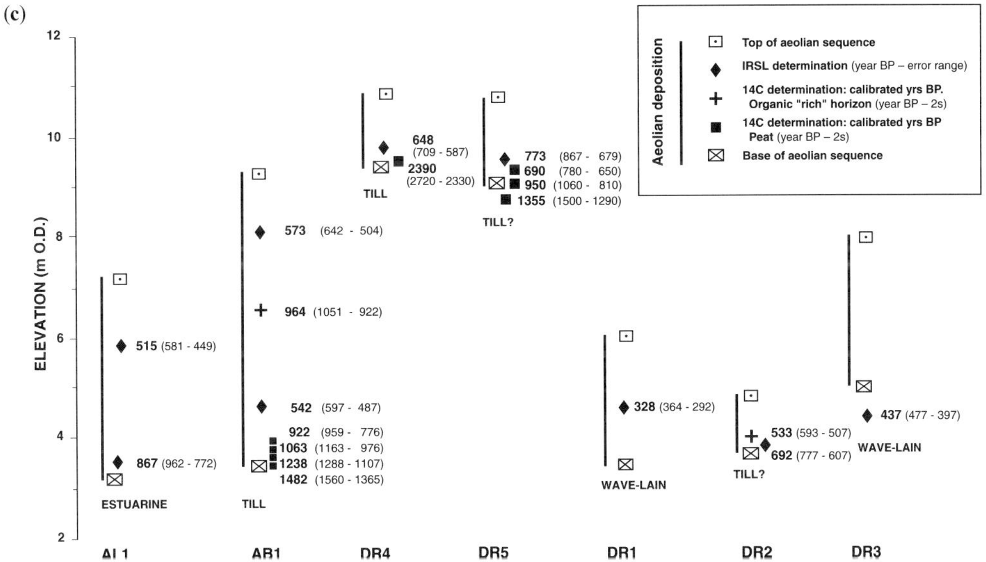

Fig. 11. (*continued*)

shoreline in centre-bay positions. The southern section also shows a greater persistence of dune building than elsewhere, but does not allow for the survival of dunes older than 3 ka BP. This may be due to either the presence of a ravinement surface associated with the transgressive shoreline eroding older dune sediments, or the fact that older dune sediments lie presently offshore.

The central section shows remnants of the oldest dune activity on the Northumberland coast (St Aidan's and Newton Links), a gap of over 2 ka of apparent aeolian inactivity and then recent last-millennium dune activity. The lack of evidence of dune activity in the 2-ka gap could equally be construed to indicate dune activity of which all trace has now been lost by erosion, as much as suggesting the lack of dune activity *per se*. The position of this central section relative to the hinge point between northern regression and southern transgression identifies the area as potentially one where extensive reworking in

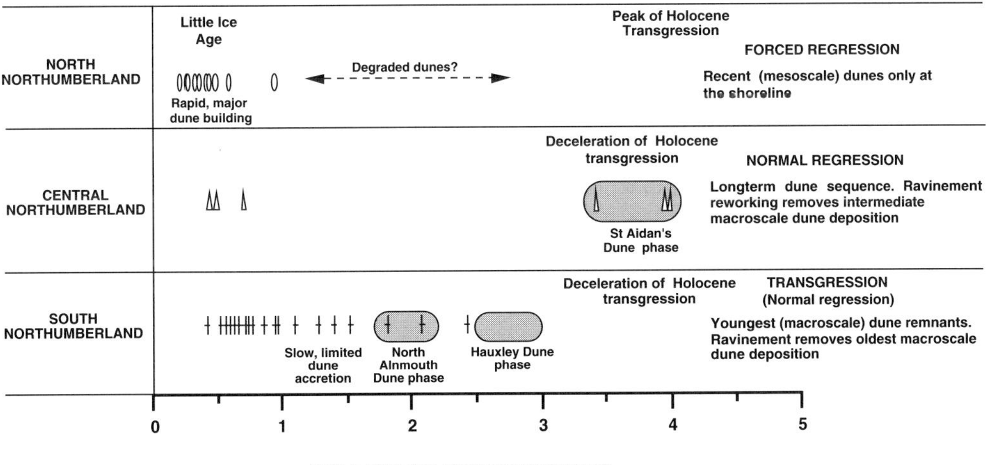

Fig. 12. Consolidated chronology of dune building in Northumberland over the last 4 ka. Hatched symbols indicate a number of dated dune advances that are not, as yet, determined as separate dune initiation events.

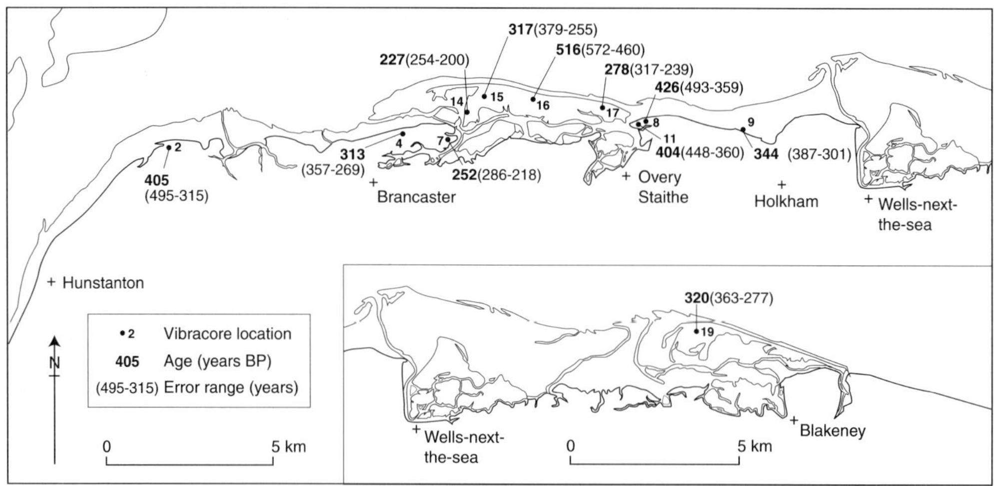

Fig. 13. Dune initiation age and error range (a BP) from 11 Norfolk cores, determined by IRSL of sediment drawn from the base of the aeolian deposition.

a limited landward spatial frame is likely. St Aidan's is another site where rock headlands have protected late–middle Holocene sedimentation from erosion. Newton Links is a site where erosion occurred due to the sediment reworking by the near-stationary late Holocene shoreline. This was followed by prograded dune building during the current millennium.

All Northumberland sites show the initiation and build-up of major dunes between 600–200 a BP, with a development maximized during the Little Ice Age (LIA) (Lamb 1995). Although not explicitly related to the initiation of dunes, it is important to note that the greatest volume of recovered dune sediment is linked by IRSL dates to the LIA, while most of the single major dune ridges identified at all sites were also formed during this relatively narrow time window.

Norfolk

All the dated cores from Norfolk show the progradation model of dune initiation as a consequence of a transgressive shoreline leading to normal regressive deposition through high sediment supply. Figure 13 shows the limited temporal spread of dune initiation dates (all within the last 600 years) compared with the millennia spread of Northumberland dates. This could be the result of a rising sea-level, by which any earlier evidence of aeolian deposition had been eroded. However, Pethick (1980) mentions that a pre-Roman artefact had been found in the landward dunes at Holkham, while in the same area (between sites 8 and 9, Fig. 13), Bristow *et al.* (in press) have recognized with the aid of ground penetrating radar, a major, steep, seaward dipping erosional bounding plane cutting across internal dune beds, suggesting that some landward dunes are stratigraphically older than the LIA element. However, there is an absence of dune dates that specify an earlier age than LIA to the lower unit. Elsewhere, a degraded sand ridge lying on top of the marshes at Stiffkey (Stiffkey Meals between Wells and Blakeney) has been suggested by Boomer & Woodcock (1999) as a possible barrier, with dune remnant driven over the marsh surface by a higher than present-day sea-level *c.* 1–0.8 ka BP (the date produced by a palaeoecological method). Andrews *et al.* (1999) discuss the development of the depositional shoreface along this coastline since 3 ka BP, underlining the possibility of older dune remnants pre-existing before the LIA. However, it is important to recognize the very limited nature of this evidence for dune development prior to the LIA.

The spatial spread of dates from Norfolk supports a model of longshore progradation of Scolt Head barrier during the early part of this millennium (average progradation *c.* 4–5 m a^{-1}). Net progradation is seen in the recurve extensions of Scolt Head and the development of dunes on these recurves. Dunes would also have been built on the barrier extensions between the recurves as long as the barrier's sediment budget was in positive balance. Eventually in this westerly extension the sediment budget must have shifted into a negative budget balance, such that the barrier would have moved into a cannibalization phase leading to a barrier showing spatially variable barrier thinning. During this stage, a barrier becomes prone to the development of positions where wave-generated erosion-dominated perturbations occur, as wave–sediment cells self-organize along the barrier length (cf. Jennings *et al.* 1998). These sediment deficit positions can lead to barrier breaching, the development of tidal dominated entrances–exits and ebb-tidal deltas. A subsequent phase of dune development is associated with inlet mouth modifications by wave refraction around the ebb-tide delta, moving sediment preferentially into inlets (Fig. 7). This sediment is also directed by wave refraction into the beach ridge series, forming inlet recurves, which then acted as the core to aeolian dunes as the beach face and intertidal areas were deflated. In the case of north Norfolk, this activity phase appears to have been aided by two processes:

(a) anthropogenic reduction of the back-barrier tidal prisms by reclamation of the upper tidal marsh areas from the seventeenth century onwards;
(b) continuing longshore depletion of barrier–dune sediment by erosion under a transgressive shoreline.

The latter process generated sediment for dune building, while the former process defined and controlled the spatially limited depositional sinks for sand along the upper barrier over the last 300 years.

Conditions for past dune initiation

Dune development needs a source sediment volume (traditionally the beach face), a wind of sufficient velocity to entrain the sediment and a consistent wind direction that moves the sediment onshore. In the case of Northumberland and Norfolk there are some basic inconsistencies to these simple requirements that need some exploration.

Clearly a number of sites where there has been LIA dune development, have had a source sediment that has been coarser than normal on beach faces, or has experienced an enhanced entrainment capacity to move coarser material onshore. There is also a need to consider from where the sediment volume sufficient for dune formation appeared. It may be that there has always been an excess sediment volume in the beach face storage sufficient for dune building throughout the Holocene, but only entrained under a temporally limited opportunistic set of conditions such as during the LIA. Contrary to that view, the contemporary inner-reflective–outer-dissipative status of beaches in Northumberland suggests that there is no great sediment volume available on the beach for a further period of dune building. Likewise the ebb-delta domination of the north Norfolk coast and the adjacent beach face exposure of sub-barrier peat beds underline the lack of sediment on the beachface. Given this lack of sediment, the question is posed as to whether the beach face supply is now exhausted (apart from contemporary shoreline reworking) and without major sediment renewal (via sea-level change), leading to the proposition that the sediment currently in the dunes reflects a major 'one-off' loss to the shoreline. If this is the case, then under what conditions could this sediment volume have been exposed to aeolian processes for beach–dune exchange?

The contemporary nature of wind direction recorded at Boulmer (Northumberland) and Hemsby (Norfolk) during 1987–1991 is shown in Fig. 14 and reinforces the longer-term aggregate data identified by HMSO (1940) for east coast UK. The signal is consistent for both coasts in that under contemporary conditions, wind direction is dominated (60%) by west–south offshore vectors with limited (<10%) wind in any of the onshore vectors. Onshore vectors show a tendency for concentration during April and May (HMSO 1940) but with limited wind velocity (<10% chance of wind

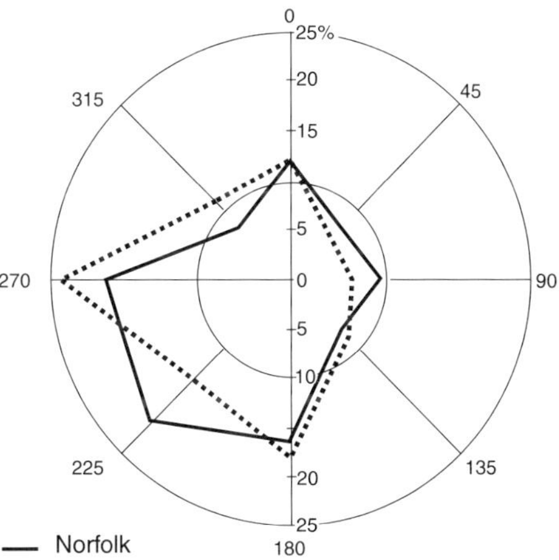

Fig. 14. Annual percentage distribution of wind by direction in Northumberland and Norfolk. Source: UK Meteorological Office's Daily Weather Record for 1988–1991.

exceeding Force 4). This lack of onshore wind in either coastal site reflects the limited contemporary dune development in either study area and underlines the need for some shift in atmospheric forcing to be sufficient to account for major dune formation in the past.

Lamb (1995) has compiled a substantial set of proxy data by which he has documented climate change in the British Isles and western Europe over the last 3 ka. In this piecing together of multitudinous evidence, Lamb outlined as one aspect of his work the nature of atmospheric warming and cooling that led to a strengthening or weakening of the Atlantic thermal gradient and its effect on cyclonic activity and sea-level around the British Isles. A chronological synthesis of Lamb's evidence on forcing conditions, together with our evidence of dune phases from Northumberland and Norfolk, is presented in Fig. 15. Lamb (1995) identifies two distinctive phases of higher sea-levels: around 1.7–1.6 ka BP and 0.8–0.7 ka BP, both of which are associated with a relative warming of the atmosphere and a 'deflection' northwards of westerly storm tracks. Note that higher sea-levels are considered only as relative to preceding lower sea-level. After the warming periods, Lamb suggests cooling phases in which either intensification of cyclonic activity over the Atlantic and/or increasing anticyclone activity over Fenno-Scandinavia occurred. The difference relates to the dominant air mass control of the period. In the former case the low-pressure westerlies are dominant over the British Isles due to the enhancement of a wet, cooling maritime low pressure over the Greenland cell, while in the latter case the expansion of an anticyclonic, cold stable high pressure from polar air over Scandinavia, forces westerlies both further north and south leading to an increase in southeast and northwest winds, respectively, over the east coast of England.

Figure 15 identifies the close proximity of the semicontinuous phase of dune building in south Northumberland during the period (1.5–1 ka BP), which followed the shift to a relatively lower sea level, as well as identifying a second possible relative sea-level fall (amount unknown, though Lamb mentions *c.* 0.5 m) following the 'climatic optimum' of the thirteenth century, which could be associated with the commencement of the dune deposition that accelerates during the Little Ice Age. This suggests that falling sea-level was a prior condition by which intertidal sediment was exposed, dried and deflated in sufficient volume to allow beach–dune exchange to current dune positions. Figure 15 also shows that around the LIA there was an earlier, though subdued, phase of dune initiation in Northumberland compared with the later burst of Norfolk dune building. This could

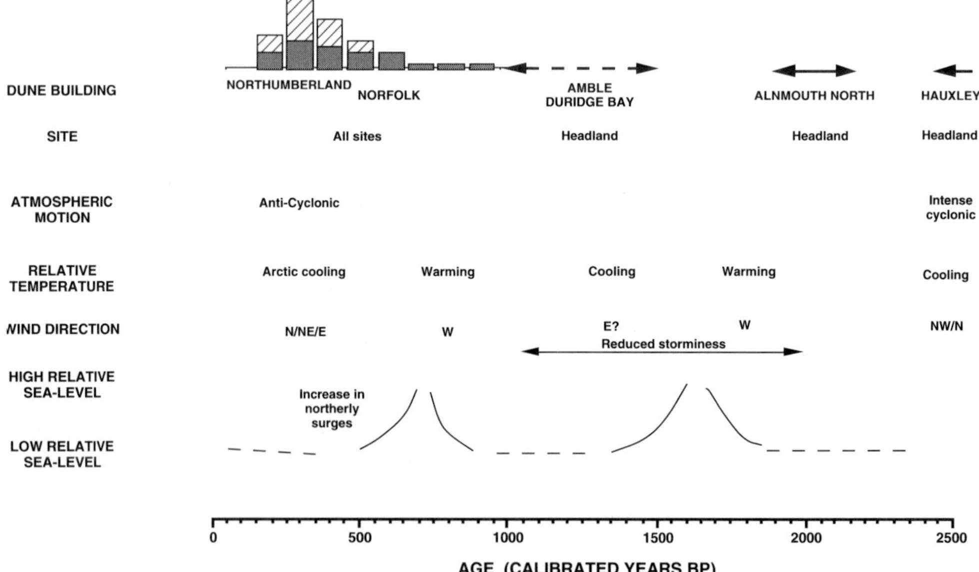

Fig. 15. Comparison of late Holocene atmospheric and sea-level data from Lamb (1995) and dune building phases in Northumberland and Norfolk over the last 2.5 ka to illustrate possible sea-level controls on dune building phases. The bar graph identifies the multiplicity of dates availble for the two coastal sites during the last millennium.

relate to a diachronous period of Arctic cooling whose effects reached Norfolk later, or had to contend with a stronger transgressive signature in Norfolk, which had to be reduced or even reversed. Lamb (1995) comments that during the fifteenth century Norfolk suffered an above average phase of inundation from northerly directed surges. These, Lamb suggested, were due to a stronger northerly component to this cooling phase, which again would have been coincidental with the start of LIA dune building in Norfolk. These same northerly surges may account for some of the reworking of the central barrier dunes and the development of the sediment supply to the ebb-deltas and the estuary mouth side-bars. However, they do not account for the conditions in which the original dune line along the seaward edge of the barrier were built, and reinforce the requirement of a pre-fifteenth century dune building phase following the 'climatic optimum'.

The relationship between falling sea-level and dune building is equivocal. Figure 15 also identifies the dune phases at Hauxley and Alnmouth, which are not associated directly with sea-level changes; though Innes & Frank (1988) believe that the Hauxley dune development was associated with a relative regression. The initiation of St Aidan's dunes (the oldest dunes in Northumberland) is coincident with a major deceleration of the relative Holocene transgression rather than a reversal of sea-level tendency and might identify that when a generally transgressive shoreline is forced to slow down, then normal regressive tendencies generated through sediment supply can come into play. In brief, the transgressive rate is reducing to a threshold below which the rate of relative sea-level rise is insufficient to rework sediment progradation.

An argument can be made in which dune formation is dependent on rising sea-level generating sediment volume by shoreline erosion (Pye 1984). This seems unlikely in the context of central and southern Northumberland where the dunes occupy the only existing sites of potential material that could be eroded, i.e. the source areas are the sinks. Although the same argument could be raised concerning more contemporary dune development (cf. the model of current dune development in Norfolk, Fig. 7), it would imply the need for a continuous history of autogenic dune emplacement, erosion and replacement, which does not appear evident from our data. Rather, our data are interpreted as showing discrete phases of dune building that are forced by allogenic changes in the external environment, which controlled shoreline development. Though the direction of such allogenic change in terms of sea-level and atmospheric forcing is signalled here, there is still a need to establish the scale of such changes as well as the processes involved.

Conclusions

The chronostratigraphy of dune development in Northumberland and north Norfolk has been obtained for the first time, based on vibracore sediments dated by IRSL and ^{14}C methods. Dune development in the two study areas reflects varying coastal responses to spatially varying Holocene sea levels. Presently observed coastal dune morphology is mainly Little Ice Age in origin. Dunes do not appear to be a repetitive element of the Holocene transgression due to the intermittency of onshore (north–east) wind during this period. The oldest dunes are potentially to be found in two possible contexts:

(a) areas with forced regression (north Northumberland), but likely to be degraded such that they are not clearly observed via their morphological attributes;
(b) an area of marginal normally regressive coastline (central Northumberland) that is spatially stable.

Only the latter has been observed at *c.* 4 ka BP. LIA dunes appear, by association, to have required a regressive sea-level tendency in order to develop. However, some LIA dunes may well be a function of excess local sediment supply. Most post-LIA dunes in Norfolk are found around the mouth of tidal inlets and probably reflect anthropogenic induced changes to the estuary tidal prism due to major estuary reclamation in the eighteenth century. Finally, the study indicates the time dependency of dune formation along the North Sea coast and indicates the potential fallacy of assuming that coastal dunes are a sustainable element of coastal defences in eastern England.

This work was funded by the UK Natural Environment Research Council (contract GST/02/789) through their LOIS programme. The vibracoring programme was planned with the logistical assistance of the Coastal Research Group, British Geological Survey (Keyworth) and physically undertaken by the British Geological Survey (Edinburgh and Keyworth). Facilities for storing and logging cores was extended by BGS (Keyworth). Our thanks go to P. Balson, H. Glaves, J. Rees and J. Ridgeway for their assistance in this initial phase of the work, to F. Musson for laboratory measurement of the IRSL samples and to G. Alexander and M. Pringle for diagram work. Comments from two referees were appreciated. We also extend our thanks to the numerous landowners and land-owning non-governmental organizations that allowed us access to the dunes in both Northumberland and Norfolk. This is LOIS Publication No. 587.

References

ANDREWS, J. E., BOOMER, I., BAILIFF, I., BALSON, P., BRISTOW, C., CHROSTON, P. N., FUNNELL, B. M., HARWOOD, G. M., JONES, R., MAHER, B. A. & SHIMMIELD, G. B. 2000. Sedimentary evolution of the north Norfolk barrier coastline in the context of Holocene sea-level change. *This volume.*

BOOMER, I. & WOODCOCK, L. 1999. The nature and origin of Stiffkey Meals. *Bulletin of the Geological Society of Norfolk*, **48**, in press.

BRISTOW, C. S., CHROSTON, N. P. & BAILEY, S. The structure and development of coastal dunes: insights from ground penetrating radar (GPR) surveys, Norfolk, England. *Sedimentology*, in press.

——, HORN, D. P. & RAPER, J. F. 1993. Evolution of a barrier island and recurved spits on a macrotidal coast. *In*: LIST, J. (ed.) *Large Scale Coastal Behaviour '93.* United States Geological Survey Open File, Report, **93–381**, 9–12.

CARTER, R. W. G. 1988. *Coastal Environments.* Academic Press, London, 617pp.

——1990. The geomorphology of coastal dunes in Ireland. *In*: BAKKER, T. W. M., JUNGERIUS, P. D. & KLIJN, J. A. (eds) *Dunes of the European Coasts.* Catena Supplement, **18**, 31–39.

—— & WILSON, P. 1990. The geomorphological, ecological and pedological development of coastal foredunes at Magilligan Point, Northern Ireland. *In*: NORDSTROM, K. F., PSUTY, N. P. & CARTER, R. W. G. (eds) *Coastal Dunes: Form and Process*, J. Wiley, Chichester, 129–157.

—— & ——1993. Aeolian processes and deposits in north-west Ireland. *In*: PYE, K. (ed.) *The Dynamics and Environmental Context of Aeolian sedimentary systems.* Geological Society Special Publications, **72**, 173–190.

CHANG, S. C. & EVANS, G. 1992 Source of sediment and sediment transport on the east coast of England: significance or coincidental phenomena? *Marine Geo*logy, **107**, 283–288.

CHRISTIANSEN, C., DALSGAARD, K., MØLLER, J. T. & BOWMAN, D. 1990. Coastal dunes in Denmark. Chronology in relation to sea level. *In*: BAKKER, T. W., JUNGERIUS, P. D. & KLIJN, J. A. (eds) *Dunes of the European Coasts.* Catena Supplement **18**, 61–70.

CLEMMENSEN, L. B., ANDREASEN, F., NIELSEN, S. & STEN, E. 1996. The late Holocene coastal dunefield at Vejers, Denmark: characteristics, sand budget and depositional dynamics. *Geomorphology*, **17**, 79–98.

FLEMMING, B. W. & ANTIA, E. E. 1990. Interaction between storm-generated resonance and nearshore morphodynamics along the Frisian barrier island coast, southern North Sea. *Abstracts of the Thirteenth International Sedimentology Congress*, p. 171.

FRANK, R. B. 1982. A Holocene peat and dune-sand sequence on the coast of northeast England – a preliminary report. *Quaternary Newsletter*, **36**, 24–32.

FUNNELL, B. M. & PEARSON, I. 1989. Holocene sedimentation on the north Norfolk barrier coast in relation to relative sea-level change. *Journal of Quaternary Science*, **4**, 25–36.

GERRITSEN, F & DUNSBERGEN, D. 1998. Morphological–empirical relationships for ebb-tidal deltas as a tool in dynamic–empirical modelling. *Journal of Coastal Research*, Special Issue, **26**, 273–281.

HARDY, J. A. 1964. The movement of beach material and wave action near Blakeney Point, Norfolk. *Transactions of the Institute of British Geographers*, **34**, 53–69.

HER MAJESTY'S STATIONERY OFFICE 1940. *Weather in Home Waters and North-Eastern Atlantic*, **II**, Part 5, *The North Sea.* 202 pp.

HESP, P. A. 1988. Morphology, dynamics and internal stratification of some established foredunes in south-east Australia. *Sedimentary Geology*, **55**, 17–41.

HUNTLEY, D. J., GODFREY-SMITH, D. I. & THEWALT, M. L. W. 1985. Optical dating of sediments. *Nature*, **313**, 105–107.

INNES, J. B. & FRANK, R. M. 1988. Palynological evidence for Late Flandrian coastal changes at Druridge Bay, Northumberland. *Scottish Geographical Magazine*, **104**, 14–23.

JENNINGS, S., ORFORD, J. D., CANTI, M., DEVOY, R. J. N. & STRAKER, V. 1998. The role of relative sea-level rise and changing sediment supply on Holo-cene gravel barrier development: the example of Porlock, Somerset, UK. *The Holocene*, **8**, 165–181.

JONES, R. 1966. Long term coastal changes. *Proceedings of the MAFF Conference of River and Coastal Engineers*, Univ. Keele, July 1966, Section **8.4**. Ministry of Agriculture and Fisheries, London, 1–9.

KLIJN, J. A. 1990. The younger dunes in The Netherlands; chronology and causation. *In*: BAKKER, T. W., JUNGERIUS, P. D. & KLIJN, J. A. (eds) *Dunes of the European Coasts.* Catena Supplement, **18**, 89–100.

KNIGHT, J., ORFORD, J. D., WILSON, P., WINTLE, A. G. & BRALEY, S. 1998. Facies, age and controls on recent coastal sand dune evolution in north Norfolk, eastern England. *Journal of Coastal Research*, Special Issue, **26**, 154–161.

LAMB, H. H. 1995. *Climate, History and the Modern World.* 2nd Edition, Routledge, London.

OLLERHEAD, J., HUNTLEY, D. J. & BERGER, G. W. 1994. Luminescence dating of sediments from Buctouche Spit, New Brunswick. *Canadian Journal of Earth Sciences*, **31**, 523–531.

ORFORD, J. D., COOPER, J. A. G., JACKSON, D., MALVAREZ, G. & WHITE, D. 1999. Extreme storms and thresholds on foredune stripping at Inch Spit, south-west Ireland. *Coastal Sediments '99*, American Society of Civil Engineers, 1852–1866.

PEARSON, I., FUNNELL, B. M. & MCCAVE, I. N. 1989. Sedimentary environments of the sandy barrier/tidal marsh coastline of north Norfolk. *Bulletin of the Geological Society of Norfolk*, **39**, 3–44

PETHICK, J. S. 1980. Salt-marsh initiation during the Holocene transgression: the example of the North Norfolk marshes, England. *Journal of Biogeography*, **7**, 1–9.

PLATER, A. J. & SHENNAN, I. 1992. Evidence of Holocene sea-level change from the Northumberland coast, eastern England. *Proceedings of the Geologists' Association*, **103**, 201–216.

POSAMENTIER, H. W., ALLEN, G. P. & JAMES, D. P. 1992. High resolution sequence stratigraphy, the East Coulee delta, Alberta. *Journal of Sedimentary Petrology*, **62**, 310–317.

PSUTY, N. P. 1992. Spatial variation in coastal foredune development. *In*: CARTER, R. W. G., CURTIS, T. F. G. & SHEEHY-SKEFFINGTON, M. (eds) *Coastal Dunes: Geomorphology, Ecology and Management for Conservation*. Balkema, Rotterdam, 3–13.

PYE, K. 1982. Negatively skewed aeolian sands from a humid tropical coastal dunefield, Northern Australia. *Sedimentary Geology*, **31**, 249–266.

——1983. Formation and history of Queensland coastal dunes. *Zeitschrift fur Geomorpologie*, Supple. Bd, **45**, 175–204

——1984. Models of transgressive coastal dune building episodes and their relationship to Quaternary sea level changes: a discussion with reference to evidence from eastern Australia. *In*: CLARKE, M. (ed.) *Coastal Research: UK Perspectives*. Geo Books, Norwich, 81–104.

—— & NEAL, A. 1993. Late Holocene dune formation on the Sefton Coast, northwest England. *In*: PYE, K. (ed.) *The Dynamics and Environmental Context of Aeolian Sedimentary Systems*. Geological Society Special Publications, **72**, 207–217.

——, STOKES, S. & NEAL, A. 1995. Optical dating of aeolian sediments from the Sefton coast, northwest England. *Proceedings of the Geologists' Association*, **106**, 281–292.

—— & TSOAR, H. 1990. *Aeolian Sand and Sand Dunes*. Unwin Hyman, London, 396pp.

SHENNAN, I., LAMBECK, K., HORTON, B., INNES, J. B., LLOYD, J., MCARTHUR, J. & RUTHERFORD, M. 1999. Holocene crustal movements and relative sea-level changes on the east coast of England. *This volume*.

SHERMAN, D. J. & BAUER, B. O. 1993. Dynamics of beach-dune systems. *Progress in Physical Geography*, **17**, 413–447.

——, JACKSON, D. W. T., NAMIKAS, S. L. & WANG, J. 1998. Wind-blown sand on beaches: an evaluation of models. *Geomorphology*, **22**, 113–133.

SHORT, A. D. & HESP, P. A. 1982. Wave, beach and dune interactions in southeastern Australia. *Marine Geology*, **48**, 259–284.

STEERS, J. A. 1989. The physical features of Scolt Head Island and Blakeney Point. *In*: ALLISON, H. & MORLEY, J. (eds) *Blakeney Point and Scolt Head Island*. 5th Edition, The National Trust, Norfolk, 14–27.

STUIVER, M. & REIMER, P. J. 1993. Extended ^{14}C database and revised CALIB 3.0 ^{14}C age calibration program. *Radiocarbon*, **35**, 215–230.

TAYLOR, B. J., BURGESS, I. C., LAND, D. H., MILLS, D. A. C., SMITH, D. B. & WARREN, P. T. 1971. *British Regional Geology: Northern England*. Her Majesty's Stationery Office, London, 121pp.

TERWINDT, J. H. T. & KROON, A. 1993. Theoretical concepts of parameterisation of coastal behaviour. *In*: LIST, J. (ed.) *Large Scale Coastal Behaviour '93*. United States Geological Survey, Open File Report, **93–381**, 193–196

WILSON, P. 1990. Coastal dune chronology in the north of Ireland. *In*: BAKKER, T. W., JUNGERIUS, P. D. & KLIJN, J. A. (eds) *Dunes of the European Coasts*. Catena Supplement, **18**, 71–79.

——1996. Morphological and chemical variations of a buried palaeocatena in late Holocene beach-ridge sands at Magilligan Foreland, Northern Ireland. *Journal of Coastal Research*, **12**, 605–611.

—— & MCKENNA, J. 1996. Holocene evolution of the River Bann estuary and adjacent coast. *Proceedings of the Geologists' Association*, **107**, 241–252

WINTLE, A. G. 1993. Luminescence dating of aeolian sands: an overview. *In*: PYE, K. (ed.) *The Dynamics and Environmental Context of Aeolian Sedimentary Systems*. Geological Society, Special Publications, **72**, 49–58.

——, CLARKE, M. L., MUSSON, F. M., ORFORD, J. D. & DEVOY, R. J. N. 1998. Luminescence dating of recent dunes on Inch Spit, Dingle Bay, southwest Ireland. *The Holocene*, **8**, 331–339.

Sedimentary evolution of the north Norfolk barrier coastline in the context of Holocene sea-level change

J. E. ANDREWS,[1] I. BOOMER,[1,6] I. BAILIFF,[2] P. BALSON,[3] C. BRISTOW,[4] P. N. CHROSTON,[1] B. M. FUNNELL,[1] G. M. HARWOOD,[1]† R. JONES,[1] B. A. MAHER[1] & G. B. SHIMMIELD[5]

[1] *School of Environmental Sciences, University of East Anglia, Norwich NR4 7TJ, UK (e-mail: j.andrews@uea.ac.uk)*

[2] *Environmental Research Centre, Department of Archaeology, University of Durham, Durham DH1 3LE, UK*

[3] *British Geological Survey, Kingsley Dunham Centre, Keyworth, Nottingham NG12 5GG, UK*

[4] *Department of Geology, Birkbeck College, Malet Street, London WC1E 7HX, UK*

[5] *Dunstaffnage Marine Laboratory, PO Box 3 Oban, Argyll PA34 4AD, UK*

[6] *Present address: Department of Geography, Daysh Building, University of Newcastle, Newcastle-upon-Tyne NE1 7RU, UK*

Abstract: Holocene sediments of the north Norfolk coast (NNC) between Weybourne and Hunstanton have been studied using geophysical, sedimentological, biofacial and dating techniques. New cores and refraction seismic data have defined the topography of the pre-Holocene surface and show that the NNC sediment prism is underlain by an east–west trending Quaternary trough, probably a palaeo-river-valley. The age of the Holocene fill has been dated using radiocarbon and luminescence dates, while sedimentation rates were constrained by, and compared with, modern rates using radionuclide data. The Holocene sediments are divided into a sandy-barrier lithofacies association (LFA), and a muddy–silty–peat back-barrier LFA. The oldest Holocene sediments are peats, formed on an undulating till surface. These peats were forming by 11–10 cal. ka BP and continued to form until at least 7 cal. ka BP in a number of places. As Holocene sea-level rose, marine mudflat and saltmarsh environments began to form between 7 and 6 cal. ka BP east of Holkham and around 6 cal. ka BP or younger west of Holkham. A marked erosion surface between the barrier and back-barrier LFA in the Holkham to Burnham Overy area is imperfectly dated at <3 cal. ka BP, but suggests the sediment prism has thinned by about 3 km over 6 to *c.* 3 cal. ka BP. This surface probably records the westward progress of laterally migrating tidal channels that caused back-barrier sediment erosion, along with shoreface processes, as sea-level rose. Small-scale regressive and transgressive saltmarsh sequences occur at different elevations along strike but cannot be correlated, suggesting that the control on saltmarsh and mudflat development is autocyclic rather than allocyclic. Generally, transgressive and regressive events are related to disposition of coastal barriers and these are superimposed on a general facies evolution governed by regional sea-level change. Predictions about how this barrier coastline might respond to increased rates of regional sea-level change caused by global warming, or climatic events like increased storminess, require an understanding of how specific segments of the coastline have responded over millennial time-scales. This longer-term evolution provides the baseline information for decision making and management strategy. It is likely that sandy sediment supply is limited on the NNC and this implies that the barriers will continue to move landward, probably at increased rates relative to today, suggesting that parts of the NNC will become more vulnerable to erosion and flooding.

† Deceased.

From: SHENNAN, I. & ANDREWS, J. (eds) *Holocene Land–Ocean Interaction and Environmental Change around the North Sea*. Geological Society, London, Special Publications, **166**, 219–251. 1-86239-054-1/00/$15.00

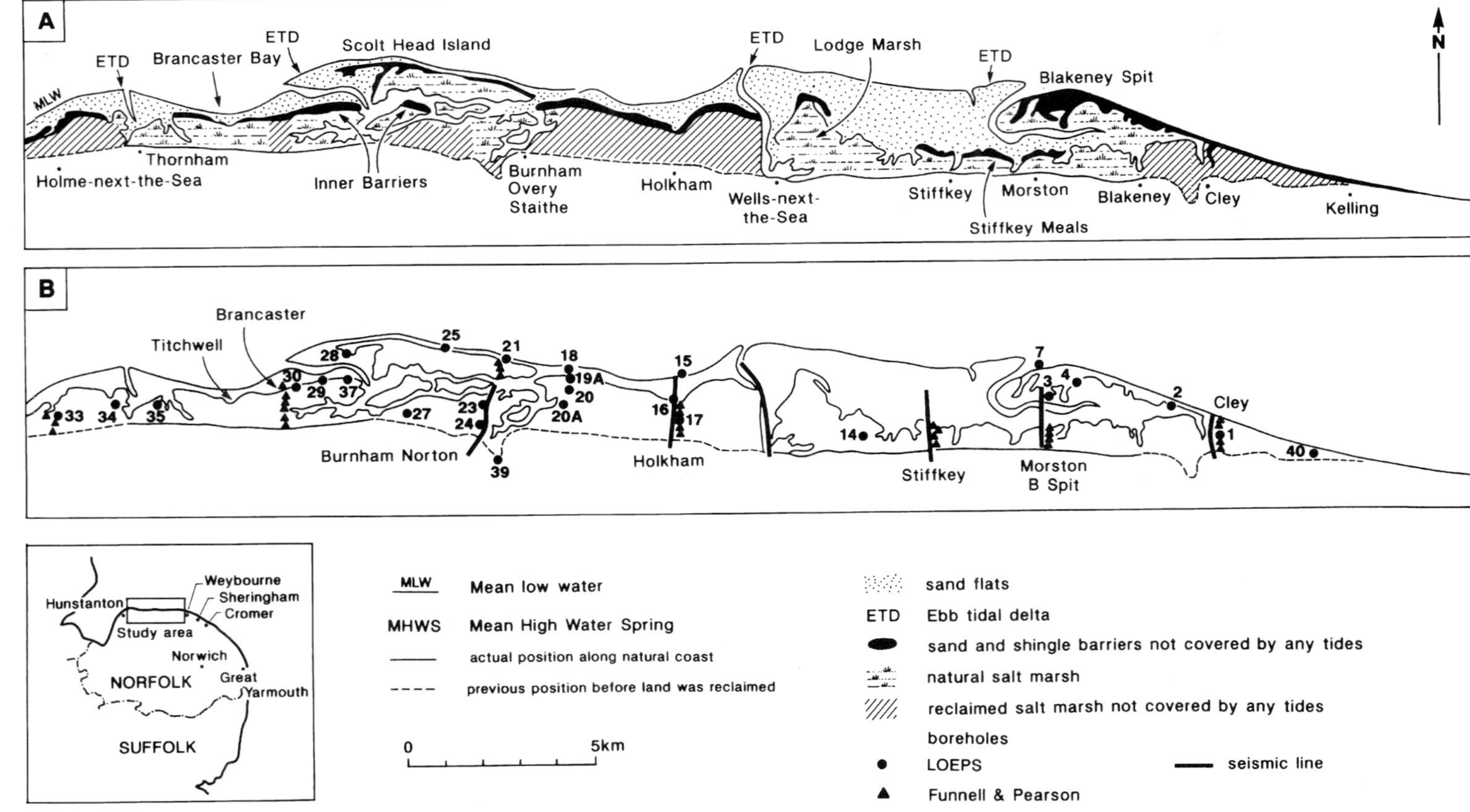

Fig. 1. Location map of the study area. **(a)** Schematic map of the dominant geomorphological composition of the NNC adapted from Funnell (1992) and Spencer & French (1992). ETD, ebb-tidal delta. **(b)** Position of the LOEPS boreholes (40, NNC40, etc.), earlier borehole transects by Funnell & Pearson (1989) and the location of the land seismic lines (details in Chroston *et al.* 1999).

An objective of the Land–Ocean Evolution Perspective Study (LOEPS) of the Land–Ocean Interaction Study (LOIS) was 'to determine, through study of the Holocene sedimentary record and changing coastal disposition ... the regional history of sediment fluxes, sources and sinks at the river–atmosphere–coast–study (RACS) site' (LOIS 1992). It was emphasized that the role of sea-level change would be an important component influencing these parameters. Special Topic 32 (the study described here) proposed a study of the barrier coastline of north Norfolk (NNC) in the south of the RACS site (Fig. 1). The objective of the study was to use combined sedimentological, geophysical, biofacial and dating techniques to: (a) define the regional sedimentary framework and establish the Holocene sedimentary evolution of the north Norfolk coast; and (b) determine rates of environmental change and help predict future coastal changes during sea-level rise.

These objectives have direct relevance to future management of the NNC, a coastline designated either under domestic or international legislation as a Site of Special Scientific Interest (SSSI), an Area of Outstanding Natural Beauty (AONB), as a Special Protection Area (SPA) or as a Biosphere Reserve or Ramsar (wetland of international importance) site (see also Funnell 1992). Titchwell (Fig. 1), for example, is the Royal Society for the Protection of Birds' (RSPB) most visited reserve, while the Norfolk Wildlife Trust reserve at Cley (Fig. 1) is protected under international legislation as an SPA under the Birds Directive. Shoreline management plans (SMPs) are supposed to provide a co-ordinated strategic approach for making decisions on coastal defence, but these typically consider only quite short term environmental change and fail to account for the longer-term (greater than centennial) evolution of the coastal zone addressed here.

The barrier coastline of north Norfolk stretches from Weybourne in the east to Hunstanton in the west (Fig. 1). Previous work on the subsurface record in this Holocene sediment prism has been limited. The most comprehensive information is in Pearson (1986) and Funnell & Pearson (1989), who studied a series of power auger transects along NNC between Holme and Cley (Fig. 1), and in Allison (1985), who presented power auger information from Scolt Head Island (Fig. 1). Other studies of the Holocene record include local investigations in the Brancaster (Murphy & Funnell 1980) and Titchwell (Fig. 1) areas (Wymer & Robins 1994). Based on this existing information, especially the transects of Funnell & Pearson (1989), we proposed the siting of 28 new cored boreholes (part of the LOEPS core programme).

Background to the north Norfolk coast

Modern coastal geomorphology

The low-lying barrier coast of north Norfolk (localities mentioned in the text are shown on Fig. 1) is today being shaped by a macrotidal to high mesotidal regime (Admiralty tide tables show the mean spring tidal range is 6.6 m at Hunstanton and 4.7 m at Cromer). However, because of the mainly shallow water offshore, wave energy is low to moderate with an average wave height of 0.3–0.4 m between Blakeney and Sheringham and 0.2–0.3 m between Scolt Head Island and Blakeney (Pye 1992). The combination of high tidal range and the shallow, gently inclined offshore slope has resulted in the development of extensive intertidal areas, generally comprising protected back-barrier mudflats and saltmarsh (see overview in Pye 1992) and a wide beach or wave-influenced sandflat in front of the sand and gravel barriers (Fig. 1). Blakeney Spit and Scolt Head Island are the only portions of the barrier coast fronted by relatively steeply shelving beaches immediately seaward.

The overall wave climate is complex; however, the dominant wave direction is from the northeast and longshore wave transport from Sheringham to Scolt Head Island is westward (Steers 1927, Hardy 1964, Vincent 1979). This longshore energy transports some sand and shingle westward, as shown by the recurved lateral ridges on Blakeney Spit and Scolt Head Island (Steers 1929, 1964). The potential for westward sand transport past Blakeney has been calculated as $350 \times 10^3 \, m^3 \, a^{-1}$ (Vincent 1979), although the actual transport rate is much lower, and may be zero (Clayton & O'Riordan 1995). Wave observer data at Blakeney showed significant net eastward drift in places (Clayton 1976), suggesting that the eastern end of the spit is realigning to keep pace with cliff retreat east of Weybourne. At the western end of the study area, between Holme and Brancaster, the dominant longshore transport path is locally reversed, i.e. eastward, as shown by the orientation of some recurved spits, e.g. Gore Point near Holme (Steers 1929; Roy 1967). Grain size data from intertidal sands between Cromer and Hunstanton indicate some (but not strong) preferential winnowing in the direction of longshore transport (McCave 1978), suggesting that at least some sand supply might be from the north or northwest (Shih-Chiao &

Evans 1992), where much the shallow floor of the north Sea is composed of glacial sediments or veneered by up to 1 m of sand (Pye 1992). The source of shingle, especially at Scolt Head Island, is problematic (Steers 1989), and it is not at all clear that *new* sediment is being supplied to these beaches; the sand and shingle may simply be recycled between the beaches, ebb-tidal deltas (see below) and offshore banks.

The sand and gravel barriers occur in a variety of forms along the coast (Fig. 1).

(1) Between Weybourne and Morston the shingle spit of Blakeney Point is accreting and migrating westwards and the gravel beach at its trailing eastern edge is rolling landwards by about 1 m a^{-1} (Clayton 1976).
(2) Between Morston and Wells, extensive and apparently stable intertidal sand flats are backed by a narrow, vegetated dune ridge.
(3) Between Wells and Burnham Overy, the sandflats are prograding. Here, vegetated aeolian dune ridges are fronted 500 m seawards by a foredune ridge along the beach platform with incipient saltmarsh forming behind.
(4) The area between Burnham Overy and Brancaster Staithe is dominated by Scolt Head Island, a barrier island with an accreting spit at its western end and erosion in its central part; west of the island, intertidal sandflats backed by aeolian dune ridges are being eroded in Brancaster Bay.
(5) Between Holme and Brancaster the beaches are generally stable or accreting, but locally eroding, and protect extensive mudflats and saltmarshes.

The tidal prism enters and drains the back-barrier areas through tidal inlets at Thornham, Brancaster, Burnham Overy, Wells and Blakeney. The tidal prism has been severely reduced over the last few hundred years following the reclamation of around 50% of the back-barrier saltmarshes (Pye 1992; Clayton 1995), including the loss of a tidal inlet at Holkham. The form of other tidal inlets, e.g. the Wells Channel, has changed by straightening, embankment and dredging, decreasing both the ebb and flood tidal current in channels. Wave-modified ebb tidal deltas are accreting north of the tidal inlets at Thornham, Brancaster, Burnham Overy, Wells and Blakeney (Fig. 1). The tidal deltas are sediment sinks and form areas of sediment convergence where beaches are accreting. The deltas divide the coast into a series of coastal cells and intervening cell divides may mark zones of erosion such as at Brancaster Bay (Fig. 1).

Regional sea-level change

A detailed record of the sea-level changes impacting the Holocene evolution of the NNC have not so far been available (but see Shennan *et al.* this volume), although the broad positions of the palaeocoastline in the early Holocene were suggested by Jelgersma (1979) and modelled by Lambeck (1995) and in now in detail by Shennan *et al.* (this volume). Holocene changes in relative sea-level in nearby Fenland and Broadland have been detailed by Shennan (1982, 1986) and Shennan & Woodworth (1992). These data sets suggest mean crustal subsidence of 0.9 mm a^{-1} in Fenland over the last 6.5 ^{14}C ka, and that since 3 ^{14}C ka sea-level has risen by about 1 mm a^{-1}, while in Broadland, sea-level rise over the past 7.5 ^{14}C ka averages at about 0.6 mm a^{-1}. It is likely that sea-level rise in the earlier part of the Holocene was faster than this, but reliable records do not extend much beyond 6–7 ^{14}C ka BP. Funnell & Pearson (1989) related the depositional record of the NNC Holocene sediment prism to this overall sea-level history, and also proposed more local-scale positive and negative sea-level tendencies, based on the position of transgressive and regressive overlaps.

The following sections demonstrate how we have built upon existing knowledge to give a more complete view of the history of NNC Holocene sedimentation. This paper is an overview of progress made during the LOEPS and in places the reader is referred to specific papers resulting from LOEPS that contain the details of methodology and results.

Methods

Shell and auger percussion cores were recovered in 1-m-long rigid plastic liners; recovery was virtually 100% in muddy and peaty sediments (Ridgway *et al.* this volume) but much lower and sometimes zero in sandy sediments. High water content frequently caused liquefaction in sandy sediments, destroying most sedimentary structures. All core sites were levelled to Ordnance datum (OD) with closure error <2 mm (see details in Ridgway *et al.* this volume). Sedimentological, lithofacies and biofacies studies were based on the cores and associated hand auger profiles (<5 m depth).

Multi-channel seismic reflection trials revealed a strong reflection from the top of the Chalk formation; however, within the Holocene sediments resolution was poor and structures could not be resolved. Refraction seismology was thus adopted for the main survey on land (Chroston *et al.* this volume) and 80 separate refraction spreads were established mostly along five principal traverses (Fig. 1). Two hundred kilometres of marine single-channel continuous reflection profiling (boomer source) were obtained offshore, augmented by a further 15 km in creeks and channels. Despite the shallow inshore waters, multiples were often weak allowing identification of sub-bottom reflectors.

Radiocarbon dates from 48 samples (42 *in situ* peats, one wood, five $CaCO_3$ shells) were provided as part of the LOEPS (core programme) allocation at the NERC facility (East Kilbride). All samples were scrutinized and approved by the LOEPS radiocarbon committee and dates were calibrated to calendar years BP using the appropriate calibrations (including the reservoir effect for marine shell carbonate when necessary) in the CALIB 3 program (Stuiver & Reimer 1993). We did not use dates from bulk organic matter in muddy sediments (see e.g. dates in Allison 1985) as these show anomalous age–depth relation and are clearly contaminated by reworked older carbon. In this paper, calibrated radiocarbon ages are generally given as the central value of the (2σ) range, although when direct comparisons with other dates are made the range values (2σ) are quoted. The full data for specific dates mentioned are given in Table 1.

Luminescence dating measurements were made on samples from five of the cored boreholes. Samples were taken from opaque lined cores under subdued red light and sealed in black plastic containers to preserve water content. (Six infrared-stimulated luminescence (IRSL) ages were produced for basal tidal sands overlying mudflats and 18 IRSL ages for silts.) Ages were obtained via the age equation in its simplest form, i.e. luminescence age = palaeodose/annual dose. Evaluation of the palaeodose was achieved by the application of luminescence techniques to quartz and feldspar mineral fractions (details of laboratory procedures are given in Bailiff & Tooley this volume). Two techniques were used: (a) with the mineralogically undifferentiated silt-sized fraction, IRSL, which selectively stimulates luminescence in feldspars; and (b) with separated and acid-treated quartz grains from sand fractions, visible light (450–550 nm) stimulated luminescence (OSL). Annual dose was determined via various analytical techniques (Bailiff & Tooley this volume). Although OSL measurements on quartz extracted from three cores (NNC16, NNC18 and NNC19) were unsuccessful because of inherently low luminescence sensitivity, the recovery of small amounts of feldspar allowed application of IRSL. Ages are quoted with their 1σ confidence limits.

High-resolution records of saltmarsh accumulation rates were made using natural and anthropogenic radionuclides (^{210}Pb and ^{137}Cs). Their presence in fine-grained sediment is ascribed to adsorption onto detrital sediment particles. By collecting cores from a range of sites including pioneer and lower saltmarsh to mature upper saltmarsh at Stiffkey, Wells and Scolt Head Island, average accumulation rates of sediment can be calculated over much longer time-scales than conventional surveying techniques. Both ^{210}Pb, ^{226}Ra and ^{137}Cs were counted simultaneously by high resolution gamma spectroscopy and quantified using spiked sediment samples of known activity. The procedure was validated with international reference standards. Results are in bequerels per kilogram dry weight of sample and the errors represent 1σ deviation on the counting statistics.

Foraminifera and ostracoda from 150 sediment samples (43 000 specimen counts) were used to identify four separate intertidal environments, i.e. upper saltmarsh, lower saltmarsh, upper mudflat, lower mudflat (Boomer 1998). Modern analogues from the same coastline identified four assemblages each defined in terms of altitude/flooding frequency. The upper two are well constrained altitudinally, allowing precise sea-level reconstructions; the second two less so. Knowing the relationship between modern microfaunal assemblages and the tidal levels at which they live, enabled us to determine tidal levels in core samples based on the microfossil assemblages. Using this information and the regional sea-level curve (Lambeck 1995), we were able to assign approximate deposition dates for core samples. Despite uncertainties to be expected in such a method, 86 'microbiofacies dates' were obtained (Funnell & Boomer 1998). The derived microbiofacies 'dates' are proxy radiocarbon dates, and with an assumed error, can be calibrated in the same way as radiocarbon dates. Errors for individual microbiofacies 'dates' quoted in this paper are tabulated in Funnell & Boomer (1998): the largest errors are $<\pm 320$ years, the smallest around ± 40 years.

Underlying structure and pre-Holocene surface

The thickness of the Holocene sediments is determined by the elevation of an underlying glacial and post-glacial erosion surface, which cuts into underlying glacial sediments and Chalk (see cross sections and discussion in Funnell & Pearson 1989). Coring and geophysics (Fig. 2a) indicate that Holocene and glacial sediments

Table 1. *Radiocarbon ages and calibration data for dates quoted in the text*

Site	Material	Laboratory code	^{14}C age (a BP)	error ($\pm 1\sigma$)	Calibrated age (a BP)*			Altitude (m OD)
					Max.	Mean	Min.	
Cley NNC1	Amorphous peat	AA22702	8770	60	9918	9 767	9532	−8.04
Blakeney NNC2	Clayey peat	AA23462	5625	60	6527	6410	6296	−4.47
Blakeney NNC2	Clayey peat	AA22703	5725	50	6664	6492	6409	−4.55
Blakeney Point NNC4	Amorphous peat	AA23463	6705	65	7627	7540	7397	−11.49
Blakeney Point NNC4	Amorphous peat	AA22704	7770	55	8599	8500	8406	−11.73
Warham Marsh NNC14†	*Scrobicularia* shell *in situ*	AA27230	5115	55	5585	5460	5311	−5.50
Warham Marsh NNC14	Decomposed peat	AA27231	6585	65	7540	7410	7296	−12.41
Warham Marsh NNC14	Decomposed peat	AA27232	6860	85	7880	7640	7529	−12.47
Warham Marsh NNC14	Clayey peat	AA27233	7420	90	8369	8160	7973	−14.35
Warham Marsh NNC14	Amorphous peat	AA22686	7530	100	8483	8330	8026	−14.38
Warham Marsh NNC14	Decomposed peat	AA27588	9265	90	10531	10250	10029	−14.64
Holkham NNC16	Decomposed peat	AA22679	7250	95	8174	8025	7833	−8.04
Holkham NNC16	Decomposed peat	AA22680	9280	100	10772	10255	10024	−8.57
Holkham NNC17†	*Scrobicularia* shell *in situ*	AA22707	2715	70	2672	2360	2271	−1.61
Holkham NNC17	Amorphous peat	AA22681	5930	100	7006	6750	6494	−6.36
Holkham NNC17	Decomposed peat	AA23465	6375	60	7384	7240	7173	−6.53
Holkham NNC17	Amorphous peat	AA22682	7760	95	8943	8490	8346	−6.70
Burnham Overy NNC 18†	*Cerastoderma* shell	AA22705	3250	45	3193	3059	2925	−4.78
Burnham Overy NNC 18	Amorphous peat	AA22683	6800	100	7790	7580	7432	−5.68
Burnham Overy NNC 18	Amorphous peat	AA22684	7155	95	8124	7930	7725	−5.73
Burnham Overy NNC19A†	*Hydrobia* shells	AA22706	5525	45	5984	5900	5822	−5.33
Burnham Overy NNC19A	Amorphous peat	AA22685	9410	95	10889	10380	10140	−6.23
Burnham Overy NNC19A	Sandy peat	AA22700	9680	60	10992	10940	10488	−6.52
Burnham Overy NNC20	Decomposed peat	AA22698	3770	60	4347	4120	3930	−0.31
Burnham Overy NNC20	Decomposed peat	AA22699	4260	50	4871	4835	4645	−0.49

Scolt Head NNC21†	*Scrobicularia* shell *in situ*	AA22708	1565	40	1 200	1 107	1 005	0.43
Burnham Deepdale NNC27	Silty humified peat	AA23458	3455	85	3 921	3 690	3 473	0.02
Burnham Deepdale NNC27	Amorphous peat	AA23459	5795	60	6 742	6 605	6 448	–1.71
Scolt Head NNC28	Poorly humified peat	AA23464	5985	60	6 988	6 825	6 677	–6.53
Scolt Head NNC28	Laminated clayey peat	AA23460	9470	75	10 902	10 470	10 227	–6.80
Brancaster NNC29	Amorphous peat	AA22687	2700	40	2 861	2 776	2 749	–1.68
Brancaster NNC29	Amorphous peat	AA22688	3260	45	3 576	3 467	3 372	–1.94
Brancaster NNC29	Dark brown peat	AA22689	5845	55	6 786	6 670	6 496	–5.87
Brancaster NNC29	Dark brown peat	AA22690	9870	70	11 470	11 000	10 959	–6.19
Thornham NNC35	Decomposed peat	AA22691	2640	55	2 848	2 760	2 717	–1.11
Thornham NNC35	Decomposed peat	AA22692	3785	50	4 344	4 121	3 985	–1.97
Thornham NNC35	Silty peat	AA22701	4380	80	5 286	4 925	4 738	–3.17
Thornham NNC35	Decomposed peat	AA22693	4690	55	5 582	5 390	5 299	–3.37
Thornham NNC35	Decomposed peat	AA22694	8750	75	9 921	9 745	9 497	–3.98
Brancaster NNC37	Decomposed peat	AA23461	4105	55	4 826	4 570	4 426	–2.14
Salthouse NNC40	Decomposed peat	AA22696	3940	50	4 517	4 408	4 232	–3.64
Salthouse NNC40	Decomposed peat	AA22697	4495	50	5 305	5 149	4 877	–3.85
Salthouse NNC40	Silty peat	AA22695	6145	55	7 175	7 010	6 881	–5.92
Titchwell Foreshore§	Decomposed peat	AA28179	4560	50	5 443	5 292	5 042	–2.85
Titchwell Foreshore§	Decomposed peat	AA28178	6495	55	7 471	7 380	7 239	–3.14
Brancaster Marsh‡,§	Decomposed peat	UB-3998	3416	64	3 833	3 660	3 474	0.16
Titchwell Foreshore‡,¶	Decomposed peat	UB-3978	3506	40	3 874	3 774	3 639	–1.19
Titchwell Foreshore‡,¶	Rooted wood (*in situ*)	UB-3979	3017	39	3 340	3 210	3 071	–1.03

* The calibrated ages shown are the age ranges which contain 95.4% of the area under the probability curve. Ages in this table were calibrated with CALIB 3.0 using the bidecadal atmospheric curve (Stuiver & Reimer 1993; and references therein) except for samples marked.
† Which were calibrated with the marine calibration dataset (Stuiver & Braziunas 1993).
‡ Traditional radiocarbon date made at the University of Belfast (data courtesy of Brian Funnell & Ian Boomer).
§ Sampled from gouge auger.
¶ Sampled from foreshore exposure.

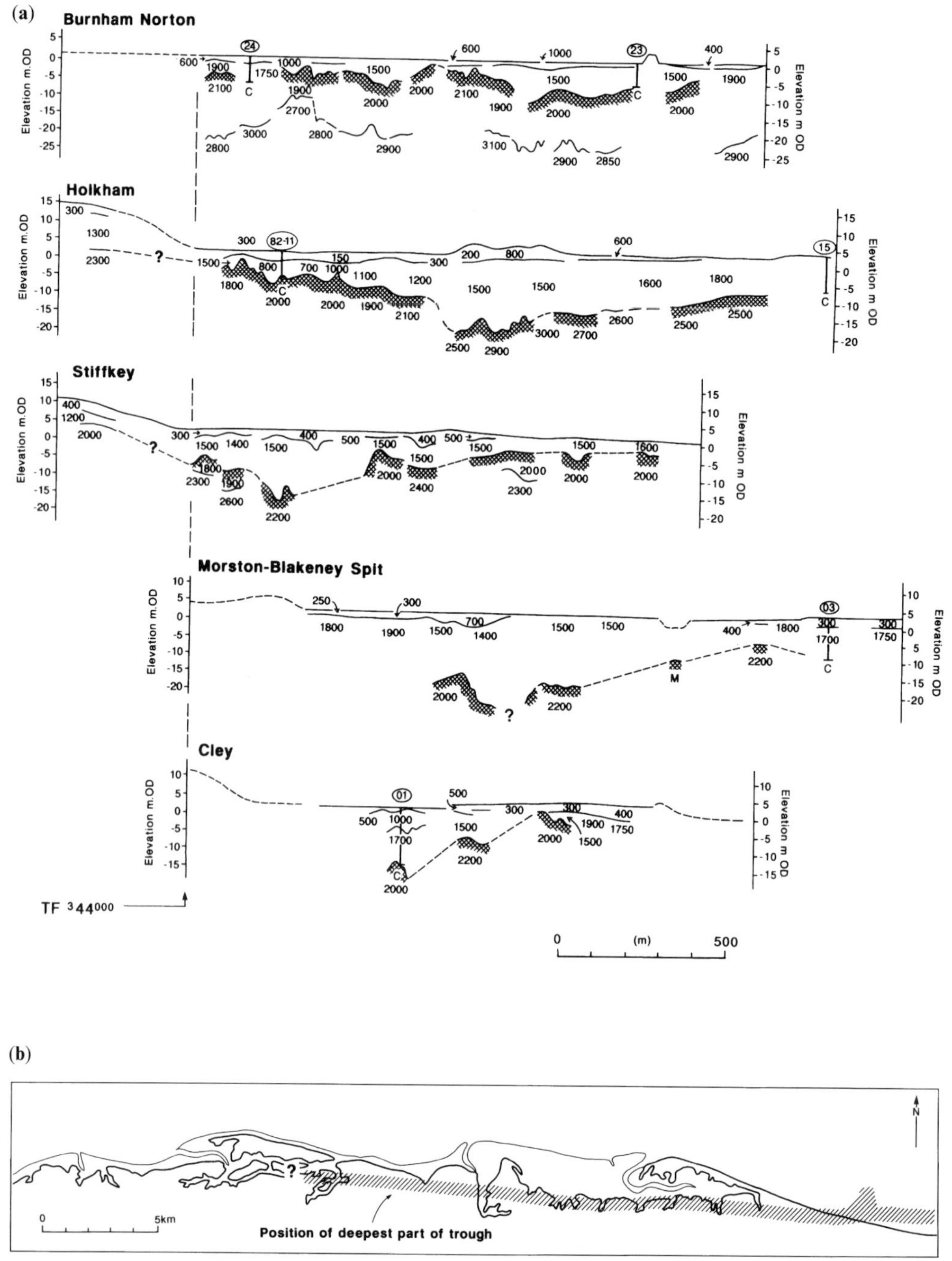

(a)

Burnham Norton

Holkham

Stiffkey

Morston-Blakeney Spit

Cley

TF 344000

0 (m) 500

(b)

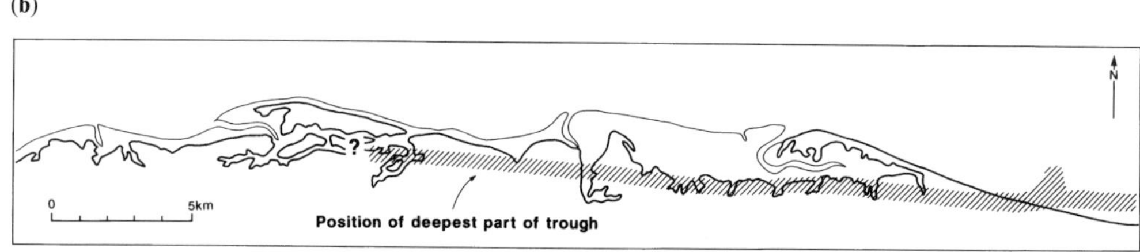

Fig. 2. (**a**) Seismic cross sections at Holkham, Stiffkey and Cley (Fig. 1), showing depth to the top of the Chalk. Borehole cores at 03 (i.e. NNC3, etc.) are shown where they prove the depth of Chalk (C). Note the trough shaped feature in the Chalk surface. Four seismic units are identified: unit 1 is a surface low velocity layer (200–550 m s^{-1}); unit 2 is Holocene and glacial sediments (800–1900 m s^{-1}); unit 3 is weathered chalk (1750–2150 m s^{-1}) and unit 4 is fresh chalk (2200–3000 m s^{-1}); after Chroston (*et al.* 1999). The transects are coast normal and are anchored on the [344 000] northing grid line. (**b**) Map showing position of the deepest part of the Quaternary trough identified by geophysical and borehole evidence (after Chroston *et al.* 1999). Reproduced by permission of Cambridge University Press.

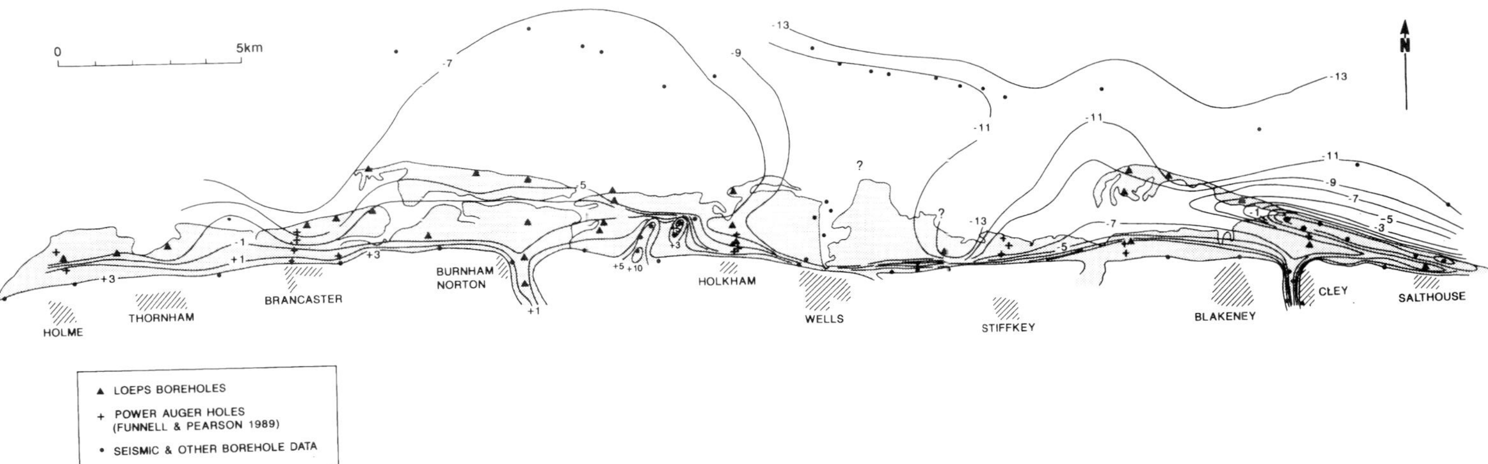

Fig. 3. Contour map of the pre-Holocene surface relative to Ordnance Datum based on borehole, auger and geophysical evidence. Dots mark position of borehole or geophysical spot data. The +3 m landward contour is taken as the depositional edge of the Holocene deposits. The shaded area represents the morphology of the modern coastal zone. Glacial 'eyes' occur as positive areas in the coastal zone between Cley and Salthouse and also west of Holkham.

occur in a west–east trough defined by the surface of the Chalk and interpreted as a palaeovalley (Chroston *et al.* 1999). The trough plunges gently east from Holme to Weybourne; beneath Blakeney spit and then offshore from Salthouse and Weybourne, with seismic profiles showing the trough continuing eastward (Fig. 2b and Chroston *et al.* 1999). The trend of the palaeovalley may have been determined by west–east faulting in the sub-Chalk 'basement' (Chroston *et al.* 1999). The within-trough Holocene sediments generally thicken from west to east.

Base Holocene surface

From the new and existing borehole and seismic data we reconstructed the topography of the Late Glacial surface, which was transgressed by rising sea-level during the Holocene. At many localities the basal Holocene sediment is terrestrial peat, formed as the water table rose in advance of the marine transgression; absence of basal peat suggests either subsequent erosion or non-deposition. The Holocene sequence rests variously on the Chalk, till, or unconsolidated glacial sands and gravels. In some cores it is difficult to distinguish between glacial deposits and Holocene barrier sands and gravels although work in progress is helping refine the provenance and age of these sediments (Funnell & Corbett pers. comm. 1998). Data defining the pre-Holocene palaeotopography was initially contoured using the surface plotting program SURFER v.5.02 (Golden Software Inc. 1994) and then adjusted manually to remove 'edge' effects (Fig. 3).

The depth to the base of the Holocene sequence ranges from about +3.00 m OD, where modern upper saltmarsh abuts the Chalk Formation/glacial sediment cliff-line on the landward edge of the sedimentary wedge, to about −14.50 m OD in borehole NNC14 (Warham Marsh). Key features observed from east to west in Fig. 3 include

(1) A ridge of glacial sediment underlying the eastern half of Blakeney Spit with a parallel trough just landward of it, now occupied by the main tidal channel. This glacial sediment ridge breaks the surface in places and can be seen in the Clay area forming the glacial 'islands' of Cley Eye, Blakeney Eye, Little Eye and Gramborough Hill (seen in section from [TG 043 453–086 443]), which are otherwise surrounded by back-barrier saltmarsh and barrier gravels.
(2) A basinal area to the west of the Blakeney glacial sediment ridge to about the Wells Channel including the thickest Holocene sequence at Warham in NNC14. Most of the basal Holocene along this stretch of coast lies at −7.00 to −11.00 m OD.
(3) An area west of the Wells Channel to the westernmost end of Scolt Head Island, where the base of the Holocene appears to form a plateau (at about −5.00 to −7.00 m OD) extending some kilometres out to sea. Towards the landward limit, some isolated glacial sediment highs are recorded between Wells and Burnham Norton rising to more than +5.00 m OD.

The westernmost section from Brancaster to Holme (Fig. 3) lacks onshore geophysical data and the depths are based entirely on borehole evidence. The basal Holocene is generally at a similar depth to that in the adjacent sector, although the −7.00 m OD contour comes close to the shore in Brancaster Bay. Evidence for the presence of the west–east trending channel at Holme (Funnell & Pearson 1989) was reconfirmed at borehole site NNC33.

Lithofacies and lithofacies associations of Holocene sediments

The modern barrier coastline sedimentary environments include: (a) intertidal sandflats with megaripples and beach bars, of variable width (up to 1.75 km) typically seaward of the main barriers (Pearson *et al.* 1990); (b) barrier and spit systems, such as Blakeney Spit and Scolt Head Island, typically composed of gravel and coarse sand (Steers 1929; 1964) and often capped by aeolian dunes (Knight *et al.* 1998); (c) back-barrier saltmarsh and intertidal muds and silts (Pye 1992); and (d) sandy tidal channel deposits with small amounts of gravel (Pearson *et al.* 1990).

These modern environments are easily recognizable in the Holocene subsurface sediments at the lithofacies association scale (see below). The Holocene sediments were first classified into seven lithofacies based on the information in Table 2. These included LF1, peat; LF2, fine-grained back-barrier sediments (saltmarsh and intertidal muds and silts); LF3, muddy sand; LF4, pebbly sand; LF5, rooted sand; LF6, interbedded sand; and LF7, gravel. Peat biofacies were studied in detail for sea-level studies (see LOEPS examples in Shennan *et al.* this volume *a*), and in general represent freshwater deposits, in places younging to marine saltmarsh peats. These LF1 peats, particularly the saltmarsh peats, have no clear modern counterparts except for small patches of *Phragmites* reed

Table 2. *Lithofacies data, descriptions and interpretations*

Lithofacies	Description	Process interpretation	Depositional environment
LF1 Peat	Black–dark brown humic peats and organic-rich muds. Upper and lower contacts can be sharp (erosional) or gradational	Accumulation of largely autochthonous plant-derived organic material, typically under anoxic conditions	Freshwater swamp to alder carr. Rare saltmarsh peats in some cores
LF2 Back-barrier muds & silts	Muds and silts, often laminated on millimetre-scale, locally bioturbated and shelly	Settling of clay–silt grade material from suspension at high tide, routed via creeks and tidal channels	Intertidal mudflats, lower and upper saltmarsh
LF3 Muddy sand	Sharp-based sand with mud-flakes/pebbles, mud matrix	Fluctuating energy flows, mud clasts rapidly deposited, minimal transport	Tidal channel/creeks
LF4 Pebbly sand	Clean sand, scattered pebbles, occasionally rooted	Consistently high energy, wave influenced environment. Some storm or subaerial deposition	Exposed, wave influenced environment, washovers
LF5 Rooted sand	Vertical rootlets, no pebbles, some mud/sand	Subaerial, stable setting; gradual accumulation	Vegetated windblown dunes
LF6 Interbedded sand	Thin interlaminated sands, muds, sometimes rooted	Mixed energy shallow water environment, periodic higher energy events	Mixed tidal flat
LF7 Gravel	Clast-supported flint gravel with sand matrix	Consistent high energy	Upper beach

beds, probably associated with springs on the landward limit of the Holocene sediment prism.

LF2 fine-grained clastic sediments were assigned to precise depositional environments using biofacies information. Four foraminiferal assemblages were recognized (details in Boomer 1998) each comprising a number of species.

(1) An upper saltmarsh assemblage (USM), representing highest astronomical tide–mean high water (HAT–MHW) tidal levels, of agglutinating taxa dominated by *Trochammina inflata* and *Jadammina macrescens*, which together usually constitute at least 99% of the total number of individuals.
(2) A lower saltmarsh assemblage (LSM), representing MHW–MHW neap tidal levels, transitional between upper saltmarsh (USM) and upper mudflat (UMF) assemblages (see below), where the highest abundances of *Miliammina fusca* and *Elphidium williamsoni* are recorded, commonly reaching 20% each. The assemblages generally comprise a mixture of agglutinating and calcareous walled taxa.
(3) An upper mudflat assemblage (UMF), representing MHWN to about mean sea-level (MSL) tidal levels, dominated by (up to 95%) *Haynesina germanica*, *Haynesina depressula* and *Ammonia beccarii*. The exact composition is dependent on a number of factors, particularly salinity variation.
(4) A lower mudflat assemblage (LMF), representing MSL and lower tidal levels, which is similar to the UMF except that the *Ammonia* and *Haynesina* species generally constitute a smaller percentage of the total, with species of *Elphidium* (except *E. williamsoni*) and the genus Miliolina becoming much more common. Shelf genera (e.g. *Asterigerinata*, *Cibicides*, *Glabratella*, *Lagena* and *Planorbulina*) are common, albeit in low abundance, indicating proximity to marine open water conditions.

Ostracod data were used as salinity indicators and as supporting information for the foraminiferal assemblages. Three assemblages were recognized (details in Boomer 1998).

(1) A brackish ostracod assemblage *of Cyprideis torosa*, *Leptocythere castanea*, *Leptocythere porcellanea*, and *Loxoconcha elliptica*. These species live in brackish water (typically 10–25‰), probably mainly saltpans, ditches and small water bodies flooded on highest spring tides. These species can probably tolerate strong evaporation, but not desiccation.

(2) A lower saltmarsh/euryhaline ostracod assemblage where species such as *Hemicythere rubida*, *Leptocythere lacertosa*, *Hirschmannia viridis*, *Leptocythere psammophila*, *Cytherois fischeri*, *Leptocythere baltica* and *Elofsonia baltica* are capable of withstanding significant changes of both tidal inundation and salinity (probably 15–35‰) on a daily basis.
(3) A marine ostracod assemblage of *Pontocythere elongata*, *Semicytherura* sp., *Leptocythere pellucida*, *Loxoconcha rhomboidea* and *Hemicythere villosa*. These species are not tolerant of sustained salinity reduction and most of them are common in the sublittoral zone, probably indicating deposition around or below mean sea level.

The Holocene sedimentary facies are grouped into two lithofacies associations. LFA-A, is a coarse-grained barrier and tidal channel association (LF3–7) restricted to a zone beneath the modern barrier environments. LFA-B, is a fine-grained back-barrier mudflat and saltmarsh association dominated by LF2 fine-grained sediments, but often containing a basal LF1 peat.

An important aspect of this work has been attempting to understand the spatial and temporal relationship between LFA-A and LFA-B on decadal to centennial and millennial time-scales. This has required integration of our work on both lithofacies analysis and chronology.

Establishing a chronology

Radiocarbon and microbiofacies dates

The new NNC radiocarbon data, augmented by previously published dates from peats in Funnell & Pearson (1989) and Wymer & Robins (1994) are plotted against depth in Fig. 4. Addtional reliable dates on carbonate shells (without accurate elevation data) are also given in Allison (1985) and Funnell & Pearson (1989). Ages older than 7 ^{14}C ka BP (Fig. 4) come from the bases or middle parts of basal terrestrial peats (the oldest LOEPS NNC date measured is 9680 ± 60 ^{14}C a BP in NNC19A at Burnham Overy). The age–depth trajectories for these samples are very flat, indicating slow or discontinuous sedimentation. The tops of peats at >−6.0 m OD

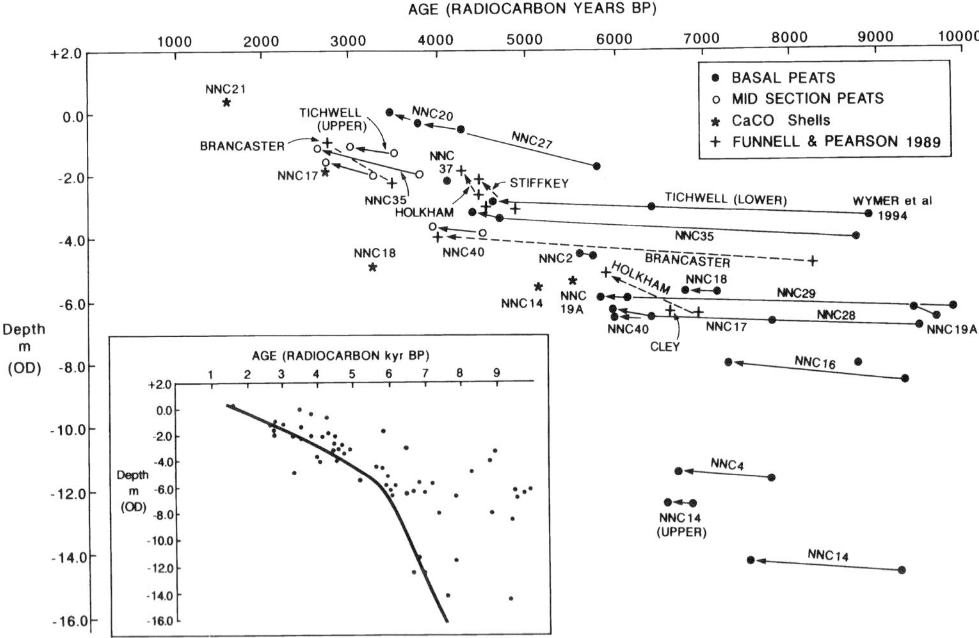

Fig. 4. Radiocarbon ages versus depth plot. Several samples taken from a single peat (e.g. base, middle and top) are joined by tie lines with the arrowhead pointing to the top sample). Black dots represent basal peats, open dots represent mid section peats and stars represent dates on $CaCO_3$ shell material (mainly *Scrobicularia plana*, but NNC19A on *Hydrobia* sp. and NNC18 on *Cerastoderma* sp.) Data from Funnell & Pearson (1989) are plotted for comparison (crosses and pecked tie lines). The inset is a simplified plot of the data with a schematic sea-level curve (see details in Shennan *et al.* this volume).

have ages between 6 and 7.5 ^{14}C ka BP and probably define the change from terrestrial to marine conditions (see below). Ages on peats younger than 6 ^{14}C ka BP show a reasonably systematic age–depth relationship, and on the whole have trajectories from bases to tops that are similar to the regional sea-level rise at this time, indicating that peat accumulation kept pace with sea-level and base-level rise. The mid-section peats of *c.* 4.5 ^{14}C ka BP identified by Funnell & Pearson (1989) between Holkham and Cley may have formed as a response to lowered sea-level rise, accompanied by greater groundwater inflow at the landward edge of the Holocene sediments or rising ground-water base-level.

Dates on shell material from back-barrier sediments are broadly consistent with the age–depth pattern (Fig. 4), except the date on a *Scrobicularia plana* at −5.50 m OD in NNC14 (Warham, Fig. 5), which appears to be 1 m deeper than its age suggests (Fig. 4). Assuming the date is reliable, this might indicate that the shell had burrowed deep into the saltmarsh sediment; compaction could have also had an effect. The date on a *Cerastoderma* shell at −4.78 m OD in NNC18 (Burnham Overy) appears anomalously deep for its age (Fig. 4; see also Fig. 7a); however, this is the only date from a shell in the barrier sands. The sample comes from just above a marked erosion surface (see below) and may be an important date in constraining the age of downcutting and barrier emplacement in this area. There is only one radiocarbon date post-2.5 ^{14}C ka BP, leaving a chronological gap to the present day, although this is in part addressed by radionuclide chronology (below).

Combined sedimentological and microfossil data (Shennan *et al.* this volume *a*) from the dated peats were used to establish as precisely as possible the timing of marine inundation at a

Table 3. *Luminescence ages*

Sample depth (m OD)	Sample reference	Material examined*,†	Luminescence age (a ×1000)	±Overall error (a ×1000)	±Random error (a ×1000)
0.58	NNC-14-5	PFG	2.25		0.15
−0.73	NNC-14-1	PFG	3.0	0.3	0.15
−3.20	NNC-14-4	PFG	5.3	0.45	0.3
−5.41	NNC-14-2	PFG	5.6	0.5	0.3
−13.23	NNC-14-3	PFG	9.9	1.0	0.6
−3.56	NNC-16-1	Feldspars, Na + K	Weak luminescence		
−2.11	NNC-16-2	Feldspars, K	3.0	1.0	1.0
−6.59	NNC-16-3^{I}	Feldspars, Na + K	1.9	0.5	0.5
−6.59	NNC-16-3^{II}	Feldspars, K	2.1	0.7	0.6
−7.09	NNC-16-4	PFG	5.3	0.5	0.2
−7.87	NNC-16-5	PFG	5.9	0.65	0.4
0.54	NNC-17-3	PFG	2.3	0.2	0.15
−1.44	NNC-17-2	PFG	2.8	0.3	0.2
−2.94	NNC-17-5	PFG	4.7	0.4	0.3
−3.00	NNC-17-6	PFG	4.5	0.4	0.2
−4.88	NNC-17-4	PFG	5.7	0.7	0.4
−6.10	NNC-17-1	PFG	5.9	0.6	0.3
3.62	NNC-18-1	Feldspars, K	Weak luminescence		
−4.61	NNC-18-2	Feldspars, K	4.2	1.2	1.2
−4.84	NNC-18-3	PFG	4.6	0.45	0.35
−5.53	NNC-18-4	PFG	5.0	0.4	0.3
2.62	NNC-19-1	Feldspars	Weak luminescence		
1.21	NNC-19-2	Feldspars, K	1.0	0.4	0.4
0.14	NNC-19-3	Feldspars, Na + K	Weak luminescence		
−0.59	NNC-19-4	Feldspars, K	3.0	0.9	0.8
−0.70	NNC-19-5	PFG	2.5	0.35	0.3
−2.43	NNC-19-6	PFG	2.1	0.4	0.35
−6.02	NNC-19-7	PFG	6.1	0.6	0.4

* PFG refers to the polymineral fine-grained samples that were prepared for IRSL dating measurements using an additive dose fine-grained technique. The uncertainties are standard errors given at the 68% level of confidence.
† Feldspars K & Na: 2.53–2.58 gm cm^{-3}; feldspars K: <2.53 gm cm^{-3}; both 200–300 m sieved fractions.

site, and therefore former sea-level. These provide a local history of relative sea-level change for the past 6.5 ^{14}C ka and indicate a near linear relative sea-level rise of 1.3 mm a^{-1} since that time (Shennan *et al.* this volume *a*, fig. 23). The microbiofacies 'dates' are tied to pre-existing interpretations of sea-level relative to Ordnance Datum and therefore are bound to have a consistent age–depth relationship.

Luminescence ages

The IRSL ages (Table 3) for LF2 silts in cores NNC14, NNC16, NNC17, NNC18 & NNC19A (Fig. 1) appear reasonable in so far as: (1) the ages increase with increasing depth without reversals; and (2) with one exception the oldest IRSL, ages in samples above basal peats are younger than the top of the basal peat itself. (In NNC14 (Warham) the luminescence age of 9900 ± 1000 years at −13.23 m OD only overlaps the date on the peat upper surface of 8160 cal. a BP (2σ range 8369–7973 cal. a BP) at −14.34 m OD within 2σ limits; Fig. 5.) The IRSL ages can be also be directly compared with radiocarbon dates on shell material (Fig. 5). In NNC14, a *Scrobicularia plana* shell apparently in life position at −5.5 m OD yielded an age of 5460 cal. a BP (2σ range 5585–5311 cal. a BP) in

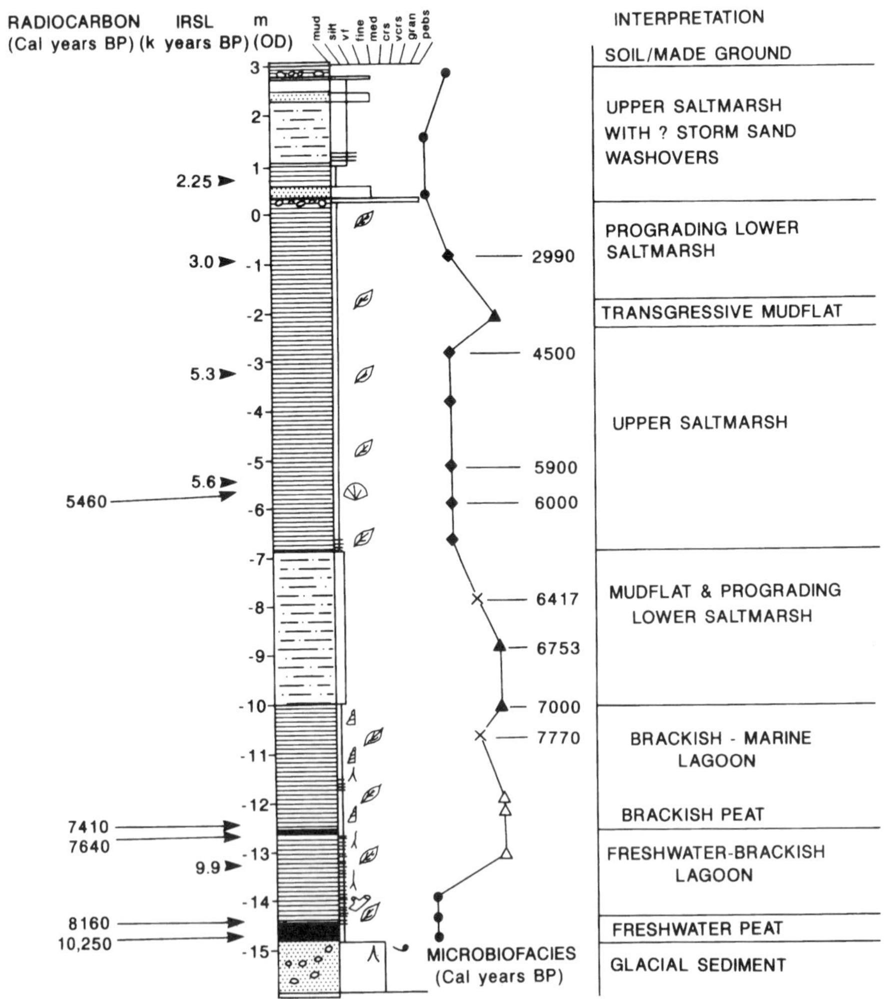

Fig. 5. Sedimentological and biofacies log of the Warham core NNC14 (Fig. 1, [TF 9484 4430]) showing radiocarbon, IRSL and microbiofacies chronologies and palaeoenvironmental interpretations.

agreement with an IRSL age of 5600 ± 500 years at −5.41 m OD. A similar comparison can be made in NNC17 (Holkham) where a *S. plana* shell in life position at −1.61 m OD was dated at 2360 cal. a BP (2σ range 2672–2271 cal. a BP), just within error of the IRSL age of 2800 ± 300 years at −1.44 m OD (Fig. 6a). Given the slight differences in sample depths, uncertainties and error bands in both dating methods (e.g. see Aitken 1990), uncertainties in radiocarbon calibration (Aitken 1990) and uncertainties in the burrowing activities of *Scrobicularia* (as discussed above), the level of agreement in dating is reasonable. Similarly, a date of 5900 cal. a BP (2σ range 5984–5822 cal. a BP) on *Hydrobia* sp. shells within LF2 mudflat–saltmarsh sediments at −5.33 m OD in NNC19A (Burnham Overy) compares well with an IRSL age of 6100 ± 600 years at −6.02 m OD (Fig. 7b).

Comparison between IRSL ages for LF2 silts with microbiofacies 'dates' from similar horizons is easiest to do with reference to specific cores (e.g. Figs 5–7); however, the important points are: (1) that IRSL ages in the age range 4.2–6.3 ka are consistent with the microbiofacies 'dates' in cores NNC14 and NNC17, but are consistently younger than the corresponding microbiofacies 'dates' by some 1 ka in NNC16, NNC18 and the upper section of NNC19A; (2) that IRSL ages younger than 3 ka in NNC14 and NNC17 are within 0.5 ka of the microbiofacies age estimates.

In two cores, NNC14 (Warham) and NNC17 (Holkham), a conventional approach to estimating the plausibility of the IRSL ages can be established by calculating mean sedimentation rates for LF2 muddy units without obvious sedimentological breaks or biofacies changes. In NNC14 the 2.21-m-thick sequence between −3.20 and −5.41 m OD (Fig. 5) represents some 300 years of upper saltmarsh sedimentation based on the differences between the central values of the IRSL ages (5600 ± 300 and 5300 ± 300 years). The calculation of the apparent sedimentation rate ($8 \pm 9\ \mathrm{mm\,a^{-1}}$), assuming a constant accretion rate, is beyond the resolution of the technique. However, a sedimentation rate for a similar depth interval in this core (between −2.80 and −5.86 m OD), based on microbiofacies 'dates' of 4.5 and 6 cal. ka BP

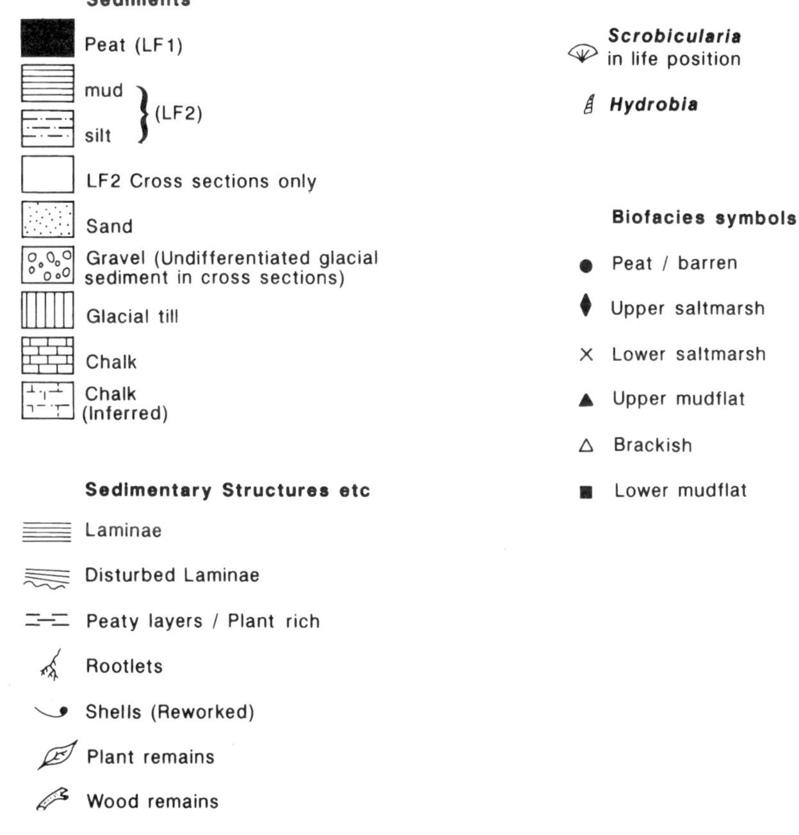

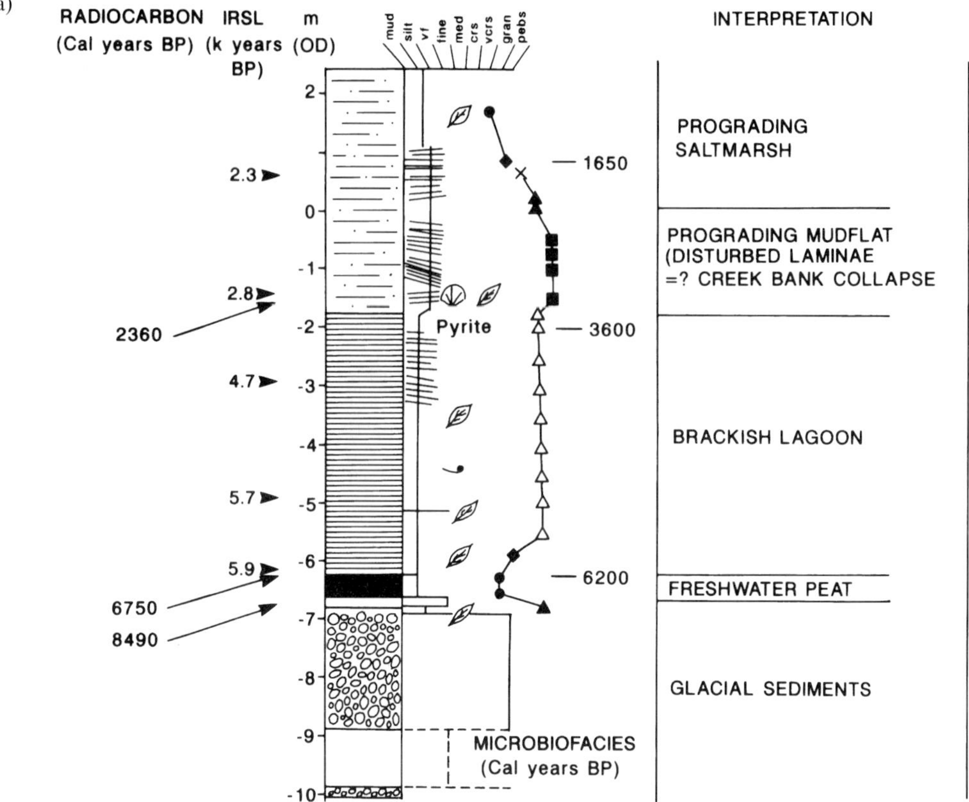

Fig. 6. Sedimentological and biofacies logs of the Holkham cores NNC17 (**a**) and NNC16 (**b**) (Fig. 1, [TF 8913 4455] and [TF 8903 4500]) showing radiocarbon, IRSL and microbiofacies chronologies and palaeoenvironmental interpretations. The positions of these cores in the Holkham transect are shown in Fig. 12. (See legend in Fig. 5.)

(Fig. 5) yields an apparent sedimentation rate of about 2 mm a^{-1}. Comparison of these sedimentation rates with modern upper saltmarsh accumulation rates (see discussion below) suggests that the 2 mm a^{-1} values are plausible. For NNC17, microbiofacies information suggests that LF2 muds between −2 to −6 m OD represent deposition in brackish–marine lagoonal conditions without any obvious sedimentological breaks. IRSL ages of 4.7 and 5.9 ka at −3.0 and −6.1 m OD, respectively, yield a sedimentation rate of 2.6 ± 1.8 mm a^{-1}, compared with 1.6 mm a^{-1} based on microbiofacies 'dates' (Fig. 6a). Given the errors involved, these sedimentation rates are plausible when compared with measurements of modern sedimentation rates (see below).

Overall, the IRSL method for LF2 sediments gives method-consistent ages although on the whole the ages appear to be young when compared to age–depth relations based on microbiofacies 'dating', especially for ages older than 4 ka BP. Experience with IRSL ages in other LOEPS cores from Fenland (Bailiff & Tooley this volume) suggest that a systematic underestimate relative to calibrated radiocarbon dates may not apply to all age calculations. It is possible that loss of moisture during coring has affected some age determinations, since the calculation is very sensitive to assumed changes in water content during the burial period. This problem needs further investigation involving new coring experiments, and currently the differences remain unexplained.

The dating of LFA-A sands based on the measurement of IRSL in feldspars presented the greatest experimental uncertainties because very few feldspar grains were present and the calculation of the annual dose rate was subject to potentially large uncertainties. There was also the risk that these larger-sized grains were not fully bleached before deposition. Tests for this

(b)

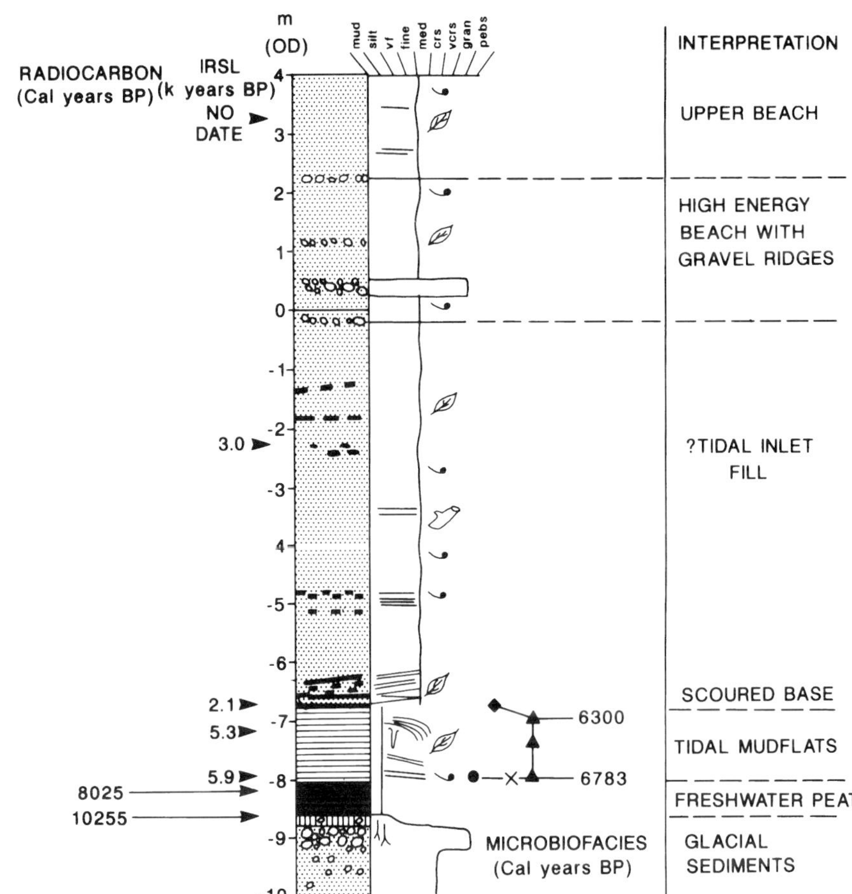

factor were constrained by the amount of material recovered. The validity of these LFA-A IRSL ages (Table 3) is therefore difficult to assess. The quartz fraction of the sands was also tested using OSL (450–550 nm), but the signals were not of sufficient intensity.

In core NNC16 (Holkham, Fig. 6b), the age of sands at −2.1 m OD is *c.* 1 ka *older* than the age at −6.6 m OD, although the ages overlap within 1σ error limits. Further tests are necessary with the shallower sample to examine for the possibility of incomplete bleaching at deposition. The feldspars within this sample were sparse and much larger quantities of core material would be required to test this (alternatively other ways may yet be found to stimulate stronger OSL from the quartz fraction). In core NNC18 (Burnham Overy, Fig. 7a) the stratigraphic consistency is more encouraging: the IRSL age of 4200 ± 1200 years for the basal LFA-A sand at −4.6 m OD overlaps with the age of LF2 sediments (4600 ± 450 years) immediately underlying the facies boundary, within experimental error limits (and similar overlaps occur in core NNC19A, Burnham Overy, Fig. 7b).

Comparison of the IRSL age of 4200 ± 1200 years for the basal LFA-A sand at −4.6 m OD in NNC18, can also be directly compared with a radiocarbon date of 3059 cal. a BP (2σ range 3193–2925 cal. a BP) on a *Cerastoderma* shell at −4.78 m OD (Figs 7a and 4). This sample comes from just above a marked erosion surface and, if reliable, indicates the maximum age of downcutting and barrier emplacement in this area.

Radionuclides: modern sediment accretion rates

^{137}Cs profiles from saltmarshes at Stiffkey, Wells and Scolt Head Island (Figs 8a and 9a) are broadly similar. Around the UK coastline, ^{137}Cs is found in both sediment and water as a

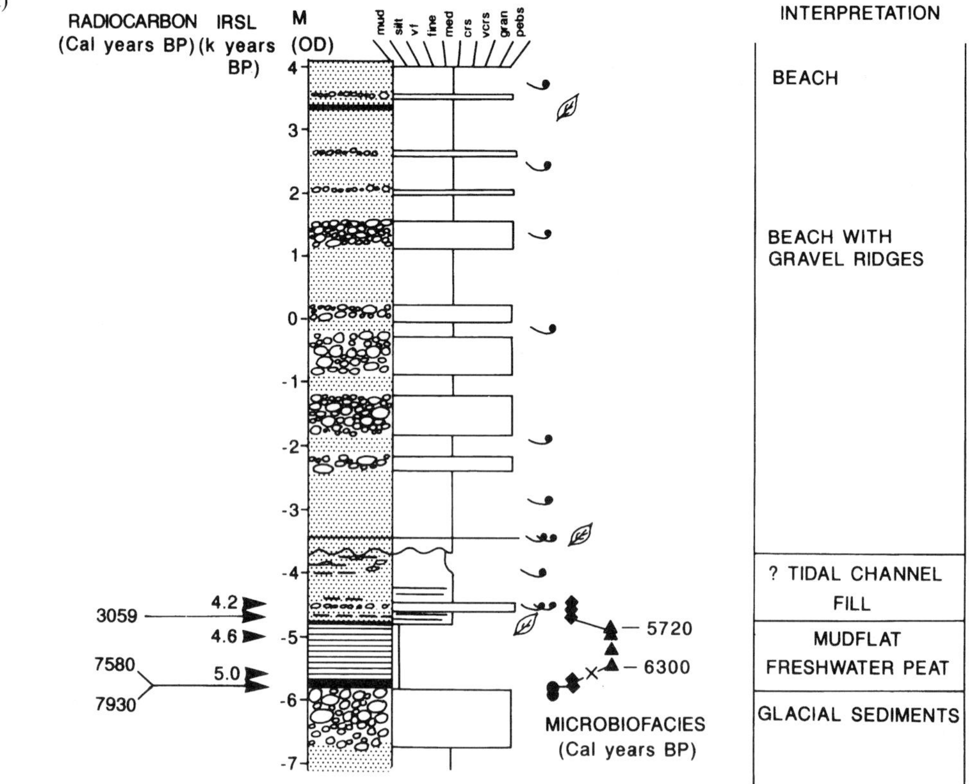

Fig. 7. Sedimentological and biofacies logs of the Burnham Overy cores NNC18 (**a**) and NNC19A (**b**) (Fig 1, [TF 85765 45898] and [TF 85814 45644], respectively) showing radiocarbon, IRSL and microbiofacies chronologies and palaeoenvironmental interpretations. The positions of these cores in the Burnham Overy transect are shown in Fig. 13. (See legend in Fig. 5.)

consequence of: (1) discharges from Sellafield and Cap de la Hague nuclear fuel reprocessing plants (peak Sellafield discharge in 1975); (2) potentially from the Chernobyl accident fallout (1986); and (3) from residues of atmospheric weapons testing in the 1950s and early 1960s. While bioturbation, physical mixing and Cs diffusion might influence the down-core distribution, the very similar profiles (and the accompanying ^{210}Pb, Figs 8b and 9b) suggest that redistribution has been minimal. The characteristics of the ^{137}Cs profile are: (1) absent ^{137}Cs at depth; (2) exponential increase in activity from 20–30 cm depth; towards (3) an inflection point midway to the peak maximum (manifest as a discrete peak in the Stiffkey cores); (4) a peak maximum that is often a doublet; and (5) exponential decrease towards the surface, but never returning to low or zero background.

Linear sedimentation rates and mass accumulation rates for these marshes (Table 4) have been calculated assuming that the peak maximum in ^{137}Cs is derived predominantly from a Sellafield discharge source (in contrast to the interpretations of Callaway *et al.* 1996), with a six-year transit time to the NNC. The rationale for this interpretation is that: (1) signal matching with average ^{210}Pb chronology both in Norfolk and Humberside compares favourably to residence time models for the North Sea (Prandle 1984); (2) similar profiles between marshes, and within the same marsh, argue for uniform sources of ^{137}Cs; (3) specific activity of ^{137}Cs and total ^{137}Cs inventory is consistent with Sellafield origin; (4) a pulse tracer input, such as Chernobyl fallout, would produce a much sharper, well-defined peak with lower activity between the 1963 weapons test peak and the fallout maximum in the absence of sediment reworking, and (5) the minimal fluvial input to the NNC would limit a catchment-derived Chernobyl fallout signal. (To be definitive, further tracers of sediment history,

(b)

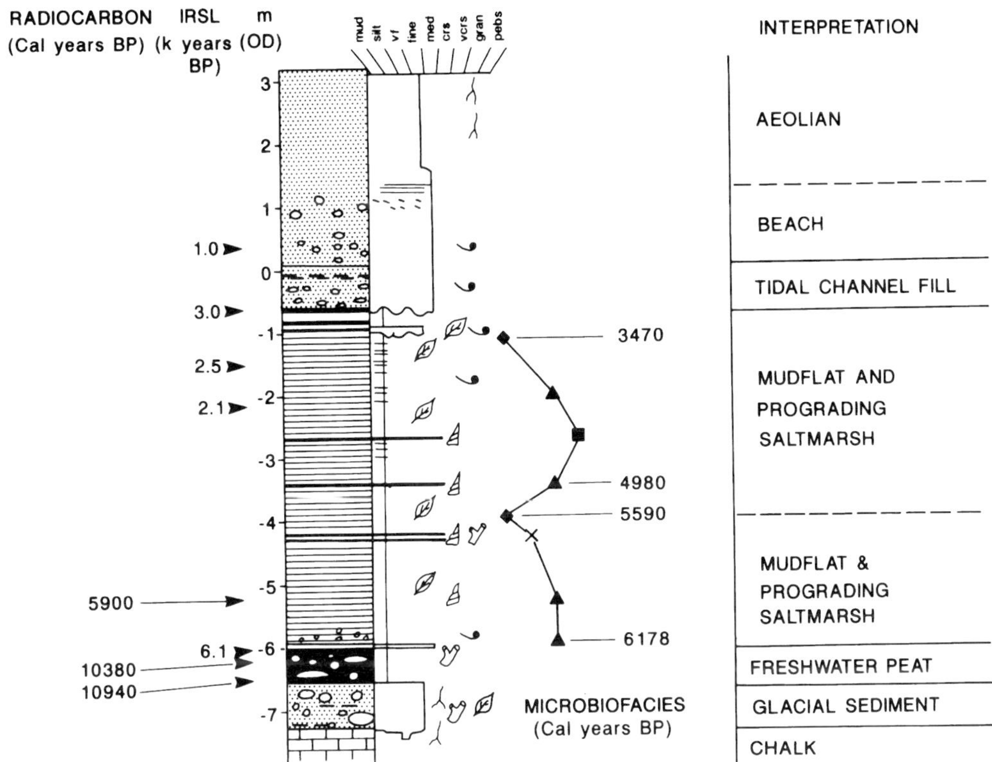

e.g. lead isotopes or large volume analysis for trace ^{134}Cs are needed.)

Irrespective of the origin of the ^{137}Cs (and therefore the precise date assigned to peak maxima), the data show that sediment accumulation in lower saltmarsh environments (e.g. the *Spartina* saltmarshes at Stiffkey, core 1, and Missel Marsh on Scolt Head Island, core 5) is rapid (mass accumulation rates 0.46–0.49 $g\,cm^{-2}\,a^{-1}$, Table 4), while other upper saltmarsh environments (the other Stiffkey and Scolt Head Island cores, Table 4) generally have lower mass accumulation rates (0.1–0.35 $g\,cm^{-2}\,a^{-1}$). These mass accumulation rates translate to linear sedimentation rates of around 6 $mm\,a^{-1}$ for lower saltmarshes and 2–4 $mm\,a^{-1}$ for upper saltmarshes. Pioneer mudflat/samphire-*Spartina* marsh forming in Holkham Bay has a linear accumulation rate of 1.8 $mm\,a^{-1}$ based on a ^{210}Pb profile. These rates are consistent with previous studies of saltmarsh accumulation from the NNC (based largely on levelling techniques) where short term linear accretion rates are generally between 1 and 8 $mm\,a^{-1}$ (Pye 1992; French & Spencer 1993) but spatially variable on upper saltmarshes depending on proximity to tidal creeks. Pye (1992) in a synthesis of earlier data suggests that decadal and longer-term accretion rates on upper saltmarshes are around 1.5 $mm\,a^{-1}$, in good agreement with the lower radionuclide linear accretion rates.

These modern accumulation rates constrain likely accumulation rates in Holocene LF2 back-barrier muds and silts assuming roughly similar rates of sea-level rise, allowing us to establish the validity of mean linear sedimentation rates derived from luminescence and microbiofacies methods (described above). In the upper part of core NNC14 (Warham, Fig. 5) the sedimentation rate of 2 $mm\,a^{-1}$ for upper saltmarsh sediments based on microbiofacies 'dates' appears reasonable. In the brackish–marine lagoonal LF2 sediments in the middle part of NNC17 (Holkham, Fig. 6a), sedimentation rates of 2.6 ± 1.8 and 1.6 $mm\,a^{-1}$ based on IRSL and microbiofacies ages, respectively, are consistent with modern lower saltmarsh sedimentation rates in the 2–4 $mm\,a^{-1}$ range.

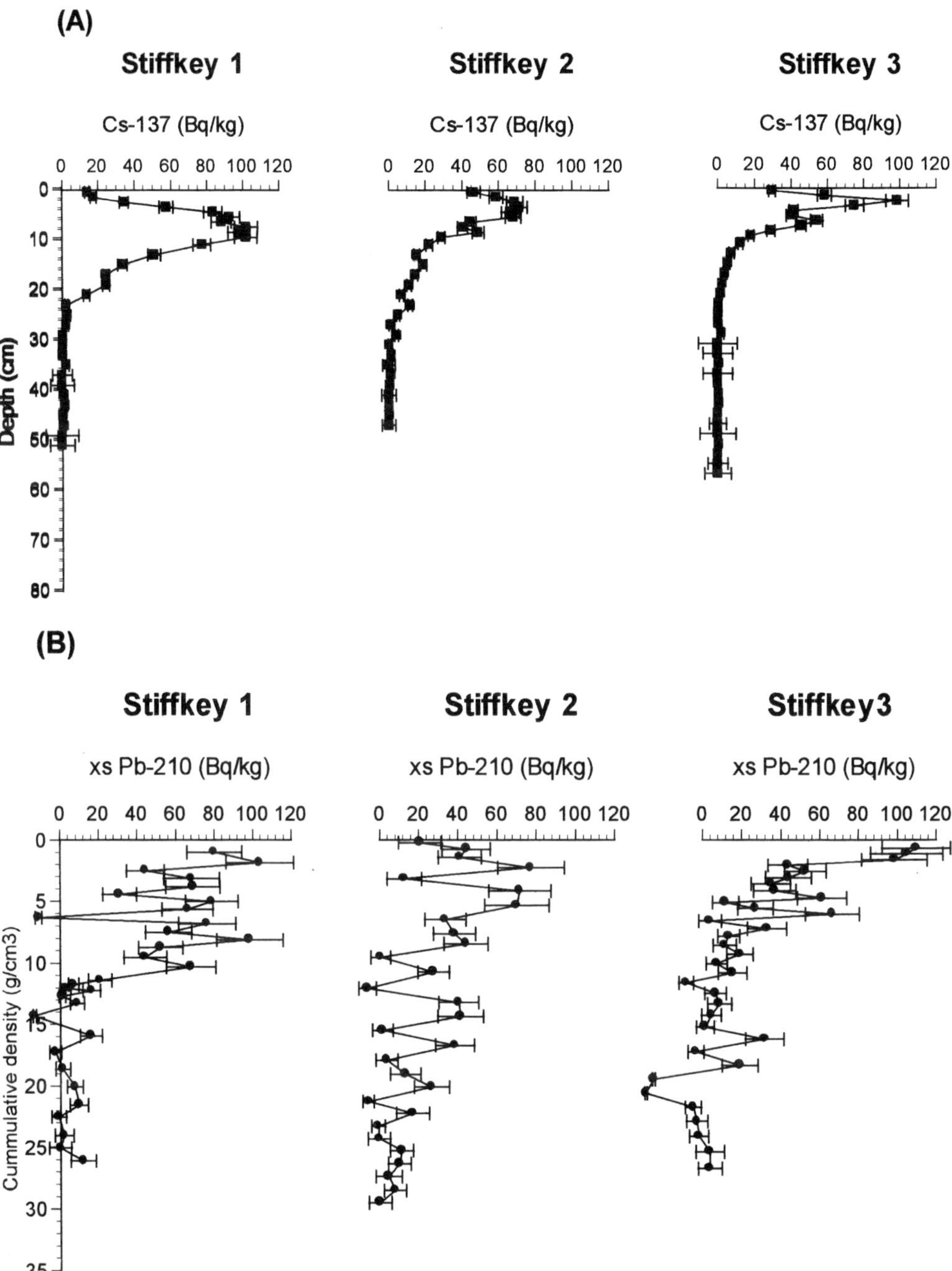

Fig. 8. (**a**) ^{137}Cs profiles plotted against depth and (**b**) excess ^{210}Pb profiles plotted against cumulative sediment density at Stiffkey (Fig. 1). Stiffkey 1 (lower *Spartina* saltmarsh [TF 965 448]), Stiffkey 2 (mature saltmarsh 100 m south of Stiffkey Meals [TF 965 446]), Stiffkey 3 (mature saltmarsh 300 m north of Green Lane car park [TF 964 444]).

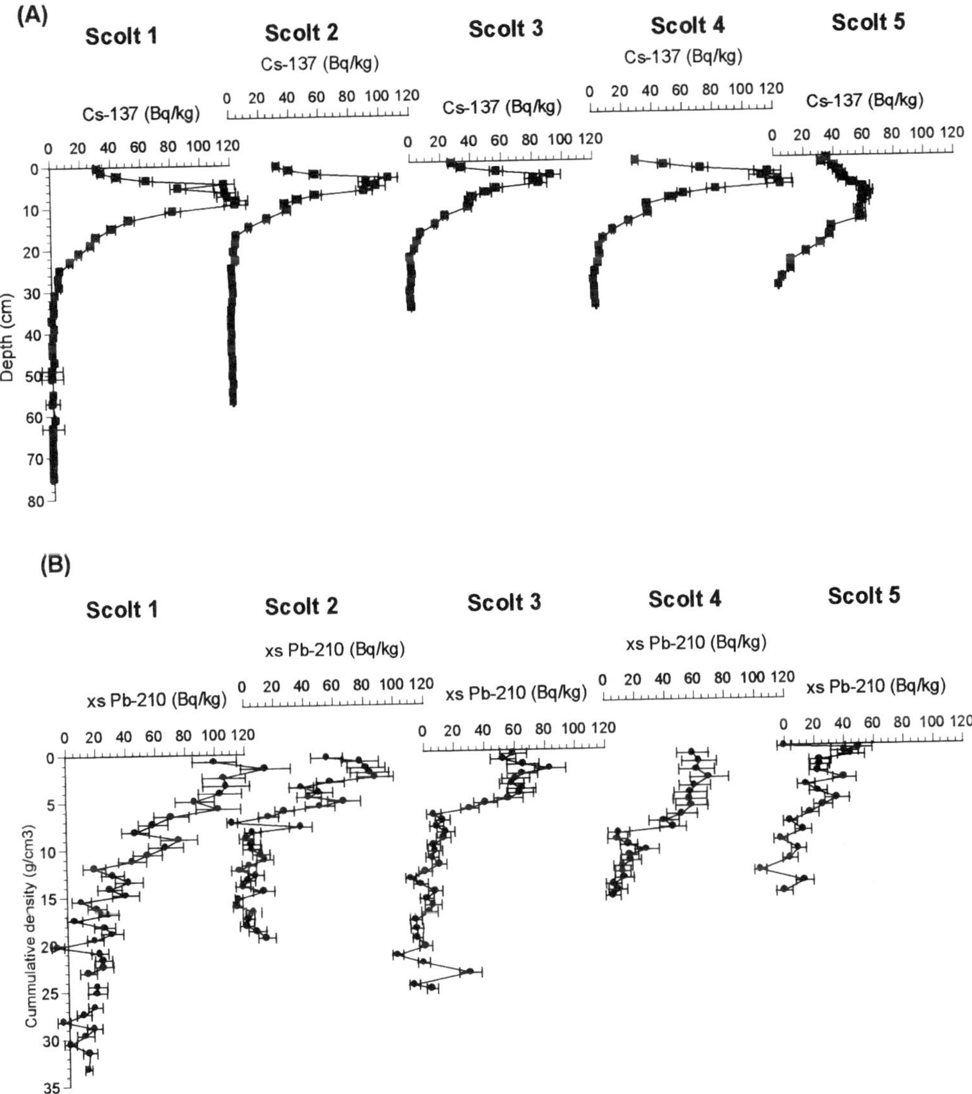

Fig. 9. (**a**) ^{137}Cs profiles plotted against depth and (**b**) excess ^{210}Pb profiles plotted against cumulative sediment density at Scolt Head Island (Fig. 1) Scolt 1 (upper saltmarsh, Great Aster Marsh [TF 839 459]), Scolt 2 (upper saltmarsh, Plantago Marsh [TF 830 459]), Scolt 3 (upper saltmarsh, Plover Marsh [TF 821 459]), Scolt 4 (upper saltmarsh, Hut Marsh [TF 812 463]), Scolt 5 (lower *Spartina* saltmarsh, Missel Marsh [TF 805 460]).

Evolution of Holocene sedimentation

The lithofacies evolution of this coast on a millennial time-scale, records the chronology of events of filling available accommodation space as sea-level rose following the last glacial maximum. It is likely that a system of sand and gravel barriers developed some way, possibly a number of kilometres, offshore in the early Holocene, protecting saltmarsh and mudflat environments on the landward side. We know that these barriers have moved shoreward, towards their present position as sea-level has risen, because we have evidence of old LF2 back-barrier sediments *overlain* by LFA-A barriers sands in seaward cores at a number of localities, e.g. in NNC18 at Burnham Overy, in NNC16 at Holkham (Fig. 6b) and in the Blakeney Spit cores

Table 4. *Saltmarsh accumulation rates based on ^{137}Cs profiles, assuming peak maximum from Sellafield in 1976 with six-year lag to north Norfolk*

Location, grid reference and approximate surface elevation	Linear sedimentation rate ($mm\,a^{-1}$)	Mass accumulation rate ($g\,cm^{-2}\,a^{-1}$)
Stiffkey Marshes		
S1 *Spartina* marsh, [TF $^{5}9640$ $^{3}4480$] (+2.50 m OD)	6.4	0.49
S2 Mid marsh, [TF $^{5}9640$ $^{3}4460$] (+2.85 m OD)	3.6	0.19
S3 Upper marsh (near bridge), [TF $^{5}9640$ $^{3}4440$] (+2.70 m OD)	2.1	0.12
Scolt Head Island		
SH1 Great Aster Marsh, [TF $^{5}8399$ $^{3}4588$] (+2.3 m OD)	5.4	0.47
SH2 Plantago Marsh, [TF $^{5}8302$ $^{3}4595$] (+2.94 m OD)	3.9	0.22
SH3 Plover Marsh, [TF $^{5}8215$ $^{3}4592$] (+2.89 m OD)	3.2	0.18
SH4 Hut Marsh, [TF $^{5}8117$ $^{3}4628$] (+2.9 m OD)	3.9	0.35
SH5 Missel Marsh, [TF $^{5}8047$ $^{3}4599$] (+2.9 m OD)	7.9	0.46

(locations in Fig. 1). However, shoreface erosion has completely removed any traces of the former back-barrier sediments seaward of these cores.

Soon after the Devensian ice retreated from north Norfolk, sedimentation in the area of the present day Holocene sediment prism was dominantly fluvial, the rivers flanked by marshy and peaty deposits that formed on an undulating till or gravel surface (found at the bases of many of the NNC cores). These terrestrial peaty deposits were forming by 11–10 cal. ka BP (9.6 ^{14}C ka) from Thornham in the west to Cley in the east (Fig. 4). Peat formation was probably not continuous, as changes in sample elevation of a few centimetres can give dates separated by thousands of years (Fig. 4); although sedimentation lasted until at least 7 cal. ka BP (about 6 ^{14}C ka) in many places (Fig. 4).

Back-barrier lithofacies association (LFA-B)

The first records of marine sedimentation, typically in mudflat or saltmarsh environments, and the transition from brackish to marine biofacies (details in Funnell *et al.* in press) are dated using the radiocarbon dates from the tops of peats and the microbiofacies dates in the marine sediments. Marine sedimentation began at around 7 cal. ka BP at Warham and Holkham (NNC14 and NNC17, Figs 5 and 6), 6.4 cal. ka BP at Cley (NNC1, Fig. 10b) and between 7 and 6 cal. ka BP at Salthouse (NNC40, Fig. 11). West of Holkham the first records of marine sedimentation are around 6 cal. ka BP at Burnham Overy (NNC18 and NNC19A, Fig. 7), but at younger dates in more westerly located boreholes. These estimates of the timing of initial transgression suggest that the sea first entered the Holocene trough in its central or eastern part from the east (Shennan *et al.* this volume *b*), finally flooding the area west of Burnham Overy sometime after 5.5–5 cal. ka BP (based on dates in NNC29 at Brancaster and NNC35 at Thornham).

The history of sedimentation following initial transgression is simplest to summarize in the Holkham to Cley area landward of the modern Blakeney Spit and barrier beaches, where four NNC boreholes augment published information from transects at Holkham, Stiffkey, Morston and Cley (Funnell & Pearson 1989). At Warham, where the marine LFA-B muddy sequence is about 13 m thick, 7 cal. ka BP mudflat sediments gradually give way to lower saltmarsh sediments at *c.* 6.4 cal. ka BP (Fig. 5), followed by the establishment of upper saltmarsh deposits at about 6 cal. ka BP and continuing more or less to 1.95 cal. ka BP (Fig. 5). This thick sequence is comparable with transects from Stiffkey and Morston marshes, 3 and 5 km to the west (Funnell & Pearson 1989), where up to 10 m of sediment typically begins with mudflat sediments that are replaced upward by upper saltmarsh deposits. Funnell *et al.* (in press) have recently

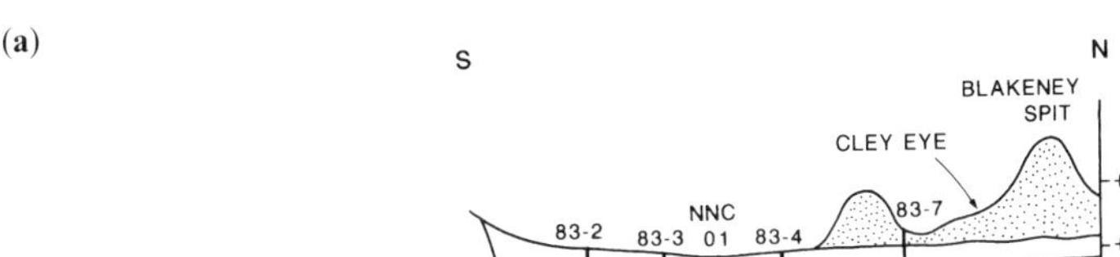

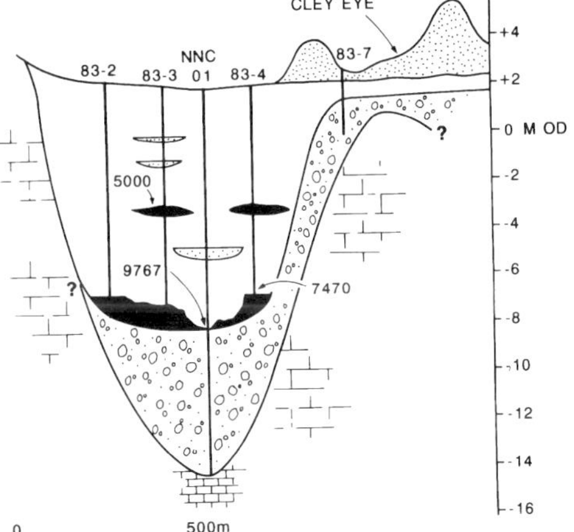

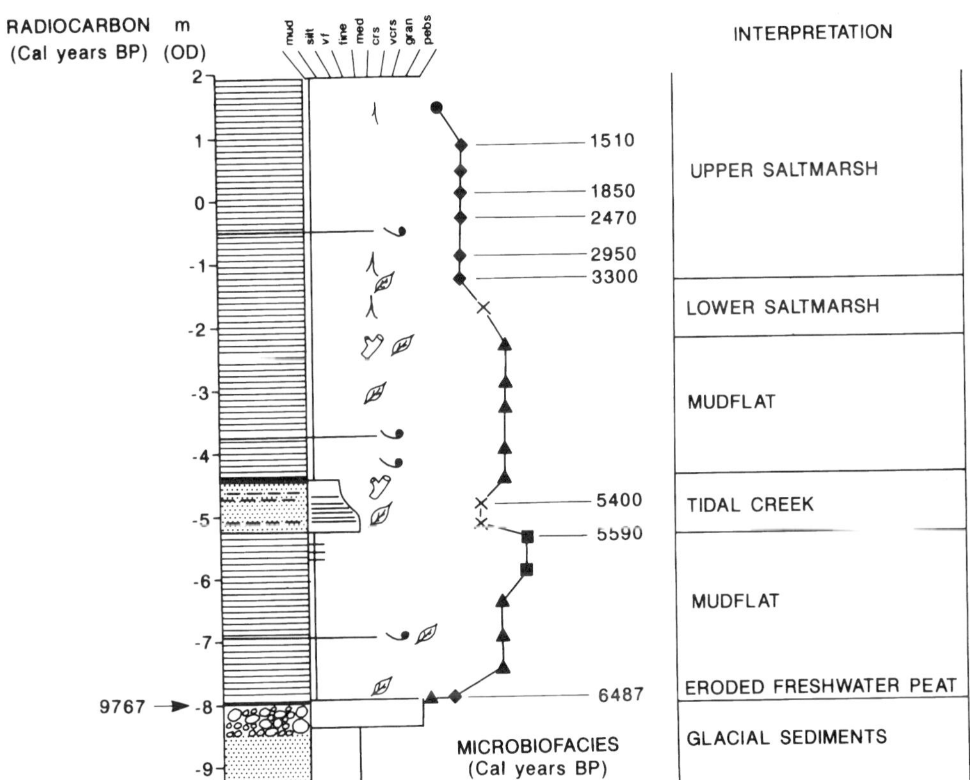

Fig. 10. (**a**) Coast normal cross-section of the Holocene sediments at the Cley transect (Fig. 1) based on borehole and power auger information (83-x numbers) from Funnell & Pearson (1989). The depth to the top of the Chalk is based on seismic data (Fig. 2) from Chroston *et al.* 1999. The landward margin is drawn to the +3 m OD contour. Note the Holocene and glacial sediments are located in a trough and thin on the Chalk high beneath the modern Blakeney Spit. Radiocarbon dates are in calibrated years before present. (**b**) Sedimentological and biofacies log of the Cley core NNC1 (Fig. 1, [TG 0391 4462]) showing radiocarbon and microbiofacies chronologies and palaeoenvironmental interpretations. (See legend in Fig. 5.)

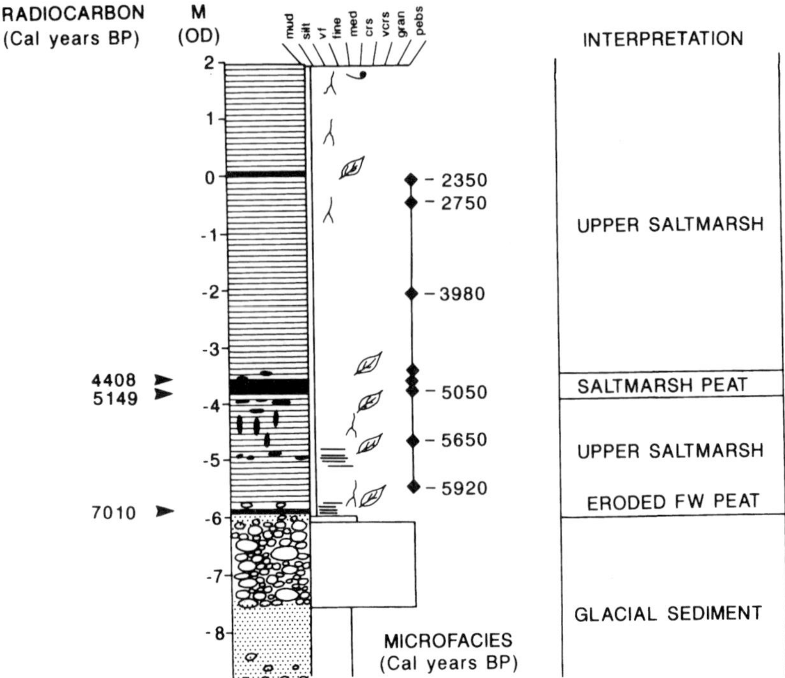

Fig. 11. Sedimentological and biofacies log of the Salthouse core NNC 40 (Fig. 1; NGR TG 8140 4400] showing radiocarbon and microbiofacies chronologies and palaeoenvironmental interpretations (See legend in Fig. 5.).

shown that the transition from mudflat to upper saltmarsh probably happened at around 4.4–5 cal. ka BP (at Stiffkey and Morston, respectively). These sequences thus seem to represent long-term maintenance of mudflat and then saltmarsh conditions south of a fairly stable barrier system. The timing of the regressive mudflat to saltmarsh change may be locally controlled by proximity to main creeks in the saltmarsh system; there is certainly evidence of a creek within the Stiffkey transect (Funnell & Pearson 1989).

In the Cley transect (NNC1 plus information in Funnell & Pearson 1989), the LFA-B muds landward of Blakeney Spit suggest marine flooding at *c.* 6.4 cal. ka BP followed by mudflat or lower saltmarsh sedimentation at least in the deepest parts of the Holocene trough until *c.* 3.3 cal. ka BP (Fig. 10b), followed by upper saltmarsh sedimentation, probably until nineteenth century reclamation for grazing marsh. It is also clear from Fig. 10a (see also Fig. 8b in Funnell & Pearson 1989), that the LFA-A barrier sands, i.e. dune sands and washover fans from Blakeney Spit, are locally overlapping these older sediments as the modern barrier rolls landward.

In our one core east of the Cley transect, NNC40 at Salthouse (Fig. 11), marine sedimentation from 6 cal. ka BP until reclamation, was wholly in upper saltmarsh conditions, punctuated by a thin saltmarsh peat (Funnell *et al.* in press) dated at the base and top at 5149 and 4408 cal. a BP, respectively. These dates are similar to the ages of mid section peats at Cley dated at 5 cal. ka BP (top of peat in borehole 83-3, Funnell & Pearson 1989), and 5610–5320 cal. a BP at Stiffkey (borehole 83-10, Funnell & Pearson 1989). These peaty deposits may represent a period of lowered sea-level rise, accompanied by greater ground-water inflow at the landward margin of the Holocene sediments (Funnell *et al.* in press).

To the west of the thick sequence at Warham, the Holkham transect (Fig. 12) again shows unbroken LFA-B sediments seaward of the modern barrier-beach. In NNC17, an early phase of saltmarsh deposition around 6 cal. ka BP gave way to brackish lagoonal conditions surrounded by mudflats that persisted until about 3.6 cal.ka BP (Fig. 6a). Palaeomagnetic data (Boomer & Maher pers. comm. 1998) and sudden changes in dates over a short vertical interval suggest a depositional hiatus at −1.70 m OD, followed by mudflat sedimentation. An *in situ Scrobicularia plana* in the mudflat sediments at −1.61 m OD was

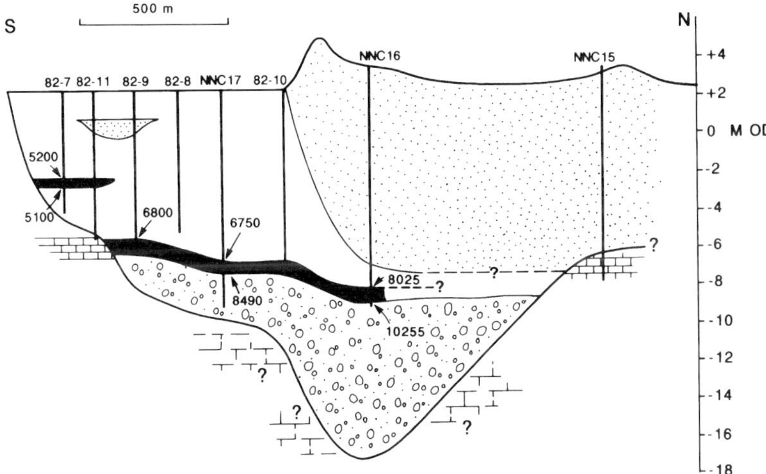

Fig. 12. Coast-normal cross section of the Holocene sediments at the Holkham transect (Fig. 1) based on borehole and power auger information (82-x numbers) from Funnell & Pearson (1989). The depth to the top of the Chalk is based on seismic data (Fig. 2) from Chroston *et al.* 1999. The landward margin is drawn to the +3 m OD contour. Note the Holocene and glacial sediments are centred on a trough on the Chalk surface. The subsurface contact between LFA-A sands and LFA-B back-barrier sediments is clear in core NNC16. Radiocarbon dates are all in calibrated years before present (cal. a BP), and show the backstepping relationship of the basal peat. Details of cores NNC16 and NNC17 are given in Fig. 6.

dated at 2360 cal. a BP (Fig. 6a) and corroborated by a similar IRSL age (2800 ± 300 years). Mudflat sediments give way to upper saltmarsh sediments at about 1650–2000 cal. a BP (based on IRSL and microbiofacies ages). On the landward edge of this transect in borehole 82-7, Funnell & Pearson (1989) again identified a *c.* 5 cal. ka BP mid-section peat, suggesting that the period of lowered sea-level rise is an event that can be correlated along the eastern half of the Holocene sediment prism between Holkham and Salthouse (Figs 10a and 12).

West of Holkham the LFA-B back-barrier sediments are rather thinner between Burnham Overy and Burnham Norton, where mudflat to upper saltmarsh sediments were accumulating in places by between 6 and 5 cal. ka BP until present, although now overstepped by LFA-A barrier sands in the more seaward boreholes (e.g. NNC19A Burnham Overy; Fig. 7b). Peats between −2.0 and 0.0 m OD in NNC20 and NNC27 on the landward edge of the Burnham Overy and Burnham Deepdale transects (Fig. 13) contain *Phragmites* debris, and represent freshwater marshes, probably fed by chalk groundwater springs, which are evident in the area today. Up to 7 m of LFA-B mudflat–saltmarsh back-barrier sediments are present in the Brancaster and Thornham boreholes (see also Funnell & Pearson 1989), and in places were probably accumulating from *c.* 4.5 cal. ka BP onwards. At various places in the Brancaster, Titchwell, Thornham and Holme area, the saltmarsh muds are intercalated with terrestrial peats, some with *in situ* birch and alder tree stumps (P. Murphy pers. comm. 1997) apparently forming between 4 and 3 cal. ka BP. This regressive episode is not recorded further east than Brancaster and may represent local barrier movement cutting off the area from seawater inundation for thousands of years. In late 1998, a human constructed wooden structure, the so called Seahenge, was discovered at Holme in peat that is probably of this age range (see e.g. Eastern Daily Press, 14 January 1999).

Barrier lithofacies association (LFA-A)

The cored boreholes in the LFA-A barrier sands are less helpful in determining the Holocene evolution, in part because the barriers are migrating shoreward with time (see below) and shoreface erosion is removing the record of earlier barriers, and also because it has proved difficult to date the age of the sands with any confidence. Fluidization of sands during coring has also obscured facies designations. Despite these problems the cores from the Blakeney,

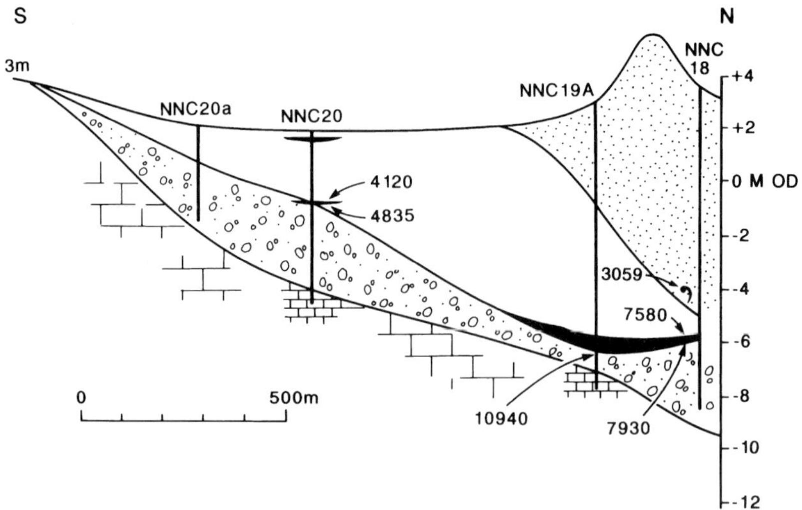

Fig. 13. Coast-normal cross section of the Holocene sediments at the Burnham Overy transect (Fig. 1) based on borehole information. The depth to the top of the Chalk is based on core data in NNC19A and NNC20 and in part inferred from seismic data at the nearby Burnham Norton transect (Fig. 1), from Chroston *et al.* 1999. The landward margin is drawn to the +3 m OD contour. The subsurface contact between LFA-A sands and LFA-B back-barrier sediments is clear in cores NNC18 and NNC19A (details of these cores are given in Figs 6b and 7b). Radiocarbon dates are in calibrated years before present (cal. a BP). Note the date of 3059 cal. a BP on the *Cerastoderma* shell in NNC18.

Holkham–Burnham Overy, Scolt Head Island and Brancaster areas (Fig. 1) are the first cores to penetrate the barrier sands and show the relationship with the back-barrier sediments. Most of these cores were spudded on beaches or at the base of aeolian dune sands. The dune sand to wave-lain sand boundary is usually marked by the uppermost prominent gravel deposit (beach ridge) and IRSL dates on dune initiation above these surfaces appears to cluster between AD 1500 and AD 750 (Knight *et al.* 1998).

Both the Blakeney and the Scolt Head Island areas (Fig. 1) are dominated by the present-day spit–barrier geomorphology. New interpretation of cartographic information for the last 400 years shows westward migration of Blakeney Spit at about 3.5 m a^{-1} and landward rollover at 1 m a^{-1} (the latter value can be compared with the decadal time-scale estimate of 0.6 m a^{-1} made by Hunter & Mottram (1925)). Recurved spits on Scolt Head Island mark the former position of the western end (Steers 1934), and record episodic westward accretion at a rate of approximately 3.5 m a^{-1} over the past 1155 years. Recent surveying over the past seven years (C. Bristow pers. comm. 1998) indicates that westward growth of Scolt Head Island is continuing at a similar rate.

The boundary between the barrier LFA-A and back-barrier LFA-B is very clear in borehole transects between Holkham and Scolt (Figs 12–14). At Holkham (Fig. 6b; NNC16), only 0.55 m of peat (10.3–8 cal. ka BP) and an overlying 1.25 m of intertidal mud was proved at depth underlying LFA-A barrier sand. The age of these muds is probably 6.3–6.8 cal. ka BP based on microbiofacies dates; IRSL ages of between 5300 ± 500 and 5900 ± 650 years appear systematically too young by about 1 ka, although they overlap within 2σ limits. Between Burnham and Holkham, these fine-grained LFA-B sediments are truncated seaward, presumably by shoreface erosion and replaced by LFA-A sands and occasional gravels. In NNC16 (Fig. 6b) the contact between LFA-A sand resting on the LFA-B muds at −6.80 m OD is sharp, presumably an erosion surface. Although there are doubts concerning the reliability of the IRSL ages for the overlying sands, the ages indicate that sand sedimentation occurred between 2 and 3 ka ago. A similar erosion surface is present in NNC18 at Burnham Overy; here the surface is at −4.80 m OD, overlying *c.* 6 cal. ka BP LFA-B back-barrier sediments. The radiocarbon date of 3059 cal. a BP on a *Cerastoderma* shell at −4.78 m OD (Fig. 7a), just above the erosion surface, if reliable, indicates that the maximum age of downcutting and barrier emplacement in this area was around 3 ka ago. Correlation between the erosion surfaces in these boreholes

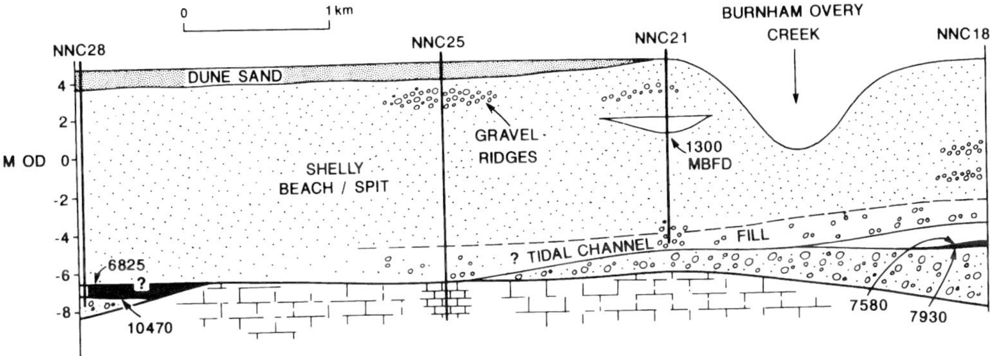

Fig. 14. Coast-parallel cross-section of the Holocene sediments on the Scolt Head Island transect (Fig. 1) based on borehole information. The transect is dominated by LFA-A sands and gravels, although eroded remnants of LFA-B back-barrier sediments are present in NNC28 and NNC18. The isolated patch of LFA-B back-barrier sediment in NNC21 at about 0 m OD is probably a saltmarsh enclosed within recurved spits. Radiocarbon and microbiofacies dates (MBFD) are all in calibrated years (cal. a BP) before present.

would imply metre-scale relief of the surface over about 3.5 km. This relief, combined with facies transitions in the LFA-A sands above the erosion surface is probably most consistent with erosion and LFA-A deposition by a laterally migrating tidal inlet active at some time after 6 cal. ka BP, probably after 3 cal. ka BP. This interpretation implies that on millennial time-scales laterally migrating tidal inlets are a significant cause of saltmarsh erosion, despite their apparent stability on decadal time-scales (see e.g. French & Stoddart 1992). Similar channels can be seen today at the eastern and western ends of Scolt Head Island (Fig. 1).

The Scolt Head Island borehole transect (Fig. 14) provided the first complete penetration of the Holocene barriers sediments in the area (earlier power augering by Allison (1985) and Funnell & Pearson (1989) had only achieved penetration to −4 m OD). The boreholes reveal a sequence dominated by LFA-A shallow marine barrier sands and gravels, with a basal erosion surface at −6.50 to −7.00 m OD cutting basal Holocene peat with a top surface age of *c.* 6.8 cal. ka BP (NNC28) but elsewhere abutting putty chalk (NNC25), and glacial sediment (NNC21). It is likely that this erosion surface is a marine ravinement surface active some time after 6.8 cal. ka BP. Because there is good cartographic evidence of westward migration of the island over historical times (Steers 1934; Allison 1985), we suspect, although cannot prove, that the LFA-A sediments are diachronous, oldest in the east and youngest in the west. The overall up core lithofacies changes are interpreted as a regressive sequence from tidal delta or tidal channel deposits, passing upward into shoreface beach and/or recurved spit deposits capped by dune sands (see below). It is therefore possible that the Scolt Head barrier initiated as a spit and extended west from Holkham within the last 3–2 ka. A radiocarbon date of 1107 cal. a BP from an *in situ* shell in a saltmarsh mud sequence at about +0.50 m OD in NNC21 (Fig. 14) suggests back-barrier marsh formation by this time, and there is an AD 1585 map that can be interpreted to support the presence of a spit (see Steers 1934, 1960). This interpretation would imply that the island has prograded over and now overlies its associated tidal delta and channel deposits (Fig. 14). If our interpretation is correct, it reopens the debate concerning the initiation and early evolution of the island. Steers (1934, 1960) eloquently summarized his preferred interpretation that the island initiated on a shingle ridge and has always been an island. Future work to better constrain the age of these LFA-A sediments should help resolve this debate.

Before Scolt Head Island began extending into the present area, i.e. at *c.* 2 cal. ka BP, the coastline between Burnham Overy and Brancaster Staithe probably lay along a line of inner barriers (now vegetated gravel ridges such as the Nod [TF 813 457], Great Ramsey [TF 834 455] and Little Ramsey [TF 820 455], Steers 1934; inner barriers south of Scolt Head Island on Fig. 1). Along this coast it is also likely that older barriers existed further offshore; partly because of the presence of peat and back-barrier muds at the bases of NNC18 and NNC21, and also because of the presence of *c.* 3–4 cal. ka BP peats outcropping on the modern beach in the Brancaster–Titchwell area.

If our model for the evolution of Scolt Head Island proves to be correct, and *if* the erosion surface between Holkham and Burnham Overy was scoured by the migrating tidal channel at the leading edge of island, then the erosion surface should not be older than about 3–2 cal. ka BP in the Burnham Overy area (NNC18) and perhaps not more than 3 cal. ka BP in the Holkham area (NNC16). These predictions are consistent with the limited radiocarbon data, but await further refinement of the luminescence dating of the LFA-A sands.

East of Holkham we were not able to core through the LFA-A barrier sediments (due to difficult access in the Warham–Stiffkey area) except in the far east of the study area at Blakeney Spit. Here the boundary between LFA-A barrier sediments and back-barrier LFA-B sediments is clear. At its eastern end the modern spit is grounded on small (<10 m high) hills ('Eyes') of glacial deposits (Funnell *et al.* in press; Figs 10a and 15), landward of which are the Cley–Salthouse back-barrier LFA-B deposits. Two kilometres northwest of Cley in NNC2 the boundary between LFA-A sands and the underlying LFA-B sediments is at −3.40 m OD. Two–three kilometres further northwest along the modern spit, cores from NNC4 and NNC7 show that the LFA-A/LFA-B boundary is much deeper at −9.5 to −10.5 m OD. An AD 1586 map suggests that these boreholes were sited where the main channel was at that time, accounting for the 13 m thickness of LFA-A sands in these cores. A radiocarbon date on a basal peat in NNC4 (Fig. 15) indicates initiation of marine Holocene sedimentation in this area sometime after 7.5 cal. ka BP, probably around 6.4 cal. ka BP, and clearly in the protection of a barrier seaward of the present one.

The westward extension of Blakeney Spit probably occurred in a similar way to Scolt Head Island. The LFA-A sediments are thus interpreted as laterally migrating tidal inlet deposits, passing upward into spit–barrier beach deposits. These deposits are currently being reworked in places by ongoing landward barrier rollover at about 1 m a^{-1}.

In the area between the western end of Blakeney Spit and the Wells channel there is no detailed subsurface information. Refraction geophysics on West Sand [TF 965 455], at the seaward end of the Stiffkey geophysical transect suggests that the top of the Chalk is at about −4 m OD suggesting that the LFA-A barrier sands are about 7 m thick here. The modern barrier complex is composed of an ebb-tidal delta at the mouth of the Wells Channel with an extensive low-gradient sandflat with attached

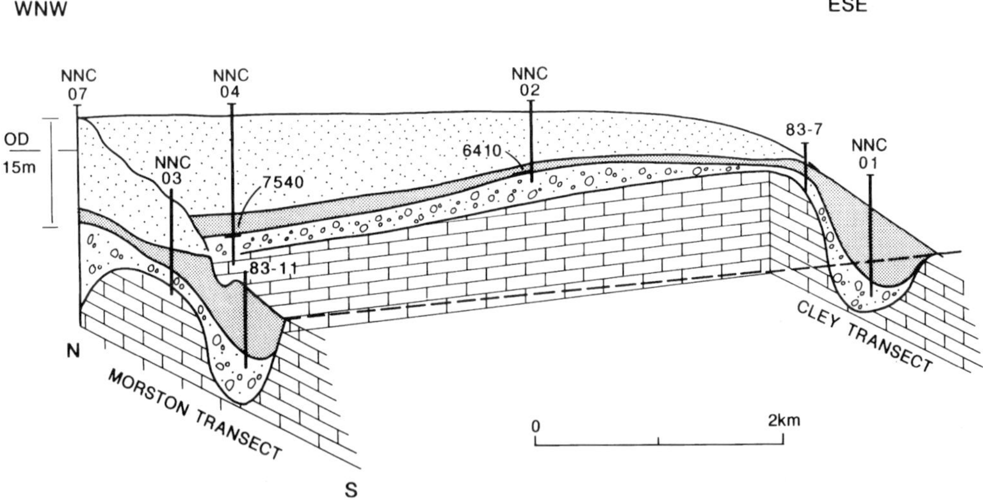

Fig. 15. Schematic fence diagram for the Blakeney area based on the Blakeney Spit, Cley and Morston transects (Fig. 1). Depths are based on borehole and seismic information. Note that the modern Blakeney Spit (line of section from 83-7 to NNC7) is grounded at its east-southeast end on a high in the Chalk subsurface topography. The LFA-A barrier sands are underlain by a thin strip of LFA-B back-barrier sediments (grey tone), removed by either tidal channel or shoreface erosion. The Quaternary trough in the Chalk is clear in the back-barrier area. The pecked line represents the modern depositional limit on the landward side (about +3 m OD).

bars and megaripple fields to the east (Pearson *et al.* 1990). Lodge Marsh, part of the Warham Marsh complex, was formed by reclamation in the late 1700s; here about 0.5 m of modern saltmarsh is underlain by sand and gravels, part of the wide sandflat to the east. Before reclamation, the LFA-A/LFA-B transition and the natural coastline probably ran along East Fleet [TF 925 452].

Between Warham and Morston, coast-parallel sand–gravel ridges, known locally a 'Meols' or 'Meals', are well developed (similar gravel ridges are also present in places to the west between Brancaster and Holme). Augering at Stiffkey has showed that these 'inner barriers' lie on LFA-B intertidal marsh deposits, which have been accumulating since at least 6 cal. ka BP (Boomer & Woodcock 1999). These inner barriers probably record the transport of coarse material across saltmarsh during a relatively short-lived emplacement episode, no more than 1000 years ago based on the microbiofacies age–depth model. Major storms would be necessary to bring sand and gravel shoreward from the outer barriers in sufficient amounts to build the 'Meals' (see also Forbes *et al.* 1991) and Boomer & Woodcock (1999) argue that this may have been caused by increased storminess during the early part of the Little Ice Age (AD 1445–1680). This idea is consistent with the interpretations of Knight *et al.* (1998) who suggested that the elevation of wave-lain gravel platforms deposited between AD 1430 and 1730 was raised by about 2 mm a^{-1}, indicating a transgressive phase of relative sea level.

Lithofacies continuity and correlation on local and regional scale

The long-term pattern of landward barrier roll-over and lateral migration observed today was probably continuous during most of the Holocene, with earlier barrier LFAs established further north of the present coastline. These sandy barrier facies belts have migrated south and aggraded more or less in pace with rising sea-level (see also Orford *et al.* 1995). Short-term landward movement of the barriers has probably been episodic and driven by storms (see e.g. Steers *et al.* 1979; Forbes *et al.* 1991), while the shoreface has eroded so that Holocene sediments have accumulated in an increasingly narrow coastal prism.

If the present rate of landward movement of the barriers of about 1 m a^{-1} has been more or less constant during the phase of steady sea-level rise between 7.4 cal. ka BP (6.5 ^{14}C ka BP, Fig. 4) and present, it is possible to propose a crude estimate for the narrowing of the coastal zone during this time. In core NNC16 at Holkham, we know that young LFA-A sands rest disconformably on *c.* 6 ka old LFA-B back-barrier sediments at −6.80 m OD (Fig. 6b). This tells us that the barrier sands were seaward of this position 6 ka ago. While we do not have a reliable age for the emplacement of the sands in NNC16, our model for the evolution of Scolt Head Island and limited radiocarbon dates from the Burnham Overy transect, suggests that the sands are unlikely to be more than 3 ka old (they could be much younger). This implies that there was no more than 3 ka to roll the barrier into this position from offshore, suggesting, at landward movement of about 1 m a^{-1}, a narrowing of the coastal zone at this time of no more than 3 km. This estimate, while making a number of assumptions, suggests that the Holocene coastal prism in north Norfolk was approximately twice its present maximum width (about 6 km): an estimate in keeping with the implied rate of cliff-line erosion to the east over the last 5 ka (Clayton 1989) and in broad agreement with seismic evidence for the position of the early Holocene coastline in the Yarmouth area (Arthurton *et al.* 1994). This estimate is also supported by the discovery of an 8360 cal. a BP (7580 ^{14}C a BP) old saltmarsh peat and intertidal clay about 5 km off the north Norfolk coast (Shennan *et al.* this volume *b*, core 52/+01/2699). Shoreface erosion has left almost no record of this wider coastal zone; however, it is likely that the sediments that made up this coast have been extensively recycled and redeposited landward as the coastal zone has thinned.

In some areas the preserved back-barrier sediments show a predominantly regressive first-order lithofacies transition from open mudflat to mature saltmarsh with time. However, at the borehole and transect level in the back-barrier facies (LFA-B) small-scale regressive and transgressive sequences are very difficult to correlate for more than a few hundred metres at most. Some of these apparently small-scale sequences might reflect very local facies changes. For example, an apparently transgressive sequence from upper saltmarsh to mudflat might simply result from the core penetrating a tidal creek that had formed in the upper part of a mature saltmarsh. However, it is likely that the general second order control on saltmarsh and mudflat development is autocyclic rather than allocyclic, probably a response to local changes in the form and disposition of tidal channels and barriers.

The onset of marine sedimentation in various NNC boreholes between about 7.8 and

5.8 cal. ka BP (7–5 ^{14}C ka BP, Fig. 4) occurs during the regionally significant Wash I/II positive sea-level tendency in nearby Fenland (Shennan 1982; Shennan *et al.* 1983). Similarly, the short fresher-water episode (mid section peats at *c.* 5.2 cal. ka BP, i.e. *c.* 4.5 ^{14}C ka BP, Fig. 4), may have occurred during a negative sea-level period, when peat (Fenland IV regressive; Shennan 1982, 1986), rather than estuarine sediments, were extensive in Fenland and freshwater, rather than estuarine deposition, was initiated in the Yare Valley of east Norfolk (Coles & Funnell 1981). However, younger terrestrial peats within the marine sequences at Thornham, Titchwell and Brancaster cannot be related to regional sea-level tendencies, although overlaps after *c.* 3.3 cal. ka BP (3 ^{14}C ka BP, Fig. 4) are transgressive and appear to correspond to Wash VI positive sea-level tendency in Fenland (Shennan 1982). The clear implication is that the 5.2–3.3 cal. ka BP (4.5–3 ^{14}C ka BP, Fig. 4) NNC transgressive and regressive units are generally not correlatable along the coast and do not directly always represent external forcing (global or regional sea-level change); it is very likely that they reflect local modification of coastal barriers superimposed upon regional sea-level change.

The future of the north Norfolk coast

The combined effort of earlier workers and ourselves has identified the main processes operating on the modern NNC and the sediment record of the Holocene history. However, to understand how this coast will respond to environmental change in the future we need to have an understanding of sediment storage and flux over Holocene time. This is not easy to address because: (a) we have very little information regarding pre-8 cal. ka BP sedimentation; (b) the absolute chronology between 8 and 2.5 cal. ka BP is only reasonably well-constrained for LF1 peats and some LF2 muddy sediments (luminescence and microbiofacies ages); (c) because of the post-2.5 cal. ka BP gap in most of these chronologies.

There is, and probably always has been over the last 10 ka, a negligible sediment supply to the NNC from the small rivers draining the till-capped Chalk hinterland, due to low runoff and low sediment load. Almost all sediment is therefore supplied from shoreline erosion and from offshore. It is clearly important to determine whether there is sufficient supply of coarser sand and gravel to maintain the spits and barriers that form the framework within which fine-grained sediment can accumulate. The net longshore transport of sandy sediment from the cliffs around Sheringham eastward is uncertain, potentially around $60 \times 10^3\, m^3\, a^{-1}$ (Vincent 1979; Clayton *et al.* 1983); however, lack of sandy sediment in the Blakeney area suggests that this potential transport is not realized (Vincent 1979), while shingle movement is complex (Clayton 1976). Clearly, longshore drift within the intertidal and shallow-subtidal surf zone transports most sediment east–west along much of the coast today, and has done for at least 2 ka, as shown by the recurved laterals on Blakeney Spit and Scolt Head Island; however, this sediment may have been supplied from offshore (Shih-Chiao & Evans 1992). An easterly directed tidal residual current transports sand from the north and east along parts of this coast in the subtidal zone and this might deliver some sand to the western part of the coastal zone. These poorly defined transport paths are presumably locally intercepted by the tidal inlets and associated ebb-tidal deltas that probably act as hydraulic groins and sinks for sand-sized sediment. The rather scant information we have suggests a long term under-supply of coarse-grained sediment to the NNC barriers, resulting in the landward migration of the barrier systems and narrowing of the coastal prism; a trend that is recognized globally (Clayton 1995). In its present configuration, our data suggest that the NNC Holocene sediment prism (measured to the Ordnance Survey low water mark) has a total volume of $685 \times 10^6\, m^3$, of which about 48% is LFA-A barrier sands and 52% is LFA-B fine-grained back-barrier sediment.

In contrast to this implied under-supply of sand and gravel, the NNC is probably well sup-plied with $<63\, \mu m$ material, predominantly from the eroding cliffs of Holderness to the north (McCave 1978), some of it recycled through the Wash (Ke *et al.* 1996), and in part from erosion of the Norfolk cliffs. The NNC is estimated to be a sink for about 104 000 tonne a^{-1} of $<63\, \mu m$ material (McCave 1978), which is consistent with our calculation of net storage of $<63\, \mu m$ material, of about 8.2×10^8 tonne over the Holocene. This, combined with information on sedimentation rates in saltmarshes (see above; also Pye 1992; French & Spencer 1993) suggests that the back-barrier marshes are presently well supplied with fine-grained sediment and are able to accrete at a similar pace to rising sea level (although the caveat of French *et al.* (1995) should be bourne in mind).

Response of the NNC to future sea-level change and management

Global warming scenarios predict a 50 cm rise in sea-level over the next 100 years (Wigley & Raper 1992). Our results confirm that the NNC saltmarshes have accreted in pace with sea-level rise over the last 6–7 ka, and one-dimensional modelling (French 1993) suggests that they might keep up with the predicted accelerated rate of sea-level rise. While fine-grained sediment does not seem to be in short supply for building back-barrier saltmarshes, the coarse clastic barrier systems may well be sediment supply limited. Moreover, if the offshore area is, or was, a significant source of sediment, rising sea-level will progressively isolate this area from strong wave action, decreasing erosion and sediment supply (Jones 1996). This suggests that the barriers will continue to roll landward, possibly at faster rates if sea-level rise is accelerated (see also Orford *et al.* 1995). This landward rollover will further narrow the coastal prism (see also French 1993), and probably change the configuration of tidal inlets and their ebb-tidal deltas.

Our borehole data shows that longshore and shoreward movement of the barriers has been a long-term feature of the evolution of this coast, and management and flood defence strategies should use this information to allow realistic forward planning (see also Spencer & French 1992). If accelerated sea-level rise were associated with increased storminess we might expect a phase of increased sand and gravel transport shorewards (e.g. see Steers 1953, 1971; Steers *et al.* 1979) and bar formation on top of the back-barrier muds, landward of the main barriers, like the ancient Stiffkey–Morston Meals examples. These ridges might potentially act as nuclei for sand dune formation (cf. Knight *et al.* 1998). However, large-scale dune-building may not occur if relative sea-level rise is so rapid that extensive offshore sandflats are not exposed, or if the dominant wind direction remains offshore as today (Knight *et al.* 1998; Orford *et al.* this volume). Thus the attractive idea that new dunes might act as a line of natural defence remains doubtful. This is unfortunate because apparently fragile systems like sand dunes stand up remarkably well to major storm surges (Steers 1953, 1971; Steers *et al.* 1979).

While natural marshes will probably be able to accrete in pace with accelerated sea-level rise (see also Spencer & French 1992), reclaimed marshes (over half the saltmarsh area) already have a deficit of up to 300 years of intertidal sedimentation, resulting in 60 cm height difference between natural and reclaimed marshes (Pye 1992). The continued raising of sea-defences to provide protection for some developed areas (e.g. some of the larger villages, such as Wells and Blakeney) may be justified (Clayton 1993, 1995). However, the continued protection of reclaimed marsh for agriculture will not justify expenditure, and the continued isolation of these former marshes from the sea will exacerbate the sediment deficit problem. The return of reclaimed marsh to tidal flooding by the re-opening of tidal creeks, will create instant accommodation space, and presumably the progressive formation of intertidal creek to saltmarsh environments. This artificially low sediment surface will undoubtedly flood during storm surges to create temporary and possibly semi-permanent lagoons. These lagoons may gradually silt up to recreate natural saltmarsh environments over many hundreds of years. They should be allowed to do so, to create large buffer areas able to absorb wave and tidal energy. However, in the meantime, and with judicious management, these lagoons might create valuable new habitats for fish spawning and bird life (see also Funnell 1992).

The present obligation to protect the reserve at Cley (SPA), which is situated on reclaimed grazing marsh landward of a barrier that has been rolling landward for at least 6 ka, is not sustainable in the long term (O'Riordan & Ward 1997), despite being economically justifiable in the short term (Klein & Bateman 1998). Opportunities must be pursued to manage new natural habitat as it evolves elsewhere along the coast, allowing the wildlife reserves to move with natural coastal changes and removing the pressure to protect what are actually artificial environments at existing 'fixed-site' nature reserves.

H. Glaves (British Geological Survey) helped with the levelling of borehole sites and H. Roberts (Durham) assisted with the IRSL analyses. Geophysical fieldwork and hand augering would not have been possible without the stalwart help of L. Cartwright, J. Stevenson, A. Goillau, B. Makin and T. Hardman (all UEA). J. Smith (University of Edinburgh) helped with short coring and did radionuclide analyses. S. Davies and P. Judge (UEA) helped prepare the final diagrams. We very much appreciate the permissions to work on private and publicly owned land and especially thank the National Trust, the Norfolk and Norwich Naturalists' Trust and the Holkham Estate. The authors would like to dedicate this paper to the memory of G. Harwood who died on 12 March 1996. This research was supported by NERC LOIS Special Topic allocation GST/02/737and the paper is LOIS publication number 586.

References

AITKEN, M. J. 1990. *Science-based Dating in Archaeology*. Longman, Harlow.

ALLISON, H. M. 1985. *The Holocene Evolution of Scolt Head Island, Norfolk*. PhD thesis, University of Cambridge.

ARTHURTON, R. S., BOOTH, S. J., ABBOTT, M. A. W. & WOOD, C. J. 1994. *Geology of the Country Around Great Yarmouth*. British Geological Survey, HMSO, London.

BAILIFF, I. K. & TOOLEY, M. J. 2000. Luminescence dating of fine-grain Holocene sediments from a coastal setting. *This volume*.

BOOMER, I. 1998. The relationship between meiofauna (Ostracoda, Foraminifera) and tidal levels in modern intertidal environments of north Norfolk: a tool for palaeoenvironmental reconstruction. *Bulletin of the Geological Society of Norfolk*, **46**, 17–29.

—— & WOODCOCK, L. 1999. The nature and origin of Stiffkey meals. *Bulletin of the Geological Society of Norfolk*, **49**, 3–13

CALLAWAY, J. C., DELAUNE, R. D. & PATRICK, JR., W. H. 1996. Chernobyl ^{137}Cs used to determine sediment accretion rates at selected northern European coastal wetlands. *Limnology and Oceanography*, **41**, 444–450.

CHROSTON, P. N., JONES, R. & MAKIN, B. in press. Geometry of Quaternary sediments along the north Norfolk coast UK: a shallow seismic study. *Geological Magazine*, **136**, 465–474.

CLAYTON, K. M. 1976. Norfolk sandy beaches: applied geomorphology and the engineer. *Bulletin of the Geological Society of Norfolk*, **28**, 49–67.

——1993. Adjustment to greenhouse gas induced sea-level rise on the Norfolk coast; a case study. *In*: WARRICK, R. A., BARROW, E. M. & WIGLEY, T. M. L. (eds) *Climate and Sea-level Change: Observations, Projections and Implications*. Cambridge University Press, Cambridge, 310–321.

——1995. Predicting sea-level rise and managing the consequences. *In*: O'RIORDAN, T. (ed.) *Environmental Science for Environmental Management*. Longman, Harlow, 165–184.

——1989. Sediment input from the Norfolk cliffs, eastern England: a century of coast protection and its effects. *Journal of Coastal Research*, **5**, 433–422.

—— & O'RIORDAN, T. 1995. Coastal processes and management. *In*: O'RIORDAN, T. (ed.) *Environmental Science for Environmental Management*. Longman, Harlow, 151–164.

——, MCCAVE, I. N. & VINCENT, C. 1983. The establishment of a sand budget for the East Anglian coast and its implications for coastal stability, *Shoreline Protection*. Thomas Telford, Institute of Civil Engineers, 91–96.

COLES, B. P. & FUNNELL, B. M. 1981. *Holocene Palaeoenvironments of Broadland, England*. Special Publication of the International Association of Sedimentologists, **5**. Blackwell Scientific Publications, Oxford, 123–131.

FORBES, D. L., TAYLOR, R. B., ORFORD, J. D., CARTER, R. & SHAW, J. 1991. Gravel-barrier migration and overstepping. *Marine Geology*, **97**, 305 313.

FRENCH, J. R. 1993. Numerical simulation of vertical marsh growth and adjustment to accelerated sea-level rise, north Norfolk, UK. *Earth Surface Processes and Landforms*, **18**, 63–81.

—— & SPENCER, T. 1993. Dynamics of sedimentation in a tide-dominated backbarrier saltmarsh, Norfolk, UK. *Marine Geology*, **110**, 315–331.

—— & STODDART, D. R. 1992. Hydrodynamics of salt marsh creek systems: implications for marsh morphological development and material exchange. *Earth Surface Processes and Landforms*, **17**, 235–252.

——, SPENCER, T., MURRAY, A. L. & ARNOLD, A. S. 1995. Geostatistical analysis of sediment deposition in two small tidal wetlands, Norfolk, UK. *Journal of Coastal Research*, **11**, 308–321.

FUNNELL, B. M. (1992) A geological heritage coast: north Norfolk (Hunstanton to Happisburgh). *In*: STEVENS, C., GORDON, J. E., GREEN, C. P. & MACKLIN, M. G. (eds) *Conserving our Landscape: Evolving Landforms and Ice-Age Heritage*. Crewe, 59–62.

—— & BOOMER, I. 1998. Microbiofacies tidal-level and age deduction in Holocene saltmarsh deposits on the north Norfolk coast. *Bulletin of the Geological Society of Norfolk*, **46**, 31–55.

—— & PEARSON, I. 1989. Holocene sedimentation on the north Norfolk barrier coast in relation to relative sea-level change. *Journal of Quaternary Science*, **4**, 25–36.

——, BOOMER, I. & JONES, R. in press. Holocene evolution of the Blakeney Spit area of the North Norfolk coastline. *Proceedings of the Geologists' Association*.

HARDY, J. R. 1964. The movement of beach material and wave action near Blakeney Point, Norfolk. *Transactions of the Institute of British Geographers*, **34**, 53–70.

HUNTER, R. E. & MOTTRAM, W. E. 1925. A note on the occurrence of natural preservation of plant tissues. *New Phytologist*, **24**, 193–206.

JELGERSMA, S. 1979. Sea-level changes in the North Sea basin. *In*: OELE, W., SCHUTTENHELM, R. T. E. & WIGGERS, A. J. (eds) *Quaternary History of the North Sea*. **2**, Acta Universitativ Upsaliensis, Symposia Universitatis Upsaliensis Quingentesium Celebrantis, 2 Uppsala, 233–248.

JONES, R. 1996. Holocene coastal evolution in north Norfolk. *In*: JONES, P. S., HEALY, M. G. & WILLIAMS, A. T. (eds) *Studies in European Coastal Management, Coastlines 95*. Samara Publishing, Cardigan.

KE, X., EVANS, G. & COLLINS, M. B. 1996. Hydrodynamics and sediment dynamics of The Wash embayment, eastern England. *Sedimentology*, **43**, 157–174.

KLEIN, R. J. T. & BATEMAN, I. J. 1998. The recreational value of Cley marshes Nature Reserve: an argument against managed retreat? *Journal of Water and Environmental Management*, **12**, 280–288.

KNIGHT, J., ORFORD, J. D., WILSON, P., WINTLE, A. G. & BRALEY, S. 1998. Facies, age and controls on recent coastal sand dune evolution in north

Norfolk, eastern England. *Journal of Coastal Research (Special Issue)*, **26**, 154–161.

LAMBECK, K. 1995. Late Devensian and Holocene shorelines of the British Isles and North Sea from models of glacio-hydro-isostatic rebound. *Journal of the Geological Society of London*, **152**, 437–448.

LOIS 1992. *Science Plan for a Community Research Project*. Natural Environment Research Council, Swindon.

MCCAVE, I. N. 1978. Grain-size trends and transport along beaches: example from Eastern England. *Marine Geology*, **28**, M43–M51.

MURPHY, P. & FUNNELL, B. M. 1980. Preliminary Holocene stratigraphy of Brancaster Marsh. *Bulletin of the Geological Society* of *Norfolk*, **31**, 11–15.

ORFORD, J. D., CARTER, R. W. G., MCKENNA, J. & JENNINGS, S. C. 1995. The relationship between the rate of mesoscale sea-level rise and the rate of retreat of swash-aligned gravel-dominated barriers. *Marine Geology*, **124**, 177–186.

——, WILSON, P., WINTLE, A. G., KNIGHT, J. & BRALEY, S. 2000. Holocene coastal dune initiation in Northumberland and Norfolk, eastern UK: climate and sea-level changes as possible forcing agents for dune initiation. *This volume*.

O'RIORDAN, T. & WARD, R. 1997. Building trust in shoreline management: creating participatory consultation in shoreline management plans. *Land Use Policy*, **14**, 257–276.

PEARSON, I. 1986. *Holocene Evolution of the North Norfolk Coast*. PhD thesis, University of East Anglia.

——, FUNNELL, B. M. & MCCAVE, I. N. 1990. Sedimentary environments of the sandy barrier/tidal marsh coastline of north Norfolk. *Bulletin of the Geological Society of Norfolk*, **39**, 3–44.

PRANDLE, D. 1984. A modelling study of the mixing of ^{137}Cs in the seas of the European continental shelf. *Philosophical Transactions of the Royal Society of London*, **A310**, 407–436.

PYE, K. 1992. Saltmarshes on the barrier coastline of north Norfolk, eastern England. *In*: ALLEN, J. R. L. & PYE, K. (eds) *Saltmarshes: Morphodynamics, Conservation and Engineering Significance*. Cambridge University Press, Cambridge, 148–178.

RIDGWAY, J., ANDREWS, J. E., ELLIS, S., HORTON, B. P., INNES, J. B., KNOX, R. W. O'B., MCARTHUR, J. J., MAHER, B. A, METCALFE, S. E., MITLEHNER, A., PARKES, A., REES, J. G., SAMWAYS, G. & SHENNAN, I. 2000. Analysis and interpretation of Holocene sedimentary sequences: techniques applied in the Humber Estuary. *This volume*.

ROY, P. S. 1967. *The Recent Sedimentology of Scolt Head Island, Norfolk*. PhD thesis, University of London.

SHENNAN, I. 1982. Interpretation of Flandrian sea level data from the Fenland. *Proceedings of the Geologists' Association*, **93**, 53–63.

——1986. Flandrian sea-level changes in the Fenland. II: tendencies of sea-level movement, altitudinal changes, and local and regional factors. *Journal of Quaternary Science*, **1**, 155–179.

—— & WOODWORTH, P. L. 1992. A comparison of late Holocene and twentieth century sea-level trends from the UK and North Sea regions. *Geophysical Journal International*, **109**, 96–105.

——, LAMBECK, K., HORTON, B., INNES, J., LLOYD, J., MCARTHUR, J. & RUTHERFORD, M. 2000*a*. Holocene isostacy and relative sea-level changes on the east coast of England. *This volume*.

——, ——, FLATHER, R., HORTON, B., MCARTHUR, J., INNES, J., LLOYD, J., RUTHERFORD, M. & WINGFIELD, R. 2000*b*. Modelling western North Sea palaeogeographies and tidal changes during the Holocene. *This volume*.

——, TOOLEY, M. J., DAVIS, M. J. & HAGGART, B. A. 1983. Analysis and interpretation of Holocene sea-level data. *Nature*, **302**, 404–406.

SHIH-CHIAO, C. & EVANS, G. 1992. Source of sediment and sediment transport on the east coast of England: significant or coincidental phenomena? *Marine Geology*, **107**, 283–288.

SPENCER, T. & FRENCH, J. R. (1992) Geomorphologically informed conservation of coastal depositional landforms and associated habitats. *In*: STEVENS, C., GORDON, J. E., GREEN, C. P. & MACKLIN, M. G. (eds) *Conserving our Landscape: Evolving Landforms and Ice-Age Heritage*, Crewe, 69–73.

STEERS, J. A. 1927. The East Anglian Coast. *Geographical Journal*, **69**, 24–43.

——1929. The physiographical evolution of Scolt Head Island. *Transactions of the Norfolk and Norwich Naturalists Society*, **12**, 230–238.

——1934. Scolt Head Island. *Geographical Journal*, **83**, 479–494.

——1953. The east coast floods, January 31–February 1 1953. *Geographical Journal*, **119**, 280–298.

——1960. *Scolt Head Island*. W. Heffer & Sons, Cambridge.

——1964. *Blakeney Point and Scolt Head Island*. The National Trust, London.

——1971. The east coast floods 31 January–1 February 1953. *In*. STEERS, J. A. (ed.) *Applied Coastal Geomorphology*. Macmillan, London, 198–223.

——1989. The physical features of Scolt Head Island and Blakeney Point. *In*: ALLISON, H. & MORLEY, J. P. (eds) *Blakeney Point and Scolt Head Island*. The National Trust, Blickling, Norfolk, 14–27.

——, STODDART, D. R., BAYLISS-SMITH, T. P., SPENCER, T. & DURBIDGE, P. M. 1979. The storm surge of 11 January 1978 on the east coast of England. *Geographical Journal*, **145** 192–205.

STUIVER, M. & BARZIUNAS, T. F. 1993 Modeling atmospheric ^{14}C influences and ^{14}C ages of marine samples back to 10 000 BC. *Radiocarbon*, **35**, 137–189.

—— & REIMER, P. J. 1993. Extended ^{14}C data base and revised CALIB 3.0 ^{14}C age calibration program. *Radiocarbon*, **35**, 215–230.

VINCENT, C. E. 1979. Longshore sand transport rates: a simple model for the East Anglian coastline. *Coastal Engineering*, **3**, 113–136.

WIGLEY, T. M. L. & RAPER, S. C. B. 1992. Implications for climate and sea level of revised IPCC emissions scenarios. *Nature*, **357**, 293–300.

WYMER, J. J. & ROBINS, P. A. 1994. A long blade flint industry beneath boreal peat at Titchwell, Norfolk. *Norfolk Archaeology*, **42**, 13–37.

Holocene sedimentary evolution and palaeocoastlines of the Fenland embayment, eastern England

DAVID S. BREW,[1] TINA HOLT,[2] KEN PYE[2] & RHONDA NEWSHAM[3]

[1] *Coastal and Engineering Geology Group, British Geological Survey, Keyworth, Nottingham, NG12 5GG UK (e-mail: d.brew@bgs.ac.uk)*
[2] *Postgraduate Research Institute for Sedimentology, University of Reading, PO Box 227, Whiteknights, Reading RG6 6AB, UK*
[3] *Geospatial Information Systems Group, British Geological Survey, Keyworth, Nottingham NG12 5GG, UK*

Abstract: The Holocene sedimentary facies of the Fenland are described using a lithological database and new cores recovered as part of the Land–Ocean Interaction Study. Landward, the Holocene sequence is dominated by mud facies with intercalated peat layers, whereas the seaward areas are sand-dominated. The sedimentological characteristics of the mud facies are homogeneous and are similar for the whole sequence. Glacial deposits located north of The Wash are thought to be the main sediment sources. The sand facies generally fines upwards and the chemistry reflects this change. However, elemental ratios show only slight variations between the two facies implying a general constancy of sediment provenance. The evolution of the Fenland has been dominated by three main events. Firstly, the initial post-glacial transgression, which started *c.* 7850 cal. BP. Secondly, the sedimentary infilling of the embayment with rising sea-level; deposition of intertidal clastic sediments alternating with peat accumulation. Thirdly, renewed expansion of tidal flat areas between *c.* 2750 and 1500 cal. BP forming the final clastic fill.

The Fenland embayment, located on the English North Sea coast, is one of the largest sinks of Holocene sediment in the UK. The Fenland has been created through land-claim and is separated from The Wash by seawalls (Fig. 1). The present-day Wash is characterized by a macrotidal regime with a spring-tide range of about 6.5 m. The Fenland is drained by four main rivers; the Great Ouse, Nene, Welland and Witham (Fig. 1). Since marine inundation of the area in the early Holocene, clastic sediments and peats have accumulated over an area of about 3400 km^2 and to thicknesses of up to 30 m.

Research on the Fenland Holocene sequence began in earnest with the litho- and biostratigraphic (mainly pollen) work of Sir Harry Godwin and his co-workers (Godwin 1940, 1975, 1978; Godwin & Clifford 1938; Godwin & Vishnu-Mittre 1975). Later, Shennan (1982, 1986*a*, *b*) approached the evolution of the Fenland from a chronostratigraphic viewpoint using well-constrained radiocarbon dates from peat–clastic boundaries to develop a model for the sea-level history of the area (see also Shennan *et al.* 1999). Waller (1994*a*) utilized both litho- and chronostratigraphic data to provide a first attempt at Fenland-wide palaeogeographical reconstructions (the extent of marine and freshwater sedimentation) at various times through the Holocene.

Although these previous studies incorporated information from the basal and intercalated peat units and their boundaries, few detailed data are available from the lithological and sedimentological components of the clastic units. The recovery of 35 new cores (Fig. 1) as part of the National Environmental Research Council (NERC's) Land–Ocean Interaction Study (LOIS) has now made this possible. In addition, a new database of Holocene lithological data for the Fenland has been compiled as part of LOIS. The aim of this paper, therefore, is to provide an improved understanding of the Holocene sedimentary evolution of the embayment. This aim will be addressed by combining three methods: (a) interrogation of the database to define the geometry of the Holocene lithofacies; (b) sedimentological analyses of the lithofacies recovered in LOIS cores; (c) combination of the above data with lithostratigraphic data and

From: SHENNAN, I. & ANDREWS, J. (eds) *Holocene Land–Ocean Interaction and Environmental Change around the North Sea*. Geological Society, London, Special Publications, **166**, 253–273. 1-86239-054-1/00/$15.00

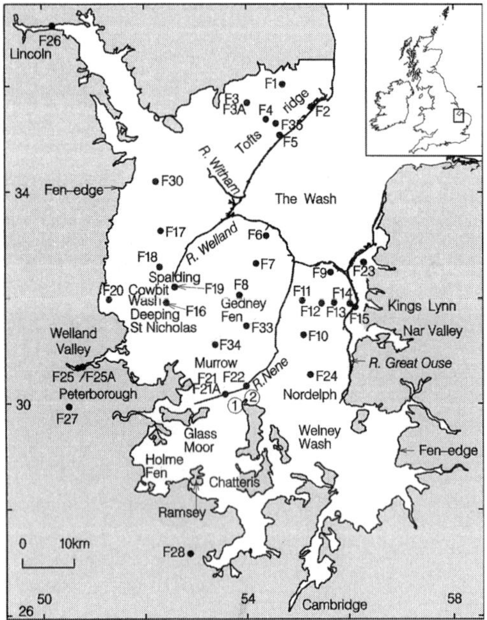

Fig. 1. Fenland: locations mentioned in the text and the position of the LOIS cores. 1, Adventurers' Land; 2, Guyhirn.

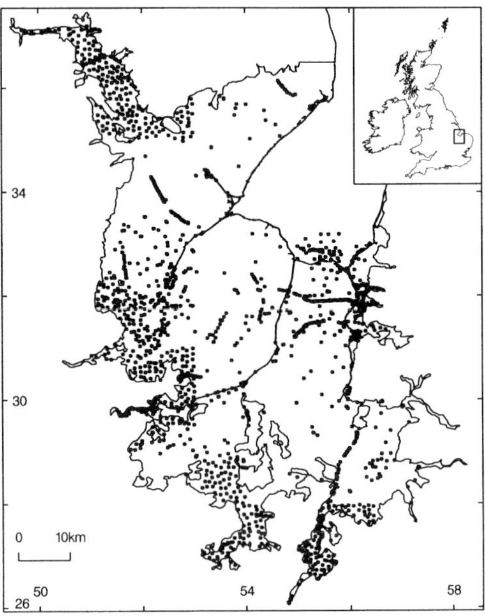

Fig. 2. Fenland: positions of the boreholes held in the lithological database.

radiocarbon dates to provide an alternative interpretation of the coastline positions through the Holocene.

Methodology

Database and modelling

The lithological information is held in a Microsoft ACCESS database and contains over 2000 boreholes (logs archived at the British Geological Survey and including LOIS cores) which have their tops levelled relative to Ordnance datum (OD) Newlyn (*c.* 0.5 m above mean sea-level in this area). Contour maps of the surface topography and pre-Holocene surface, and an isopachyte map of the Holocene sequence (all constructed using Bentley Microstation^{TM} Terrain Analyst software) are described in this paper. As Fig. 2 demonstrates, there is local paucity of borehole coverage in certain areas of the Fenland, with poor distribution in parts of the southeast and northwest. Some of the holes in the north-central part of the Fenland did not reach the base of the Holocene sequence, reducing the number of holes in this area used to model the pre-Holocene surface.

Particle size, major element chemistry and clay mineralogy

High-resolution particle size data were obtained in the 0.1–850 μm range using a Coulter LS-130 laser granulometer, which relies on scattering of laser light by the sediment sample in suspension. The recorded angle of diffraction was converted to particle volume using calculations based on the Fraunhofer model. From this, the equivalent spherical diameter was obtained and the results plotted as a differential particle size distribution curve. Bulk chemistry was analysed by X-ray fluorescence using Philips PW2400 and PW1480 spectrometers controlled by Philips X40 and SuperQ software, respectively (Ridgway *et al.* 1998). The major elements analysed were TiO_2, Al_2O_3, Fe_2O_3, MnO, MgO, CaO and K_2O. SiO_2 was not analysed, but estimated by assuming that the unanalysed component (subtracting percentage loss on ignition) is mainly SiO_2. Clay mineralogy was analysed by X-ray diffraction using a Siemens D5000 diffraction system and Siemens Diffrac Plus processing software. The $<16\,\mu$m fraction, rather than the $<2\,\mu$m fraction, was used to characterize all the clay minerals in the fine silt and clay fraction (to supplement whole rock mineralogical analyses of this size fraction in another study). Since the samples analysed share an almost identical

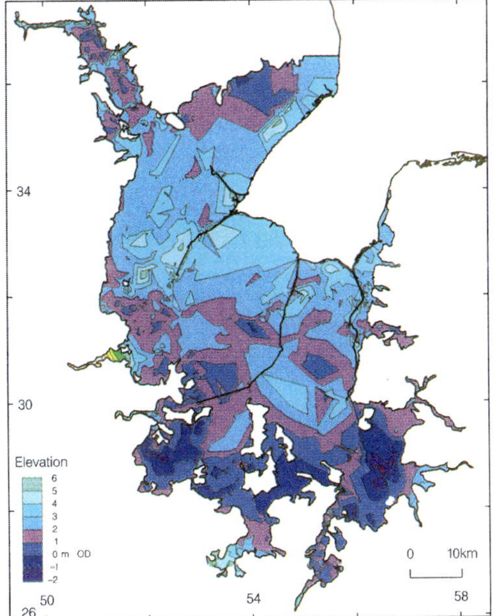

Fig. 3. Elevation model of the Fenland ground surface at 1 m intervals relative to OD.

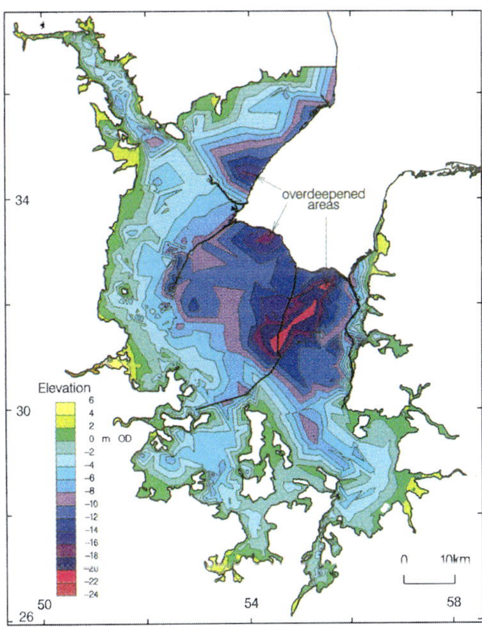

Fig. 4. Elevation model of the pre-Holocene surface at 2 m intervals relative to OD.

particle size distribution between 0.1 and 16 μm, differences in clay mineralogy due to variations in particle size will be minimal. The procedures of Brindley & Brown (1980) were used to determine the clay minerals present, and a semi-quantitative estimation of their relative abundances made, based on the procedure of Weir *et al.* (1975). Further details of methodology can be found in Ridgway *et al.* (1999).

Modelling results

Topography

A prerequisite for the accurate calculation of the thickness of the Holocene sequence is a reliable model of the surface topography. The surface elevation model shown in Fig. 3 was created using the elevation of the boreholes in the database with any made-ground removed. Apart from isolated highs and lows, the surface of the Fenland shows a general tendency to lower towards the south before rising again in the river valleys. The highest parts of the Fenland (5–6 m OD) occur along the lower parts of the modern river systems entering The Wash, particularly the River Welland. The Tofts ridge in northwestern Fenland (Fig. 1), is also modelled, as is the low-lying area behind it.

The inland lowering of the ground surface may be explained by three factors. Firstly, the larger

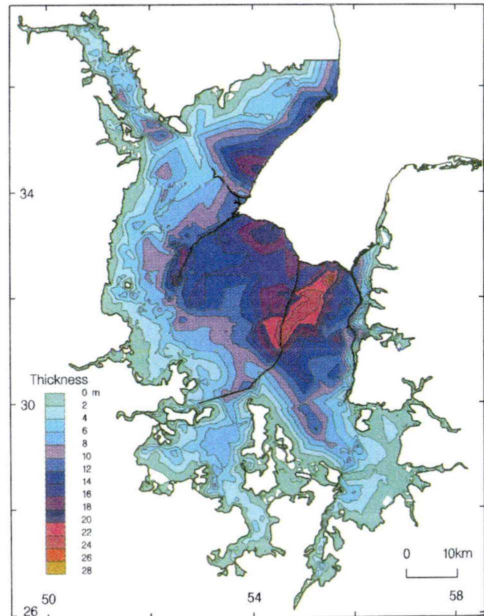

Fig. 5. Isopachyte contours of the Holocene sequence. The calculated volume of sediment preserved is approximately 24 km^3. This volume does not include the Holocene sediment of The Wash (the modern coastline is the seaward boundary of the model) and stops at northing 365 000 at the transition with the Lincolnshire Marsh sequence.

quantities of compressible mud and peat that are preserved in the Holocene sequence of the southern Fenland compared to the relatively incompressible sandier sediments of the northern Fenland (Fig. 6), has led to differential consolidation. Large areas of the southern Fenland have subsided below OD, with extreme lows in the south-eastern corner (−1 to −2 m OD) related to a subsurface composed of 100% peat. Secondly, many parts of the southern Fenland have suffered from surface peat wastage (loss). Third, the accumulation of sediment in the more seaward areas took place at a later date than sediment accumulation further landward (see later section on Fenland palaeocoastlines) and has therefore been deposited under the influence of a higher relative sea-level.

Pre-Holocene surface

The pre-Holocene surface is mainly composed of either Jurassic bedrock, Pleistocene till, sands and gravels, laminated clays, or heterogeneous sandy clays known as the Crowland Beds (Booth 1982). The surface in the Fenland deepens northeastwards (Fig. 4). Adjacent to the southeast corner of The Wash there is evidence for a southwest–northeast aligned depression, which cuts across the eastern Fenland. It is closed at both ends and reaches a maximum depth of about −24 m OD. This depression is interpreted as a former palaeovalley of an ancient Great Ouse–Nene river system, which has been modified and overdeepened by later marine tidal scour processes, creating a closed form. The shape of this depression does not, therefore, reflect the original topography in this area at the opening of the Holocene. This may also be the case for two other deep areas (up to −18 m OD) located at the seaward ends of the modern Rivers Welland and Witham (Fig. 4). Without chronological control, the timing of the scour is difficult to ascertain, a problem also encountered with similar tidally scoured hollows in The Wash (Brew 1997). Elsewhere in the Fenland the modelled pre-Holocene surface is considered to reflect the topography at the beginning of the Holocene.

The main overdeepened area represents the seaward convergence of two separate Great Ouse and Nene palaeovalleys, which extend into the southeast and southwest corners of the Fenland, respectively (Fig. 4). These valleys were important as conduits (in the form of tidal channels) for the waters and sediment of the Holocene transgression entering these areas, and have had a significant control on the lithostratigraphy.

Sediment thickness

The models of surface topography and the pre-Holocene surface have been subtracted to create a model of the total thickness of the Holocene sediment (Fig. 5). The resulting isopachs generally reflect variations in the pre-Holocene surface with the thickest sediment (up to 28 m) occurring where the pre-Holocene surface is lowest. Based on this model, approximately 24 km^3 of Holocene sediment is preserved beneath Fenland, including the riverine alluvium. This value must be considered to be a minimum estimate of the total volume of deposited sediment because it does not take into account any wastage (loss) of surface material that has occurred, either through natural or man-made processes.

Sedimentary facies

The Holocene sequence is divided into four main lithofacies; intertidal mud, intertidal-marine sand, peat and riverine alluvium. The mud facies is distributed across the landward parts of the Fenland, where it is generally less than 6 m thick (Figs 5 and 6). It is composed of mainly muddy sediments with minor sand. Palaeoecological data from LOIS cores (Brew unpublished data) show that the facies was deposited mainly in low energy intertidal mudflat or saltmarsh settings. A number of peat layers (the peat facies), including a basal peat, are intercalated within the mud. Towards the extreme edge of Fenland, all the peat layers merge to form a Holocene sequence composed entirely of the peat facies.

The north–central and northwestern Fenland are underlain by the sand facies (Fig. 6). It is generally greater than 10 m in thickness (Fig. 5) and composed of mainly sand and silty sand. Palaeoecological data from LOIS cores (Brew unpublished data) show that the facies was deposited in intertidal sandflat, tidal channel or subtidal settings. The sand facies extends into The Wash, where it has been recorded on numerous seismic profiles, which reveal periods of widespread sedimentation separated by periods of both local and regional erosion (Brew 1997). Indeed, the processes that led to the deposition of the sand facies may have eroded parts of the mud and peat facies in the central part of the Fenland. This erosion is interpreted to have been caused by channel migration. Elsewhere, the vertical and lateral relationships between the sand and mud–peat facies are complicated, although an interdigitated boundary is considered likely.

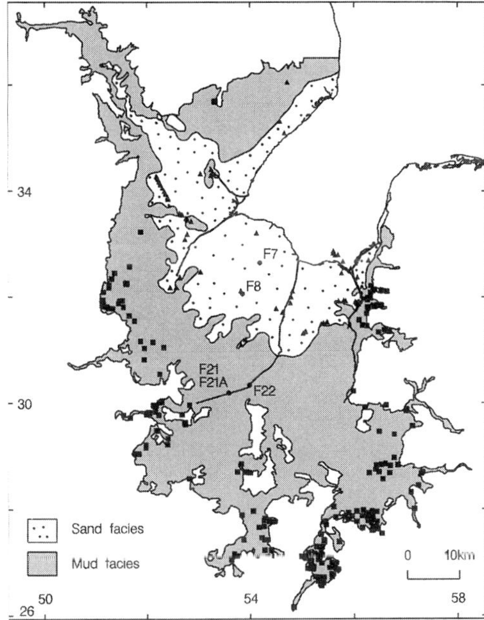

Fig. 6. Distribution of the mud and sand facies. Triangles show the locations of the boreholes where the sand facies overlies the mud facies. Squares show the locations of the boreholes where the sequence is 100% peat facies. Note that the alluvial facies is not shown on this map.

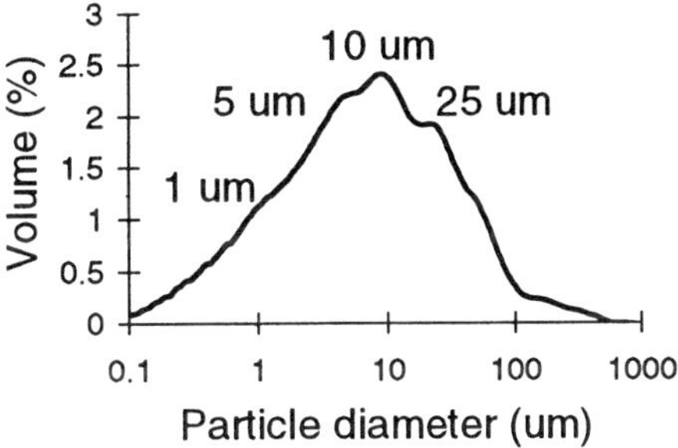

Fig 7. Example of a typical particle size distribution curve for the Holocene fine grained sediment of the Fenland. The curve is divided into 100 equal logarithmic size classes between 0.1 and 900 μm, and displays modes (maxima) and turning points (a shoulder-like portion of the curve). Modes are located where $dy/dx = 0$ and dy/dx changes from positive to negative as x increases. Turning points are located where $d\rho/dx = 0$ (ρ represents the local radius of curvature of the graph) and dy/dx remains unchanged in sign as x increases. The modes and turning points represent an enhanced concentration of particles of a specific particle size. In the fine fraction these may result from the presence of minerals of limited particle size range.

The alluvial facies occurs in the river valleys entering the Fenland and is composed of a heterogeneous mix of mud and sand with subordinate gravel. The alluvial facies was deposited by the modern rivers and will not be discussed in this paper.

Mud facies

Particle size. The mud facies is dominated by clays and fine silts, with subordinate coarse silt and minor sand. In almost all the samples from the mud facies, the fine fraction is typified by modes and turning points clustering around 1, 5, 10 and 25 μm. In many cases, the percentage volume of each mode or turning point is relatively constant, giving rise to certain common particle size distribution curves. The most common is a broad unimodal curve with the mode at 10 μm, turning points of roughly equal height on the falling limbs at 5 and 25 μm, and a third turning point at 1 μm (Fig. 7). A slight variation occurs where the 25 μm feature is larger and equals the mode at 10 μm, whereby the curve becomes bimodal. Core F21A at Adventurers' Land in the Nene Valley (Fig. 1) illustrates these particle size distribution curve types (Fig. 8) and displays the main recurring features that are present in muddy samples elsewhere in the Fenland. In this core, the particle size of the Holocene sequence is very uniform, excepting slight changes immediately adjacent to the peat layers.

By averaging the frequency curve data for groups of cores (mainly Holocene sediments, but also including some pre-Holocene samples), composite curves have been created to indicate the most common particle size categories (Fig. 9). It should be noted that these curves do not provide a quantitative representation of the total sediment column, but indicate the main regional trends. As expected, muddy sediments are concentrated in landward areas (compare with Fig. 6), and, in addition to the features described at Adventurers' Land, include layers of coarse silt with a composite mode (or turning point) at 58 μm (Fig. 9). The commonly recurring particle size characteristics of the mud facies indicate that the fine silts and clays were not sorted as they were transported in suspension (probably as flocs).

Major element chemistry. The chemistry of the Fenland sediment is largely dependent on

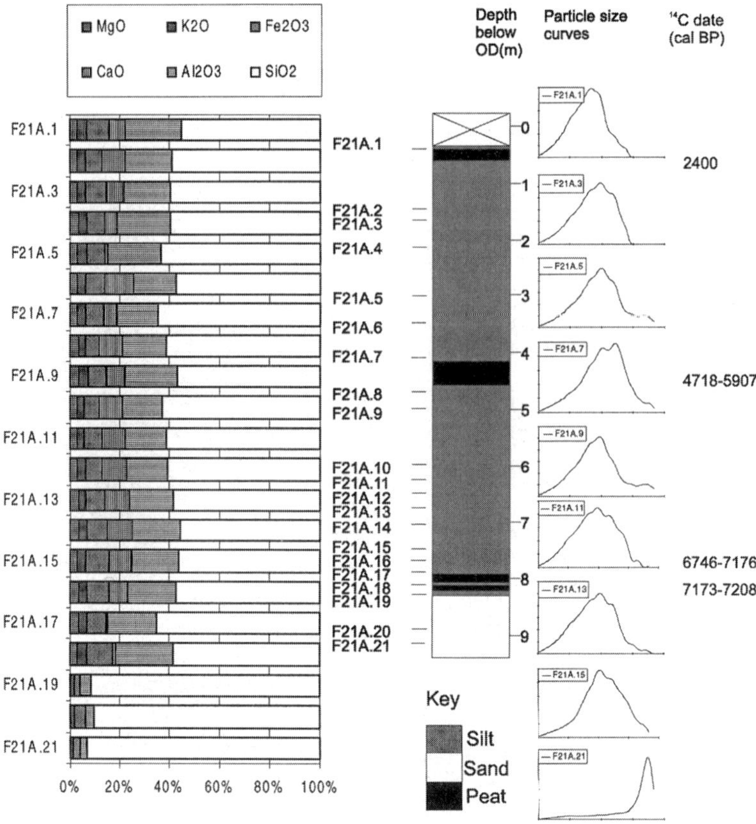

Fig. 8. Core F21A at Adventurers' Land showing particle size distribution curves for selected samples in the mud facies, and down-core changes in major element oxide composition. Curves have same axis scales as Fig. 7. Location of the core is shown in Figs. 1 and 6.

particle size: sandy sediments contain a high proportion of silica, whereas clays and silts are enriched in alumina (and other elements associated with clay minerals). For this reason, the mud facies has a low silica : alumina ratio, which appears to demonstrate little vertical variation within the Holocene sequence (Fig. 10a). Higher silica : alumina ratios are found in sands and till directly underlying the basal peat, because of their greater sand component. Similar vertical homogeneity can be seen in the concentrations of MgO, K_2O and Fe_2O_3 (e.g. Adventurers' Land, Fig. 8). However, the percentage of CaO varies because of decalcification (commonly near to peat horizons) and enhancement of shelly material in some layers.

Clay mineralogy of the <16 μm fraction. The mud facies in cores F21A and F21 (a few metres apart at Adventurers' Land in the Nene Valley, Fig. 1) are dominated by illite (45–59%) with subordinate expandable clays (15–29%), kaolinite (13–21%) and minor chlorite (6–13%) (Fig. 11). Because the proportion of expandable clays can vary depending on the degree of weathering, illite : kaolinite ratios were also examined. Three samples from above the lowest intercalated peat have illite : kaolinite ratios of about 3 : 1 (Fig. 12e). Given the uniformity of the chemistry and particle size of the Holocene sequence in core F21A, it seems likely that the clay mineralogy of these samples is representative of a large part of the mud facies at Adventurers' Land. Their composition lies between that of the Devensian tills to the north of the Fenland and the Anglian tills to the east, with the Devensian component having the stronger influence (Fig. 12a, b). Similar illite : kaolinite ratios of 3 : 1 were obtained from mud directly beneath the basal peat in core F21, and also on

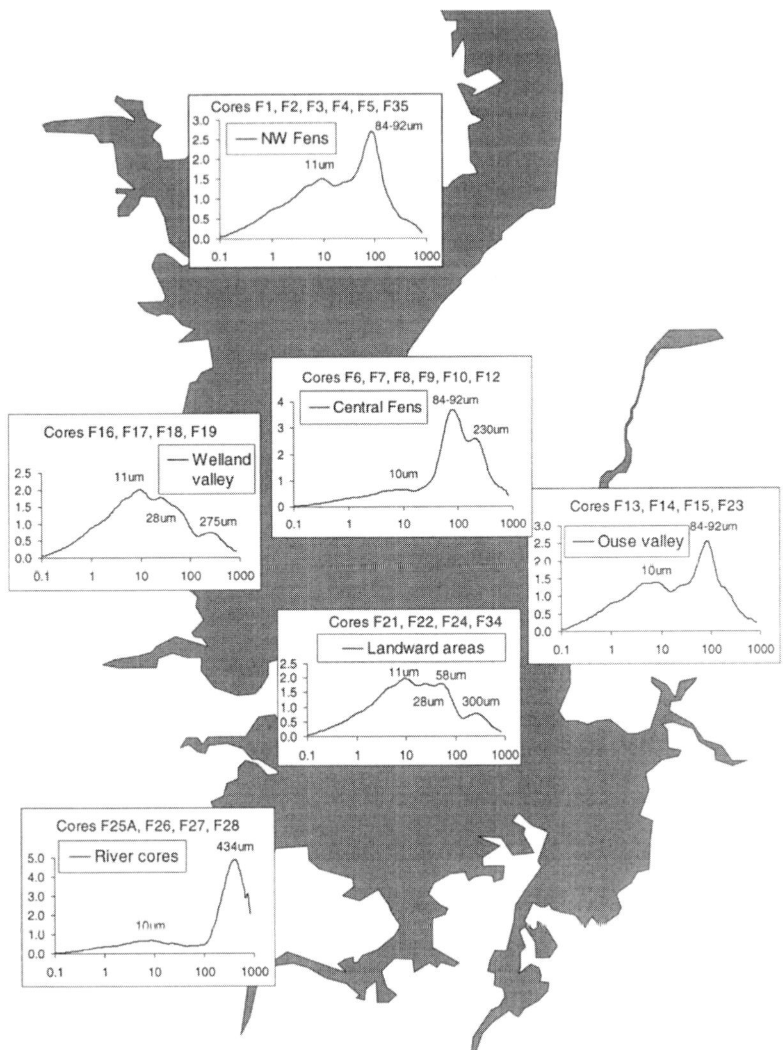

Fig. 9. Composite particle size curves for five Fenland regions and a sixth group of combined size data from beyond the tidal limits in the four river valleys (cores F25A, F26, F27, F28). Locations of the cores are shown in Fig. 1.

the northern flank of the Welland Valley (core F17) and Gedney Fen (core F8) (Fig. 12d). The mud facies was probably derived from a mix of till sources exposed along the adjacent coasts and offshore areas, and from erosion of the pre-Holocene surface in the embayment itself.

In contrast to the rest of the mud facies at Adventurers' Land, a thin mud layer between the basal peat and the lowest intercalated peat (Figs 8 and 12e) has a relatively low illite : kaolinite ratio. This is also the case for the lowest parts of the mud facies in core F16 in the Welland Valley (Fig. 12f). Even lower ratios are found from mud samples directly beneath the basal peat in certain parts of the Welland Valley (Fig. 12d). The Oxford Clay in the upper reaches of the Welland Valley (F25A, Fig. 1) has a very low illite : kaolinite ratio of about 1:1 (Fig. 12c). These data suggest that early in the Holocene, the mud facies in the landward parts of the Nene and Welland palaeovalleys represents a combination of glacial sources (with higher illite : kaolinite ratios) and local fluvially reworked and/or *in situ* Oxford Clay sources (with lower ratios). It should be noted that these lower mud facies sediments (Fig. 12e, f) also have similar

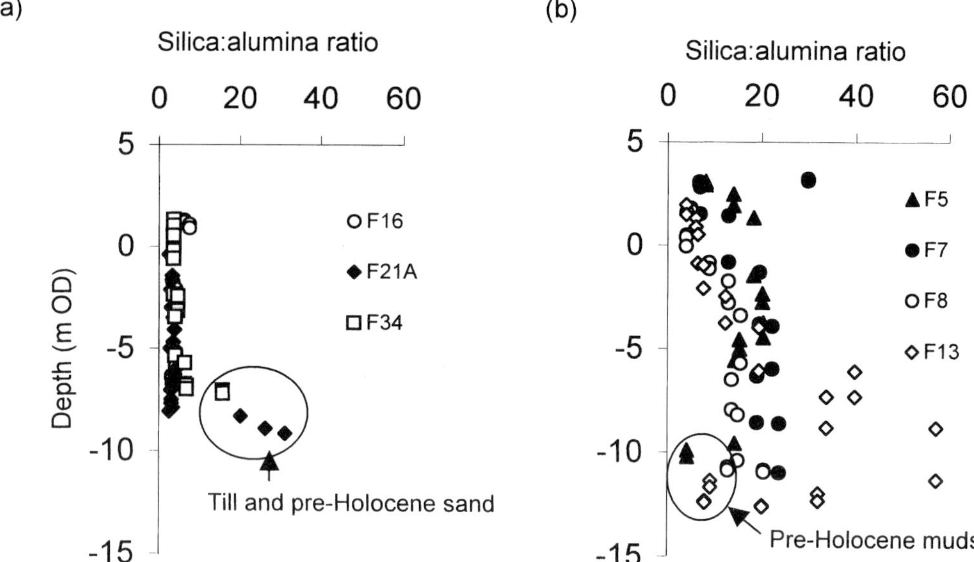

Fig. 10. Silica : alumina ratio of (**a**) landward cores F16, F21A and F34; (**b**) seaward cores F5, F7, F8 and F13. Pre-Holocene samples are circled. Locations of the cores are shown in Fig. 1.

illite : kaolinite ratios to selected tills from the Holderness coast (Fig. 12a). The latter deposits would have been well mixed in the North Sea sediment pool prior to deposition in the Fenland, and it is more likely that a contribution of Oxford-Clay-derived sediment has controlled local variations in the illite : kaolinite ratios.

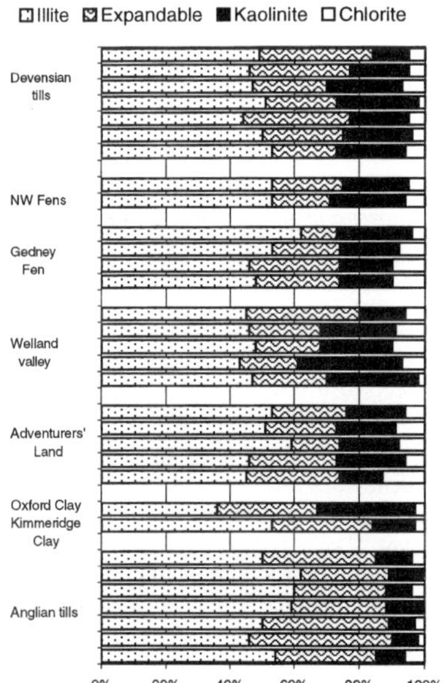

Fig. 11. Clay mineralogy of the <16 μm fraction of Fenland Holocene and pre-Holocene samples.

Sand facies

Particle size. The sand facies is dominated by clean, well-sorted sands with rare mud laminae. Composite particle size curves reflect the dominance of sand-grade sediment in the north-central Fenland (Fig. 9). On the eastern and northwestern margins of the Fenland the curves indicate that the sand facies has a mode between 84 and 92 μm (very fine sand). The more central areas have an additional mode at 230 μm (coarse end of the fine sand range), which relates to the lower part of the facies. In contrast, the sands from the river valley cores are mainly medium-grained (430 μm, Fig. 9).

West of the River Nene (Fig. 1), the sand facies tends to fine upwards with few obvious erosional surfaces or breaks in sedimentation. However, east of the River Nene, erosional horizons, associated with shelly lags, mud clasts, organic matter, scattered pebbles and gravel bands, are more common. Where the sequence is disrupted by erosive episodes, a layer of coarser sediment is commonly found, generally followed by a fining upward sequence until the next erosive episode. Despite these complexities, the overall particle size trend is fining upwards, from

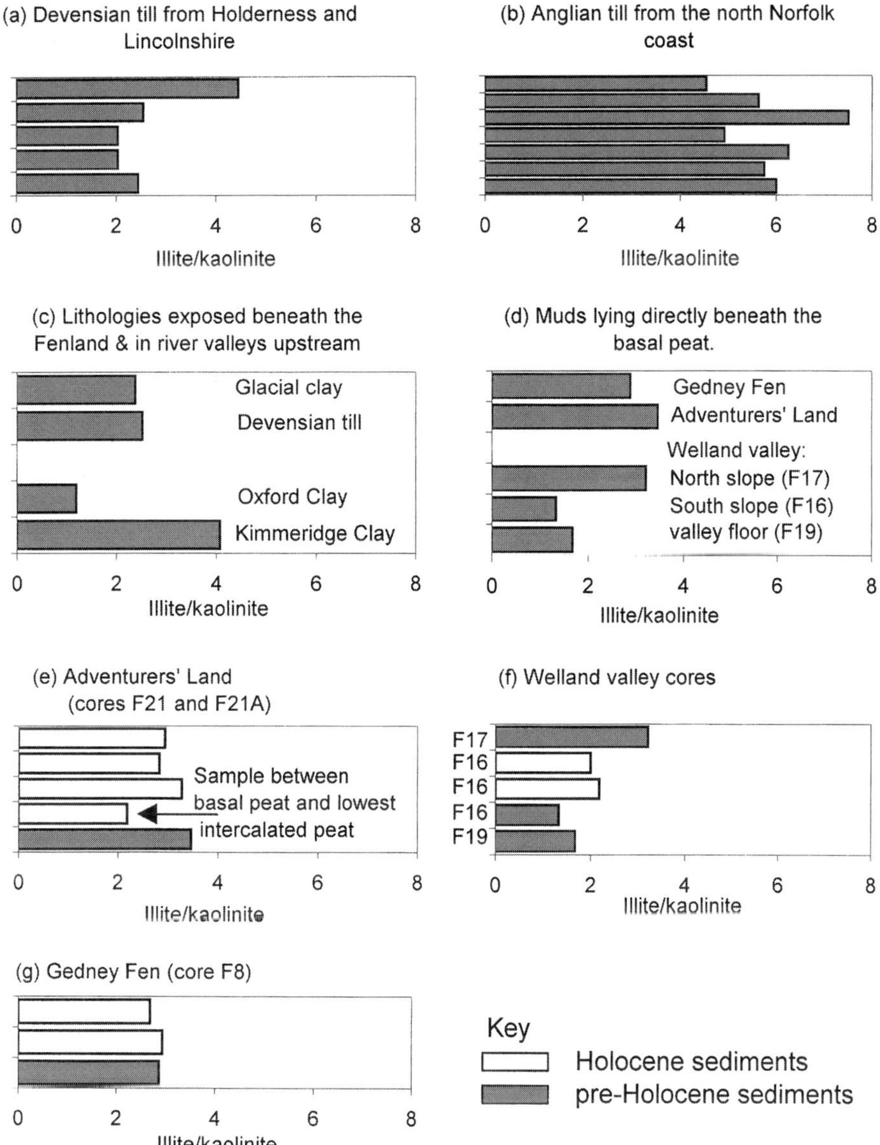

Fig. 12. Illite : kaolinite ratios of the <16 μm fraction of Fenland Holocene and pre-Holocene samples. Note that all the pre-Holocene muds shown in (d) are duplicated in (e), (f) and (g) so both local and regional comparisons can be made.

medium or coarse sand at the base of the Holocene sequence to very fine sand and coarse silt near the surface (Figs 13 and 14). Across most of the northern Fenland, the sand facies passes into silt-dominated sediments near the surface (Fig. 14).

Major element chemistry. The sand facies is primarily composed of quartz with minor feldspar, and hence low Al_2O_3 (alumina), K_2O, MgO and Fe_2O_3 contents. As a result, the silica : alumina ratio is relatively high compared with that in the mud facies (Fig. 10b). Towards the top of the sand facies, the chemistry alters as the dominant grade of sediment changes from sand to silt. At Gedney Fen (core F8, Fig. 1), increasing alumina towards the surface corresponds with a decrease in silica (Fig. 14). This is

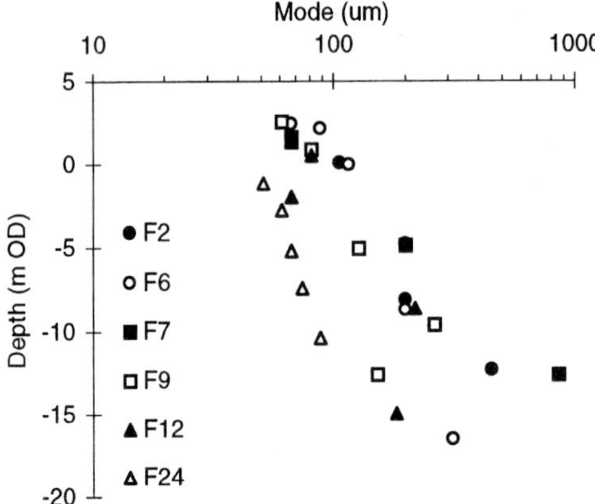

Fig. 13. Modal particle size of cores from the seaward areas showing a fining upwards trend. Note that core F24 is mainly mud facies composed of coarse silt, which can be hydraulically sorted. Locations of the cores are shown in Fig. 1.

accompanied by a decrease in CaO due to decalcification of the surface silts, and an increase in Fe_2O_3.

The effects of hydraulic sorting dominate the chemical variations within the Holocene sequence of the Fenland. To compare the chemical signatures of the mud- and sand-sized sediments together, particle size sorting effects were reduced by comparing appropriate elemental ratios, such as aluminium : titanium (Al : Ti). Figure 15 shows that the bulk of the sediments have Al : Ti ratios that cluster between 10 and 20, suggesting homogeneity between the mud and sand facies. It would appear, therefore, that the bulk of the sediment has been well mixed (tidally) prior to its deposition, and that the mud and sand facies are simply facies with a lateral transition. The few anomalous Al : Ti values (Fig. 15) are generally located in the river valleys around the Fenland margin, where it is believed the river sediments are poor in titanium bearing minerals.

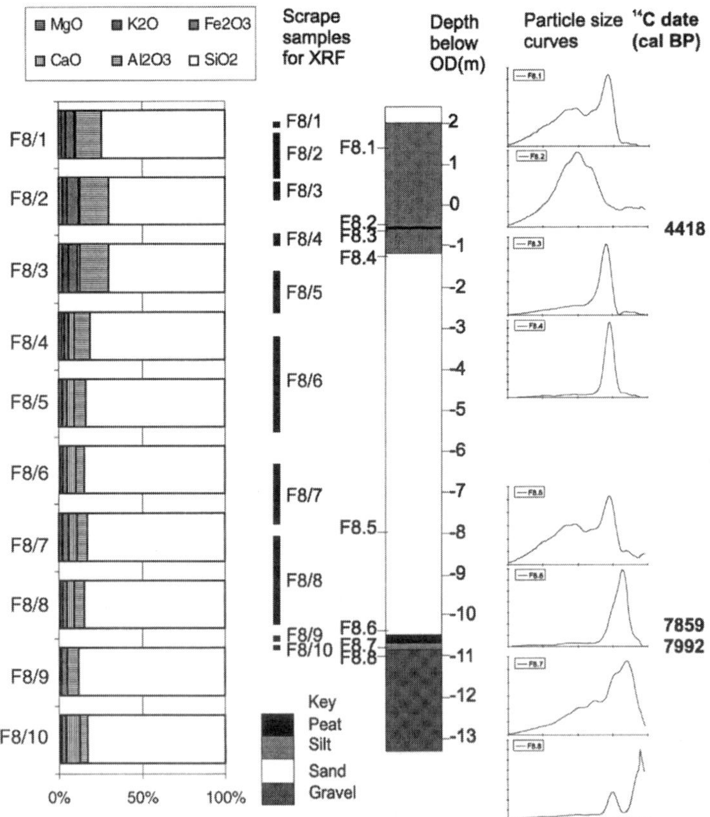

Fig. 14. Core F8 at Gedney Fen showing particle size distribution curves for selected samples in the sand facies, and down-core changes in major element oxide composition. Curves have same axis scales as Figs 7 and 8. Location of the core is shown in Figs 1 and 6.

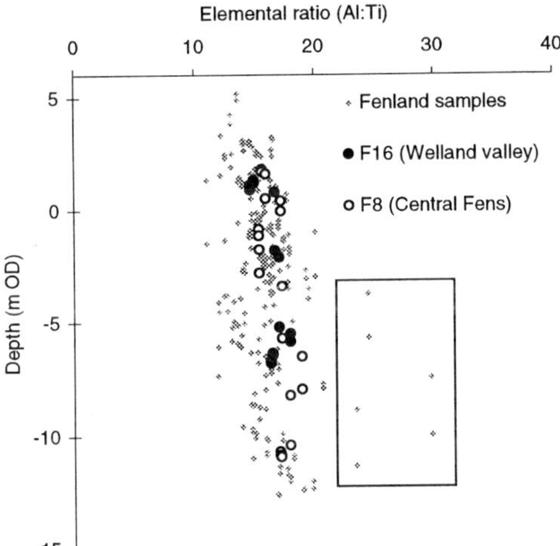

Fig. 15. The ratio of Al : Ti in Fenland cores, highlighting the similarity between muddy landward sequences (e.g. F16) and sandy seaward sequences (e.g. F8). Note that anomalous values (within rectangle) are primarily from marginal sandy horizons, and probably contain fluvially derived sediment.

Clay mineralogy of the <16 µm fraction. At Gedney Fen (core F8), muds in the sand-dominated Holocene sequence have illite : kaolinite ratios of about 3:1 (Fig. 12g). In this respect, they are similar to the ratios in the bulk of the mud facies at Adventurers' Land (F21A, Fig. 12e). This relative uniformity of much of the sequence at both Gedney Fen and Adventurers' Land implies that, on a regional scale, clay-grade sediment entering the embayment during the Holocene has maintained a constant composition. The main source lithologies supplying this sediment (till deposits eroded from the North Sea coast and sea bed) have not changed. However, in the lower parts of the sequence, variations exist due to early fluvial inputs and/or reworking of local contrasting lithologies on the Fenland floor.

Peat facies

The peat facies is found intercalated within the mud facies and occasionally within the sand facies. The number, lateral extent and elevation of peat layers varies across the Fenland making lithostratigraphic correlation difficult (Wheeler & Waller 1995). However, the peat layers are important as they have been radiocarbon dated (Table 1) to provide a chronology for the evolution of the Fenland into which the sedimentary characteristics of the clastic units are fitted.

Basal peat. The basal peat (Fig. 16) is widespread around the western and southern Fenland. In contrast, it is very patchy along the eastern edge of the Fenland and in the central Fenland. Mean calibrated dates (BP) of the upper surface of the peat (Fig. 16) show that the sea entered the Fenland embayment *c.* 7850 cal. a BP and quickly flooded the central and eastern sectors. It is possible that this rapid flooding did not allow significant accumulation of the basal peat in these areas (Waller 1994*c*). After *c.* 4400 cal. a BP the sea finally flooded the western and southern sectors, where the basal peat is well preserved (possibly through development over a longer period).

Intercalated peats. A number of intercalated peats occur in the western Fenland (between the Rivers Welland and Nene) at elevations between *c.* −2 and −5 m OD. Within what is a complicated stratigraphy, a laterally consistent peat layer (*c.* −3 to −5 m OD) occurs at the landward end of the Nene palaeovalley. The base of this peat appears to be diachronous and has dates between 5907 and 4847 cal. a BP and the top between 4722 and 4528 cal. a BP (Fig. 17).

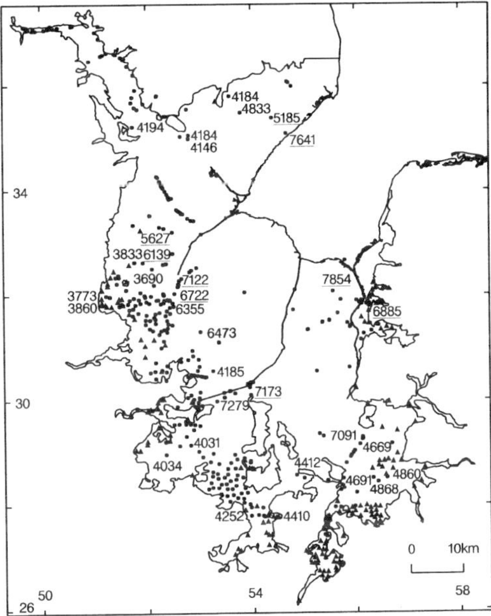

Fig. 16. Location of boreholes proving the basal peat, including merged peats (triangles), at the extreme edge of Fenland. The mean calibrated dates (BP) of the upper contact of the peat are also shown. These dates are compiled from the *good* positive sea-level tendencies of Waller (1994*d*) and nine new LOIS dates, which are underlined.

Table 1. *LOIS radiocarbon dates from the Fenland*

	Laboratory code	^{14}C age BP ± 1σ (a)	Calibrated age (a BP)			Altitude (m OD)	Tendency of sea-level	Change in mean sea-level from present (nsl (m) ± error)
			Min.	Median	Max.			
Sea-level index points*								
Adventurers' Land, F21	AA22362	6310 ± 65	7026	7208	7367	−7.88	+	−11.68 ± 0.20
Adventurers' Land, F21	AA22669	6255 ± 55	7010	7173	7233	−7.76	+	−11.36 ± 0.20
Adventurers' Land, F21	AA22668	6265 ± 50	7018	7176	7232	−7.69	−	−11.67 ± 0.20
Adventurers' Land, F21	AA22361	5925 ± 65	6576	6746	6890	−7.56	−	−11.16 ± 0.20
Adventurers' Land, F21	AA22360	5130 ± 60	5736	5907	5989	−4.54	−	−8.52 ± 0.20
Adventurers' Land, F21	AA22359	4165 ± 55	4459	4718	4839	−4.23	+	−7.83 ± 0.20
Adventurers' Land, F21	AA26362	2435 ± 50	2343	2400	2720	−0.33	−	−4.13 ± 0.40
Clenchwarton, F13	AA22355	7215 ± 70	7840	7960	8127	−12.55	+	−16.35 ± 0.20
Clenchwarton, F13	AA22354	7035 ± 65	7668	7854	7935	−12.41	+	−16.01 ± 0.20
Cowbit, F16	AA22358	6080 ± 60	6785	6899	7157	−6.59	+	−10.39 ± 0.20
Cowbit, F16	AA26357	5875 ± 75	6491	6722	6874	−6.5	+	−10.10 ± 0.20
Cowbit, F16	AA22357	5635 ± 65	6296	6412	6610	−6.27	−	−10.28 ± 0.20
Cowbit, F16	AA22356	5095 ± 55	5723	5819	5939	−5.92	+	−9.52 ± 0.20
Eastville, F3A	AA26356	3180 ± 65	3219	3378	3549	0.12	−	−3.82 ± 0.81
Gedney Fen, F8	AA22675	3985 ± 45	4298	4418	4534	−0.56	−	−4.35 ± 0.40
Gedney Hill, F34	AA26372	5775 ± 65	6414	6592	6737	−5.65	+	−9.45 ± 0.20
Gedney Hill, F34	AA26371	5695 ± 65	6314	6473	6665	−5.59	+	−9.19 ± 0.20
Gosberton, F17	AA26361	4845 ± 55	5340	5592	5702	−3.05	+	−6.85 ± 0.20
Gosberton, F17	AA26360	4895 ± 55	5490	5627	5736	−2.99	+	−6.59 ± 0.20
Guyhirn, F22	AA26359	2080 ± 50	1896	2007	2145	−0.3	−	−4.28 ± 0.20
Guyhirn, F22	AA26358	1220 ± 60	976	1161	1280	1.07	+	−2.53 ± 0.20
Pinchbeck, F18	AA22671	5890 ± 55	6564	6728	6856	−5.35	+	−9.15 ± 0.20
Oinchbeck, F18	AA22670	5345 ± 55	5949	6139	6283	−5.17	+	−8.77 ± 0.20
South Lynn, F15	AA26366	6045 ± 75	6730	6885	7155	−6.1	+	−9.90 ± 0.20
South Lynn, F15	AA26364	2925 ± 50	2885	3067	3213	−0.57	−	−4.55 ± 0.20
South Lynn, F15	AA26363	1940 ± 60	1719	1874	1995	0.25	+	−3.85 ± 0.20
Spalding, F19	AA22364	6270 ± 70	7004	7178	7279	−8.4	+	−12.20 ± 0.20
Spalding, F19	AA22363	6230 ± 80	6895	7122	7265	−8.2	+	−11.80 ± 0.20
Thorpe Culvert, F1	AA26355	2920 ± 60	2870	3065	3248	−0.17	−	−3.96 ± 0.20
Wrangle Bank, F4	AA26373	4495 ± 85	4864	5185	5445	−2.5	+	−5.45 ± 0.20
Wrangle Bank, F4	AA26374	4735 ± 60	5311	5526	5596	2.58	−	−5.90 ± 0.20
Wrangle Lowgate, F5	AA22366	6860 ± 70	7540	7641	7791	−9.89	+	−12.84 ± 0.40
Wrangle Lowgate, F5	AA22365	6920 ± 75	7558	7682	7898	9.54	−	−12.69 ± 0.40

* Locations of the cores are shown in Fig. 1.

† The tendency of sea-level describes whether the point records an increase or decrease in water level: +ve, increase, −ve, decrease.

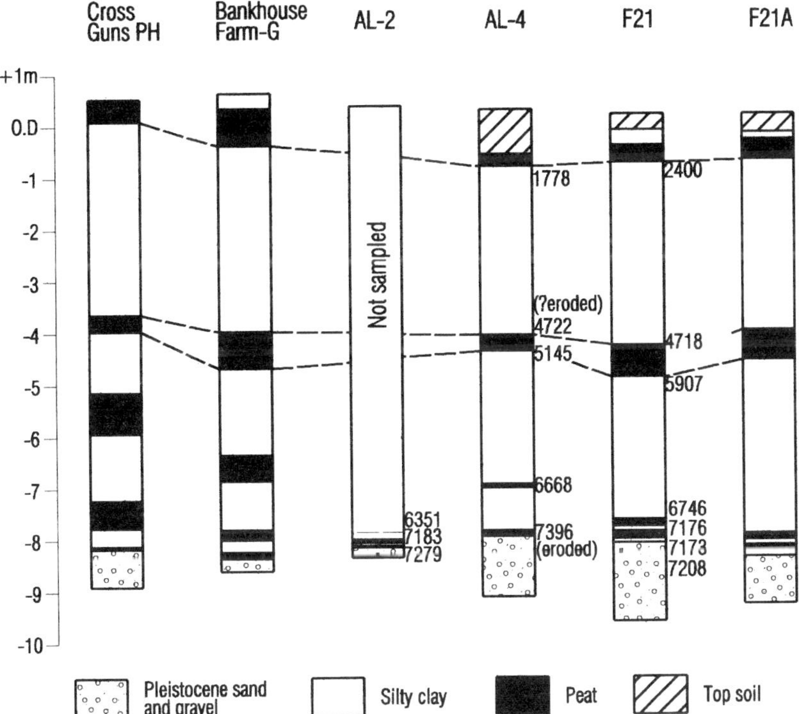

Fig. 17. Logs for six holes in the River Nene palaeovalley at Adventurers' Land illustrating the elevation and mean calibrated dates (BP) of the intercalated and basal peats. F21 and F21A are LOIS holes. AL-2 and AL-4 are described in Shennan (1986*a*). Cross Guns PH and Bankhouse Farm-G are described in Godwin & Clifford (1938).

Peat layers at the landward end of the Welland palaeovalley occur at slightly higher elevations (*c.* −2 to −3 m OD). Reliable dates for these peats are presently unavailable. However, Shennan and Alderton (1994) suggest a possible correlation with the Nene palaeovalley peats based on a poorly validated date of 5013 cal. a BP for the top of a peat at Spalding. If this correlation is correct, then it is possible that the varying elevations of the peat may be due to differences in consolidation across the area post-5 cal. ka BP.

The uppermost and most widespread intercalated peat layers occur between *c.* −2 and +2 m OD across most of the southern Fenland and parts of the northwestern Fenland (Fig. 18). Interrogation of the database shows that in any one core only a single layer of peat occurs between these levels. This peat layer has been lithostratigraphically defined as the Nordelph Peat by British Geological Survey (1978) and Gallois (1979) based on its distribution along the eastern Fen-edge. According to Wheeler & Waller (1995) this definition is flawed for two main reasons. Firstly, chronological connotations are implicit in the definition, and, secondly, the definition is not easily adapted to other areas of the Fenland.

Mean calibrated dates from the lower contacts of the −2 to +2 m OD peats vary from *c.* 4.4 to 2 cal. ka BP (Fig. 18). Accumulation began earliest, between *c.*4.4 and 3.9 cal. ka BP, in eastern and southern Fenland, followed a few hundred years later, *c.* 3650–3050 cal. a BP, in western and northwestern Fenland, with the latest formation in central Fenland between *c.* 2.8 and 2 cal. ka BP. The ages of the top of these peats (Fig. 19) suggest a complicated pattern of resubmergence starting at *c.* 2750 cal. a BP in the western Fenland. A cluster of dates between *c.* 2250 and 1950 cal. a BP in central and southeastern Fenland suggest a later, but fairly rapid, reflooding of these areas.

Holocene palaeocoastline reconstruction

The timing of the post-glacial transgression into the Fenland is recorded from dating of the

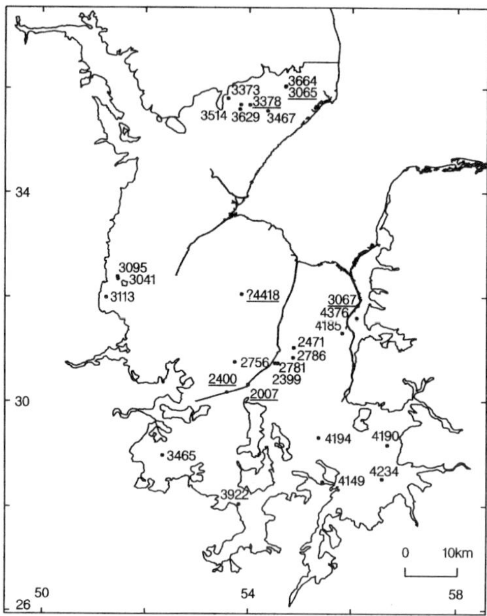

Fig. 18. Mean calibrated dates (BP) of the lower contacts of intercalated peats at elevations between *c.* −2 and +2 m OD. These dates are compiled from the *good* negative sea-level tendencies of Waller (1994*d*) and six new LOIS dates, which are underlined.

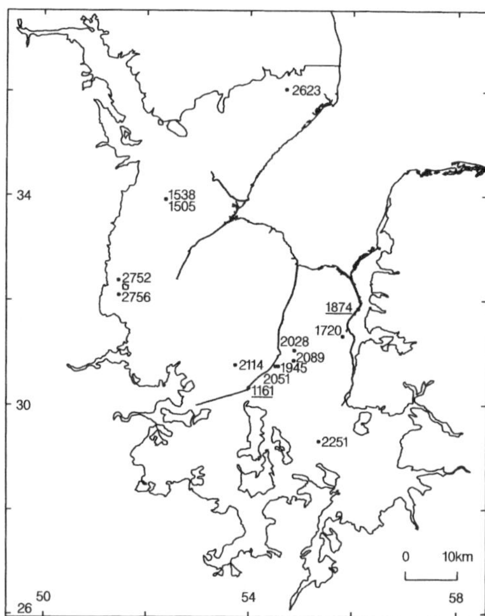

Fig. 19. Mean calibrated dates (BP) of the upper contacts of intercalated peats at elevations between *c.* −2 and +2 m OD. These dates are compiled from the *good* positive sea level tendencies of Waller (1994*d*) and two new LOIS dates, which are underlined.

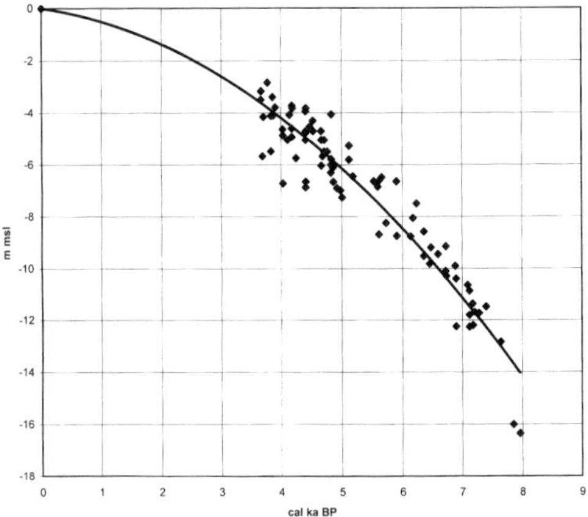

Fig. 20. Sea-level curve for the Fenland relative to mean sea-level (MSL) using dates from basal peats only (see also Shennan *et al.* 1999).

upper contact of the basal peat prior to its inundation by clastic sediments. The basal peat is widespread and differs both in elevation and age across the embayment (Fig. 16). The pre-Holocene surface model (Fig. 4) is important in the reconstruction of palaeocoastlines across the Fenland because it is against this surface that the surface indicating relative sea level at any particular time is intersected, so giving the form of the embayment at that time. Using a high-resolution sea-level curve generated for the Fenland (Fig. 20) (Shennan *et al.* 1999) in combination with the modelled pre-Holocene surface, lithostratigraphical and sedimentological data, palaeocoastline positions for the embayment at 1-ka 'snap-shots' through the Holocene, are presented. The maps shown have been constructed using mean calibrated dates (Table 1), which have been calculated from the original ^{14}C dates by using the CALIB program 3.0.3 of Stuiver & Reimer (1993).

Palaeocoastlines are taken to be located at about mean high water spring tide (MHWST). The sea-level curve (Fig. 20), which is relative to mean sea-level, was corrected to MHWST relative to OD using an average correction factor computed for three tidal stations in the inner Wash (Tabs Head, Kings Lynn and Boston, Hydrographer of the Navy 1997). The average MHWST for these three stations is 3.83 ± 0.06 m above OD and the average mean tide level (*c.* mean sea-level) 0.51 ± 0.11 m above OD. A correction factor of +3.32 m is therefore applied to mean sea-level to locate the position of MHWST relative to OD. The reconstructions of the palaeocoastlines assume no change in tidal range through the Holocene.

8–7 cal. ka BP

One of the successes of the LOIS radiocarbon dating campaign was the recovery of ten reliable new dates (between 8 and 5 cal. ka BP) from deeper basal peats, enabling a better understanding of the early Holocene evolution of the embayment. Some of these dates (Fig. 16) add new information on the position of the coastline between 8 and 7 cal. ka BP. By 7 cal. ka BP (Fig. 21) most of the eastern and north-central Fenland had been flooded, reaching as far south as Welney Wash along the palaeovalley of the River Great Ouse and Adventurers' Land along the palaeovalley of the River Nene. It is likely that during this period the tidal scour processes began, which led to the overdeepening of the seaward part of the Great Ouse–Nene palaeovalley system. In northwestern Fenland the marine incursion had reached just beyond the position of the modern coastline. At Adventurers' Land there is evidence for the early stages of a local regressive phase where an intercalated peat has been recorded which started accumulating *c.* 7180 cal. a BP (Figs 17 and 21). In all the LOIS cores where a deep, pre-7 cal. ka BP basal peat has been found (F5, F13, F19, F21A, Figs 1 and 16) the immediately overlying clastic sediments are fine-grained and indicative of low energy environments.

7–6 cal. ka BP

The coastline at 6 cal. ka BP is shown in Fig. 22. The initial transgressive episode that was still affecting most parts of the Fenland was interrupted by several local regressive phases during this period. At Welney Wash, an intercalated peat has been recorded which accumulated between *c.* 6690 and 6480 cal. a BP (Waller *et al.* 1994). At Spalding and Cowbit Wash, an intercalated peat has been recorded, which accumulated between *c.* 6410 and 5820 cal. a BP (Figs 22

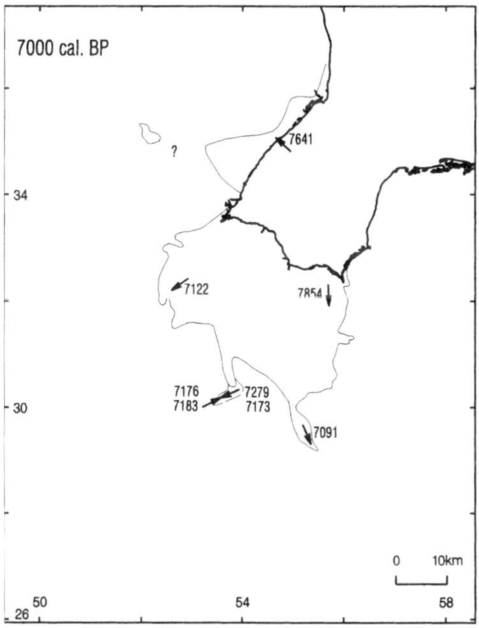

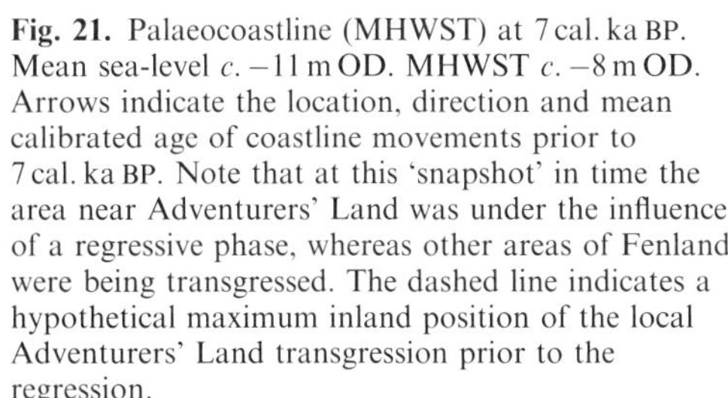
Fig. 21. Palaeocoastline (MHWST) at 7 cal. ka BP. Mean sea-level *c.* −11 m OD. MHWST *c.* −8 m OD. Arrows indicate the location, direction and mean calibrated age of coastline movements prior to 7 cal. ka BP. Note that at this 'snapshot' in time the area near Adventurers' Land was under the influence of a regressive phase, whereas other areas of Fenland were being transgressed. The dashed line indicates a hypothetical maximum inland position of the local Adventurers' Land transgression prior to the regression.

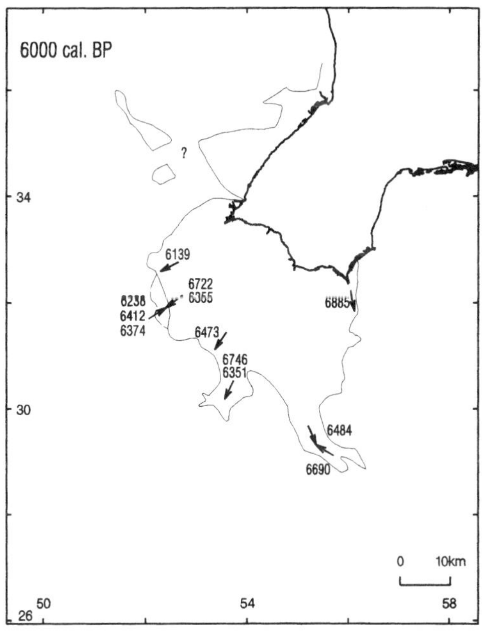

Fig. 22. Palaeocoastline (MHWST) at 6 cal. ka BP. Mean sea-level *c.* −8.5 m OD. MHWST *c.* −5 m OD. Arrows indicate the location, direction and mean calibrated age of coastline movements between 7 and 6 cal. ka BP. The regressive phases at Adventurers' Land and Welney Washes had finished and transgression had resumed. However, at Cowbit and Spalding, regression was ongoing. The dashed line indicates the maximum extent of the local Cowbit Wash–Spalding transgression before peat accumulation began.

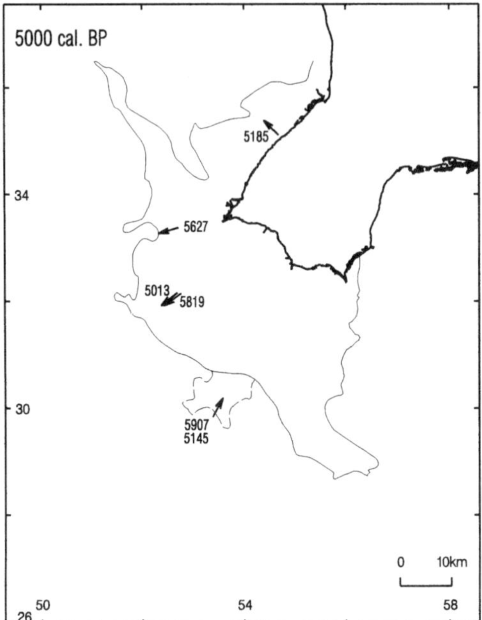

Fig. 23. Palaeocoastline (MHWST) at 5 cal. ka BP. Mean sea-level *c*. −6 m OD. MHWST *c*. −3 m OD. Arrows indicate the location, direction and mean calibrated age of coastline movements between 6 and 5 cal. ka BP. The area near Adventurers' Land was under the influence of a regressive phase, whereas other areas of Fenland were being transgressed. The dashed line positions the coastline at its maximum extent at Adventurers' Land prior to the regression.

and 23). The regressive phase at Adventurers' Land continued until it ended between *c*. 6750 and 6350 cal. a BP. The lateral extent of these three regressive episodes is difficult to establish because only site specific data are available. However, they are all recorded at the landward ends of the main palaeovalleys defined by the pre-Holocene surface model (Fig. 4) and it is possible that differential fluvial input of sediment into these valleys may have contributed to their timing. Indeed, the clay mineralogical evidence from cores F21 and F21A in the Nene palaeovalley (Fig. 1) suggests an enhanced fluvial contribution to the system between accumulation of the basal and the lowest intercalated peats. It may be that the fluvial sediment supply to these landward areas at this time was greater than the increase in accommodation space created through sea-level rise, leading to a regressive tendency.

The position of the coastline in southwestern Fenland at this time represents a departure from the position proposed by Waller (1994*a*). The Waller (1994*a*) model suggested that the sea penetrated into the southwestern corner as far as Chatteris and Ramsey (p. 68, map 4). The lithostratigraphic evidence from this study suggests that the coastline did not progress any further inland than the Adventurers' Land area (Figs 1 and 22).

6–5 cal. ka BP

During this period most of the Fenland was under the influence of a transgressive episode (Fig. 23). However, there is evidence, both from Shennan (1986*a*, *b*) and this study, of a regressive phase at Adventurers' Land between *c*. 5910 and 4720 cal. a BP (Figs 23 and 24), which led to the accumulation of an intercalated peat (possibly the 'Middle' Peat in the Peterborough area of Horton 1989) (Fig. 17). As with the earlier regressive phase that affected this area, it was probably confined to the Nene palaeovalley. At Murrow

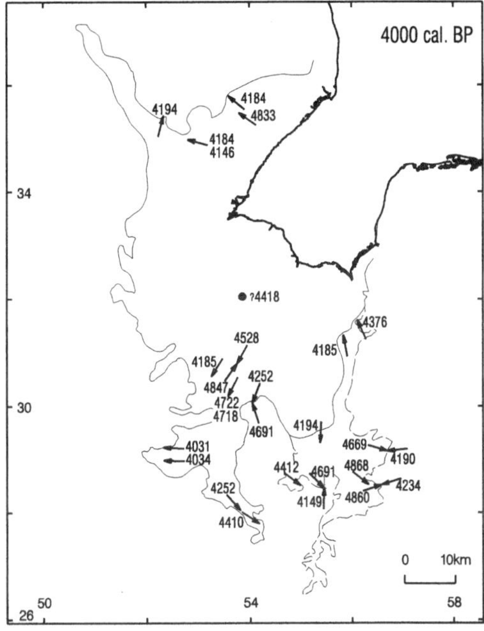

Fig. 24. Palaeocoastline (MHWST) at 4 cal. ka BP. Mean sea-level *c*. −4 m OD. MHWST *c*. −1 m OD. Arrows indicate the location, direction and mean calibrated age of coastline movements between 5 and 4 cal. ka BP. The southeastern Fenland was under the influence of a regressive phase, whereas other areas of Fenland were being transgressed. The dashed line indicates the position of the maximum inland limit of the coastline in southeastern Fenland prior to this regressive phase.

(more seaward in the palaeovalley) the same peat has been recorded slightly higher in the sequence and dated a little later between *c.* 4850 and 4530 cal. a BP (Waller 1994*b*) (Fig. 24). Waller (1994*b*) and Shennan & Alderton (1994) suggested that this regressive phase may have had a more regional signature in areas to the north and northwest (possibly the Lower Peat 'upper leaf' of Booth 1983).

5–4 cal. ka BP

During this period the transgression reached its maximum extent in southeastern Fenland with the sea finally invading beyond Welney Wash between *c.* 4870 and 4410 cal. a BP (Fig. 24). The marine incursion resumed in southwestern Fenland reaching beyond Adventurers' Land between *c.* 4410 and 4030 cal. a BP, and the inland edge of northwestern Fenland was being approached *c.* 4170 cal. a BP. However, during the later part of the period eastern and southeastern Fenland were under the influence of a regressive phase with development of an intercalated peat layer between *c.* 4380 and 4150 cal. a BP (Figs 18 and 24). It would appear, therefore, from the distribution of the radiocarbon dates that a situation where the entire Fenland was flooded simultaneously was unlikely to have occurred.

4–3 cal. ka BP

The early part of this period is important because across most of the Fenland (apart from the southeastern sector where it occurred earlier) the transgression reached its maximum extent (Fig. 25). The western Fen-edge was the last area to be inundated between 3860 and 3690 cal. a BP. However, by *c.* 3 cal. ka BP essentially the whole of the Fenland was at various stages of the regressive phase that had begun earlier in the southeastern corner (Figs 24 and 25), and which led to continued accumulation of intercalated peats. Peat formation began between *c.* 3920 and 3470 cal. a BP in southwestern Fenland, between *c.* 3660 and 3070 cal. a BP in northwestern Fenland and finally between *c.* 3110 and 3040 cal. a BP along the western Fen-edge. A transgressive phase (*c.* 3360–2820 cal. a BP, Figs 25 and 26) briefly interrupted peat accumulation in parts of northwestern Fenland (Brew *et al.* in press).

The coastline of Waller (1994*a*) at about this time (p. 76, map 9) is postulated to have been further seaward than the coastline postulated in this study. The differences in position are likely to be due to interpretation of different lithological databases associated with each study.

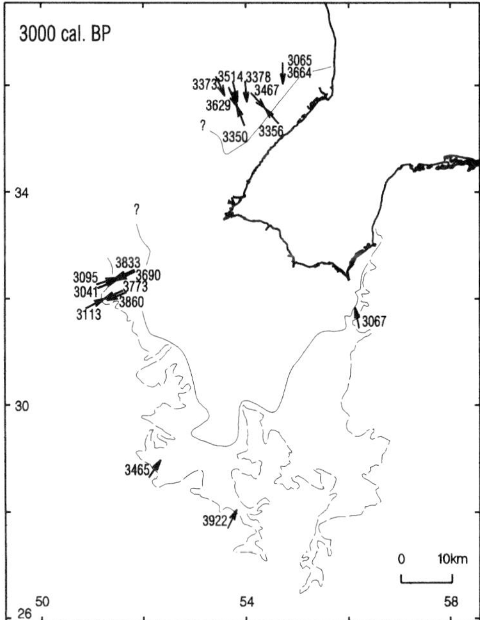

Fig. 25. Palaeocoastline (MHWST) at 3 cal. ka BP. Mean sea level *c.* –2.5 m OD. MHWST *c.* +0.5 m OD. Arrows indicate the location, direction and mean calibrated age of coastline movements between 4 and 3 cal. ka BP. The whole of the Fenland was under the influence of a regressive phase with widespread formation of peat. The dashed line indicates the position of the maximum inland limit of the coastline prior to 3 cal. ka BP (based on the lithostratigraphic database).

3–2 cal. ka BP

This period saw the continued seaward expansion of freshwater conditions, which led to later peat accumulation (*c.* 2790–2010 cal. a BP) in the south-central Fenland (Fig. 26). However, at a similar time, *c.* 2750 cal. a BP, marine–brackish sedimentation was resumed along the western Fen-edge resulting in only 300–350 years of peat accumulation in that area. In the Welney Wash area marine conditions became re-established post-2250 cal. a BP, but did not extend into the far southeastern corner of the Fenland (Fig. 27) where peat is exposed at the surface. By *c.* 2 cal. ka BP (Fig. 26) most of the Fenland was at various stages of the transgressive phase.

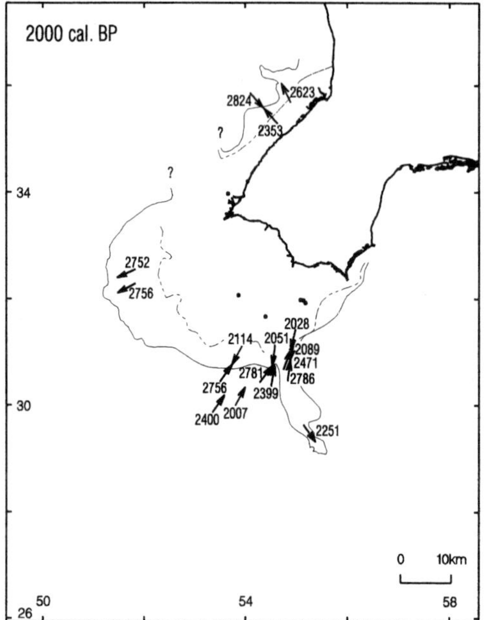

Fig. 26. Palaeocoastline (MHWST) at 2 cal. ka BP. Mean sea-level c. −1.5 m OD. MHWST c. +2 m OD. Arrows indicate the location, direction and mean calibrated age of coastline movements between 3 and 2 cal. ka BP. Fenland was again under the influence of a transgression. The dashed-dotted line indicates the hypothesized maximum seaward limit of the coastline during the previous regressive phase. The dots indicate cores where 'outliers' of intercalated peat have been found.

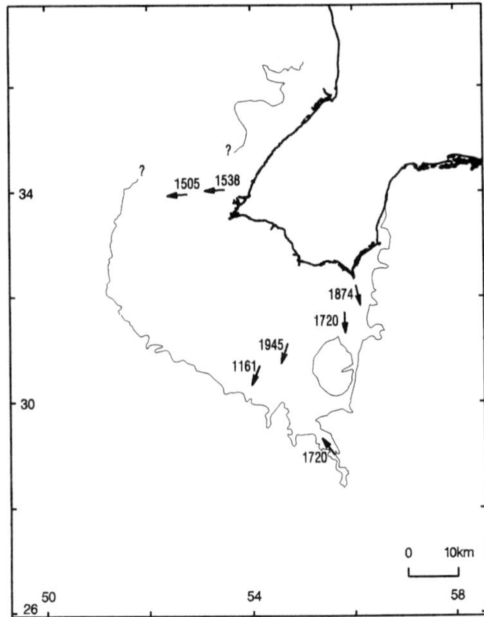

Fig. 27. Map showing the maximum inland limit of the latest transgressive phase (renewed expansion of tidal flat areas).

Post-2 cal. ka BP

Marine–brackish conditions returned to the eastern Fen-edge between c. 1870 and 1720 cal. a BP, a couple of hundred years later than in south-central Fenland, marking a landward movement of marine conditions (Fig. 27). This transgressive phase led to a more landward marine inundation in this area than the earlier transgressive phase with clastic sediments recorded well up the eastern valleys (see also Smith 1985). However, in the southern Fenland the opposite is true (compare Figs 25 and 27). Fig. 27 shows an area along the eastern Fen-edge north of Nordelph, where marine–brackish sediments were not deposited as part of this later transgressive phase (British Geological Survey 1995). This area corresponds with a distinct low in the topography (Fig. 3), which has been caused by wastage and removal of peat, which was previously exposed at the surface. Shell marls located in isolated meres that developed in the southern Fenland were also deposited post-2 cal. ka BP.

Discussion

The evolution of the Fenland has been dominated by three main events. Firstly, the initial post-glacial transgression over the pre-Holocene surface. Secondly, the sedimentary infilling of the embayment with rising sea-level; deposition of clastic sediments alternating with peat accumulation, with the final stages dominated by peat formation between c. 4.4 and 2 cal. ka BP. Thirdly, renewed expansion of tidal flat areas between c. 2750 and 1500 cal. a BP forming the final clastic fill.

The earliest intercalated peat accumulations occurred in the palaeovalleys of the Rivers Welland, Nene and Great Ouse. For example, a local fluctuation in the position of the coastline in the Nene palaeovalley at Adventurers' Land took place c. 7200 cal. a BP (Fig. 21). Given the overlap of the ages for the top of the basal peat (7279–7173 cal. a BP) and the base of the lowest intercalated peat (7183–7176 cal. a BP) (Fig. 17), it seems probable that the muddy layer between the peats represents deposition over a short period of time. This mud contains a fluvial component, which differs in clay mineralogy from sources local to the embayment itself or the North Sea. A similar situation occurs (although not as distinct as in the Nene) in the Welland palaeovalley where the coastline shifted locally early in the Holocene (Figs 22 and 23). Although the clay mineralogy indicates that the mud between the basal peat and the lowest intercalated peat is dominated by tidally transported sediment, there is a detectable fluvial component. It would appear that early Holocene fluctuations in the position of the coastline may have been controlled in part by the amount of fluvial sediment supplied to the system. This contrasts with the modern system where fluvial sediment input is negligible compared to marine inputs (Wilmot & Collins 1981; Evans & Collins 1987; Dugdale *et al.* 1987).

The bulk of the Holocene sediments of the Fenland have been deposited from a well-mixed pool of sediment transported by tidal currents into the embayment from the North Sea. Deposition of the mud and sand facies has taken place under the influence of a rising sea-level which has gradually slowed towards the present-day (Fig. 20). The particle size of the sediments might be expected, therefore, to show a coarsening-upwards trend. This is shown not to be the case, particularly with respect to the sand facies, where the sands are generally fining-upwards, becoming very silty towards the top (Figs 13 and 14). The mud facies has a uniform particle size throughout the Holocene (Fig. 8). This indicates that sedimentation kept pace with, or exceeded, sea-level rise. Higher energy environments were replaced by lower energy environments by lateral progradation and vertical accretion with the available accommodation space being filled due to the ample sediment supply. With a constant supply of sediment available from offshore sources and the slowing down of the rate of sea-level rise, the embayment silted-up. This led to the eventual accumulation of large areas of intercalated peat, which began in different areas between *c.* 4.4 and 2 cal. ka BP (Fig. 18). The radiocarbon data suggest that the tidal channels that drained the southeastern Fenland were the first to become silted-up and reduced.

The mechanisms that led to the renewed expansion of tidal flat areas (*c.* 2750–1500 cal. a BP) are difficult to establish with the data available. The Fenland embayment is postulated to have remained 'open' to the North Sea with no protective barriers behind which major formation of intercalated peats could have occurred (Shennan 1986*a*, *b*). Unpublished seismic profiles towards the entrance to The Wash support this view with no seismostratigraphic evidence for barriers or remnants of barriers. Therefore, the formation and subsequent destruction of migrating barriers (similar to those that formed along the Dutch coast, Vos & van Heeringen 1997) cannot be used as an explanation.

A natural process of inundation is tentatively put forward here, which is based on the hypothesis of Beets *et al.* (1994) for the later Holocene evolution of the Dutch coastal embayments. Between *c.* 7 and 3 cal. ka BP, the supply of sediment from marine sources to the Fenland embayment kept pace with, or exceeded, sea-level rise, leading to sedimentary infilling and progradation. However, beyond *c.* 3 ka BP the continued removal of sandy sources from offshore led to a deficit in supply relative to sea-level rise (even at a reduced rate), and, to compensate, new sources had to be found. The new supply came from reworking of the previously deposited Holocene sediments of the embayment itself, leading to shoreface erosion and renewed landward migration of the tidal system. Indeed, the modern sea bed in the northern parts of The Wash and immediately offshore is composed of a till or chalk planation surface with very little overlying Holocene sediment. Further into The Wash, the origin of a major seismic reflector (RS-2 of Brew 1997) within the sand facies may be related to this phase of erosion.

Conclusions

The Fenland Holocene sequence comprises three main sedimentary facies. The mud facies is composed of mud with minor sand, deposited in intertidal mudflat or saltmarsh environments and occurs in the inner parts of the Fenland. This facies contains intercalated peat layers forming the peat facies. The sand facies is composed of sand, silty sand and sandy silt, deposited in intertidal sandflat or subtidal environments (mainly tidal channels) and occurs in the north-central Fenland and The Wash.

The sediments of the mud and sand facies were derived from mainly marine sources, with some reworking of pre-Holocene sediments in the embayment during the post-glacial transgression. There is evidence for locally enhanced fluvial sediment supply to the Fenland during the early Holocene, although the volume of this sediment is minor compared to the large, later, supply of sediment from marine sources.

The Holocene evolution of the Fenland began *c.* 7850 cal. a BP with the onset of marine conditions in the northern Fenland associated with post-glacial sea-level rise. Over the next 3–4 cal. ka marine–brackish sedimentation gradually spread landwards, only briefly interrupted by accumulation of peat. The uniformity of particle size in the mud facies in the southern Fenland and the fining-upward sequences of the sand facies in the north-central Fenland suggest that deposition kept pace with, or exceeded, sea-level rise. The particle size data and the formation of peat layers indicate that sediment supply was sufficient both for vertical accretion in response to sea-level rise and seaward progradation of the coastline. Between *c.* 4.4 and 2 cal. ka BP, this mechanism finally led to widespread accumulation of intercalated peat layers. Partial marine flooding of these peats after *c.* 2750 cal. a BP may have been caused by starvation of sediment supply from offshore leading to reworking of previously deposited sediments, erosion of the shoreface and landward migration of the tidal system.

This study has highlighted several issues which require further research.

(1) The reconstruction of later Holocene coastlines is subject to some uncertainty. The main difficulty lies in constructing an accurate sea-level curve after 4 cal. ka BP, when the transgression reached its maximum extent. After this time the accumulation of basal peat no longer occurred, and because intercalated peats and the underlying muds are subject to significant consolidation and deformation, their original elevation is difficult to ascertain. This makes them unsuitable for use in the construction of a sea-level curve, unless the likely consolidation history can be ascertained.

(2) The mechanisms that drove the renewed expansion of tidal flat areas post-2750 cal. a BP are unproven. A provisional hypothesis related to reworking and erosion of the shoreface after the sediment supply from offshore had diminished relative to sea-level rise, is presented here. However, without further sedimentological research this remains only a tentative suggestion.

(3) The contribution and impacts of fluvial sediment supply into the Fenland palaeovalleys during the early Holocene need to be quantified. This study has demonstrated an enhanced fluvial sediment input to Adventurers' Land at around 7.2–7 cal. ka BP. In the lower reaches of the Welland Valley, the clay composition suggests that Oxford Clay was contributed fluvially or by local tidal scour early in the Holocene. This study forwards the hypothesis that some of the early regressive phases identified in the landward parts of the Fenland may, in part, have been controlled by this supply factor. Further research is now required on the early Holocene sediments in other parts of the Fenland to identify and quantify the fluvial components.

The authors wish to thank the staff of Site Investigation Services for their collection of the LOIS cores. Radiocarbon dates were provided by University of Arizona NSF facility after preparation at the NERC Radiocarbon Laboratory, East Kilbride. T. Myers and P. Bell are thanked for their IT support and J. Ridgway for permission to publish his XRF data. The contribution of T. Holt was made as part of a PhD (NERC Studentship GT12/94/LOIS/55, grant number GST/02/0771). B. Moorlock and S. Booth reviewed earlier manuscripts of this paper. M. Waller and C. Baeteman are thanked for their useful suggestions as the external referees. This is LOIS Publication Number 580 of the LOIS Community Research Programme carried out under a Special Topic Award from the NERC. D. S. Brew and R.Newsham recognize publication with permission of the Director, British Geological Survey (NERC).

References

Beets, D. J., Van der Spek, A. J. F. & Van der Valk, L. 1994. *Holocene ontwikkeling van de Nederlandse kust.* Rijks Geologische Dienst, rapport, **40.016**, Haarlem.

Booth, S. J. 1982. *The Sand and Gravel Resources of the Country Around Whittlesey, Cambridgeshire.* Description of 10.5: 25 000 sheets TF20 and TL29. *Mineral Assessment Report of the Institute of Geological Sciences*, **93**, Her Majesty's Stationery Office, London.

——1983. *The Sand and Gravel Resources of the Country Between Bourne and Crowland, Lincolnshire.* Description of 1:25 000 sheet TF11 and parts of TF01 and TF21. *Mineral Assessment Report of the Institute of Geological Sciences*, **130**, Her Majesty's Stationery Office, London.

Brew, D. S. 1997. The Quaternary history of the subtidal central Wash, eastern England. *Journal of Quaternary Science*, **12**, 131–141.

——, Evans, G., Horton, B. P., Innes, J. B. & Shennan, I. in press. Holocene palaeoenvironmental evolution and sea level history of the northwestern Fenland, eastern England. *The Holocene.*

Brindley, G. W. & Brown, G. 1980. (eds) *Crystal Structures of Clay Minerals and their X-ray Identification.* Mineralogical Society Monograph, London, **5**.

British Geological Survey. 1978. *Kings Lynn and The Wash.* Institute of Geological Sciences 1:50 000 sheet 145 and part of 129, British Geological Survey, Keyworth.

——1995. *Solid and Drift Geology*. Wisbech 1:50000 sheet 159. British Geological Survey, Keyworth.

DUGDALE, R., PLATER, A. & ALBANAKIS, K. 1987. The fluvial and marine contribution to the sediment budget of The Wash. *In*: DOODY, P. & BARNETT, B. (eds) *The Wash and its Environment*. Nature Conservancy Council, Peterborough, 37–47.

EVANS, G. & COLLINS, M. B. 1987. Sediment supply and deposition in The Wash. *In*: DOODY, P. & BARNETT, B. (eds) *The Wash and its Environment*. Nature Conservancy Council, Peterborough, 48–63.

GALLOIS, R. W. 1979. *Geological Investigations for the Wash Water Storage Scheme*. Report of the Institute of Geological Sciences, **78/19**, Her Majesty's Stationery Office, London.

GODWIN, H. 1940. Studies of the post-glacial history of British vegetation. III. Fenland pollen diagrams. IV. Post-glacial changes of relative land and sea level in the English Fenland. *Philosophical Transactions of the Royal Society of London*, **B230**, 239–303.

——1975. *History of the British Flora*, 2nd edition, Cambridge University Press, Cambridge.

——1978. *Fenland: Its Ancient Past and Uncertain Future*, Cambridge University Press, Cambridge.

—— & CLIFFORD, M. H. 1938. Studies of the post-glacial history of British vegetation., I. Origin and stratigraphy of Fenland deposits near Woodwalton, Hunts. II. Origin and stratigraphy of deposits in southern Fenland. *Philosophical Transactions of the Royal Society of London*, **B229**, 323–406.

—— & VISHNU-MITTRE 1975. Studies of the post-glacial history of British vegetation. XVI. Flandrian deposits of the Fenland margin at Holme Fen and Whittlesey Mere, Hunts. *Philosophical Transactions of the Royal Society of London*, **B270**, 561–604.

HORTON, A. 1989. *Geology of the Peterborough District*. Memoirs of the Geological Survey, England and Wales, Sheet 158. Her Majesty's Stationery Office, London.

HYDROGRAPHER OF THE NAVY. 1997. *Admiralty Tide Tables*, **1**, *1998 United Kingdom and Ireland Including European Channel Ports*. Hydrographer of the Navy, Taunton.

RIDGWAY, J., REES, J. G., GOWING, C. J. B., INGHAM, M. N., COOK, J. M., KNOX, R. W. O'B., BELL, P. D., ALLEN, M. A. & MOLINEAUX, P. J. 1998. *Land–Ocean Interaction Study (LOIS). Land–Ocean Evolution Perspective Study (LOEPS) Core Programme. Geochemical Studies – 1: Methodology*. British Geological Survey, Technical Report, **WB/98/55**.

——, ANDREWS, J. E., ELLIS, S., HORTON, B. P., INNES, J. B., KNOX, R. W. O'B., MCARTHUR, J. J., MAHER, B. A., METCALFE, S. E., MITLEHNER, A., PARKES, A., REES, J. G., SAMWAYS, G. M. & SHENNAN, I. 2000. Analysis and interpretation of Holocene sedimentary sequences in the Humber estuary. *This volume*.

SHENNAN, I. 1982. Interpretation of Flandrian sea-level data from the Fenland, England. *Proceedings of the Geologists' Association*, **93**, 53–63.

——1986*a*. Flandrian sea-level changes in the Fenland. I: The geographical setting and evidence of relative sea-level changes. *Journal of Quaternary Science*, **1**, 119–154.

——1986*b*. Flandrian sea-level changes in the Fenland. II: Tendencies of sea-level movement, altitudinal changes, and local and regional factors. *Journal of Quaternary Science*, **1**, 155–179.

—— & ALDERTON, A. 1994. Western fen edge (Lincs). *In*: WALLER, M. (ed.) *The Fenland Project, Number 9: Flandrian Environmental Change in Fenland*. Cambridge Archaeological Committee, 268–282.

——, LAMBECK, K., HORTON, B. P., INNES, J. B., LLOYD, J. L., MCARTHUR, J. J. & RUTHERFORD, M. M. 2000. Holocene crustal movements and relative sea-level changes on the east coast of England. *This volume*.

SMITH, M. V. 1985. The compressibility of sediments and its importance on Flandrian Fenland deposits. *Boreas*, **14**, 1–18.

STUIVER, M. & REIMER, P. J. 1993. Extended ^{14}C database and revised CALIB 3.0 ^{14}C age calibration program. *Radiocarbon*, **35**, 215–230.

VOS, P. C. & VAN HEERINGEN, R. M. 1997. Holocene geology and occupation history of the Province of Zeeland (SW Netherlands). *In*: FISCHER, M. M. (ed.) *Holocene Evolution of Zeeland (SW Netherlands)*. **59**, Mededelingen Nederlands Instituut voor Toegepaste Geowetenschappen TNO, 5–109.

WALLER, M. 1994*a*. III. Data synthesis: palaeogeography. *In*: WALLER, M. (ed.) *The Fenland Project, Number 9: Flandrian Environmental Change in Fenland*. Cambridge Archaeological Committee, 60–81.

——1994*b*. West-central Fens (Cambs). *In*: WALLER, M. (ed.) *The Fenland Project, Number 9: Flandrian Environmental Change in Fenland*. Cambridge Archaeological Committee, 198–226.

——1994*c*. Eastern Fen Edge (Norfolk). *In*: WALLER, M. (ed.) *The Fenland Project, Number 9: Flandrian Environmental Change in Fenland*. Cambridge Archaeological Committee, 251–267.

——1994*d*. (ed.) *The Fenland Project, Number 9: Flandrian Environmental Change in Fenland*. Cambridge Archaeological Committee.

——, PEGLAR, S. & ALDERTON, A. 1994. South-eastern Fens (Cambs/Norfolk/Suffolk). *In*: WALLER, M. (ed.) *The Fenland Project, Number 9: Flandrian Environmental Change in Fenland*. Cambridge Archaeological Committee, 111–155.

WEIR, A. H., ORMEROD, E. C. & EL MANSLEY, M. I. 1975. Clay mineralogy of sediment of the western Nile Delta. *Clay Minerals*, **10**, 369–386.

WHEELER, A. J. & WALLER, M. P. 1995. The Holocene lithostratigraphy of Fenland, eastern England: a review and suggestions for redefinition. *Geological Magazine*, **132**, 223–233.

WILMOT, R. D. & COLLINS, M. B. 1981. Contemporary fluvial sediment supply to The Wash. *In*: NIO, S. D., SCHUTTENHELM, R. T. E. & VAN WEERING, T. C. E. (eds) *Holocene Marine Sedimentation in the North Sea Basin*. Special Publication of the International Association of Sedimentologists, **5**. Blackwell Scientific Publications, Oxford, 99–110.

Holocene isostasy and relative sea-level changes on the east coast of England

I. SHENNAN,[1] K. LAMBECK,[2] B. HORTON,[1] J. INNES,[1] J. LLOYD,[1] J. McARTHUR[1] & M. RUTHERFORD[1]

[1] *Environmental Research Centre, Department of Geography, University of Durham, Durham DH1 3LE, UK (e-mail: ian.shennan@durham.ac.uk)*

[2] *Research School of Earth Sciences, The Australian National University, Canberra ACT 0200, Australia*

Abstract: Analysis of sea-level data from the east coast of England identifies local-scale and regional scale factors to explain spatial and temporal variations in the altitude of Holocene sea-level index points. The isostatic effect of the glacial rebound process, including both the ice (glacio-isostatic) and water (hydro-isostatic) load contributions, explains regional-scale differences between eight areas: *c.* 20 m range at 8 cal. ka BP and by 4 cal. ka BP relative sea-level in Northumberland was above present, whereas in areas to the south relative sea level has been below present throughout the Holocene. Estimates for pre-industrial relative sea-level change range from 1.04 ± 0.12 mm a^{-1} in the Fenland to -1.30 ± 0.68 mm a^{-1} (i.e. sea-level fall) in north Northumberland, although this may overestimate the current rate of sea-level fall. Isostatic effects will produce similar relative differences in rates of sea-level change through the twenty-first century. The data agree closely with the patterns predicted by glacio- and hydro-isostatic models, but small systematic differences along the east coast await testing against new ice models. Local scale processes identified include differential isostatic effects within the Humber Estuary and the Fenland, tide range changes during the Holocene, and the effects of sediment consolidation. These processes help explain the variation in altitude between sea-level reconstructions derived from index points taken from basal peats and those from peats intercalated within thick sequences of Holocene sediments.

Analysis of data from estuaries and coastal lowlands around Great Britain reveals different relative sea-level changes owing primarily to the spatially variable consequences of glacio- and hydro-isostasy (e.g. Shennan 1983, 1989; Lambeck 1993*a*, *b*, 1995; Peltier 1998). These isostatic factors combine with eustatic changes in ocean volume to produce regional-scale relative sea-level changes. Great Britain lies beyond the limit of Scandinavian ice at the last glacial maximum, but close enough to result in isostatic movement caused by that ice-sheet. Although small in comparison, the British ice-sheet is the primary cause of the spatial variation of regional-scale relative sea-level changes along the east coast of England. Both observations and model predictions (e.g. Shennan 1989; Shennan *et al.* 1995; Lambeck 1993*a*, *b*, 1995) in areas under thick ice-cover from relatively small ice-sheets such as the British ice-sheet typically record a fall–rise–fall relative sea-level from the Late Devensian to the present (Fig. 1). The age and altitude of the early Holocene minimum and mid-Holocene maximum varying according to geographic location relative to the centre of thickest ice. In contrast, areas close to the limit of the ice, and beyond, record a general trend of rising relative sea-level (RSL) throughout the Holocene (Fig. 1).

Although the interaction of eustasy and isostatic factors produces the general pattern of relative sea-level changes various factors operate at the coast and within an estuary that influence the registration of relative sea-level changes in the sedimentary record. These local-scale factors include modifications in the tidal regime along the estuary and the relationship between the freshwater table and tide levels. Furthermore, changes in elevation of the sediment recording a past sea-level since the time of deposition must be taken into account. Such changes in elevation may include consolidation due to the accumulation of overlying sediments and consolidation due to land drainage.

From: SHENNAN, I. & ANDREWS, J. (eds) *Holocene Land–Ocean Interaction and Environmental Change around the North Sea*. Geological Society, London, Special Publications, **166**, 275–298. 1-86239-054-1/00/$15.00

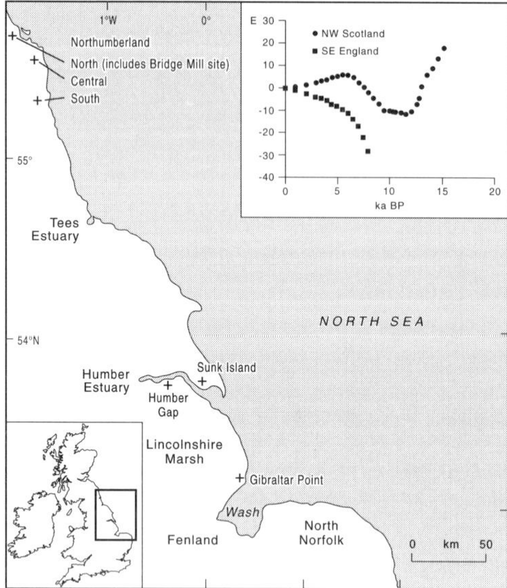

Fig. 1. Site location map with inset of schematic relative sea-level curves from northwest Scotland and southeast England based on glacio- and hydro-isostatic rebound models (after Lambeck 1995).

For each site the change in RSL ($\Delta\xi_{rsl}$) at time τ, and location φ can be expressed schematically as

$$\Delta\xi_{rsl}(\tau,\varphi) = \Delta\xi_{eus}(\tau) + \Delta\xi_{iso}(\tau,\varphi) + \Delta\xi_{local}(\tau,\varphi) \quad (1)$$

where $\Delta\xi_{eus}(\tau)$ is the time-dependent eustatic function; $\Delta\xi_{iso}(\tau,\varphi)$ is the total isostatic effect of the glacial rebound process, including both the ice (glacio-isostatic) and water (hydro-isostatic) load contributions; and $\Delta\xi_{local}(\tau,\varphi)$ is the total effect of local processes within the estuary. In order to use observations from the sedimentary record to reconstruct sea-level change the local factors can be expressed schematically

$$\Delta\xi_{local}(\tau,\varphi) = \Delta\xi_{tide}(\tau,\varphi) + \Delta\xi_{sed}(\tau,\varphi) \quad (2)$$

where $\Delta\xi_{tide}(\tau,\varphi)$ is the total effect of tidal-regime changes and the elevation of the sediment with reference to tide levels at the time of deposition, and $\Delta\xi_{sed}(\tau,\varphi)$ is the total effect of sediment consolidation since the time of deposition.

This paper is one of a series arising mainly from two projects (Project 316, Modelling Holocene depositional regimes in the western North Sea at 1000-year time intervals, and Project 313, Differential crustal movements within the river–atmosphere–coast study (RACS) site, Berwick-upon-Tweed to North Norfolk) within the Land–Ocean Interaction Study (LOIS). These two projects address different aspects of reconstructing relative sea-level change and crustal movements since the last glacial maximum (e.g. Shennan *et al.* this volume, in press *a*, *b*). Other LOIS projects contributed data to this paper (see Andrews *et al.* 1999; Brew *et al.* 1999; Metcalfe *et al.* 1999; Orford *et al.* 1999; Plater *et al.* 1999). Previous modelling studies (e.g. Lambeck 1993*a*, *b*, 1995, 1996; Lambeck *et al.* 1998) demonstrated the general trends in Holocene RSLs for different parts of the east coast and indeed the previously available data from eastern England were used to develop and validate the earth and ice models. These studies provide robust parameters for the earth model (e.g. Lambeck 1996) and the eustatic function (Fleming *et al.* 1998). Lambeck *et al.* (1998) updated the model of the Scandinavian ice sheet compared to the earlier papers and a separate paper arising from the current LOIS projects discusses a modified model of the British ice-sheet (Shennan *et al.* this volume). Therefore, the present paper does not discuss the detail of the ongoing glacio- and hydro-isostatic modelling. The major discrepancies between observations and predictions occur mostly in those areas under the thickest ice or where ice limits and volumes are less well known, and for the Late Devensian and early Holocene times, i.e. before 8 cal. ka BP, for which there are few data points from the east coast of England.

The aims of this paper are threefold. Firstly to distinguish and quantify the local-scale factors, $\Delta\xi_{tide}(\tau,\varphi)$ and $\Delta\xi_{sed}(\tau,\varphi)$, for separate sections of the east coast of England between Berwick-upon-Tweed, north Northumberland, and north Norfolk; secondly, to quantify the regional-scale differences (i.e. between the separate sections of coast) in RSL caused by the interaction of eustasy and isostasy; and finally, to quantify the pre-twentieth century rate of sea-level rise for the different coastline sections and compare them to the Intergovernmental Panel on Climate Change (IPCC) predictions of future sea-level change.

Data

A number of recent projects in eastern England (e.g. Waller 1994; van de Noort & Ellis, 1995, 1997; Long *et al.* 1998) and the LOIS programme have greatly enhanced the database available for reconstructing Holocene sea-level change since the analyses of Shennan (1989) and Lambeck (1995). Collectively, they provide a large database of reliable index points for reconstructing Holocene sea-level changes. The procedures for evaluating individual samples as relative sea-level index points are routinely explained in many publications (e.g. Shennan 1986*a*, *b*; van de Plassche 1986; Long *et al.* 1998). In brief, lithostratigraphic and biostratigraphic data are used to quantify, with an error term, the water-level at which the sample forms in relation to tidal regime and identify the tendency of sea-level movement represented by the sample. A positive tendency is defined as an increase in marine influence and a negative tendency is a decrease in marine influence. The age of the indicator comes from calibrating the radiocarbon age of the sample, expressed as the mean calibrated age and the 95% probability range (from 'method A' of Stuiver & Reimer 1993). The pollen data also provide a coarse-scale chronological check on the accuracy of the radiocarbon ages.

Much of the sea-level data used in this analysis existed before the LOIS project began. The radiocarbon database held at the Environmental Research Centre (ERC), University of Durham, consisted of 790 validated sea-level index points of which 225 came from the east coast between Berwick-upon-Tweed and north Norfolk. The temporal and spatial distributions of the index points were uneven and limited the value of earlier analyses. Most of the recorded index points were of mid–late Holocene age, an imbalance reflecting the field distribution of surviving sediments suitable for field investigation. Sediments younger than about 2 cal. ka BP have been destroyed by recent agricultural and industrial development in the coastal zone. Index points of earlier Holocene age often lie at very great depth or seaward of the current coastline. In addition, other important factors such as sediment compaction and tidal-range variations were not quantified in the existing database. Therefore the sampling strategy within the LOIS projects aimed, where possible, to fill gaps in the geographical distribution (Fig. 2) of sea-level index points and to extend the temporal range of the data set (Fig. 3). The scatter shown in the age–altitude distribution of sea-level index points (Fig. 3b) indicates the magnitude of the sum of the spatially dependent components, $\Delta\xi_{iso}(\tau, \varphi)$, $\Delta\xi_{tide}(\tau, \varphi)$ and $\Delta\xi_{sed}(\tau, \varphi)$, that are to be explained.

In general the spatial improvement of the data set and its extension towards the earlier Holocene was very successful. The database now available from the east coast of England between Berwick-upon-Tweed and north Norfolk comprises 388 sea-level index points quantitatively related to a past tide level together with an error estimate, and a further 71 data points that provide limits on the maximum altitude of the contemporary local sea-level. Less successful was the recovery of samples more recent than 2 cal. ka BP, due mainly to the great scarcity of surviving deposits of that age. New data

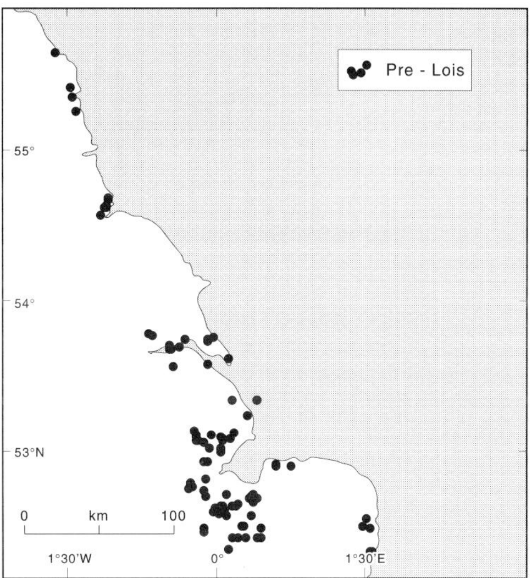

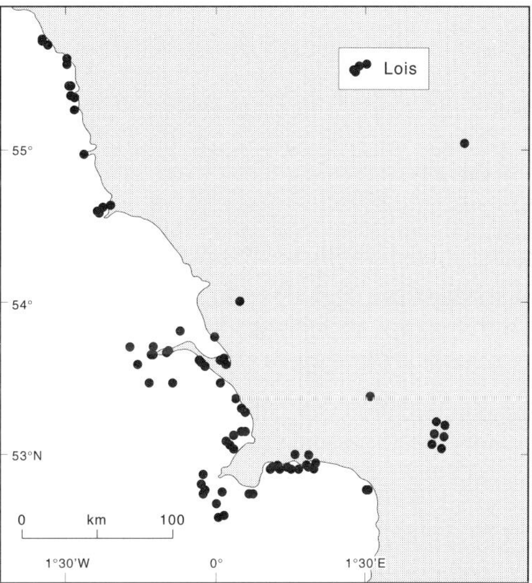

Fig. 2. Location of sea-level index points from the east coast of England and western North Sea. The dots represent study sites; each site may contain several sea-level index points.

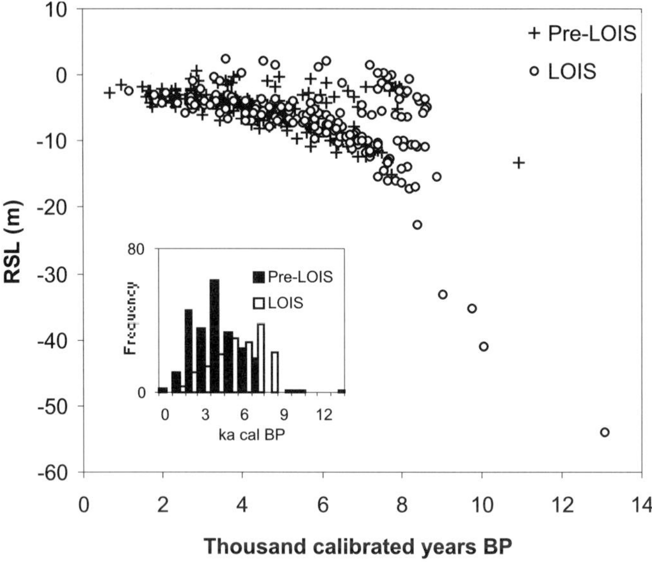

Fig. 3. Age–altitude plot, and frequency distribution (inset), of sea-level index points from the east coast of England and western North Sea.

collection also targeted basal peats, which, being less susceptible to compaction, should provide more reliable altitude data and allow an assessment of compaction effects on intercalated peats of comparable age. Where possible, sampled cores were placed within a sedimentary context by the addition of transects of cores, prior to the selection of the core to be analysed. The original sampling design identified six geographical regions, each with different pre-existing data and thus sampling requirements. The final sampling programme also had to conform to the requirements of the other LOIS projects (see other papers in this volume).

Examples of data collected

In the following section we present litho-, chrono- and biostratigraphic records from two sites that illustrate the types of sea-level index point and the limiting data used to reconstruct relative sea-level changes. Pollen and spores are shown as percentages of a total land pollen sum of at least 300, which does not include aquatics and spores. Foraminifera and diatom data are shown as numbers counted, as at some levels only small though significant totals were achieved due to poor preservation. Stratigraphic symbols follow a simplified Troels-Smith (1955) scheme.

Bridge Mill (BM95/7A), north Northumberland

This site in northern Northumberland (Fig. 1) lies in a small area of coastal plain below 10 m OD to the west of Holy Island. Coring transects across this area show the Holocene sediments to consist almost everywhere of silt, clay and sand, but localized organic intercalations occur within the clastic sequence in a sheltered area in the lee of the upland to the west. A thin upper peat occurs in several cores, and impersistent lower organic units are also present, although usually with a high clay fraction. Two distinct peat layers are best preserved at the hillslope foot in core BM95/7A [NU 0407 4567], where the lower peat rests upon sand, which could not be penetrated. The two organic units in BM95/7A are overlain and separated by slightly organic blue-grey silt clay, the intercalated clay layer including a highly organic horizon. The lithostratigraphy is shown as part of Fig. 4 which presents detailed radiocarbon and biostratigraphic data from the core.

Pollen levels were selected near lithostratigraphic unit boundaries, which represent changes in depositional environment, as well as at intervals throughout the units. Representative levels from each of the lithostratigraphic units were analysed for foraminifera and diatoms to prove their depositional origin and, if intertidal,

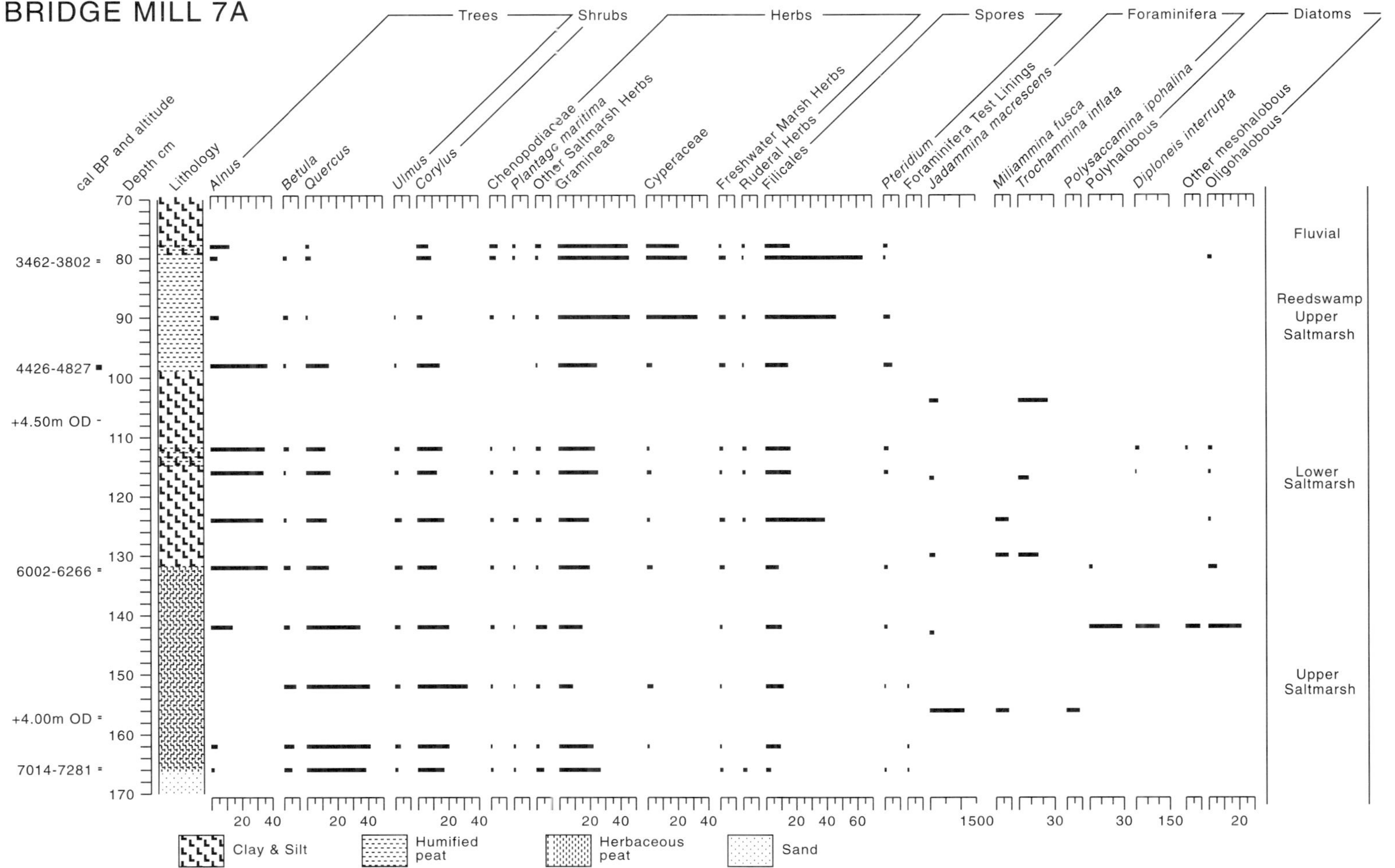

Fig. 4. Bridge Mill, Northumberland: microfossil diagram, with interpretative log at right edge. Pollen expressed as per cent total land pollen; spores as per cent Σ(land pollen + spores); foraminifera as raw counts; diatoms as raw counts and grouped according to salinity classes polyhalobous, mesohalobous and oligohalobous. Calibrated radiocarbon ages, altitudes (metres above Ordnance Datum (OD)) and depth (centimetres) down-core shown to the left of the lithology column. The sediment legend is drawn according to Troels-Smith (1955).

to assess their reference water-level within the tidal cycle. The lower peat is shown to be mid-Holocene in age by a rise in *Alnus* pollen. The *Alnus* pollen and a fall in *Ulmus* above the upper contact supports the radiocarbon date of 5290 ± 60 BP (6002–6266 cal. BP) at +4.25 m OD. The date of 6285 ± 65 BP (7014–7281 cal. BP) at + 3.89 m OD on the base of this peat is also supported by the pollen data. Gramineae dominates local wetland taxa through this peat and into the overlying clay. Saltmarsh pollen such as Chenopodiaceae and *Plantago maritima* are present throughout, but increase in frequency and range in the clay. Foraminifera test linings are present in the pollen preparations and in the separately prepared samples saltmarsh foraminifera *Jadammina macrescens*, *Miliammina fusca* and *Trochammina inflata* are common in both peat and clay. Diatom preservation was poor except for a single sample, from the upper part of the peat, with a varied saltmarsh assemblage dominated by *Diploneis interrupta*. This lower peat–clay couplet represents intertidal deposition, passing up-core from upper to lower saltmarsh.

The upper peat and surface clay are less clearly saltmarsh, containing no foraminifera and a few freshwater diatoms. Although the full range of saltmarsh pollen appears, increased Cyperaceae and Gramineae frequencies suggest deposition in a reedswamp with regular tidal input. The lower and upper contacts of the upper peat are dated 4105 ± 55 BP (4426–4827 cal. BP) at +4.59 m OD and 3360 ± 60 BP (3462–3802 cal. BP) at +4.77 m OD.

The sediments at Bridge Mill represent deposition in the upper intertidal zone under changing conditions of water-level and salinity. Table 1 illustrates how these data are used to reconstruct past RSLs. All four contacts are validated sea-level index points with a known altitude, date and indicative meaning, which is linked to a past tide level. The indicative meaning, of a coastal sample is the relationship of the local environment in which it accumulated to a contemporaneous reference tide level (van de Plassche 1986). The indicative meaning can vary according to the type of evidence and it is commonly expressed in terms of an indicative range and a reference water-level. The former is a vertical range within which the coastal sample can occur and the latter a water-level to which the assemblage is assigned, for example, mean high water spring tide (MHWST), mean tide level (MTL), etc. (van de Plassche 1986). The base of the lower peat records saltmarsh peat formation at MHWST and the upper-three dated contacts record changes between clastic and organic sedimentation occurring around MHWST. The lower peat records positive sea-level tendencies and the upper peat records negative then positive tendency.

South Farm, Sunk Island (HMB8), Humberside

This site exemplifies stratigraphic data that cannot be referred to a past tidal level and so are not valid sea-level index points *per se*, but which nevertheless provide valuable limiting data that constrain the position of past sea-levels by providing maximum altitudes for

Table 1. *Sea-level index points from Bridge Mill*

Laboratory code	AA24223	AA24224	AA24225	AA24226
^{14}C age ± 1σ (a)	3360 ± 60	4105 ± 55	5290 ± 60	6285 ± 65
Calibrated age (cal. a BP)				
Max.	3802	4827	6266	7281
Median	3606	4565	6101	7200
Min.	3462	4426	6002	7014
Altitude (m OD)	+4.77	+4.59	+4.25	+3.89
Reference water-level*,†	MHWST −0.20	MHWST +0.09	MHWST −0.20	MHWST
Indicative range‡	±0.20	±0.20	±0.20	±0.20
Tendency	+	−	+	+
Change in RSL from present§	+2.57 ± 0.20	+2.10 ± 0.20	+2.05 ± 0.20	+1.49 ± 0.20

Change in relative sea-level (RSL) is calculated as altitude minus the reference water level.
* The reference water-level is given as a mathematical expression of tidal parameters plus/minus an indicative difference. This is the distance from the mid-point of the indicative range to the reference water-level.
† Local mean high water spring tide (MHWST) is 2.40 m OD.
‡ The indicative range (given as a maximum) is the most probable vertical range in which the sample occurs.
§ The RSL error range is calculated as the square root of the sum of squares of altitudinal error, sample thickness, tide-level error and indicative range.

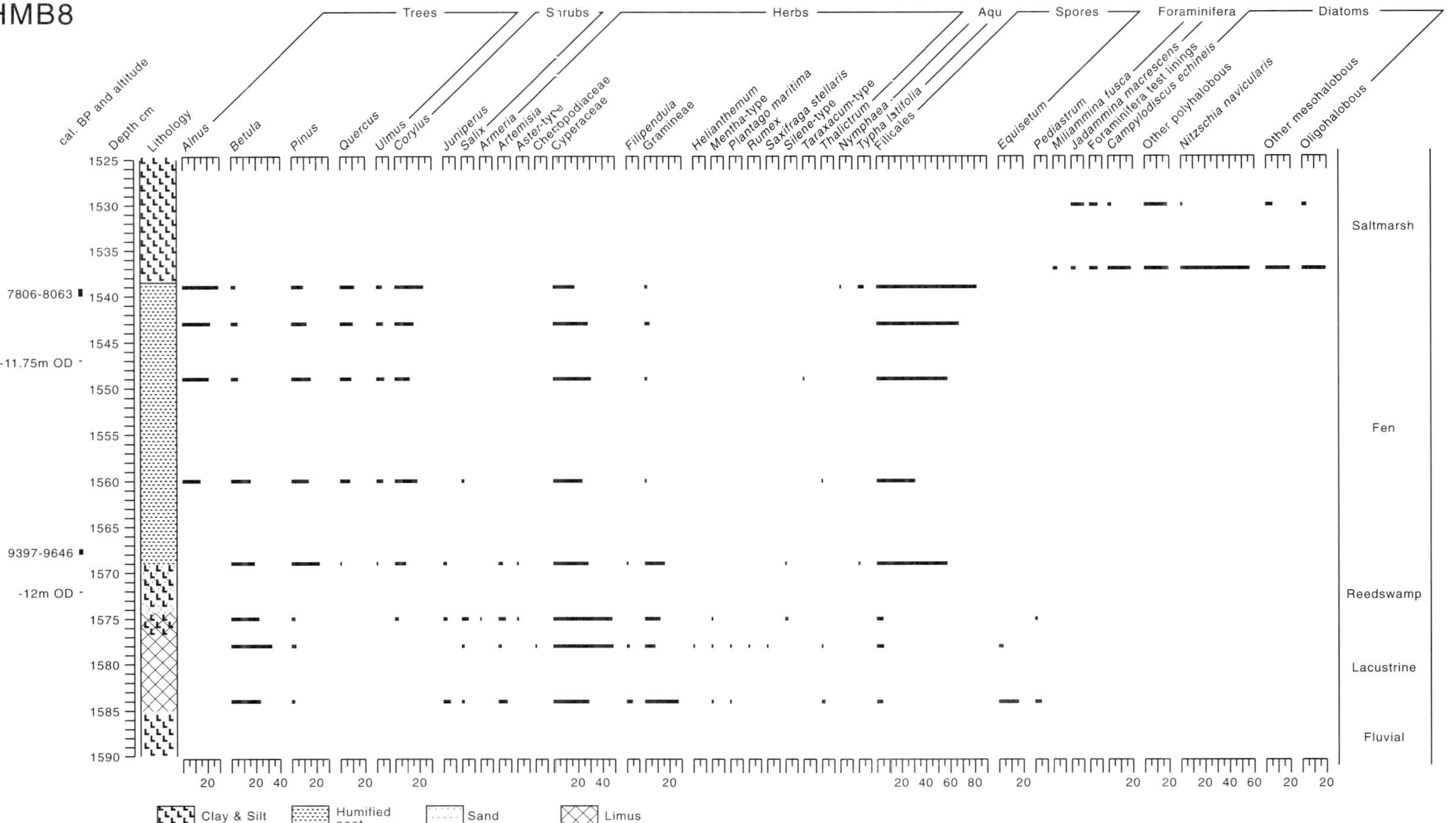

Fig. 5. South Farm, Sunk Island (HMB8), Humberside: microfossil diagram with interpretative log at right edge. Pollen expressed as per cent total land pollen; spores as per cent Σ(land pollen + spores); foraminifera as raw counts; diatoms as raw counts and grouped according to salinity classes polyhalobous, mesohalobous and oligohalobous. Calibrated radiocarbon ages, altitudes (metres below Ordnance datum (OD)) and depth (centimetres) down-core shown to the left of the lithology column. The sediment legend is drawn according to Troels-Smith (1955).

highest tidal levels. HMB8 [TA 2570 1760] lies on Sunk Island, a reclaimed area near the northern shore of the outer Humber Estuary (Fig. 1). More than 10 m of marine sands cover a layer of organic clay which contains *Phragmites* remains. This clay overlies a peat, organic clay, limnic mud and silt clay sequence, which rests on the pre-Holocene sand diamicton. The contact between the peat and the *Phragmites*-rich organic clay is very sharp and represents an erosive break in the sediment pile. The lower part of the succession is shown in Fig. 5, which presents the biostratigraphic analyses.

The limnic mud and silt clay units contain a pollen record that represents transitional Late Devensian and early Holocene vegetation communities. The lowest level near the base of the limnic mud is characterized by *Betula* and *Juniperus*, with some Gramineae and Cyperaceae. *Pediastrum* algal colonies support the interpretation of a lacustrine origin. At the start of clay deposition, above the limnic unit, many tundra-type open habitat herbs like *Helianthemum*, *Rumex* and *Thalictrum* appear. Within the clay and in the peat above, the open ground herb flora is replaced firstly by *Juniperus* and *Betula* again, followed by *Corylus* and then deciduous trees *Ulmus*, *Quercus* and *Alnus*. This succession is typical of the sequence from Late Devensian Interstadial to Loch Lomond (Younger Dryas) Stadial and then to Holocene. The change from clay to peat formation is dated 8555 ± 65 BP (9397–9646 cal. BP), which agrees with pine, birch and hazel pollen assemblages from other regions. The peat below the sharp upper contact with the overlying clay is dated 7145 ± 60 BP (7806–8063 cal. BP) at −11.67 m OD, which again supports the mid-Holocene pollen data with high *Alnus*. A rich marine diatom and saltmarsh foraminiferal assemblage occurs within this upper clay, suggesting a marine origin, but the peat and other sediments below this marine clay are barren of diatoms and foraminifera. There are no saltmarsh pollen indicators in the upper level of the peat, which apparently formed under completely freshwater conditions. These biostratigraphic data confirm an erosive hiatus between the peat and marine clay, of unknown duration. This hiatus means that the dated peat–upper clay contact cannot be accepted as a sea-level index point related directly to a past tidal level. Nevertheless, the age and altitude (Table 2) give a limiting value because at that time freshwater peat was forming and tidal influence must have operated at or below that altitude. Such peats may form in a wide vertical range depending on local palaeogeography, especially their relationship to active tidal channels and the groundwater table. Therefore, the indicative range for these dates is from above MHWST, where freshwater peat formation is controlled by tidal influence, to just below MTL, where groundwater level is the controlling factor (Godwin 1940; Shennan 1982). Subsequent marine inundation truncated the peat profile, and then laid down marine clays unconformably over the peat.

The analyses described in the following section use both sea-level index points of the types described from Bridge Mill, with quantified

Table 2. *Limiting dates from Sunk Island*

Laboratory code	AA25581	AA25582
^{14}C age ± 1σ (a)	7145 ± 60	8555 ± 65
Calibrated age (cal. a BP)		
Max.	8063	9646
Median	7923	9481
Min.	7806	9397
Altitude (m OD)	−11.67	−11.95
Reference water-level*,†	≥[MHWST + MTL]/2	≥[MHWST + MTL]/2
Indicative range‡	±1.58	±1.58
Tendency	Limiting	Limiting
Change in RSL from present§	≤ −14.61 ± 1.58	≤ −14.89 ± 1.58

Change in relative sea-level (RSL) is calculated as altitude minus the reference water-level.
* The reference water-level is given as a mathematical expression of tidal parameters plus/minus an indicative difference. This is the distance from the mid-point of the indicative range to the reference water-level.
† Local mean high water spring tide (MHWST) is 3.32 m OD.
‡ The indicative range is the most probable vertical range in which the sample occurs, although the sample could occur above that range.
§ The RSL error range is calculated as the square root of the sum of squares of altitudinal error, sample thickness, tide level error and indicative range.

relationships to tide levels (e.g. Table 1), and the limiting type of data points illustrated from Sunk Island (e.g. Table 2).

Analysis

Differences in RSL along the east coast due to $\Delta\xi_{eus}(\tau)$ and $\Delta\xi_{iso}(\tau, \varphi)$ produce a continuum within the range shown in Fig. 3b. In order to summarize the information in graphical form the data are grouped into geographical units and plotted against a single curve for that region (Fig. 6). The curve is the prediction from Lambeck (1995) for one location in the region and does not indicate any predicted within-region variations in RSL.

The ice model and earth models of Lambeck (1995) predict RSL slightly below the observations from Northumberland, the Tees Estuary, the Lincolnshire Marshes and, to a lesser extent, north Norfolk and the Humber. For the Fenland the predictions lie within the upper range of the observations. Lambeck *et al.* (1998), Fleming *et al.* (1998) and Shennan *et al.* (this volume, in press *a*, *b*) describe improvements in the ice and earth models and the eustatic factor and these are not discussed in detail in this paper. Lambeck *et al.* (1998) discuss changes to the Scandinavia ice model and Shennan *et al.* (this volume, in press *a*, *b*) evaluate earth-model parameters and revisions to the ice model for Great Britain, but at this stage they cannot identify a unique solution for the ice model. Furthermore, modelling of tidal ranges in the North Sea during the Holocene (Shennan *et al.* this volume) predicts changes of the opposite sign needed to produce a better fit between the

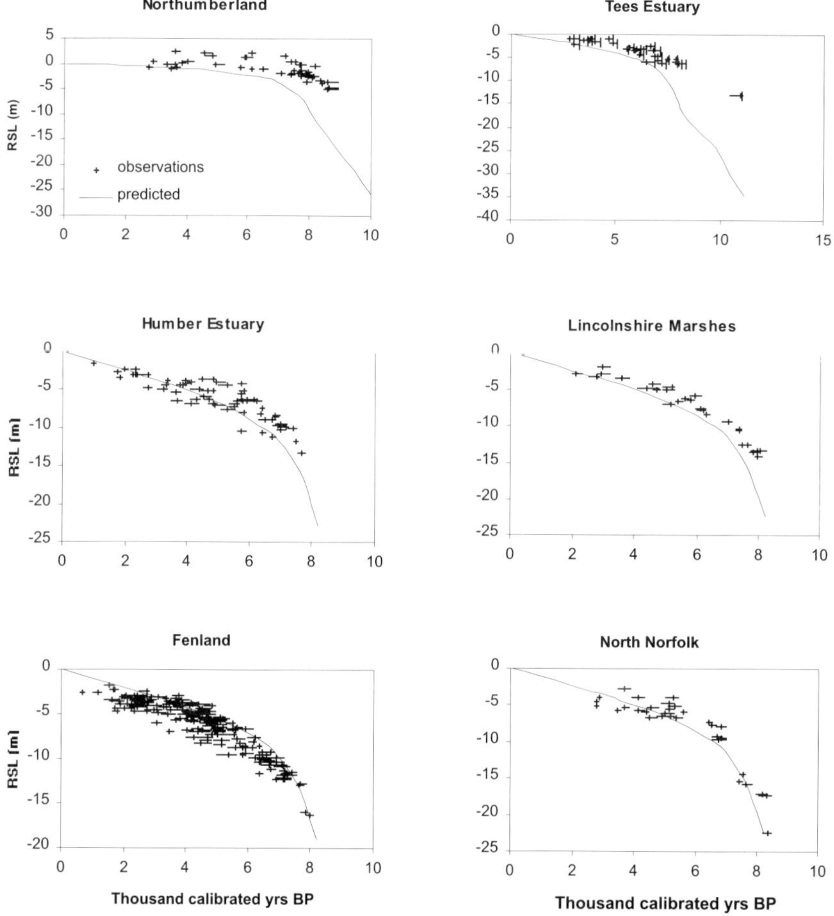

Fig. 6. Relative sea-level observations (validated index points) and predicted sea-level curves for the initial six regions.

Lambeck (1995) predictions and the observations. Also, all of the previous analyses do not consider the possible effects of sediment consolidation and tidal-range changes through time. Paul & Barras (1998) illustrate how sediment consolidation can be estimated, but this requires quantitative information on the lithology of the sequences above and below the dated sample, which is not available for a large majority of the sea-level index points.

Therefore, we have adopted the following exploratory approach based on analysis of the residuals between the observations and the summary relative sea-level curve. For each geographical region in the following section the sea-level predictions from the model of Lambeck (1995) are summarized with a best-fit polynomial. The linear term is then modified to produce a best-fit solution to the index points from basal peats only, subject to the condition that the same solution does not conflict with the set of limiting dates. The reasoning for starting from a solution for the basal peats is that they are probably less influenced by sediment consolidation than index points from peats intercalated between thick Holocene clastic sediments (e.g. Jelgersma, 1966; van de Plassche 1986; Törnqvist *et al.* 1998). Where sea-level index points on basal peats are taken from the bottom of the peat, compaction is not a factor. If sea-level index points are taken from the top of the basal peat, however, these too will be subject to some compaction.

Modification of the linear term has a similar net effect for the areas and time periods under consideration to changing an earth- or ice-model parameter (e.g. Lambeck 1993*a*, *b*) or the effect of a change in tidal range through the Holocene. The distribution of residuals against time evaluates the validity of changing the linear term. Comparable analyses were undertaken for each region. To illustrate the approach we present details of the analysis of the Humber data and then provide summary illustrations from the other areas.

Detailed example: the Humber Estuary

In relation to its size very little is known about the Holocene evolution of one of Britain's most important estuaries (Gaunt & Tooley 1974; van de Noort & Ellis 1995, 1997; Long *et al.* 1998). As a result almost 40 cores were allocated to the Humber, distributed along the estuary and in the several major river lowlands that drain into it. Thirty-six new index points were added to the database and the record now extends to almost 8 cal. a BP. Further details are given in other papers in this volume (Andrews *et al.*; Rees *et al.*; Ridgway *et al.*; Metcalfe *et al.* this volume).

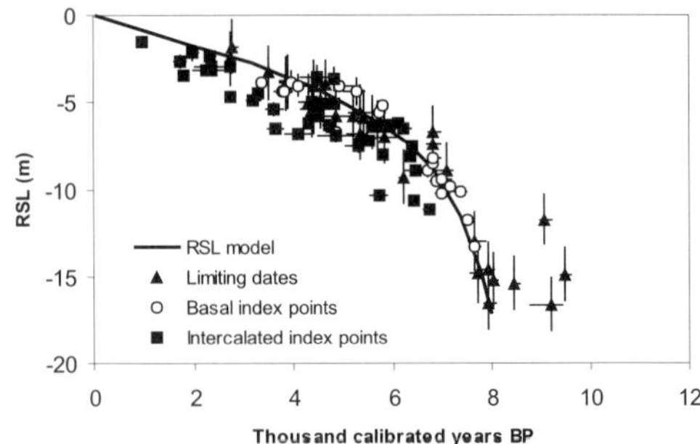

Fig. 7. Humber Estuary relative sea-level index points. Error terms are calculated as described in the text for each index point, but only appear when they extend beyond the size of the symbol.

Regional factors

The plot of all validated index points from the Humber Estuary confirms the upward trend of Holocene RSL typical of an area at or beyond the margins of the last British ice-sheet (Fig. 7). The sea-level predictions of Lambeck (1995) for the Humber Estuary are modified to produce a revised regional RSL curve (compare Figs 6 and 7). Departures from the revised regional RSL curve (Fig. 7) suggest the importance of local processes, such as sediment consolidation, changes in palaeotidal range or within-region differences in the isostatic effects. Determining the relative importance of each of these is difficult at this stage in the analysis, but careful examination of the residuals (i.e. the difference between the RSL summary curve and observations) provides an indication of their potential importance.

Local factors

Palaeotidal changes. Major changes in coastal configuration have occurred during the Holocene as changes in the rate of relative sea-level rise, sediment supply and catchment inputs of sediment and water have varied. Accompanying these will have been changes in tidal range, caused by variations in tidal prism as well as estuary configuration. As a first analysis, the RSL summary assumes no such changes and is based on present tidal variations within the Humber estuary. This hypothesis is now tested, initially

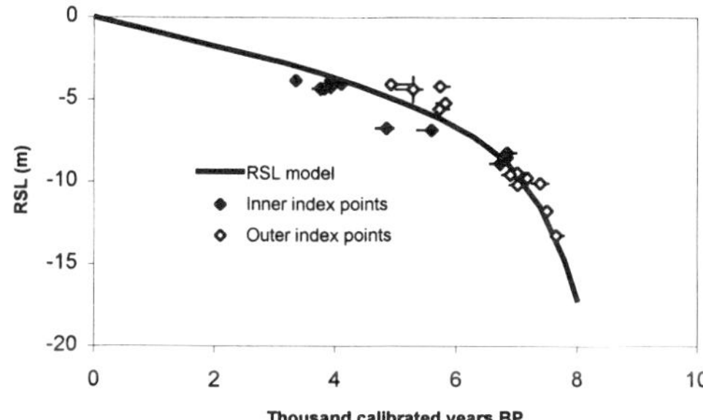

Fig. 8. Humber Estuary relative sea-level index points from basal peats: inner and outer estuary.

by dividing the sea-level index points into those from the inner estuary, west of the Humber Gap and outer estuary, east of the Humber Gap (Fig. 8).

Index points younger than 6 cal. ka BP from the outer estuary plot above the regional relative sea-level curve, whereas those from the inner estuary plot below (Fig. 8). This contrast must be qualified by the clustered distribution of the data through time, and more data are required to test this hypothesis further. Nevertheless these differences suggest that palaeotidal changes or differential isostatic movements may have occurred within the Humber Estuary.

The separation into inner and outer estuary data is an oversimplification because as the tidal wave penetrates an estuary it is modified by changes in width and depth, river flow and increased friction, which together cause non-linear tidal distortions of the higher tidal harmonics. In the Humber Estuary this results in an increase in the altitude of tide levels up estuary. This relationship is used in the subsequent analyses by using the up-estuary distance from an arbitary reference point, Grimsby, based on our reconstructions of palaeogeography at the time of sediment deposition. This still does not take into account possible changes

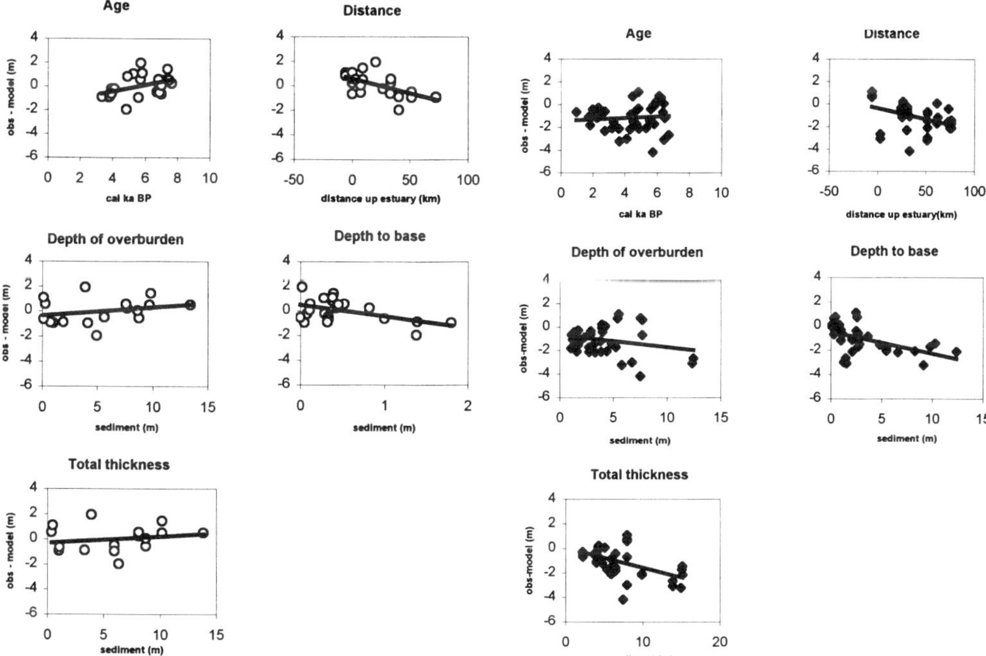

Fig. 9. Humber Estuary: scatter plots for sea-level index points from basal peats showing the relationships between residuals (observed versus modelled relative sea level) and five parameters (age, distance up estuary, depth of overburden, depth to base of Holocene sequence and total thickness of Holocene sequence) that may indicate local-scale processes. Table 3 gives the correlation coefficients.

Fig. 10. Humber Estuary: scatter plots for sea-level index points from intercalated peats showing the relationships between residuals (observed versus modelled relative sea-level) and five parameters (age, distance up estuary, depth of overburden, depth to base of Holocene sequence and total thickness of Holocene sequence) that may indicate local-scale processes. Table 3 gives the correlation coefficients.

in tidal prism through time due to the changing sea-level, bathymetry and palaeogeography within the Humber Estuary.

When the residuals are plotted against distance there is a strong relationship for basal (Fig. 9) and intercalated (Fig. 10) peats. The correlation coefficients ($r_{\text{basal}} = -0.61$; $r_{\text{intercalated}} = -0.43$) are above the critical value at both the 5 and 1% significance levels (Table 3). The residuals for both basal and intercalated sea-level index points have weaker positive relationships with age (Figs 9 and 10). One hypothesis to explain the trend with age is that tidal range within the whole estuary was less in the past than at present. The trends seen with both age and distance suggest that in the past the tidal range in the inner estuary had decreased to a greater extent than the outer estuary compared with the present day (i.e. the increase in tidal amplitude up estuary seen at the present day was reduced in the past). This suggestion is not unexpected, given the progressive infilling of the Humber Estuary during the Holocene. An additional consideration is differential isostatic movement within the estuary. The isolines of predicted mean sea level for 10, 8 and 7 ^{14}C ka BP lie almost north–south across the Humber Estuary (Lambeck 1995) and an element of the trend of residuals with distance up estuary could result from such isostatic movement. In order to discriminate between differential isostatic movement and tidal range changes through time within an area that lies across the isolines of isostatic model predictions (e.g. Lambeck 1995) it is necessary to analyse residuals from RSL predictions based on the geographic position of each index point, rather than the RSL summary for the large area used here. Until the revised glacio- and hydro-isostatic modelling described earlier is completed this preferred approach is not possible.

Sediment consolidation. Following deposition, sediment consolidation will lower index points from their original elevation and, unless corrected for, will lead to an over-estimate of the rate and magnitude of relative sea-level rise in the Humber Estuary. The effects of consolidation can be especially severe for index points that are found above and below considerable thicknesses of Holocene sediment (such as those from intercalated peats). This problem is reduced, but not entirely removed, by using index points from basal peats, which rest at, or close to, the base of the Holocene sediment column. Subdivision of the index points into basal and intercalated peats provides an initial assessment (Fig. 7), and suggests that the influence of sediment consolidation may indeed be significant. Most of the samples from intercalated peats lie below those

Table 3. *Pearson's product–moment correlation coefficients*

Area	Type†	No.	Critical value (0.05)	Critical value (0.01)	Age	Distance	Depth of overburden	Depth to base of Holocene sequence	Total thickness of Holocene sequence
Northumberland	B	6	0.71	0.83	−0.51	**−0.90**	0.34	−0.58	0.62
North	I	6	0.71	0.83	−0.15	0.60	0.57	**−0.79**	−0.56
Central	B	2	–	–	–	–	–	–	–
	I	6	0.71	0.83	0.16	**−0.87**	**−0.97**	0.01	**−0.93**
South	B	6	0.71	0.83	**0.76**	0.36	0.26	−0.70	0.07
	I	16	0.47	0.59	−0.19	**−0.48**	**−0.54**	0.25	**−0.66**
Tees	B	9	0.60	0.74	−0.03	−0.34	−0.28	**−0.68**	−0.45
	I	21	0.42	0.54	0.01	**−0.53**	**−0.59**	−0.14	**−0.48**
Humber	B	23	0.42	0.54	0.42	**−0.61**	0.27	**−0.48**	0.20
	I	40	0.30	0.39	0.08	**−0.43**	−0.23	**−0.52**	**−0.51**
Lincolnshire Marshes	B	11	0.55	0.68	0.37	**−0.60**	0.28	−0.09	0.27
	I	17	0.46	0.58	**0.54**	−0.12	−0.23	**−0.65**	−0.41
Fenland	B	85	0.21	0.27	−0.19	−0.06	**−0.35**	–	–
	I	108	0.20	0.25	−0.04	0.09	**−0.37**	–	–
North Norfolk	B	17	0.46	0.58	−0.09	0.11	−0.28	0.25	−0.23
	I	14	0.50	0.62	−0.30	−0.01	**−0.81**	−0.48	**−0.87**

* Values in bold exceed the critical value at the 0.05 significance level (two-tailed).
† B, basal peat sea-level index points; I, intercalated peat sea-level index points. Basal index points include samples from the top and bottom of peats.

from basal peats of the same age. This is not wholly unexpected and invites further analysis.

In the absence of the detailed lithological data needed for quantitative assessment of consolidation (e.g. Paul & Barras 1998) its effects on index points from peats may be considered to be dependent upon three more easily available parameters. This approach does not model the consolidation process, but gives an indication of the net effects. The parameters available for each index point are: (a) the thickness of sediment overburden; (b) the depth of sediment below to the base of the Holocene; (c) the thickness of the whole Holocene sequence (Figs 9 and 10). For the basal peat index points only the depth of sediment to the base of the Holocene is significant (Table 3), correlation coefficient $r_{basal} = -0.48$. The positive correlations with the other two variables are neither statistically significant nor of the correct sign to suggest consolidation effects.

For samples from intercalated peats, depth of sediment to the base of the Holocene and the thickness of the Holocene sequence suggest strong empirical relationships with the magnitude of residuals (Fig. 10 and Table 3). The correlation coefficients for both parameters are greater than the critical value at the 5 and 1% significance level suggesting further research in this area is justified. For example, at this stage of analysis no account has been taken of the variation in sediment types, including the proportion of different grain size distributions, organic content, water content, or of drainage histories.

Summary: regional sea level, sediment consolidation and tidal change within the Humber Estuary

The regional sea-level curve (Fig. 7) provides a summary of the Holocene sea-level data from the Humber Estuary. It is suggested that sediment consolidation, tidal changes and differential isostatic movements have played important roles in producing the scatter of relative sea-level data around the regional trend. Establishing the relative importance of these factors is not possible at this stage and requires further modelling of each process.

This procedure of data analysis is now repeated for the data collected from the five other study areas with the following amendment. Graphs of the relationships between residuals and local processes are only shown for those exceeding the critical value at the 5% significance level (Table 3). These relationships, therefore, reject the null hypothesis suggesting there is no empirical relationship between the parameters.

Northumberland

This is a critical region, as it is the most northerly in the study area and so is likely to show most clearly the effects of differential crustal movement. Only 14 pre-LOIS validated index points were available (Plater & Shennan 1992; Shennan 1992). Twenty-eight new index points have been produced with the database now extending to pre-8 cal. ka BP (Shennan *et al.* in press *b*). These index points come from individual sites, with almost 60 km between the most southerly and the most northerly. Geographically the sites fall into three clusters, labelled here north, central and south (Fig. 1, although no index points come from the most southerly parts of the Northumberland coast). Previous investigations illustrate a systematic increase in the altitude with distance northwards for sea-level index points of the same age (e.g. Plater & Shennan 1992; Shennan 1992; also Shennan *et al.* in press *b*) and the threefold division is used in the following analyses. None of the data come from large distances up estuaries, so the distance parameter used is distance from the northernmost point. Since the Northumberland coast lies along the axis of differential uplift any trend against this distance parameter will in part result from within-area differential movement.

Despite the lack of index points for the last 3 ka in all three Northumberland areas, the presence of intertidal clastic sediments above present high tide level demonstrates a fall in RSL since the time of the youngest index point. The distributions of the data from the whole of Northumberland constrain the form of the summary RSL curves (Figs 11, 13, 15) and the three differ only in the value of the linear terms in the polynomials.

The summary curve for Northumberland North (Fig. 11) indicates a mid–late Holocene maximum around 2.5 m above present. The basal peat data come from only two sites, so while the significant correlation with distance north–south may indicate differential crustal movement any local effects changing tide levels at either site will also contribute to the distribution in Fig. 12. The significant correlation with depth to base of the Holocene for index points from intercalated peats indicates a sediment consolidation effect (Fig. 12 and Table 3).

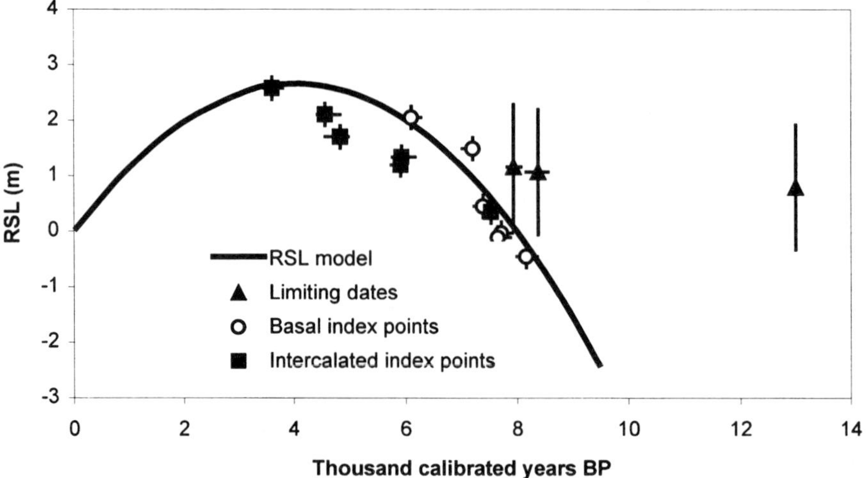

Fig. 11. Northumberland, North: relative sea-level index points. Error terms are calculated as described in the text for each index point, but only appear when they extend beyond the size of the symbol.

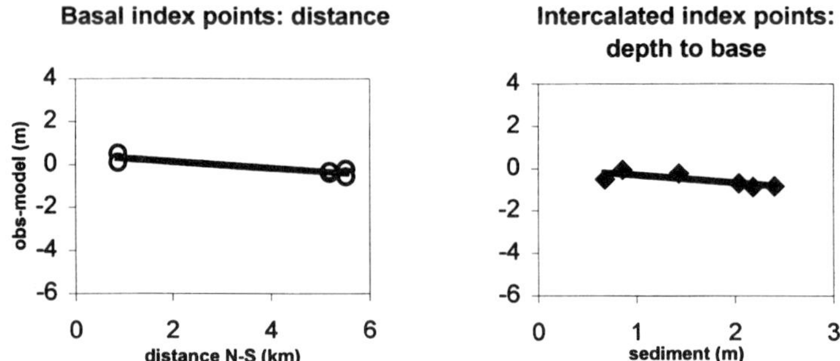

Fig. 12. Northumberland, North: scatter plots showing the statistically significant relationships from Table 3 between residuals (observed versus modelled relative sea-level) for basal and intercalated index points and parameters that may indicate local-scale processes (distance north–south and depth to base of Holocene sequence).

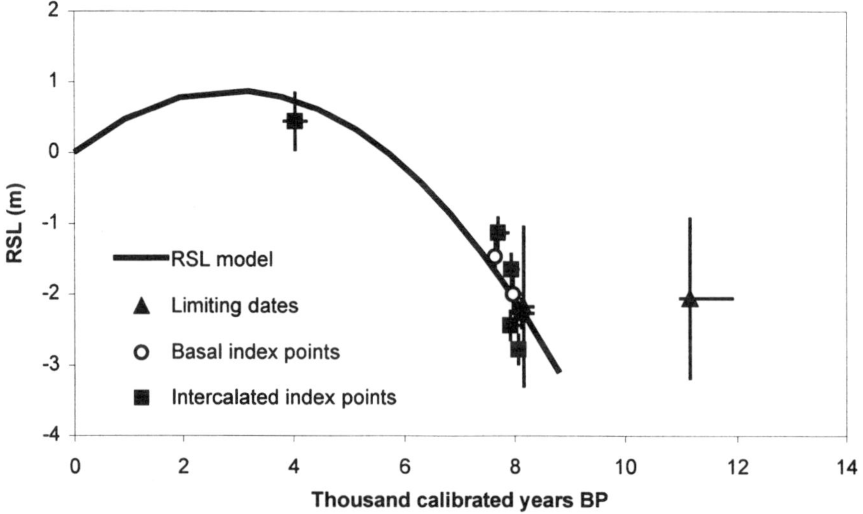

Fig. 13. Northumberland, Central: relative sea-level index points. Error terms are calculated as described in the text for each index point, but only appear when they extend beyond the size of the symbol.

The very small data set for Northumberland Central shows a maximum around 1 m above present (Fig. 13) and a within-area trend with distance (Fig. 16), which again should be qualified by the fact that the data are from few sites. Sediment consolidation effects are evident from the significant correlations with both depth of sediment overburden and total thickness of the Holocene sequence.

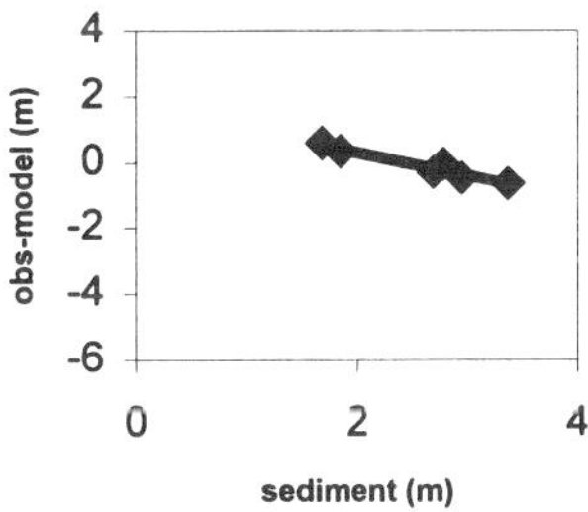

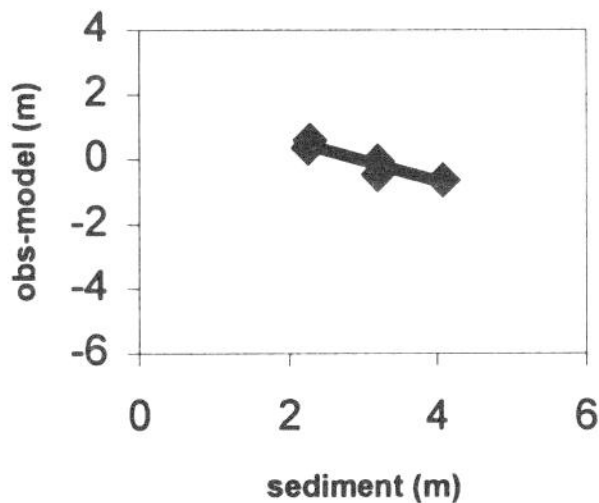

Fig. 14. Northumberland, Central: scatter plots showing the statistically significant relationships from Table 3 between residuals (observed versus modelled relative sea-level) for intercalated index points and parameters that may indicate local-scale processes (distance north–south, depth of overburden and total thickness of Holocene sequence).

Northumberland South reveals a mid–late Holocene maximum less than 1 m above present (Fig. 15). The age distribution of index points from basal peats is very restricted, and given the error estimates for both the age and altitude of residuals, no inference is made from the correlation (Fig. 16 and Table 3). The significant correlation with distance, in this case for data from intercalated peats, must be qualified with the observation that the largest negative residuals of those data younger than 4 cal. ka BP come from peat beds recently exposed on the beach at Druridge Bay, south Northumberland. These peats have been exposed following the landward migration of sand dunes, so they were once covered by many metres of dune sand, which would have caused consolidation of the peat and underlying Holocene sediment. Sediment consolidation of intercalated index points is further shown by the significant correlation with depth of overburden and sediment thickness (Table 3).

Tees Estuary

Nineteen index points were previously available from within the Tees Estuary (Fig. 1) (Tooley 1978; Shennan 1992). Eleven new index points have been gained under LOIS, which extend the Tees record to most of the period 3–11 cal. ka BP; further details in Plater *et al.* (this volume). The data indicate that sea-level did not rise above its present level (Fig. 17). Many of the samples come from thick sediment sequences and the scatter seen on the RSL plot appears to result at least in part from sediment consolidation. Significant correlations for altitude residuals against depth to base for the basal peats and depth of overburden and total Holocene thickness for the intercalated peat index points support this view (Fig. 18). The distance parameter is, like the Humber, distance up estuary. The correlation between residuals and distance is only significant at the 5% level for intercalated peats, but this may indicate a change through time of the tidal prism; although the basal peat data do not show a significant correlation coefficient.

Lincolnshire Marshes

The area between the Humber Estuary and the Fenland south of Gibraltar Point (Fig. 1) had not been studied in detail, and few index points were available despite the importance of the

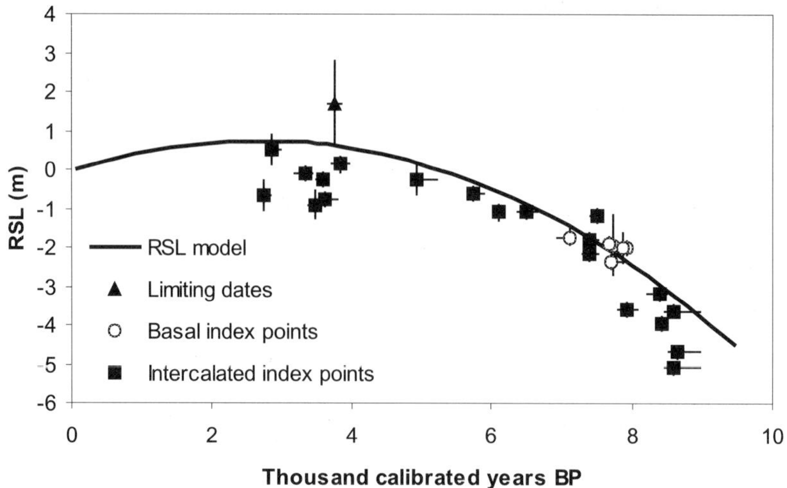

Fig. 15. Northumberland, South: relative sea-level index points. Error terms are calculated as described in the text for each index point, but only appear when they extend beyond the size of the symbol.

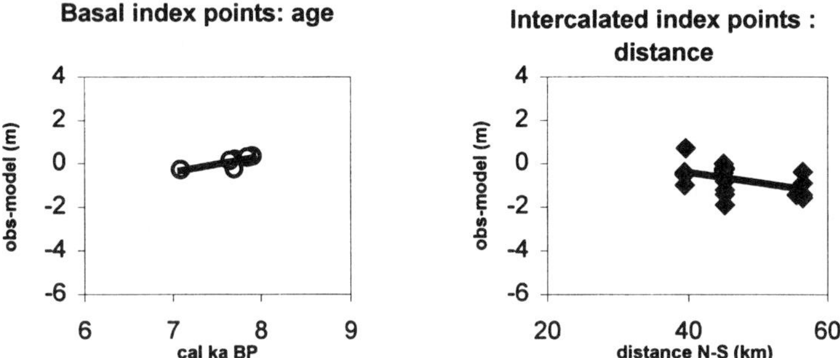

Fig. 16. Northumberland, South: scatter plots showing the statistically significant relationships from Table 3 between residuals (observed versus modelled relative sea-level) for basal and intercalated index points and parameters that may indicate local-scale processes (age and distance north–south).

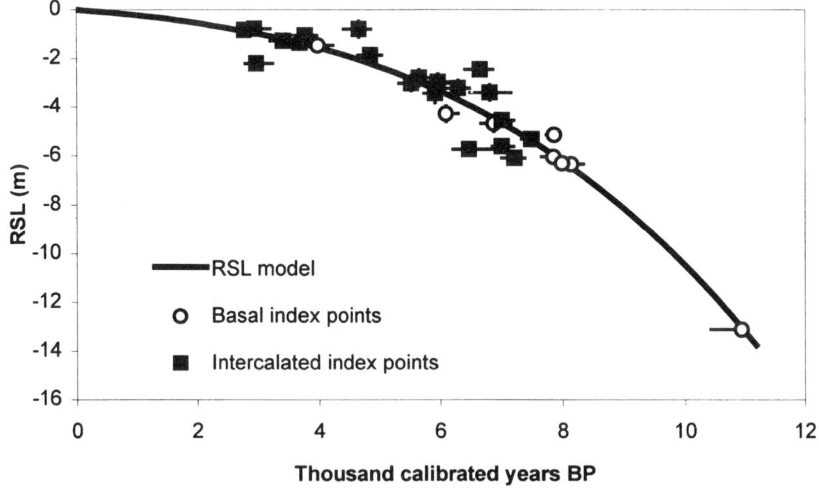

Fig. 17. Tees Estuary relative sea-level index points. Error terms are calculated as described in the text for each index point, but only appear when they extend beyond the size of the symbol.

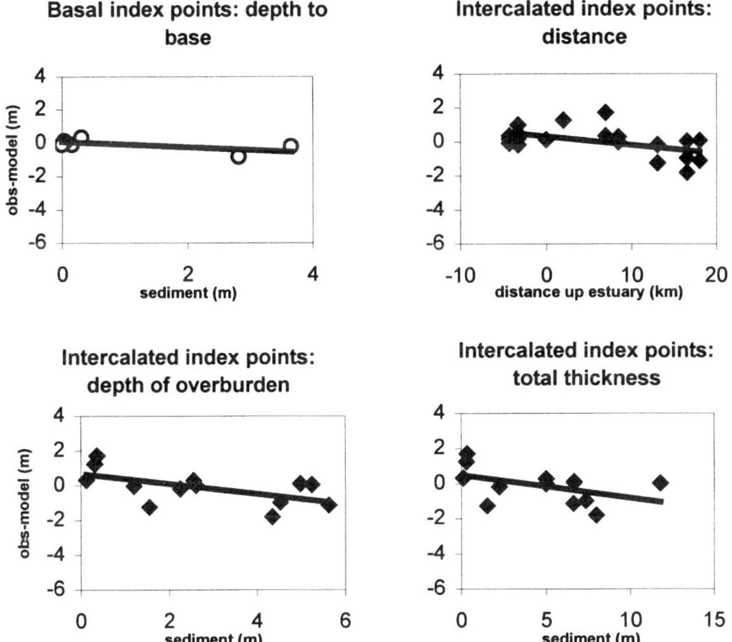

Fig. 18. Tees Estuary: scatter plots showing the statistically significant relationships from Table 3 between residuals (observed versus modelled relative sea-level) for basal and intercalated index points and parameters that may indicate local-scale processes (distance up estuary, depth of overburden, depth to base of Holocene sequence and total thickness of Holocene sequence).

Lincolnshire coast as a geological and geodynamic link between the major estuaries of the Humber and Wash (Brew 1997). There are now 28 index points for this region, including for the first time, data from calcareous material within clastic sediment units (further details in Horton *et al.* this volume). The sea-level index points record a rising curve towards present (Fig. 19),

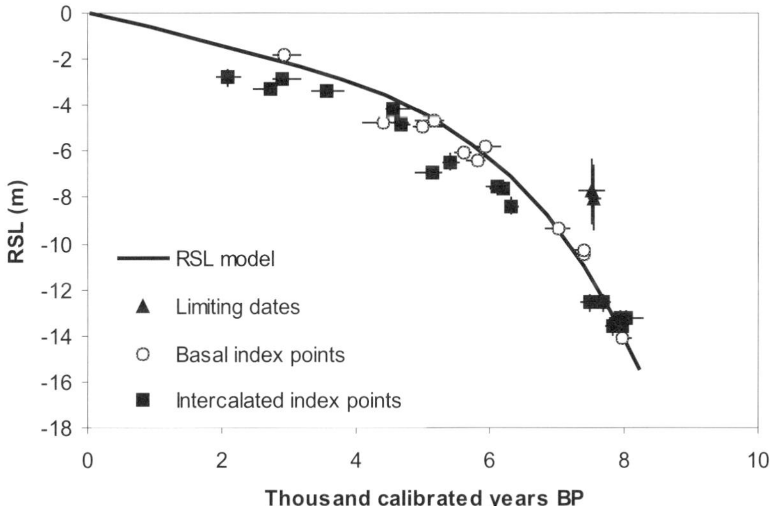

Fig. 19. Lincolnshire Marshes relative sea-level index points. Error terms are calculated as described in the text for each index point, but only appear when they extend beyond the size of the symbol.

with the majority of the samples from intercalated peats falling on or below the RSL curve. Horton *et al.* (this volume) discuss the data older than 7.5 cal. ka BP, which includes dated samples of foraminifera from clastic sediments. Differential crustal movement within the area is suggested by the significant correlation with distance, north–south along the coast as with the Northumberland analysis, for the basal peat index points (Fig. 20 and Table 3). As discussed in the Northumberland section, this could also include a tidal change effect. The cluster of samples from foraminifera discussed by Horton *et al.* (this volume) heavily influences the correlation with age (Fig. 20), and without which there is no significant trend. Sediment consolidation is suggested in the significant correlation for sediment thickness to the base of the Holocene.

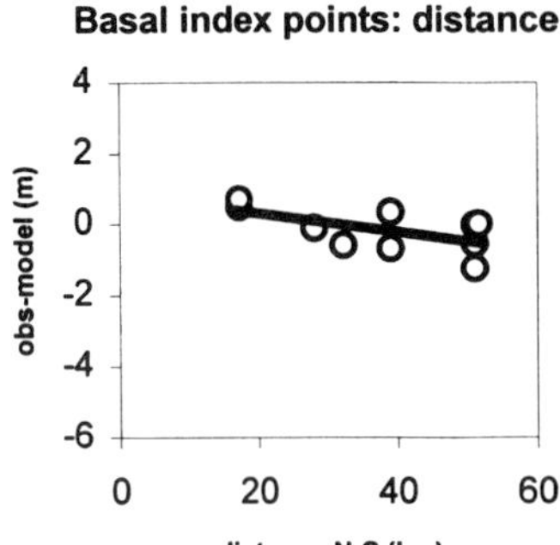

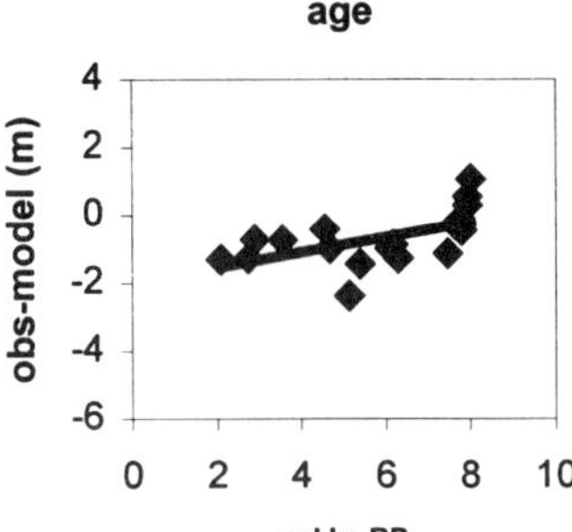

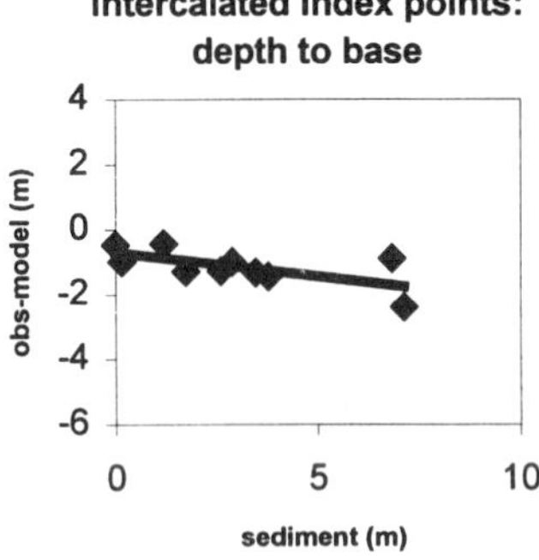

Fig. 20. Lincolnshire Marshes: scatter plots showing the statistically significant relationships from Table 3 between residuals (observed versus modelled relative sea-level) for basal and intercalated index points and parameters that may indicate local-scale processes (age, distance north–south, depth to base of Holocene sequence).

Fenland

Since a substantial mid Holocene Fenland database existed already (158 index points), LOIS sampling concentrated on deeper sediments closer to the coast where earlier index points could be recovered. In this way the Fenland database has been extended by 35, including several more index points of about 7 cal. ka BP or older. This region (Fig. 1) is large enough to hold out the possibility of recording differences in crustal history within it. Furthermore, the spread of the basal peat index points around the RSL curve (Fig. 21) is no greater than other regions with considerably smaller data sets (e.g. compare Fig. 22 with Fig. 9). The spatial distribution of samples is large, over 90 km, and previous investigations indicate separate areas within the Fenland basin showing contrasting sediment chronologies (e.g. Shennan 1986*b*; Waller 1994). The limitations of differentiating between isostatic and palaeotidal effects as outlined in the Humber analysis also occur here. The lack of a significant correlation between residuals and distance for either basal ($r_{basal} = -0.06$) or intercalated ($r_{intercalated} = 0.09$) peat index points (Table 3) should not at this stage be used to say there is no differential crustal movement within the Fenland basin. It is possible that a change in tidal range through time could have an effect of similar magnitude and opposite sign. Model results indicate that tidal range changes will have occurred in the Fenland area during the Holocene, but the predictions are poorly constrained by the resolution of the tidal models available or by the palaeogeography and palaeobathymetry models used (e.g. Hinton 1992; Shennan *et al.* this volume).

The only sediment parameter available for the majority of Fenland samples currently is the depth of overburden. For index points from both basal and intercalated peats the correlation with residual altitude is significant (Fig. 22 and Table 3), indicating the effects of sediment consolidation.

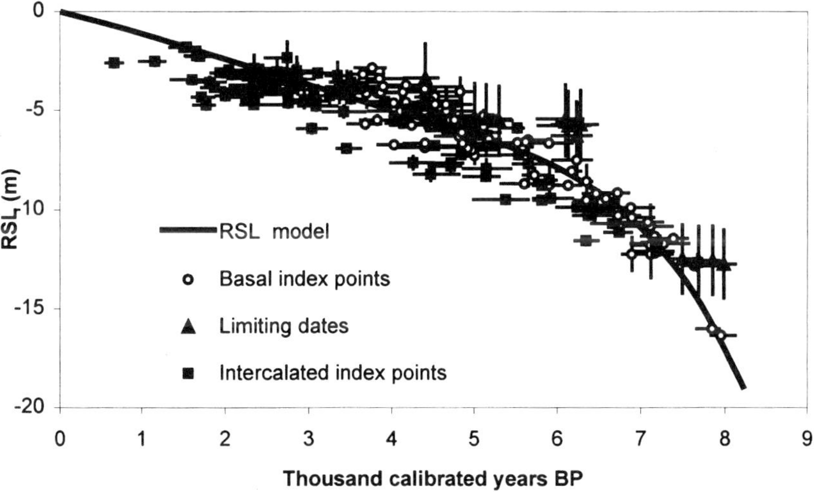

Fig. 21. Fenland relative sea-level index points. Error terms are calculated as described in the text for each index point, but only appear when they extend beyond the size of the symbol.

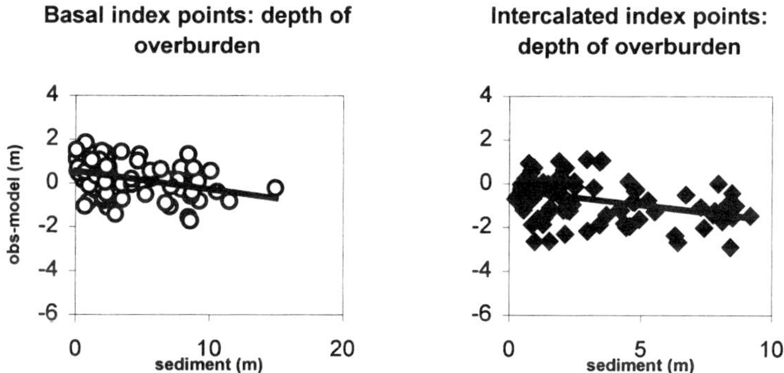

Fig. 22. Fenland: scatter plots showing the statistically significant relationships from Table 3 between residuals (observed versus modelled relative sea-level) for basal and intercalated index points and parameters that may indicate local-scale processes (depth of overburden).

North Norfolk

Locating deeper sequences and recovering several basal peats previously known but not systematically sampled enhanced the existing distribution of data points for north Norfolk (Fig. 1). Some extension along the north Norfolk coast was also achieved and 23 new index points added (Andrews *et al.* this volume). Index points from basal peats and the limiting dates from freshwater peats give a well-constrained sea-level curve, with most of the dates from intercalated peats lying below the curve (Fig. 23). Current tidal ranges change significantly east–west along the north Norfolk coast, with MHWST for Cromer, in the east, at 2.55 m OD, rising to 3.55 m OD to the west at Hunstanton. The distance parameter used in the analysis of residuals (Table 3) is distance west–east, but it shows no significant correlation with either basal or intercalated peat index points, which may have indicated an important spatially dependent change in tide levels between sites. Significant correlations with depth of overburden and total thickness of the Holocene sequence indicate the likely net effects of sediment consolidation (Fig. 24).

Regional summary: relative sea-level changes between north Northumberland and north Norfolk

The eight regional relative sea-level curves described separately in the sections above (Figs 7, 11, 13, 15, 17, 19, 21, 23) show a consistent

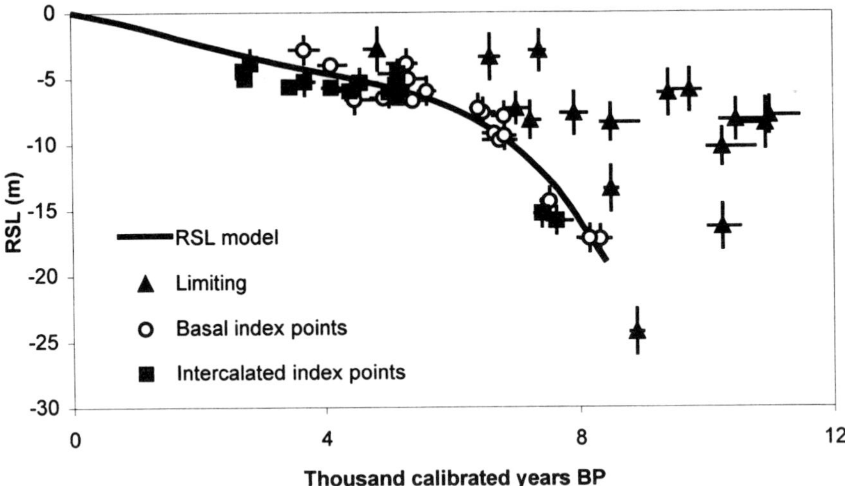

Fig. 23. North Norfolk relative sea-level index points. Error terms are calculated as described in the text for each index point, but only appear when they extend beyond the size of the symbol.

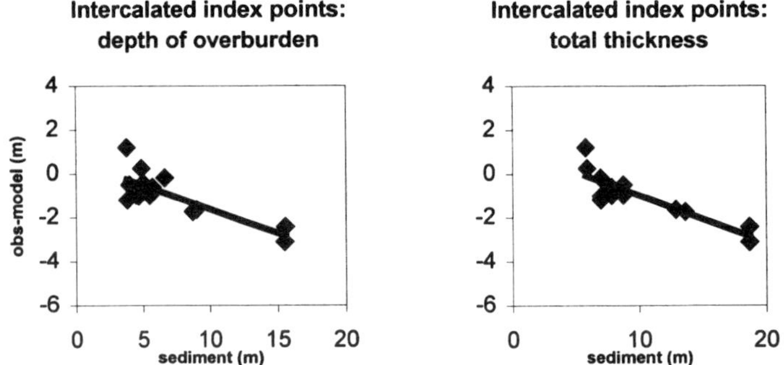

Fig. 24. North Norfolk: scatter plots showing the statistically significant relationships from Table 3 between residuals (observed versus modelled relative sea-level) for intercalated index points and parameters that may indicate local-scale processes (depth of overburden and total thickness of Holocene sequence).

sequence of increasingly higher sea level to the north at any point in time (Fig. 25). This sequence is consistent with model predictions (e.g. Lambeck 1995), but explanation of the offsets between predictions and observations (Fig. 6) awaits the results of modelling using recent analyses of the earth- and ice-model parameters (Lambeck *et al.* 1998; Shennan *et al.* this volume). Current rates of sea-level rise range from 1.04 ± 0.12 mm a^{-1} in the Fenland to -1.30 ± 0.68 mm a^{-1} (i.e. sea-level fall) in Northumberland North (Fig. 26). The scarcity of data for the last 3 ka from Northumberland, where the data sets are the smallest, gives less well-constrained statistical summaries than other regions. The approach of changing only the linear term may overestimate the current rate of sea-level fall because the polynomials are fitted to data for *c.* 3.5–8 cal. ka BP and show the rate of fall increasing from the sea-level maximum to the present (Figs 11, 13, 15). Linear solutions from the youngest index point to the present give rates of *c.* -0.66 mm a^{-1} for Northumberland North and *c.* -0.12 mm a^{-1} for Northumberland Central and South. At present there are no observations to test this further. Nevertheless, the observations of mid–late Holocene sea-level index points above present in Northumberland and below present in the Tees Estuary support the change in sign for current rates of sea-level change between these regions (Figs 25 and 26).

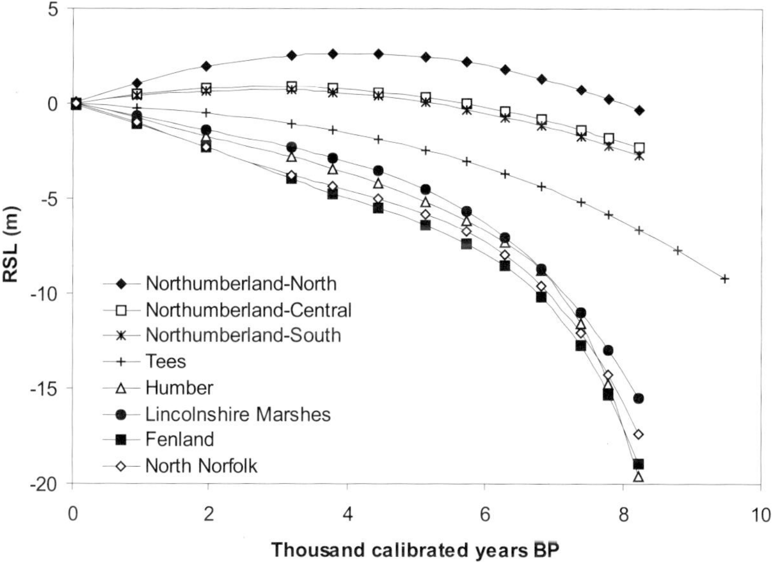

Fig. 25. Holocene relative sea-level curves for the east coast of England, based on the summary curves for each area shown in the preceding figures.

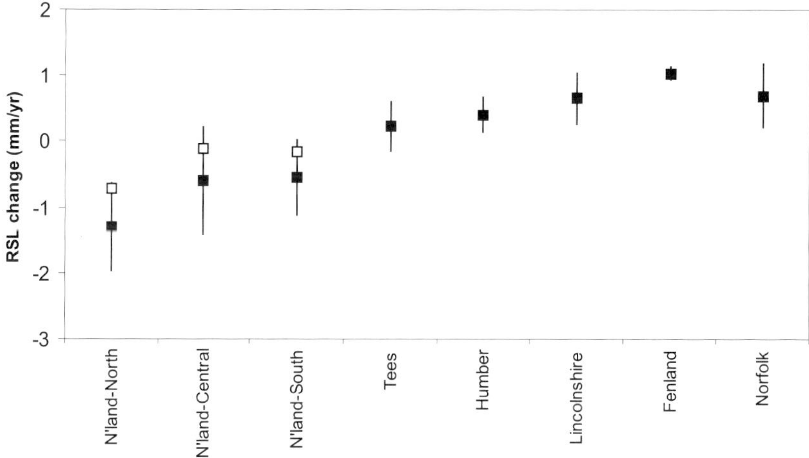

Fig. 26. Late Holocene–present rates of relative sea-level change for the east coast of England based on the extrapolation of the summary curve for each area shown in the preceding figures. The error term is the 95% regression confidence interval based on the sum of squares of the residuals from the regression. For the Northumberland areas the open square shows the rate based on the linear fit through the youngest index point only (see text for discussion).

Comparison with IPCC scenarios of future sea-level rise

The report of the Intergovernmental Panel on Climate Change 1995 (IPCC 95) summarizes a range of global mean sea-level rise scenarios (Houghton *et al.* 1996). The 'best-guess' or 'mid-range' predictions described in the technical summary and summary for policy makers in IPCC 95 refer to sea-level predictions from the IS92a prediction, which is discussed in detail by Warrick *et al.* (1996). The rise in sea-level for IS92a predicts a rise of 20 cm from 1990 to 2050, within a range of uncertainty of 7–39 cm; and 49 cm for 1990–2100, range 20–86 cm. The mean prediction translates into a rate of change in

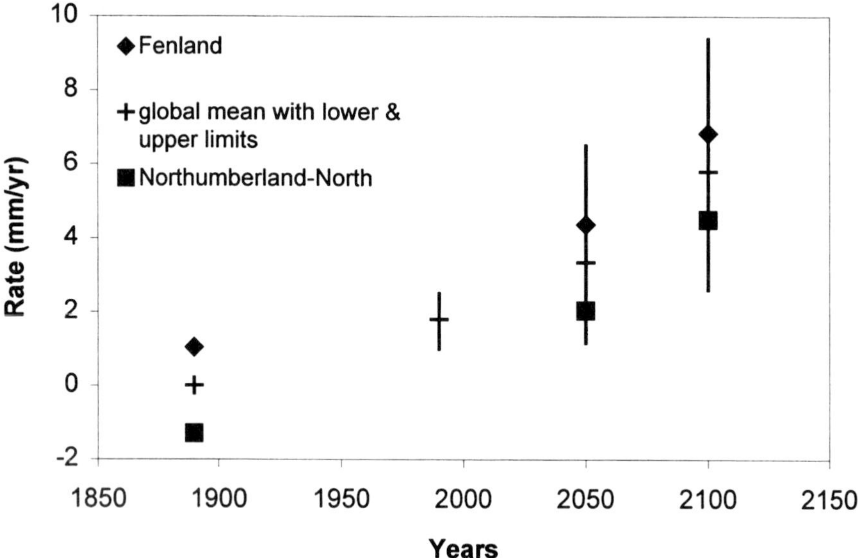

Fig. 27. Rates of sea-level change 1890–2100. Global mean data with error limits from Warrick *et al.* (1996), assuming zero eustatic sea-level rise in the pre-industrial era, shown as 1890. The late-Holocene rates for Northumberland North and Fenland (mean values from Fig. 26), shown for 1890 and added to the global mean data to give the values for 2050 and 2100.

excess of 5 mm a^{-1} by 2100 (Fig. 27). Assuming maintenance of the late Holocene relative differences between the Fenland and Northumberland North (Fig. 26) these data in combination indicate a rate of sea-level rise by 2100 in excess of 6 mm a^{-1} for the Fenland and 4 mm a^{-1} for Northumberland North. Such predictions have potentially very important implications for coastline changes and management. Detailed consideration of these is beyond the scope of this paper except to point out that Northumberland North has not experienced such rates of sea-level rise since before 4 cal. ka BP (Fig. 11), though this must be qualified by the different time-scales of the comparison, and for much of the period the coastline has developed under falling RSL. The transition sometime in the next century from relative fall to relative rise may initiate significant coastline changes. Similarly, an increase in the rate of sea-level rise in the Fenland may be sufficient to initiate a change from a regressive to transgressive phase (e.g. Shennan 1987; Allen 1990, 1995, 1997).

The uncertainties of the IPCC projections are large relative to the mean predicted sea-level rise (Fig. 27). If sea-level rise is towards the lower end of the range, the differential crustal movement or isostatic factor, i.e. the difference between Fenland and Northumberland North, is greater than the 1990–2100 increase. For sea-level predictions at the higher end of the range, the differential isostatic factor becomes proportionately less important, around 25%.

Conclusions

This analysis of sea-level data from the east coast of England has identified local-scale and regional-scale factors that explain spatial and temporal variations in the altitude of sea-level index points. The isostatic effect of the glacial rebound process, including both the ice (glacio-isostatic) and water (hydro-isostatic) load contributions, explains the differences at the regional scale, which are manifested in approximately a 20-m range for features formed 8 ka BP. By 4 ka BP, RSL in Northumberland was above present; whereas in areas to the south RSL has been below present throughout the Holocene. These differences will produce different rates of sea-level change through the twenty-first century. The data agree closely with the patterns predicted by glacio- and hydro-isostatic models, but small systematic differences along the east coast await testing against new ice models. New predictions based on the revised models will provide the basis of testing further the effects of local-scale processes identified here. These include differential isostatic effects within the Fenland, tide-range changes during the Holocene, and the effects of sediment consolidation. The paucity of data for the last 3 cal. ka BP remains a constraint on the analyses from most areas and is another priority for future research.

This is publication number 590 of the Land–Ocean Interaction Study (LOIS) Community Research Programme and the work was supported by NERC grants GST/02/0760, GST/02/0761 and GST/02/A766 under LOIS special topics 313, 316 and 348, respectively. Additional data were supplied with the collaboration of principal investigators and research staff from other LOIS projects and core programmes (details in preface of this volume). We are very grateful to the NERC radiocarbon laboratory at East Kildbride for the radiocarbon dates and to landowners for access to field sites. We thank T. Törnqvist and C. Vita-Finzi for their excellent reviews of the original version and suggestions for improvements.

References

ALLEN, J. R. L. 1990. Salt-marsh growth and stratification: a numerical model with special reference to the Severn Estuary, southwest Britain. *Marine Geology*, **95**, 77–96.

——1995. Salt-marsh growth and fluctuating sea level: implications of a simulation model for Flandrian coastal stratigraphy and peat based sea-level curves. *Sedimentary Geology*, **100**, 21–45.

——1997. On the minimum amplitude of regional sea-level fluctuations during the Flandrian. *Journal of Quaternary Science*, **12**, 501–505.

ANDREWS, J. E., BOOMER, I., BALIFF, I., BALSON, P., BRISTOW, C., CHROSTON, P. N., FUNNELL, B. M., HARWOOD, G. M., JONES, R., MAHER, B. A. & SHIMMIELD, G. B. 2000. Sedimentary evolution of the north Norfolk barrier coastline in the context of Holocene sea-level change. *This volume*.

BREW, D. S. 1997. Holocene lithostratigraphy and broad scale evolution of the Lincolnshire Outmarsh, eastern England. *East Midland Geographer*, **20**, 20–32.

——, HOLT, T., PYE, K. & NEWSHAM, R. 2000. Holocene sedimentary evolution and palaeocoastlines of the Fenland embayment, eastern England. *This volume*.

FLEMING, K., JOHNSTON, P., ZWARTZ, D., YOKOYAMA, Y., LAMBECK, K. & CHAPPELL, J. 1998. Defining the eustatic sea-level curve since the last glacial maximum using far and intermediate-field sites. *Earth and Planetary Science Letters*, **163**, 327–342.

GAUNT, G. D. & TOOLEY, M. J. 1974. Evidence for Flandrian sea-level changes in the Humber Estuary and adjacent areas. *Bulletin of the Institute of Geological Sciences*, **48**, 25–41.

GODWIN, H. 1940. Studies in the post-glacial history of British vegetation. III: Fenland pollen diagrams. IV: Post-glacial changes of relative land and sea level in the English Fenland. *Philosophical Transactions of the Royal Society of London B*, **230**, 239–303.

HINTON, A. C. 1992. Palaeotidal changes within the area of the Wash during the Holocene. *Proceedings of the Geologists' Association*, **103**(3), 259–272.

HORTON, B. P., EDWARDS, R. J. & LLOYD, J. M. 2000. Implications and applications of a microfossil transfer function in Holocene sea-level studies. *This volume*.

HOUGHTON, J. J., MEIRO FILHO, L. G., CALLANDER, B. A., HARRIS, N., KATTENBERG, A. & MASKELL, K. (eds) 1996. *Climate Change 1995: The Science of Climate Change*. Cambridge University Press, Cambridge.

JELGERSMA, S. 1966. Sea-level changes during the last 10 000 years. *In*: SAWYER, J. S. (ed.) *World Climate 8000–0 BC*. Royal Meteorological Society, London, 54–69.

LAMBECK, K. 1993*a*. Glacial rebound of the British Isles. I. Preliminary model results. *Geophysical Journal International*, **115**, 941–959.

——1993*b*. Glacial rebound of the British Isles. II. A high-resolution, high-precision model. *Geophysical Journal International*, **115**, 960–990.

——1995. Late Devensian and Holocene shorelines of the British Isles and North Sea from models of glacio–hydro-isostatic rebound. *Journal of the Geological Society of London*, **152**, 437–448.

——1996. Limits on the areal extent of the Barents Sea ice sheet in Late Weischelian time. *Palaeogeography, Palaeoclimatology, Palaeoecology*, **12**, 41–51.

——, SMITHER, C. & JOHNSTON, P. 1998. Sea-level change, glacial rebound and mantle viscosity for northern Europe. *Geophysical Journal International*, **134**, 102–144.

LONG, A. J., INNES, J. B., KIRBY, J. R., LLOYD, J. M., RUTHERFORD, M. M., SHENNAN, I. & TOOLEY, M. J. 1998. Holocene sea-level change and coastal evolution in the Humber Estuary, eastern England: an assessment of rapid coastal change. *The Holocene*, **8**, 229–247.

METCALFE, S. E., ELLIS, S., HORTON, B. P., INNES, J. B., MCARTHUR, J. J., MITLEHNER, A., PARKES, A., PETHICK, J. S., REES, J. G., RIDGWAY, J., RUTHERFORD, M. M., SHENNAN, I. & TOOLEY, M. J. 2000. The Holocene evolution of the Humber Estuary: reconstructing change in a dynamic environment. *This volume*.

ORFORD, J. D., WILSON, P., WINTLE, A. G., KNIGHT, J. & BRALEY, S. 2000. Coastal dune initiation in Northumberland and Norfolk, eastern UK: climate and sea-level changes as possible forcing agents for dune initiation. *This volume*.

PAUL, M. A. & BARRAS, B. F. 1998. A geotechnical correction for post-depositional sediment compression: examples from the Forth Valley, Scotland. *Journal of Quaternary Science*, **13**, 171–176.

PELTIER, W. R. 1998. Postglacial variations in the level of the sea: implications for climate dynamics and solid-earth geophysics. *Reviews of Geophysics*, **36**, 603–689.

PLATER, A. J. & SHENNAN, I. 1992. Evidence of Holocene sea-level change from the Northumberland coast, eastern England. *Proceedings of the Geologists' Association*, **103**, 201–216.

——, RIDGWAY, J., RAYNER, B., SHENNAN, I., HORTON, B. P., HAWORTH, E. Y., WRIGHT, M. R. & WINTLE, A. G. 2000. Sediment provenance and flux in the Tees estuary: the record from the Late Devensian to the present. *This volume.*

REES, J. G., RIDGWAY, J., ELLIS, S., KNOX, R. W. O'B., NEWSHAM, R. & PARKES, A. 2000. Holocene sediment storage in the Humber Estuary. *This volume.*

RIDGWAY, J., ANDREWS, J. E., ELLIS, S., HORTON, B. P., INNES, J. B., KNOX, R. W. O'B., MCARTHUR, J. J., MAHER, B. A., METCALFE, S. E., MITLEHNER, A., PARKES, A., REES, J. G., SAMWAYS, G. M. & SHENNAN., I. 2000. Analysis and interpretation of Holocene sedimentary sequences in the Humber Estuary. *This volume.*

SHENNAN, I. 1982. Interpretation of Flandrian sea-level data from the Fenland, England. *Proceedings of the Geologists' Association*, 93, 53–63.

——1983. Flandrian and Late Devensian sea-level changes and crustal movements in England and Wales. *In*: SMITH, D. E. & DAWSON, A. G. (eds) *Shorelines and Isostasy*. Academic Press, London, 255–283.

——1986*a*. Flandrian sea-level changes in the Fenland. I. The geographical setting and evidence of relative sea-level changes. *Journal of Quaternary Science*, **1**, 119–154.

——1986*b*. Flandrian sea-level changes in the Fenland. II. Tendencies of sea-level movement, altitudinal changes and local and regional factors. *Journal of Quaternary Science*, **1**, 155–179.

——1987. Impacts on the Wash of sea-level rise. *Research and Survey in Nature Conservation*, **7**, 77–90.

——1989. Holocene crustal movements and sea-level changes in Great Britain. *Journal of Quaternary Science*, **4**, 77–89.

——1992. Late Quaternary sea-level changes and crustal movements in eastern England and eastern Scotland: an assessment of models of coastal evolution. *Quaternary International*, **15/16**, 161–173.

——, HORTON, B. P., INNES, J. B., GEHRELS, W. R., LLOYD, J. M., MCARTHUR, J. J., RUTHERFORD, M. M. & WINGFIELD, R. in press *b*. Late Quaternary sea-level changes, crustal movements and coastal evolution in Northumberland. *Journal of Quaternary Science*.

——, INNES, J. B., LONG, A. J. & ZONG, Y. 1995. Late Devensian and Holocene relative sea-level changes in northwestern Scotland: new data to test existing models. *Quaternary International*, **26**, 97–123.

——, LAMBECK, K., FLATHER, R., HORTON, B. P., MCARTHUR, J. J., INNES, J. B., LLOYD, J. M. & RUTHERFORD, M. M. 2000. Modelling western North Sea palaeogeographies and tidal changes during the Holocene. *This volume.*

——, ——, HORTON, B. P., INNES, J. B., LLOYD, J. M., MCARTHUR, J. J., PURCELL, A. & RUTHERFORD, M. M. in press *a*. Late Devensian and Holocene records of relative sea-level changes in northwest Scotland and their implications for glacio–hydro-isostatic modelling. *Quaternary Science Reviews*.

STUIVER, M. & REIMER, P. J. 1993. Extended ^{14}C data base and revised CALIB 3.0 ^{14}C age calibration program. *Radiocarbon*, **35**, 215–230.

TOOLEY, M. J. 1978. The history of Hartlepool Bay. *International Journal of Nautical Archaeology and Underwater Exploration*, **7**, 71–75.

TÖRNQVIST, T. E., VAN REE, M. H. M., VAN 'T VEER, R. & VAN GEEL, B. 1998. Improving methodology for high-resolution reconstruction of sea-level rise and neotectonics by palaeocological analysis and AMS ^{14}C dating of basal peats. *Quaternary Research*, **49**, 72–85.

TROELS-SMITH, J. 1955. Characterization of unconsolidated sediments. *Danmarks Geologiske Undersogelse, Series IV*, **3**, 38–73.

VAN DE NOORT, R. & ELLIS, S. 1995. *Wetland Heritage of Holderness: an Archaeological Survey*. Humber Wetlands Project, University of Hull, Hull.

—— & ——1997. *Wetland Heritage of the Humberhead Levels: an Archaeological Survey*. Humber Wetlands Project, University of Hull, Hull.

VAN DE PLASSCHE, O. (ed.) 1986. *Sea-level research: a manual for the collection and evaluation of data.* Geobooks, Norwich.

WALLER, M. 1994. *The Fenland Project, Number 9: Flandrian Environmental Change in Fenland, East Anglia.* Archaeology Report Number 70, Fenland Project Committee, Cambridgeshire County Council, Cambridge.

WARRICK, R. A., LE PROVOST, C., MEIER, M. F., OERLEMANS, J. & WOODWORTH, P. L. 1996. Changes in sea level. *In*: HOUGHTON, J. T., MEIRA FILHO, L. G., CALLANDER, B. A., HARRIS, N., KATTENBERG, A. & MASKELL, K. (eds) *The Science of Climate Change: Contribution of Working Group I to the Second Assessment Report of the Intergovernmental Panel on Climate Change.* Cambridge University Press, Cambridge, 572.

Modelling western North Sea palaeogeographies and tidal changes during the Holocene

I. SHENNAN,[1] K. LAMBECK,[2] R. FLATHER,[3] B. HORTON,[1] J. McARTHUR,[1] J. INNES,[1] J. LLOYD,[1] M. RUTHERFORD[1] & R. WINGFIELD[4] †

[1] *Environmental Research Centre, Department of Geography, University of Durham, Durham DH1 3LE, UK (e-mail: ian.shennan@durham.ac.uk)*
[2] *Research School of Earth Sciences, The Australian National University, Canberra ACT 0200, Australia*
[3] *Institute of Oceanographic Sciences, Birkenhead, UK Centre for Coastal and Marine Sciences, Proudman Oceanographic Laboratory, Birkenhead L43 7RA, UK*
[4] *Coastal and Engineering Geology Group, British Geological Survey, Kingsley Dunham Centre, Keyworth, Nottingham NG12 5GG, UK*

Abstract: Analysis of cores collected from Late Devensian (Weichselian) and Holocene sediments on the floor of the North Sea provides evidence of the transgression of freshwater environments during relative sea-level rise. Although many cores show truncated sequences, examples from the Dogger Bank, Well Bank and 5 km offshore of north Norfolk reveal transitional sequences and reliable indicators of past shoreline positions. Together with radiocarbon-dated sea-level index points collected from the Holocene sediments of the estuaries and coastal lowlands of eastern England these data enable the development and testing of models of the palaeogeographies of coastlines in the western North Sea and models of tidal range changes through the Holocene epoch. Geophysical models that incorporate ice-sheet reconstructions, earth rheology, eustasy, and glacio- and hydro-isostasy provide predictions of sea-level relative to the present for the last 10 ka at 1-ka intervals. These predictions, added to a model of present-day bathymetry, produce palaeogeographic reconstructions for each time period. The palaeogeographic maps reveal the transgression of the North Sea continental shelf. Key stages include a western embayment off northeast England as early as 10 ka BP; the evolution of a large tidal embayment between eastern England and the Dogger Bank before 9 ka BP with connection to the English Channel prior to 8 ka BP; and Dogger Bank as an island at high tide by 7.5 ka BP and totally submerged by 6 ka BP. Analysis of core data shows that coastal and saltmarsh environments could adapt to rapid rates of sea-level rise and coastline retreat. After 6 ka BP the major changes in palaeogeography occurred inland of the present coast of eastern England. The palaeogeographic models provide the coastline positions and bathymetries for modelling tidal ranges at each 1-ka interval. A nested hierarchy of models, from the scale of the northeast Atlantic to the east coast of England, uses 26 tidal harmonics to reconstruct tidal regimes. Predictions consistently show tidal ranges smaller than present in the early Holocene, with only minor changes since 6 ka BP. Recalibration of previously available sea-level index points using the model results rather than present tidal-range parameters increases the difference between observations and predictions of relative sea-levels from the glacio–hydro-isostatic models and reinforces the need to search for better ice-sheet reconstructions.

The majority of the continental shelf of the North Sea was subaerial at the opening of the Holocene epoch. Relative sea-level changes, the combined result of eustatic sea-level change and glacio- and hydro-isostatic land-level change, caused the transgression of the former land areas. During this transgression there was the potential for significant erosion, transport and deposition of sediment both on the continental shelf and into the coastal zone and estuaries. Many sediment cores from the offshore have a major hiatus with much of the Holocene record missing, presumably the result of marine erosion during the transgression (Jelgersma 1961). In contrast to the

† Deceased.

From: SHENNAN, I. & ANDREWS, J. (eds) *Holocene Land–Ocean Interaction and Environmental Change around the North Sea*. Geological Society, London, Special Publications, **166**, 299–319. 1-86239-054-1/00/$15.00

present land areas, little is known in detail about the relative sea-level history of the North Sea and the configuration of the coastline at any one time. In 1994, at the start of the Land–Ocean Interaction Study (LOIS), there were very few radiocarbon-dated samples of Holocene age from the UK sector of the North Sea. Only one or two could be used to fix a past sea level, the rest were simply limiting values, either indicating fully terrestrial or fully marine conditions, with no precise relationship to sea-level or the position of the coastline at the time of deposition. Although more data are available from the rest of the North Sea (e.g. Behre *et al.* 1979; Jelgersma 1979; Ludwig *et al.* 1981) the total number of validated index points is over two orders-of-magnitude less than those available from the surrounding land areas. By integrating data from a range of recent research areas we can start to address aspects of understanding long-term environmental change that require knowledge of the Holocene evolution of the North Sea. As long-term objectives we need to understand how the changing water depths and coastline positions limit our knowledge of environmental processes and change. We also need to explain in quantitative terms the fluxes of sediments into the estuaries and coastal lowlands from the North Sea (e.g. Rees *et al.* this volume calculate that *c.* 75% of the 10km^3 of Holocene sediment in the Humber Estuary and lowlands had a coastal or marine source), the relationships between sea-level change and coastline advance or retreat, and the geographical variations in sea-level change. While the majority of samples used to reconstruct Holocene sea-level change formed close to the contemporary level of mean high water of spring tides (MHWST) rather than mean sea-level (MSL), few studies of Holocene relative sea-level changes include numerical reconstructions of tidal-range changes through time (e.g. Gehrels *et al.* 1995). Most studies standardize the data using the present tide levels, with some incorporating a tidal element within the error term of the reconstructed sea-level (e.g. Shennan 1989; Lambeck 1993*b*). As the configuration of the coastline and offshore bathymetry changed, not least with the connection of the North Sea to the English Channel, the possibility of significant changes in the tidal prism arises. Initial modelling studies with both lower and higher sea-levels predict changes in the tidal prism (e.g. Belderson *et al.* 1986; Austin, 1991; Hinton 1992, 1995, 1996; Gehrels *et al.* 1995; Scourse & Austin 1995). Temporal changes in tidal amplitudes are not only important for calibrating sea-level index points: for such changes have profound influences on shelf-sea processes, which themselves are tide-dependent, such as sand transport paths, seasonal stratification, and therefore primary productivity (Austin 1991).

Previous studies of reconstructing both the palaeocoastlines of the North Sea and the tides were limited by fundamental assumptions about the input parameters. Bathymetric maps show a low-gradient topography for much of the floor of the southern North Sea over which a relatively small change in relative sea-level would have caused a large horizontal shift in the coastline. Prior to geophysical models incorporating ice-sheet reconstructions, earth rheology, eustasy, and glacio- and hydro-isostasy (e.g. Lambeck 1991, 1993*a*, *b*, 1995; Peltier 1996, 1998) the spatial and temporal variations of relative sea-level across the North Sea were very poorly constrained (e.g. Behre *et al.* 1979; Jansen *et al.* 1979; Jelgersma 1979). This is a severe limitation on the palaeogeographic reconstructions of Jelgersma (1979) and the tidal models of Austin (1991) and Hinton (1992, 1995, 1996). Most of the tidal models simply used sea-level lowered as a plane surface by values indicated by a eustatic curve, although Hinton (1992, 1995) included a spatially variable isostatic component of up to 2 m in modelling the southwest North Sea and the Wash (Fig. 1).

The LOIS project enabled different advances to come together and re-address these issues; Shennan & Andrews (this volume) describe individual contributions from the Land–Ocean Evolution Perspective Study (LOEPS) component. In order to provide a viable database, the LOEPS core programme undertook sampling both onshore and offshore, thus providing new data for the early Holocene from many coastal locations. These new samples also allowed significant improvement of the chronostratigraphy of the western part of the North Sea. Onshore sampling concentrated on the east coast of England between Northumberland and north Norfolk (Fig. 1). The spatially variable component of relative sea-level rise shows an increasing effect back through time, reaching *c.* 30 m around 10 ^{14}C ka BP for the east coast of England (Fig. 1) and much greater in the areas of Scotland, where Late Devensian ice was thicker. This component must be included in reconstructions of both palaeocoastlines and tides. The enlarged database of sea-level index points and the possibility of identifying different tide levels from the stratigraphy (e.g. Horton *et al.* this volume; Shennan *et al.* this volume) offers the potential for testing the output from tidal models. Previous models of Holocene tides in the North Sea (e.g. Austin 1991; Hinton 1992, 1995) predict reduced tide levels compared with

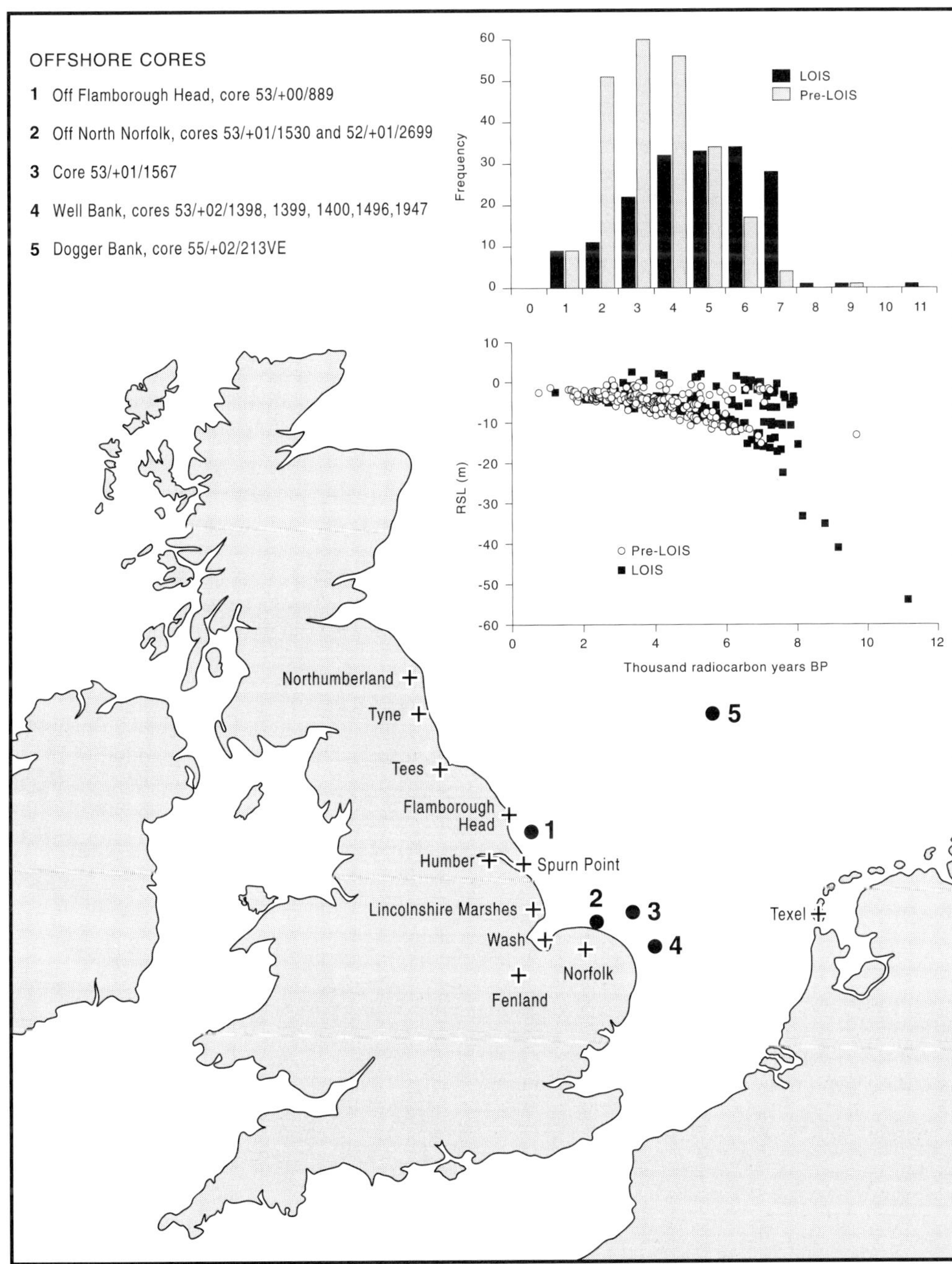

Fig. 1. Location map of the North Sea with the location of cores and insert of spatial and temporal distribution of east coast sea-level index points.

the present. These models, however, were limited in testing these predictions against observations, of sea-level index points from sedi-ment cores, that still included other possible factors such as sediment consolidation and differential isostatic movements. Lambeck (1993*a*, *b*, 1995) provided relative sea-level change predictions for each grid cell in the North Sea. These form the basis for the present paper, although recent revisions to the model parameters (Lambeck 1998), in part

Table 1. *Summary of sea-level index points and limiting data from the North Sea*

Site	Laboratory code	^{14}C age $\pm 1\sigma$	Altitude (m OD)	Reference water level*	Tendency	Change in RSL†	Reference
North Norfolk 10 52/+01/2699	AA27142	7 580 ± 70	−19.89	MHWST	+	−22.44 ± 0.54	This paper
Well Bank 4 52/+02/1398	AA27143	9 145 ± 60	−37.52	MHWST − 0.20	+	−39.29 ± 1.29	This paper
Well Bank 4a 52/+02/1399	AA27144	8 995 ± 60	−37.92	MHWST − 0.20	+	−39.69 ± 1.29	This paper
Well Bank 3 52/+02/1400	AA27145	9 045 ± 65	−38.61	MHWST − 0.20	+	−40.38 ± 1.29	This paper
53/+00/899	AA25602	11 145 ± 75	−51.40	MHWST − 0.20	+	−53.95 ± 0.20	This paper
Lemon Bank 15 53/+01/1567	AA23946	8 775 ± 70	−32.77	MHWST − 0.20	+	−35.12 ± 0.20	This paper
Well Bank 1 53/+02/1495	AA23944	9 270 ± 75	−38.53	MHWST − 0.20	+	−40.88 ± 0.20	This paper
Well Bank 7 53/+02/1496	AA27146	9 155 ± 70	−38.88	MHWST	+	−41.85 ± 1.33	This paper
Well Bank 8 53/+02/1497	AA27147	9 155 ± 75	−37.59	MHWST − 0.20	+	−39.36 ± 1.33	This paper
Dogger Bank 55/+02/213VE	AA22662	8 140 ± 55	−31.06	MHWST − 0.20	+	−33.26 ± 1.30	This paper
German Bight 235	HV7095	8 190 ± 140	−38.00	MHWST − 0.20	+	−39.80 ± 1.66	Behre *et al.* (1979)
German Bight 235	HV7094	8 485 ± 125	−38.09	MHWST − 0.20	+	−39.89 ± 1.66	Behre *et al.* (1979)
German Bight A10	HV8600	7 540 ± 80	−21.50	MHWST − 0.20	+	−23.30 ± 1.66	Behre *et al.* (1979)
German Bight A10	HV8602	7 960 ± 205	−22.60	MHWST − 0.20	+	−24.40 ± 1.66	Behre *et al.* (1979)
German Bight A10	HV8601	7 980 ± 60	−21.90	MHWST − 0.20	+	−23.70 ± 1.66	Behre, *et al.* (1979)
53/+00/889	AA27137	11 425 ± 95	−51.52	≥[MHWST + MTL]/2	Limiting	≤ −53.09 ± 1.28	This paper
North Norfolk 11 53/+01/1530	AA27148	7 975 ± 55	−22.87	≥[MHWST + MTL]/2	Limiting	≤ −23.89 ± 1.59	This paper
Well Bank 153/+02/1495	AA23945	11 325 ± 85	−38.97	≥[MHWST + MTL]/2	Limiting	≤ −40.37 ± 1.15	This paper
Elbow Formation		9 374 ± 90	−35.00	≥[MHWST + MTL]/2	Limiting	≤ −36.14 ± 1.36	Kirby & Oele (1975)
Elbow Formation		9 949 ± 120	−35.00	≥[MHWST + MTL]/2	Limiting	≤ −36.14 ± 1.36	Kirby & Oele (1975)
Geordie Trough	OXA332	9 335 ± 105	−59.00	≥[MHWST + MTL]/2	Limiting	≤ −60.57 ± 1.24	Harland & Long (1996)
German Bight 172	HV7091	8 950 ± 95	−37.27	≥[MHWST + MTL]/2	Limiting	≤ −38.34 ± 1.91	Behre *et al.* (1979)
German Bight 290		*c.* 6000–7000	−27.00	≥[MHWST + MTL]/2	Limiting	≤ −28.04 ± 1.08	Behre *et al.* (1979)
German Bight 85		*c.* 5000–6000	−17.00	≥[MHWST + MTL]/2	Limiting	≤ −18.04 ± 1.08	Behre *et al.* (1979)
German Bight 172	HV7091	8 950 ± 95	−37.27	≥[MHWST + MTL]/2	Limiting	≤ −38.34 ± 1.08	Behre *et al.* (1979)
Lower Eider 7a	HV628	7 115 ± 90	−12.00	≥[MHWST + MTL]/2	Limiting	≤ −13.04 ± 1.08	Behre *et al.* (1979)
Lower Elbe 35 7a	HV6189	8 075 ± 60	−29.00	≥[MHWST + MTL]/2	Limiting	≤ −30.04 ± 1.08	Behre *et al.* (1979)
Miele Bay 307	HV6190	6 705 ± 60	−7.00	≥[MHWST + MTL]/2	Limiting	≤ −8.04 ± 1.08	Behre *et al.* (1979)
North Sea 56/77	HV2575	7 790 ± 90	−24.90	≥[MHWST + MTL]/2	Limiting	≤ −25.94 ± 1.08	Ludwig *et al.* (1981)
North Sea 58/67	HV2143	7 720 ± 65	−24.35	≥[MHWST + MTL]/2	Limiting	≤ −25.39 ± 1.08	Ludwig *et al.* (1981)
North Sea F14	GRN5758	9 935 ± 55	−46.00	≥[MHWST + MTL]/2	Limiting	≤ −47.10 ± 1.21	Jelgersma *et al.* (1979)
North Sea F15	GRN5759	9 445 ± 80	−47.00	≥[MHWST + MTL]/2	Limiting	≤ −48.10 ± 1.21	Jelgersma *et al*, (1979)

* The reference water level is given as a mathematical expression of tidal parameters plus/minus an indicative difference. This is the distance from the mid-point of the indicative range to the reference water-level.

† Relative sea-level (RSL) is calculated as altitude minus the reference water level. The RSL error range is calculated as the square root of the sum of squares of altitudinal error, sample thickness, tide level error and indicative range. The indicative range (given as a maximum) is the most probable vertical range in which the sample occurs, although for limiting dates the sample could occur above that range.

resulting from other LOIS observations (Shennan *et al.* in press), will be used in the next generation of models. Finally, increases in computing power allow more tidal constituents to be considered. The tidal models used are essentially the same as in previous models, being those developed by Flather (e.g. 1976, 1979, 1981). Austin (1991) used only one tidal constituent and Hinton (1992) used six. Here we base our analysis on 26.

This paper represents the first stage in a new generation of modelling the Holocene palaeogeography and tidal regimes of the western North Sea. We will discuss the next stage required in this approach since the long-term aim has influenced the approach undertaken. The two broad objectives of this paper are to produce a series of palaeogeographic reconstructions of the western North Sea for the Holocene, based on spatially variable solutions of relative sea-level change, to use these reconstructions as the basis for modelling tidal regimes, and to test both the palaeogeographies and the tides against observations.

Data

Other LOEPS projects (Andrews *et al.*; Brew *et al.*; Metcalfe *et al.*; Plater *et al.*; Rees *et al.*; Shennan *et al.* this volume) describe new data from onshore locations along the east coast of England. Here we summarize new data from offshore cores. Microfossil evidence, following the methods outlined by Ridgway *et al.* (this volume) and Shennan *et al.* (this volume) indicates whether the dated sample is a sea-level index point that can be given a quantified relationship to a past tide level, with error limits. Alternatively the sample may represent a terrestrial or freshwater lacustrine deposit that indicates an age and location inland of the contemporaneous coast, and an altitude within the upper part of the tidal range or above the limit of extreme high tides. The data are summarized in Table 1 with three examples described: from the Dogger Bank; the offshore zone north of the Norfolk coast; and the Well Bank (Figs 1 and 2). The microfossil evidence reinforces the lithological interpretation that the sample is *in situ* where there is a transition between organic and minerogenic layers and hence sedimentary environments. All ages are reported as radiocarbon years, the time-scale also used in the earth rheology model and to define the history of the ice-sheets (Lambeck 1995; Lambeck *et al.* 1998). Provided that the radiocarbon time-scale differs from the calendar time-scale only in a linear way, and as long as the same time-scale is used throughout, the use of the former does not change the results. Any non-linearities in the radiocarbon time scale do not introduce significant errors into the analysis (Lambeck *et al.* 1998), provided that realistic error estimates are assumed.

Dogger Bank 55/+02/213VE

A single core from the Dogger Bank area of the central North Sea provides a reliable sea-level index point. A diamicton surface is overlain by a thin silt peat that grades into over 4 m of sand silt, which is shown to be intertidal in origin by the presence of saltmarsh foraminifera *Jadammina macrescens* and *Trochammina inflata* (Fig. 2a). The silt peat likewise formed under saltmarsh conditions as it also contains these foraminifera, abundant foraminiferal test linings, and high frequencies of saltmarsh pollen taxa *Artemisia*, Chenopodiaceae, *Plantago maritima*, *Spergularia* and *Taraxacum*-type. Saltmarsh indicators rise in abundance through the peat, although continued freshwater influence is shown by *Pediastrum* colonies. The top of the peat is a transitional transgressive contact, which forms a sea-level index point dated 8140 ± 50 BP at −31.06 m OD (Table 1). In comparison with pollen data across north-west Europe (Huntley *et al.* 1995) the high *Pinus* and *Corylus* pollen with low *Quercus* and *Ulmus* values agree with this age.

North Norfolk 52/+01/2699

This core was recovered from 15m water depth, *c.* 5 km off the north Norfolk coast, where the vibrocorer penetrated 11 cm of silt peat overlain by almost 5 m of clay. The basal part of the clay contains a rich foraminiferal assemblage (Fig. 2b) with saltmarsh (*Miliammina fusca*), estuarine (mainly *Ammonia* spp.) and shelf (e.g. *Quinqueloculina seminulum*) forms. Polyhalobous diatoms, mainly *Cocconeis scutellum* and *Opephora pacifica*, and a range of mesohalobous diatoms are also recorded, as well as dinoflagellate cysts *Operculodinium centrocarpum* and *Spiniferites* spp. The base of the clay is intertidal, also containing saltmarsh pollen Chenopodiaceae, *Artemisia*, *Taraxacum*-type and *Aster*-type. The upper levels of the silt peat also contain these saltmarsh pollen taxa and foraminifera test linings. This is a transitional transgressive con-tact and provides a sea-level index point, dated 7580 ± 70 BP at −19.89 m OD. This date is supported by the *Pinus* and *Corylus* pollen assemblage with lesser *Quercus*, *Ulmus* and

Tilia. The occurrence of *Alnus* pollen after this date also supports this chronology (Huntley *et al.* 1995).

Well Bank 53/+02/1398

This site is representative of a group of adjacent cores recovered from the Well Bank area of the southern North Sea, which all contain organic deposits at a similar depth, between −37 and −39 m OD, overlain by a variety of clastic sediments. Core 53/+02/1398 shows a silt *limus* with an increasing sand fraction in the upper 4 cm, and overlain by sand (Fig. 2c). Colonies of *Pediastrum*, freshwater bryozoans and ostracods, and a freshwater diatom assemblage dominated by *Fragilaria pinnata* with many other oligohalobous and halophobe (salt-intolerant) species indicate organic sediment accumulation in a shallow lake environment. In the upper 4 cm, coincident with the increased sand fraction, there is a rise in frequencies of estuarine and nearshore foraminifera. The foraminifera, pollen and spore assemblage shows no extensive saltmarsh flora, just single grains of *Plantago maritima* and *Artemisia* pollen, with the indicators of the freshwater lake still dominant. We interpret this sequence as the transport of sand and foraminifera into the lake as a transgression occurred. This is dated at 9145 ± 60 BP, an age supported by the tree pollen assemblage that is dominated by *Pinus* and *Corylus*. The overlying sand in contrast only contains estuarine and nearshore shelf foraminifera.

The other dates from cores from Well Bank all cluster around 9000 BP (Table 1) and have comparable microfossil assemblages that indicate the transgression of the lake. Some show a sharp upper boundary to the organic sequence, but the environmental interpretation outlined is consistent with all the litho-, bio- and chronostratigraphic data. Only core 53/+02/1497 does not have estuarine and shelf foraminifera within the top part of the freshwater limnic sequence, but the date of 9155 ± 75 BP suggests no significant erosion even though the boundary with the overlying sand is sharp.

Palaeogeographic maps

The coastline reconstructions and palaeotidal models use grids with the same spatial resolution. Computational efficiency for the tidal modelling demands use of a nested hierarchy of models. These increase in resolution at each stage by a factor of approximately three for both latitude and longitude. Table 2 gives the parameters for the nested hierarchy of hydrodynamic models used, with Fig. 3 showing their boundaries. Importing information from different sources into a geographical information system (GIS) database (Table 3) produces a model of the bathymetry for the present day at a spatial resolution higher than that required for some of the hydrodynamic models. This bathymetry was externally converted to the model resolution prior to use. Water depths at the locations of the cores from the North Sea, which provide sea-level index points, provide an independent set of observations to test the model of present-day bathymetry. Figure 4 shows a very strong correlation ($r = 0.92$), although the residuals range from −4 to +10 m, illustrating the topographic variability within a grid cell.

The earth and ice models of Lambeck (1995) provide relative sea-level predictions (metres relative to present) for each grid cell for each reconstruction and these are added to the present-day bathymetry to produce reconstructions of water depth and coastline positions (equation 3 Lambeck 1995). The earth and ice

Fig. 2. Summary of microfossil analyses. (**a**) Dogger Bank, core 55/+02/213. Pollen and spore assemblages for 406 and 408 cm expressed as per cent total land pollen, also showing foraminifera test linings preserved within the preparations. Foraminifera for 404 and 406 cm counted from separate preparations, with species expressed as total foraminifera counted per cent. Radiocarbon age and height relative to mean tide level (metres) shown at the left hand end of the figure. The sediment legend is drawn according to Troels-Smith (1955). The overlying 4 m of sand silt was not analysed. (**b**) Off north Norfolk, core 52/+01/2699. Pollen and spore assemblages for 485 and 489 cm expressed as per cent total land pollen, also showing foraminifera test linings and dinoflagellate cysts preserved within the preparations. Diatoms and foraminifera for 485 cm counted from separate preparations, with species expressed as per cent total diatom valves and per cent total foraminifera counted. Radiocarbon age and height relative to mean tide level (m) shown at the left hand end of the figure. The sediment legend is drawn according to Troels-Smith (1955). The overlying 4.8 m of clay was not analysed. (**c**) Well Bank, core 52/+02/1398. Pollen and spore assemblages for 152–162 cm expressed as per cent total land pollen, diatoms for the same levels counted from separate preparations with species expressed as per cent total diatom valves. Separate preparations for 149–152 cm show foraminifera and freshwater bryozoans and ostracods expressed as per cent total tests counted. Radiocarbon age and height relative to mean tide level (m) shown at the left hand end of the figure. The sediment legend is drawn according to Troels-Smith (1955). The overlying 1.5 m of sand was not analysed.

DOGGER BANK
55/+02/213VE
Trees
Shrubs
Herbs
Spores
Betula
Pinus
Quercus
Ulmus
Corylus
Artemisia
Chenopodiaceae
Cyperaceae
Gramineae
Plantago maritima
Silene-type
Spergularia
Taraxacum-type
Filicales
Pteridium
Pediastrum
Foraminifera Test Linings
Trochammina inflata
Jadammina macrescens
Lithology
Depth cm
age and altitude
-31.0mOD
8140±50
-31.1mOD
400
405
410
415
Silt
Sand
Humified peat
Clay

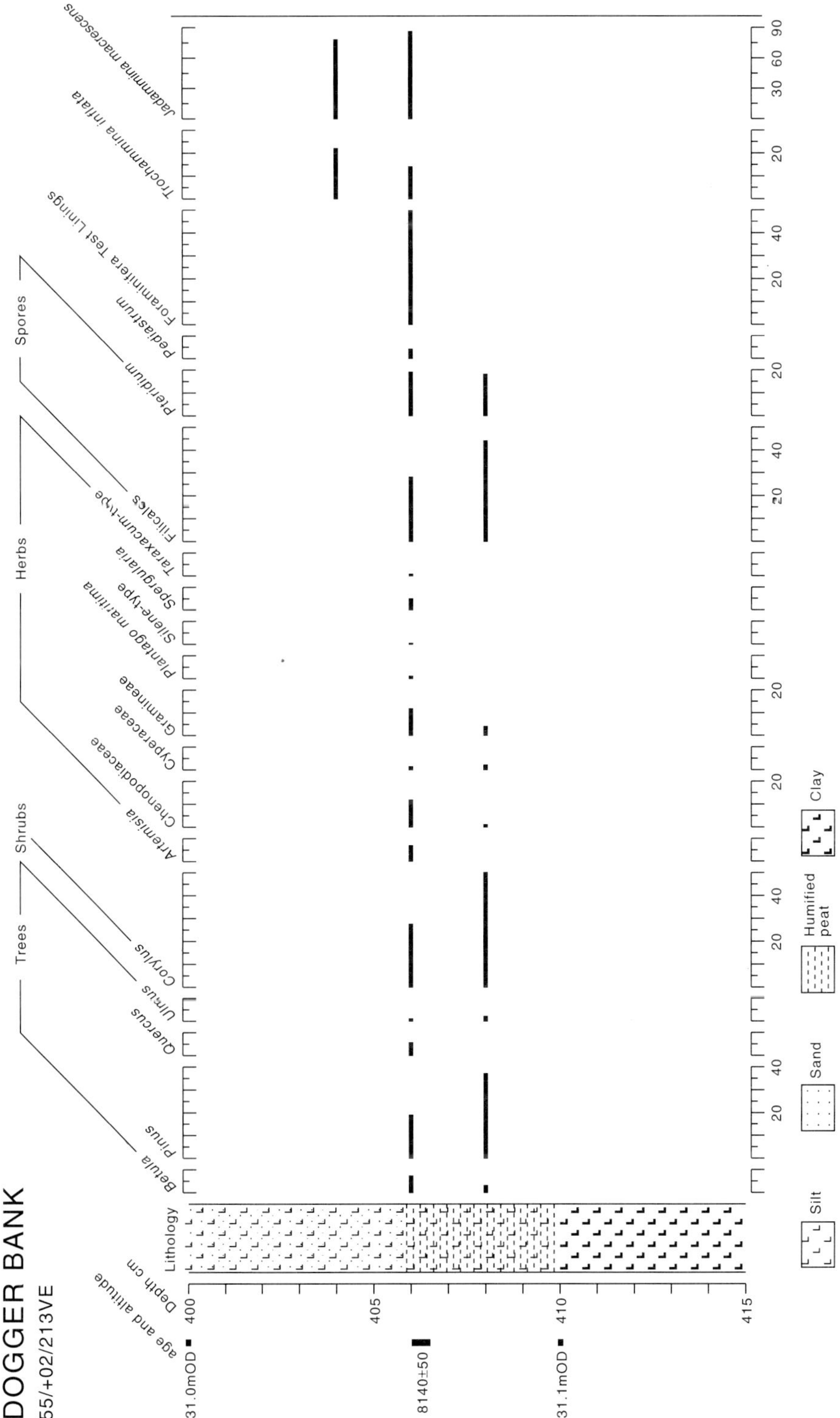

(a)

NORTH NORFOLK
52/+01/2699
Trees
Shrubs
Herbs
Spores
age and altitude
Depth cm
Lithology
Alnus
Betula
Pinus
Quercus
Tilia
Ulmus
Corylus
Artemisia
Aster-type
Chenopodiaceae
Cyperaceae
Gramineae
Taraxacum-type
Filicales
Pteridium
Pediastrum
Foraminifera test linings
Operculodinium centrocarpum
Spiniferites spp.
Cocconeis scutellum
Other polyhalobous diatoms
Mesohalobous diatoms
Oligohalobous diatoms
Miliammina fusca
Estuarine foraminifera
Nearshore shelf foraminifera
-19.8 mOD
480
481
482
483
484
485
486
487
488
489
490
491
492
493
494
495
7580 ±70
20
20 40
20
20
20 40
50
50 10 10 20
70 20
Humified peat
Silt

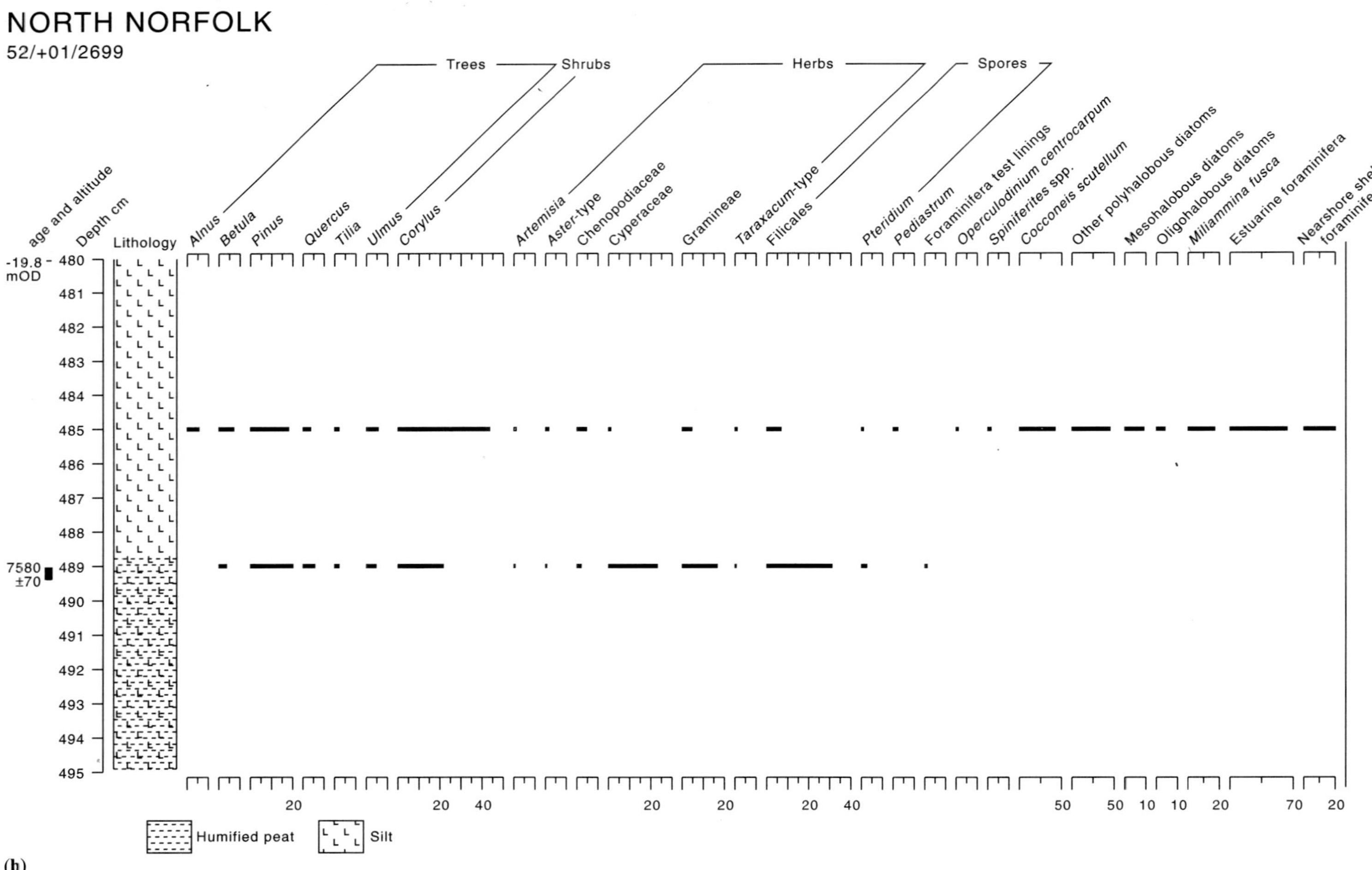

(b)

WELL BANK
52/+02/1398
Trees
Shrubs
Herbs
Aquatic
Diatoms
age and altitude
Depth cm
Lithology
Betula
Pinus
Quercus
Ulmus
Corylus
Salix
Artemisia
Cyperaceae
Gramineae
Plantago maritima
Typha angustifolia
Typha latifolia
Filicales spores
Pediastrum
Polyhalobous
Mesohalobous
Fragilaria pinnata
Other oligohalobous
Halophobe
Freshwater tests
Estuarine foraminifera
Nearshore shelf foraminifera
-37.5m OD
9145±60
-37.6m OD
-37.7m OD
145
150
155
160
165
170
20 40
20 40
20
20
20
50
70
20
90
80
30
Sand
Silt
Limus

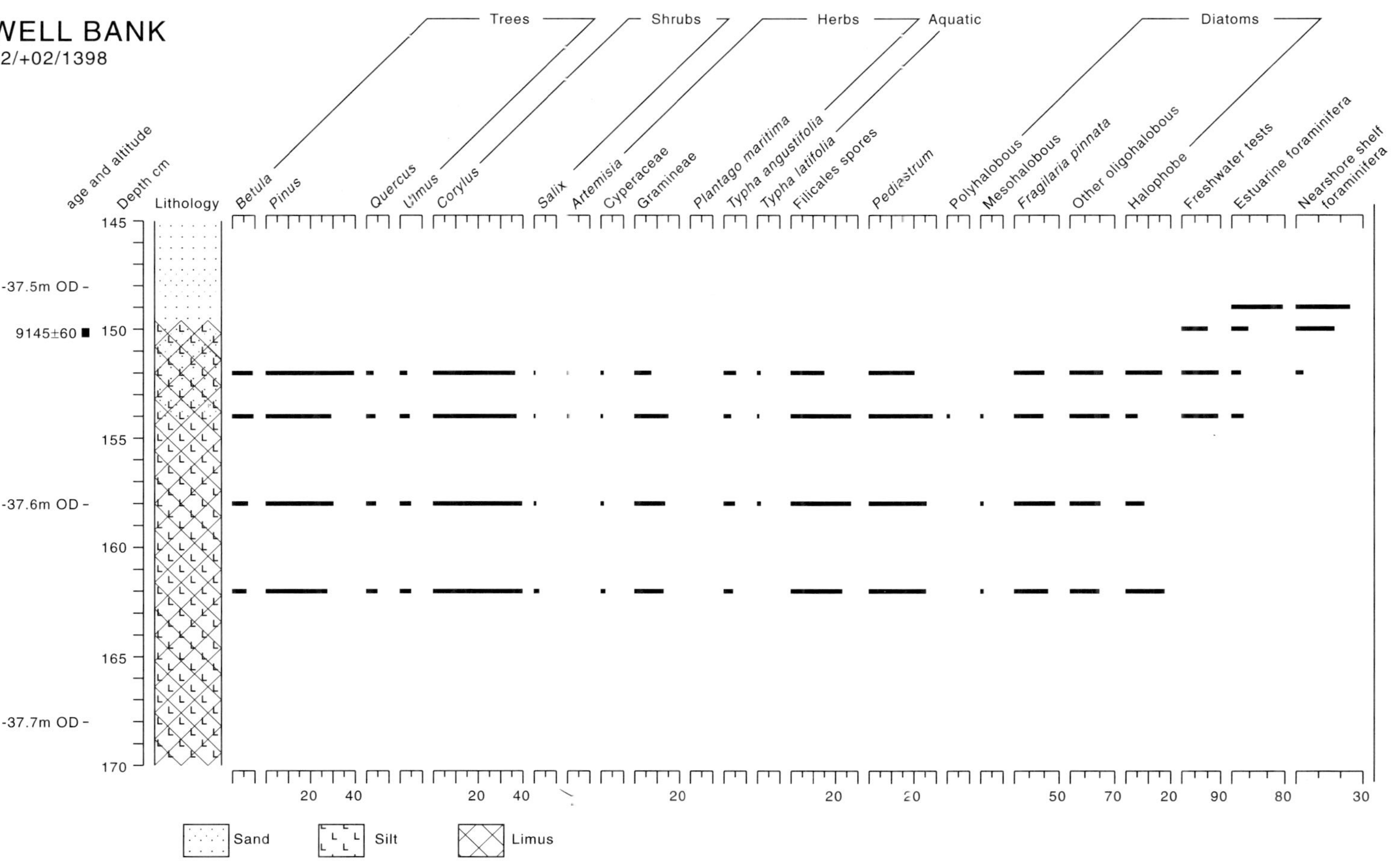

(c)

Table 2. *The set-up parameters for the nested hydrodynamic models used on the project*

Model	Northeast Atlantic NEA	Continental shelf CS3	East coast 9 EC9	East coast 30 EC30
Area	30°W–$25\frac{1}{2}$°E 37°N–$71\frac{2}{3}$°N	$11\frac{5}{6}$°W–$13\frac{2}{3}$°E $47\frac{2}{5}$°N–$62\frac{1}{5}$°N	2.5°W–6.0°E 56°N–51°N	1.0°W–1.5°E $52\frac{1}{5}$°N–$54\frac{1}{2}$°N
Grid resolution				
Longitude, deg	$\frac{1}{2}$	$\frac{1}{6}$	$\frac{1}{18}$	$\frac{1}{60}$
Latitude, deg	$\frac{1}{3}$	$\frac{1}{9}$	$\frac{1}{27}$	$\frac{1}{90}$
Approximate distance, km	36	12	4	1.2
Matrix size	106 × 113	135 × 150	135 × 139	140 × 207

models were validated against the sea-level data available at the time (Lambeck 1995), whilst the following sections assess each reconstruction against the larger database now available. At the present stage in the analysis we ignore sediment erosion and deposition but ongoing research, part of a LOIS Phase 2 project (Shennan & Andrews this volume), will address such issues. At the present stage these reconstructions are evaluated against the observation data to gauge the magnitude of the changes and the resolution of changes that can be verified by the available data. There is little justification in increasing the sophistication of models if there are no data to test their output. In the following section there is no consideration of land elevations inland of the present coastline. Discussion is confined to the predictions for the western North Sea.

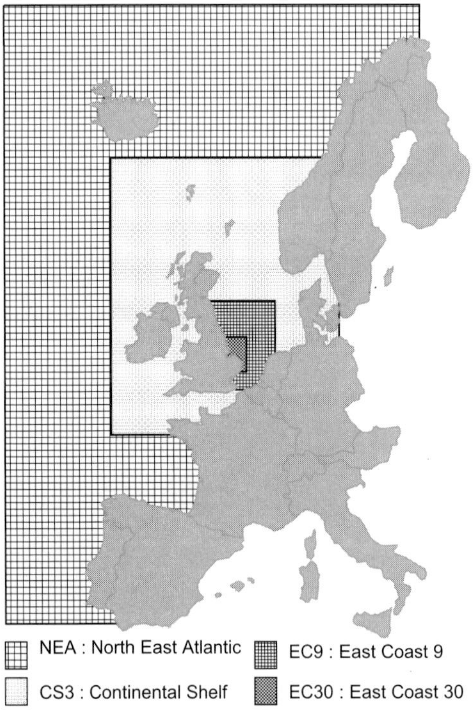

NEA : North East Atlantic
EC9 : East Coast 9
CS3 : Continental Shelf
EC30 : East Coast 30

Fig. 3. Map of the nested hierarchy of models (see Table 2 for details).

10 ka BP palaeogeography

At the start of the Holocene the North Sea coastline comprises the area of the Norwegian Trough and a western embayment extending south to the latitude of Flamborough Head (Fig. 5a). The latter did not appear in the prediction in Lambeck (1995; Fig. 3f) who used the 5′ × 5′ ETOPO-5 topographic database. This lower resolution leads to shallower estimates of water depth and in consequence, the predictions tend to lag behind those based on higher resolution bathymetry data. The coastline is only a little east of the present coast of northeast England. In most of Scotland there is the prediction of intertidal sedimentation inland of the present coast.

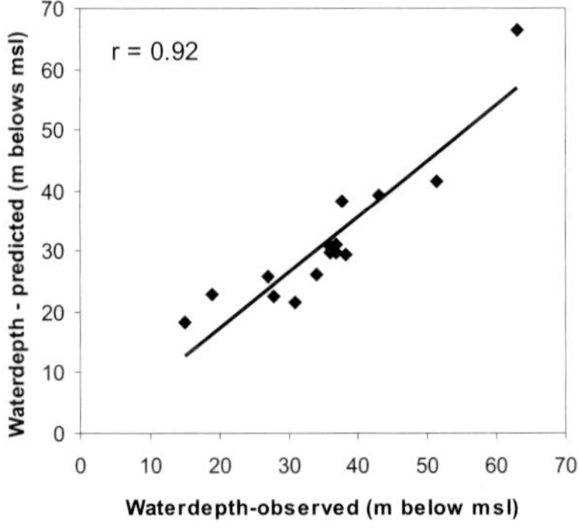

Fig. 4. Scatter plot and correlation coefficient (r) showing the relationship between observed water depths based on North Sea sea-level index points and the present-day model bathymetry.

Table 3. *Data sources for constructing the bathymetry used in the different nested models (see Fig. 3)*

Origin	Original resolution of the data longitude × latitude (deg)	Model
CCMS-POL	1/2 × 1/3	NEA
	1/18 × 1/27	EC9
	1/60 × 1/90	EC30
Lambeck (1995)	1/2 × 1	NEA
	1/4 × 1/2	CS3, EC9, EC30
British Geological Survey (BGS)	1/6 × 1/9	CS3
	1/18 × 1/27	EC9
BGS Humber LOIS project, in	200 m × 200 m	EC30
Admiralty charts	1/6 × 1/9	CS3
	1/18 × 1/27	EC9
	1/60 × 1/90	EC30
Waller (1994)*	1/60 × 1/90	EC30
Wingfield *et al.* (1978)*	1/60 × 1/90	EC30

* Data on the base of Holocene sediments.

There are few observations of this age to test this reconstruction. The sea-level index point dated 11145 ± 75 BP from core 53/+00/889 off Flamborough Head (Table 1) indicates that the western embayment (Fig. 5a) was inundated well before 10 ka BP. Marine shell material from the Geordie Trough, off northeast England, dated at 9335 ± 105 BP (Harland & Long 1996), also accord with this reconstruction. Freshwater peats from the Well Bank, east-northeast of Norfolk, with Late Devensian pollen assemblages, dated 11325 ± 85 BP in core 53/+02/1495 (Table 1), and from deep river channels in the Fenland (Waller 1994) simply concur with the prediction of those areas being inland of any tidal influence. Similarly, all the other samples that give younger ages (Table 1) lie in the land area shown in the reconstruction.

9 ka BP palaeogeography

By this time the western embayment had extended south, to off Spurn Point, and then east to produce a shallow estuary to the south of the Dogger Bank (Fig. 5b). This reconstruction is similar to that proposed by Jelgersma (1979). The earliest sea-level index point from the river Tees (Table 4) agrees with the prediction that the coastline of northeast England lay very close to the present, with tidal waters extending into the estuary. The five index points from the Well Bank, off Norfolk (cores 53/+02/1398, 1399, 1400, 1496, 1947, described above and Table 1), dated between 9155 ± 70 BP and 8995 ± 100 BP lie at the head of the bay that extends north of the Strait of Dover. Core 53/+01/1567 records the extension of the bay to the northwest at 8775 ± 70 BP as sea-level continues to rise.

8 ka BP palaeogeography

The reconstruction indicates that the North Sea was now connected to the English Channel via a narrow strait east-northeast of Norfolk and west of Texel (Fig. 5c). The Dogger Bank becomes cut off from the European mainland during high tides. It seems likely that the tidal regime around this time in the southern North Sea could be rapidly changing as the different tidal channels developed. The sea-level index point from the north side of the Dogger Bank, dated 8140 ± 55 BP in core 55/+02/213VE, (Fig. 2a) shows good agreement with the model predictions, as do the observations from Northumberland South and the River Tyne (Table 4). The model suggests that the first incursion of tidal water to the north Norfolk coast is from the east, from the deeper water eventually connecting to the English Channel. To the north there is still land above highest tides. The samples (Table 1) from cores 53/+01/1530 and 52/+01/2699 (Fig. 2b), 7975 ± 50 and 7580 ± 70 BP, respectively, and that from Warham Marshes in north Norfolk, 7530 ± 100 BP (Table 4), record the continuing transgression over the next 500 years.

This reconstruction is broadly similar to those of Jelgersma (1979) and Lambeck (1995), but these show water depths at only 20–50 m vertical resolution so the potential location of a wide intertidal zone is less easy to judge. Jelgersma (1979) suggests that the connection between the North Sea and the Channel occurred between 8.7 and 8.3 ka BP.

7.5 ka BP palaeogeography

The model predicts that by 7.5 ka BP all the estuaries of the east coast could have some tidal

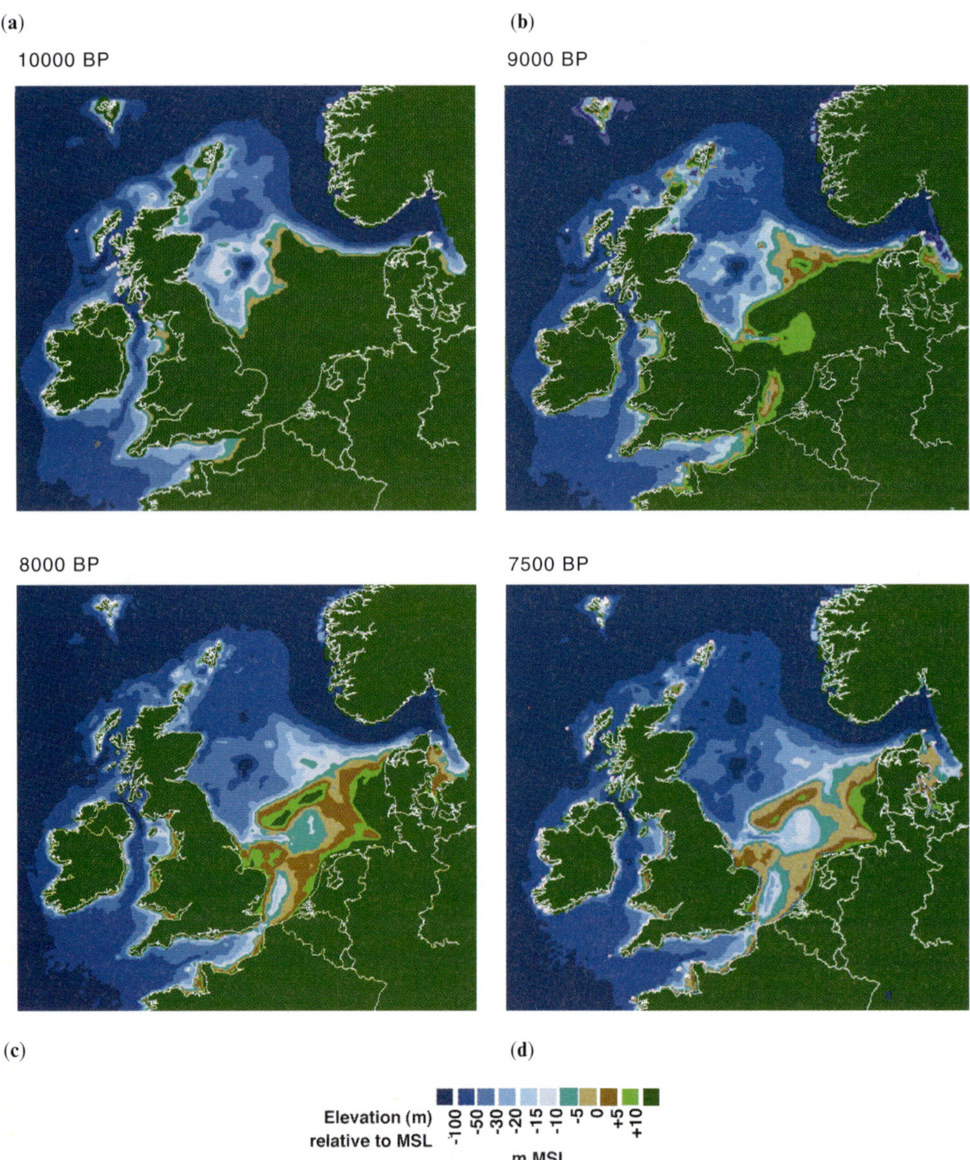

Fig. 5. Palaeogeographic reconstructions of northwest Europe. (**a**) 10 ka BP, (**b**) 9 ka BP, (**c**) 8 ka BP, (**d**) 7.5 ka BP, (**e**) 7 ka BP, (**f**) 6 ka BP, (**g**) 5 ka BP, (**h**) 4 ka BP. Elevations (metres) relative to MSL, depths below MSL are given as negative. The area covered is from 11 5/6°W–13 2/3°E to 47 2/5°N–62 1/5°N.

influence from high tides. The ages of the first sea-level index points for the Humber Estuary, Lincolnshire Marshes, Fenland and Norfolk Broads (Table 4) are from well-developed salt-marsh peats rather than indicators of the first saline water into an estuary and their younger ages are not at odds with the model. There are extensive intertidal flats from Flamborough Head to north Norfolk. The channel separating north Norfolk from mainland Europe is only 5–10 m deep at mid-tide and the channel between the Dogger Bank and mainland Europe was less than 5 m below MSL in parts (Fig. 5d).

7 ka BP palaeogeography

Wide intertidal areas are still predicted for the areas off the Humber Estuary, Lincolnshire Marshes, Fenland and north Norfolk (Fig. 5e),

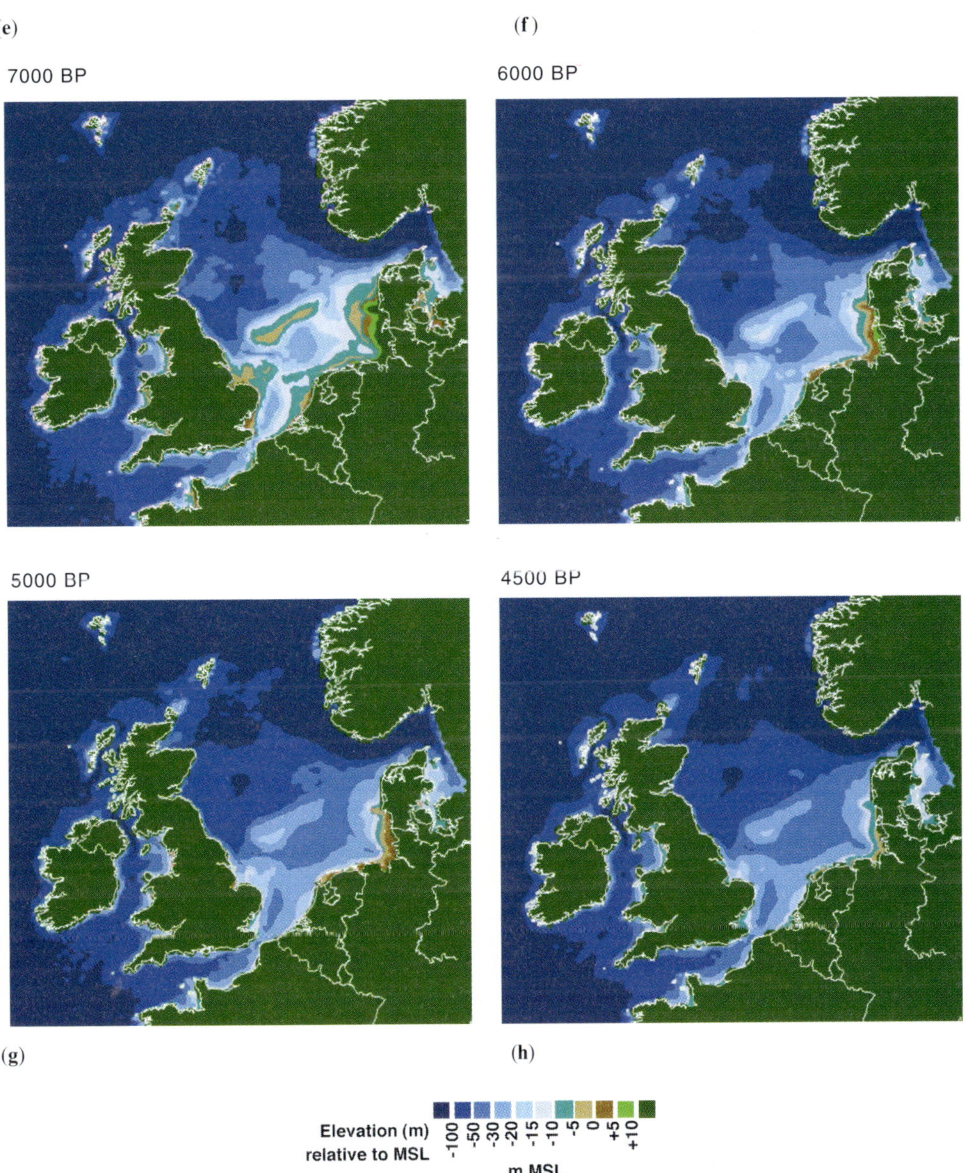

Fig. 5. (*continued*).

which agree with the sea-level index points based on saltmarsh peats for these areas (Table 4). The Dogger Bank is only exposed at low tide.

6 ka BP palaeogeography

By this time the Dogger Bank is submerged at all stages of the tide and the western margins of the North Sea are close to or inland of the present coastline (Fig. 5f), as indicated by the many index points available (Fig. 1). Waller (1994), Brew *et al.* (this volume) and Metcalfe *et al.* (this volume) make reconstructions of the palaeogeography of the Fenland and the Humber Estuary.

5, 4 and 3 ka BP palaeogeographies

From 5 ka BP to the present relative sea level increases gradually in the western North Sea

Table 4. *Earliest radiocarbon dated sea-level index points showing saltmarsh conditions for the east coast of England*

Region	Site	Code	Age BP	m OD	RSL (m)
Northumberland					
North	Beal Cast	AA23825	7420 ± 70	1.95	−0.45 ± 0.2
Central	Annstead Burn	AA27228	7355 ± 90	−0.07	−2.27 ± 0.2
South	Warkworth	AA24221	7905 ± 60	−2.54	−4.69 ± 0.2
River Tyne	Blaydon	AA23822	7795 ± 85	−3.18	−5.68 ± 0.2
Tees Estuary	Thornaby	HAR3711	9680 ± 110	−10.61	−13.11 ± 0.3
Humber Estuary	Old Den	AA27583	6875 ± 60	−10.31	−13.29 ± 0.3
Lincolnshire Marshes	Theddlethorpe	AA22667	7230 ± 55	−10.95	−14.10 ± 0.2
Fenland	Clenchwarton	AA22355	7215 ± 70	−12.55	−16.36 ± 0.2
North Norfolk	Warham Marshes	AA22686	7530 ± 100	−14.38	−17.23 ± 1.0
Norfolk Broads*	Broadland	HAR2535	7580 ± 90	−19.30	−20.04 ± 0.3

* Coles & Funnel (1981).

south of the River Tyne, but rises above present to the north to a maximum after 4 ka BP (Shennan *et al.* this volume). Such changes in water depth and coastline configuration do not show clearly at the resolution of the reconstructions illustrated (Fig. 5g–h) and reference should be made to the regional accounts in other chapters in this volume.

Other North Sea locations

Pre-Holocene data from the northern North Sea in the area of the Viking Bank (60°40′N, 1°40′E) show the marine transgression of the North Sea plateau east of Shetland by about 13 ka BP (Carlsen *et al.* 1986; Peacock 1995). Peacock & Harkness (1990) have reviewed the palaeogeographic changes in the northern North Sea over the Late Glacial period. By Late Glacial Interstadial times, about 12 ka BP, the central northern North Sea and the Shetland–Orkney area were still dry land, but a shallow marine embayment had formed that stretched along the east coast of Scotland to south of the Firth of Forth. This western extension of marine conditions agrees very well with the modelled 10 ka BP map presented in this paper, which shows this embayment extending as far south as the coast of Yorkshire by the start of the Holocene.

There are a small number of previously published Holocene data points available to test the validity of the remaining palaeogeographic maps (Table 1). Many of these published data points are based upon eroded basal peats from the coastal waters of the Dutch and German sectors of the North Sea (Jelgersma 1961, 1966; Kirby & Oele 1975; Jelgersma *et al.* 1979; Behre *et al.* 1979), which cannot be related to a past tidal altitude and so provide limiting data only. Most are early Holocene in age and prior to the time when the sea rose high enough to occupy the southeast part of the North Sea basin. They are nevertheless valuable in constraining the possible positions of palaeoshorelines, with the few more recent examples particularly so. Ludwig *et al.* (1981) have reviewed these data. Of the few examples which may be considered sea-level index points Ludwig *et al.* (1981) cite dates from *Phragmites* material within brackish sediments from site A10 in the German Bight (53°51′N, 7°51′E). The ages of between 7.5 k and 8 ka BP are in agreement with the modelled results. The modelled introduction of marine waters into the outer German Bight area by 8 ka BP is also in line with dates from German Bight core 235, which Ludwig *et al.* (1981) consider as index points. In summary, the palaeoshoreline maps are consistent with the currently published sea-level data for the North Sea.

Tidal models

The approach adopted here is very similar to previous studies whereby models developed for the present (Flather 1976) predict tides for past bathymetries and coastline configurations (e.g. Austin 1991; Hinton 1992). The major differences in comparison to previous studies is the use of bathymetries and coastlines based on modelling of differential isostatic rebound (described above) and including up to 26 tidal harmonics, rather than one or six (Austin 1991; Hinton 1992, 1995, 1996). In shallow water the progression of the tidal wave is modified by bottom friction and other processes, and a larger range of tidal harmonic constituents is necessary to represent these distortions (Pugh 1987). While six constituents may give a good approximation of MHWST (Hinton 1992, 1995, 1996), the future aim of our work is to investigate sediment movement as well as tide-level changes through time and this requires an optimum representation of shallow water distortions. Computational limitations involve a trade-off between grid resolution, matrix size, time step, number of harmonics and the length of time of the model runs.

The tidal inputs on open-sea boundaries of the northeast Atlantic (NEA) model are not altered by changes in bathymetry on the northwest European shelf. For each 1-ka time-slice, tidal inputs to CS3 and the finer grid models are provided by interpolation of harmonic constants derived from analysis of a run of the next coarser grid model, and, therefore, do account for tidal effects generated within the NEA of changes in bathymetry. During the validation process each model is run to simulate a typical six-month period after an initial five-day cold start-up. For the NEA, six months of model data are required to extract the 26 tidal constituents that are used in the other models (CS3, EC9, EC30 and Table 4). All four models were run for eleven Holocene times, 1, 2, 3, 4, 5, 6, 7, 7.5, 8, 9, 10 ^{14}C ka BP.

For computational efficiency we adopted the same approach as Hinton (1992, 1995) to model a typical 15-day period. With only six harmonic constituents, the highest recorded water-level for each cell in the 15 days gives a reasonable estimate of the MHWST. This is defined as the average throughout the year when the maximum declination is 23.5° (Admiralty Tide Tables 1997). By incorporating more harmonic constituents the maximum water-level during a 15-day period will vary much more depending on the particular 15 days chosen. The best 15-day period to minimize the difference between the model and the annual MHWST from the Admiralty tide tables varies from station to station. At the present stage of analysis, running all four models for the present and 11 Holocene reconstructions, we have kept to the 15-day model run to balance number of model runs with the output parameters required. We recognize the limitations of using the 15-day approach, and in LOIS Phase 2 aim to identify the optimum time period required for modelling sediment movement and estimating the annual average for MHWST.

Individual sea-level index points have error terms of at least ±0.2 m (e.g. Shennan 1986, Horton *et al.* this volume) and a group of index points for an area frequently show a scatter around ±1 m or more for a particular time period (e.g. Shennan *et al.* this volume). The large scatter results from a number of factors, including sediment consolidation and tidal changes. Therefore, the first aim is to model tides with a similar precision to the Holocene data, i.e. greater than or equal to ±0.2 m. The 15-day approach achieves this aim. All the model predictions are based on the astronomical conditions for the 15-day period starting 1 January 1997.

Table 5 shows the model predictions for present-day bathymetry and coastline compared to the Admiralty tide table data. The differences

Table 5. *Admiralty Tide Table data (1997) and model predictions for MHWST (m relative to msl)*

Tide Station	MHWST*	Max_t†	Max_m‡	Max_t–MHWST	Max_m–MHWST	Max_m–Max_t
Scarborough	2.32	2.42	2.23	0.10	−0.09	−0.19
Bridlington	2.55	2.65	2.26	0.10	−0.29	−0.39
Spurn	2.55	2.65	2.26	0.10	−0.29	−0.39
Grimsby	2.97	3.17	2.69	0.20	−0.28	−0.48
Bull Sand Fort	2.85	3.05	2.66	0.20	−0.19	−0.39
Skegness	3.00	3.20	2.86	0.20	−0.14	−0.34
Boston	3.37	3.57	3.28	0.20	−0.09	−0.29
Tabs Head	3.45	3.65	3.42	0.20	−0.03	−0.23
Kings Lynn	3.15	3.35	3.51	0.20	0.36	0.16
Hunstanton	3.30	3.50	3.43	0.20	0.13	−0.07
Burnham	2.35	2.55	2.60	0.20	0.25	0.05
Wells	2.50	2.70	2.69	0.20	0.19	−0.01
Blakeney	2.65	2.85	2.77	0.20	0.12	−0.08
Range				0.10	0.65	0.64
Standard deviation				0.04	0.22	0.20

* MHWST, annual mean high water spring tides (Admiralty Tide Tables 1997).
† Max_t, maximum tide level based on the astronomical conditions for the 15-day period starting 1 January 1997 (Admiralty Tide Tables 1997).§
‡ Max_m, maximum EC30 model predictions based on the astronomical conditions for the 15-day period starting 1 January 1997.
§ All Admiralty tide table values for MHWST and max_t have overall errors of ±0.1 m.

between the model (max_m) and Admiralty tide tables (max_t) data for the same 15-day period range from 0.16 to −0.48 m, a standard deviation of 0.20 m. The model (max_m) and the annual Admiralty tide tables (MHWST) predictions differ from 0.36 to −0.29 m with a standard deviation of 0.22 m. Both these comparisons are superior to previous model data (Hinton 1992, 1995). For seven tide stations Hinton (1992) reported differences from +0.17 to −0.89 m, for which we calculate a standard deviation of 0.44 m.

Tidal changes during the Holocene

The models predict an increase in tidal range for the western North Sea during the Holocene.

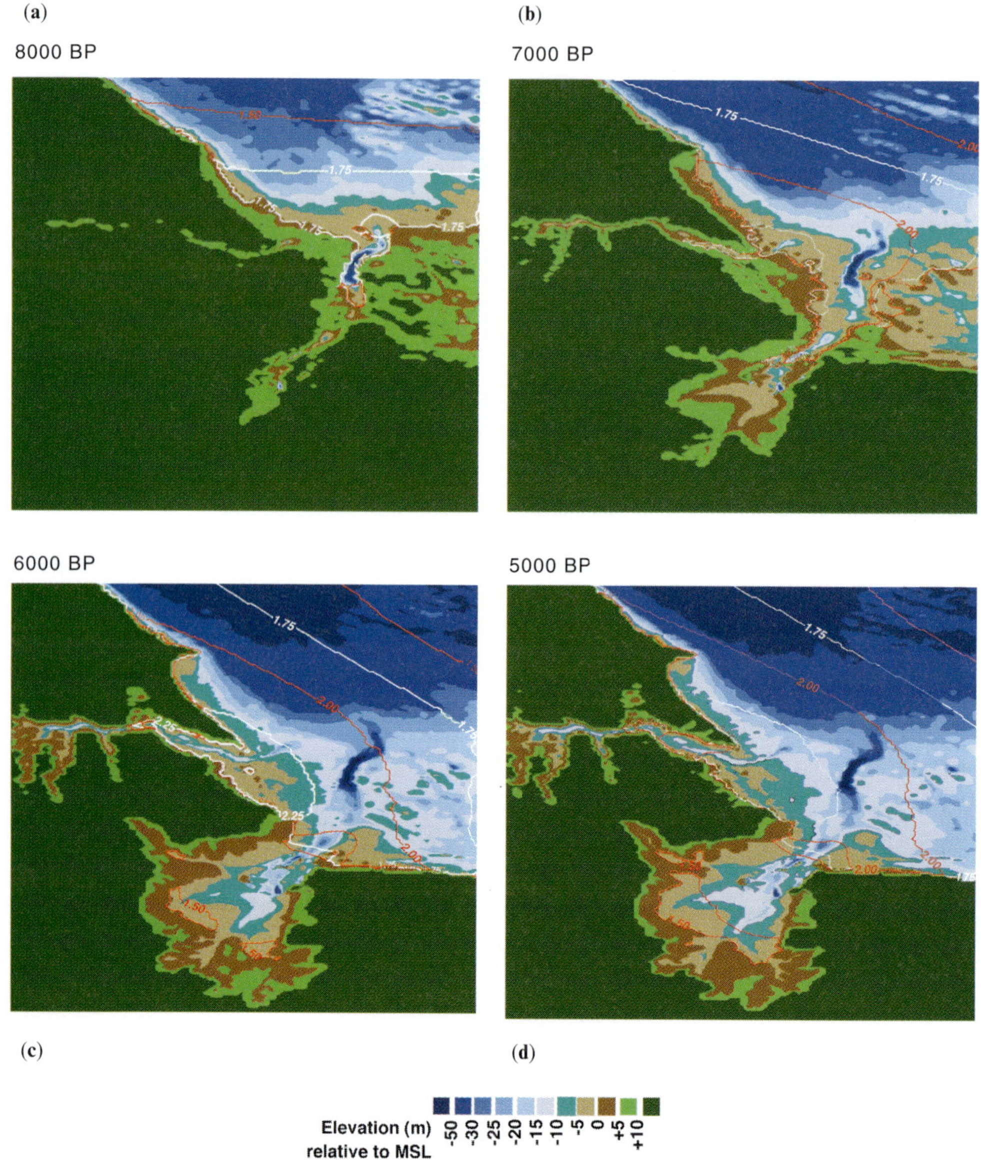

Fig. 6. Palaeogeographic reconstructions and elevations of MHWST (metres above MSL) for the southwest North Sea: **(a)** 8 ka BP, **(b)** 7 ka BP, **(c)** 6 ka BP, **(d)** 5 ka BP. Elevations (metres) relative to MSL, depths below MSL are given as negative. The area covered is from 1.0°W–1.5°E to 52.2°N–54.5°N.

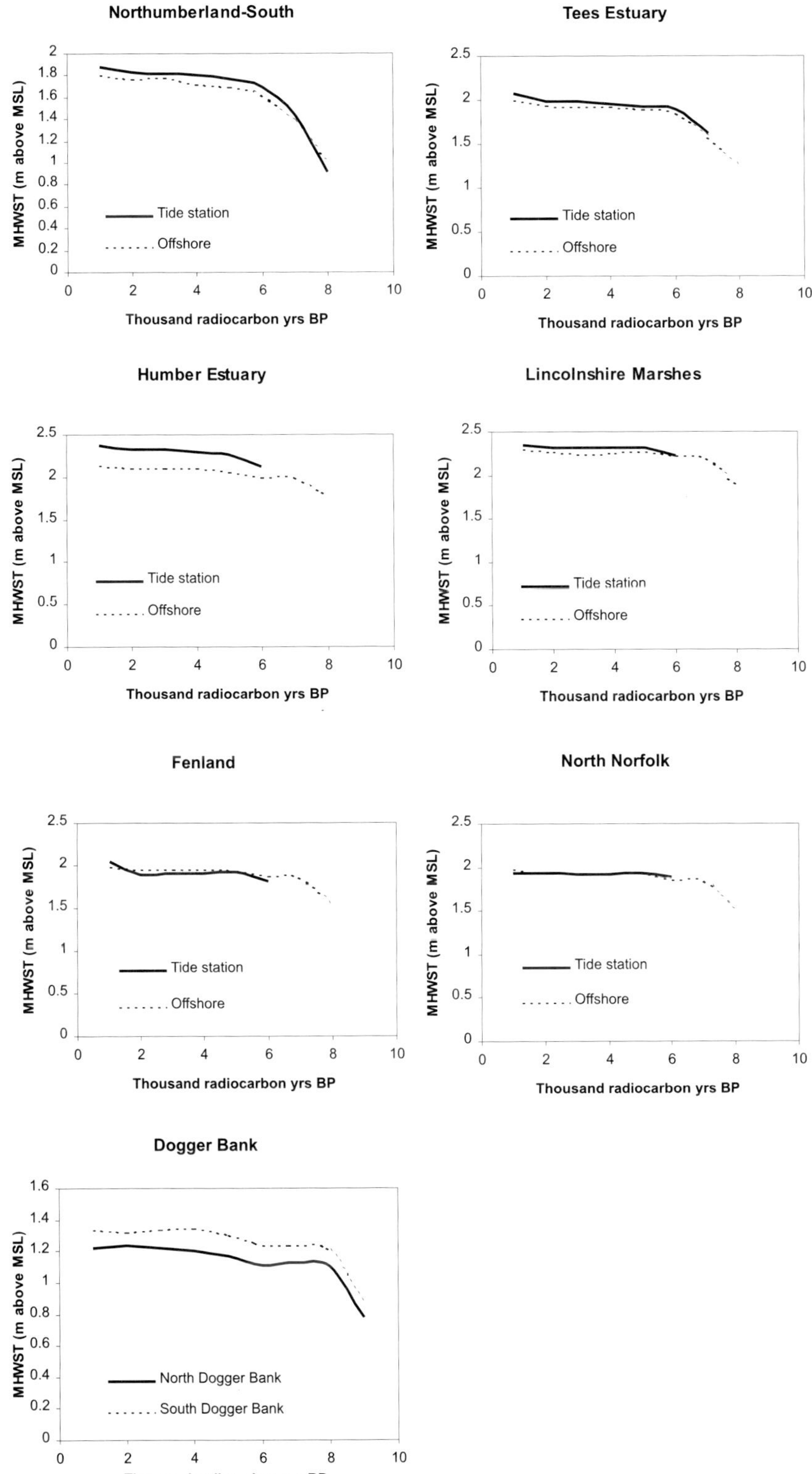

Fig. 7. Holocene changes in high tide level for selected locations in the western North Sea.

Figure 6 illustrates the spatial pattern of this increase in the area covered by EC30 for the period of most rapid change, i.e. 8–5 ka BP. This coincides with the time of major changes in palaeogeography (Fig. 5). At 8 ka BP the coastline only lay near to the present coast of Yorkshire, north of Flamborough Head, with a tidal channel extending south towards the present Wash (Figs 5c and 6a). Off Flamborough Head high tide level was *c.* 1.6 m above mean tide level at 8 ka BP, rising to 1.9 m at 7 ka BP, 2.1 m at 6 ka BP and only a little higher at 5 ka BP (Fig. 6a–d). An increasing trend occurs for all of the open-sea areas in the area covered by EC30.

Once the sea penetrates a large part of the Fenland, potentially from 6 ka BP onwards, the current versions of the models predict a reduction in the height of high tide level (Fig. 6c, d). This contrasts with the present Wash in which high tide level increases to the head of the embayment. The difference is the result of the bathymetry and the friction-dependent distortion, which is a function of bottom sediment texture. The models for the Holocene reconstructions include no model of sediment infilling of either the estuaries or the coastal lowlands above the contours for the base of the Holocene (Table 3). Thus in the model, as sea level rises the Wash/Fenland system becomes a lagoon of increasing depth and area. A lack of sediment infill in our model becomes increasingly unrealistic as the investigations of Shennan (1986), Waller (1994) and Brew *et al.* (this volume) indicate. However, there are many examples from the Fenland of clastic sediments, of 6–3 ka BP ages with brackish or marine microfossils, which comprise silt and clay-size particles and no laminations indicative of tidal currents. These may be indicative of shallow standing water or 'lagoonal' conditions in areas between tidal channels (Shennan 1986, 1994). Many sediments younger than 3 ka BP are generally coarser with more laminations and are very similar to the sediments of the present intertidal zone (Brew *et al.* this volume). A sediment accumulation model, including variations in sediment texture, is required before any robust estimates of tidal-range changes within the Fenland/Wash area can be made. A similar level of caution applies to the Humber Estuary. The present version of EC30 includes no sediment accumulation model and also no discharge from the catchments draining into the estuary. Hence there is no significant increase of high water-level up-estuary in the model. This also forms part of the LOIS Phase 2 project.

Holocene changes in high tide level for a series of locations are shown in Fig. 7. All locations record an increase in high tide level, and tidal range, through the Holocene. For example, the MHWST for Northumberland-South increases from 0.91 m above RSL at 8 ka BP to 1.88 m above RSL at 1 ka BP. By applying the best-fit line from each location in Fig. 7 to the sea-level index points from the same area, changes in RSL relative to the present are re-calculated using the modelled tidal changes (Fig. 8). This shows that the relative sea-level reconstructions corrected for tidal changes lie above those based on present tidal values. For example, index points from

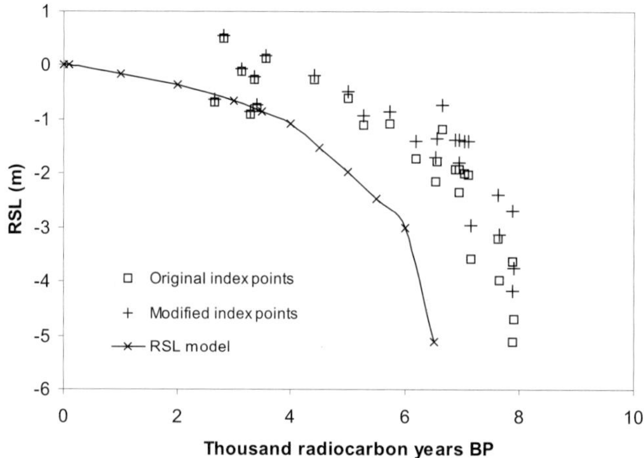

Fig. 8. Holocene relative sea-level changes and model predictions (Lambeck 1995) in Northumberland-South, including the effects of changes in high tide level.

Northumberland-South dated between 6 and 8 ka BP show an average offset of 0.64 ± 0.21 m. Shennan *et al.* (this volume) show that for virtually all of eastern England, predictions of relative sea-level change based on the ice and earth models of Lambeck (1995) lie below the observations. Corrections based on the tidal models described here increase the difference rather than reduce it. This reinforces the conclusion of Lambeck (1995; Lambeck *et al.* 1998) and Shennan *et al.* (in press) that a revised ice model is needed to explain the variations in sea-level index points from the UK, the database of which has grown significantly during the last four years, especially during LOIS.

Conclusions

Model predictions of coastline positions in the western North Sea during the Holocene show close agreement with data from both offshore and onshore cores. Even during the periods of most rapid relative sea-level rise, *c.* $0.8\,\mathrm{mm\,a^{-1}}$ (fig. 25, Shennan *et al.* this volume) in the early Holocene in the southern North Sea, coastal and saltmarsh vegetation communities formed temporarily during coastline retreat. The palaeogeographic reconstructions of coastline positions and bathymetries provide boundary conditions for tidal models for 1 ka steps through the Holocene. These models predict increases in tidal range since the early Holocene, with the major changes occurring prior to 6 ka BP. The present models omit any reconstructions of sediment accumulation within the estuaries and coastal lowlands. Incorporation of sediment transport and accumulation within the next phase of the LOIS programme will enhance the models. Incorporation of the modelled changes in tidal range reinforces the conclusions of other recent analyses that indicate the need for better models of ice sheet dimensions at the last glacial maximum and during deglaciation. Improvements in all these modelling elements will enhance the predictive capability of the modelling approach described.

Special thanks to Robin Wingfield. This is publication number 591 of the Land-Ocean Interaction Study (LOIS) Community Research Programme and the work was supported by NERC grants GST/02/0760 and GST/02/0761 under LOIS special topics 313 and 316. Additional data were supplied with the collaboration of principal investigators and research staff from other LOIS projects and core programmes. We are very grateful to the NERC radiocarbon laboratory at East Kilbride for the radiocarbon dates. We thank D. Long and J. Scourse for their excellent reviews of the original version and suggestions for improvements.

References

ADMIRALTY TIDE TABLES 1997. *European Waters Including the Mediterranean Sea.* **1**, Hydrographer to the Navy, Admiralty Hydrography Department, Taunton.

ANDREWS, J. E., BOOMER, I., BALIFF, I., BALSON, P., BRISTOW, C., CHROSTON, P. N., FUNNELL, B. M., HARWOOD, G. M., JONES, R., MAHER, B. A. & SHIMMIELD, G. B. 2000. Sedimentary evolution of the north Norfolk barrier coastline in the context of Holocene sea-level change. *This volume.*

AUSTIN, R. M. 1991. Modelling Holocene tides on the NW European continental shelf. *Terra Nova*, **3**, 276–288.

BEHRE, K. E., MENKE, B. & STREIF, H. 1979. Late Quaternary sedimentation in the North Sea. *In*: OELE, E., SCHUTTENHELM, R. T. E. & WIGGERS, A. J. (eds) *The Quaternary History of the North Sea.* University of Uppsala, 175–187.

BELDERSON, R. H., PINGREE, R. D. & GRIFFITHS, D. K. 1986. Low sea-level tidal origin of Celtic Sea sand banks: evidence from numerical modelling of M2 tidal streams. *Marine Geology*, **73**, 99–108.

BREW, D. S., HOLT, T., PYE, K. & NEWSHAM, R. 2000. Holocene Sedimentary evolution and palaeocoastlines of the Fenland embayment, eastern England. *This volume.*

CARLSEN, R., LØKEN, T. & ROALDSET, E. 1986. Late Weichselian transgression, erosion and sedimentation at Gulfaks, northern North Sea. *In*: SUMMERHAYS, C. P. & SHACKLETON, N. J. (eds) *North Atlantic Palaeoceanography.* Geological Society, Special Publications, 145–152.

COLES, B. P. L. & FUNNELL, B. M. 1981. Holocene palaeoenvironments of Broadland, England. *Special Publication, International Association of Sedimentologists*, **5**, 123–131.

FLATHER, R. A. 1976. A Tidal Model of the North West European Continental Shelf. *Memoirs, Society of Royal Scientists, Liege*, **6**, 141–164.

——1979. Recent results from a storm surge prediction scheme for the North Sea, marine forecasting. *In*: NIHOUL, J. C. J. (ed.) *Proceedings of the Tenth Liege Colloquium on Ocean Hydrodynamics.* Elsevier, Amsterdam, 385–409.

——1981. Practical surge prediction using numerical models. *In*: PEREGRINE, D. H. (ed.) *Floods Due to High Winds and Tides.* Academic Press, London, 21–44.

GEHRELS, W. R., BELKNAP, D. F., PEARCE, B. R. & GONG, B. 1995. Modelling the contribution of M2 tidal amplification to the Holocene rise of mean high water in the Gulf of Maine and the Bay of Fundy. *Marine Geology*, **124**, 71–85.

HARLAND, R. & LONG, D. 1996. A Holocene dinoflagellate cyst record from offshore north-east England. *Proceedings of the Yorkshire Geological Society*, **51**, 65–74.

HINTON, A. C. 1992. Palaeotidal changes within the area of the Wash during the Holocene. *Proceedings of the Geologists' Association*, **103**(3), 259–272.

——1995. Holocene tides of The Wash, UK: the influence of water-depth and coastline-shape changes

on the record of sea-level change. *Marine Geology*, **124**, 87–111.
——1996. Tides in the northeast Atlantic: considerations for modeling water depth changes. *Quaternary Science Reviews*, **15**, 873–894.
HORTON, B. P., EDWARDS, R. J. & LLOYD, J. M. 2000. Implications of a microfossil transfer function in Holocene sea-level studies. *This volume*.
HUNTLEY, B., BERRY, P. M., CRAMER, W. & MCDONALD, A. P. 1995. Modelling present and potential ranges of some European higher plants using climate response surfaces. *Journal of Biogeography*, **22**, 967–1001.
JANSEN, J. H. F., VAN WEERING, T. C. E. & EISMA, D. 1979. Late Quaternary sedimentation in the North Sea. *In*: OELE, E., SCHUTTENHELM, R. T. E. & WIGGERS, A. J. (eds) *The Quaternary History of the North Sea*. University of Uppsala, 175–187.
JELGERSMA, S. 1961. Holocene sea-level changes in the Netherlands. *Mededclinger van die Geologisch Stichtins*, **7**, 1–100.
——1966. Sea-level changes during the last 10 000 years. *In*: SAWYER, J. S. (ed.) *World Climate 8000–0 BC*. Royal Meteorological Society, London, 54–69.
——1979. Sea level changes in the North Sea basin. *In*: OELE, E., SCHUTTENHELM, R. T. E. & WIGGERS, A. (eds) *The Quaternary History of the North Sea*. University of Uppsala, 233–248.
——, OELE, E. & WIGGERS, A. J. 1979. Depositional history and coastal development in the Netherlands and the adjacent North Sea since the Eemian. *In*: OELE, E., SCHUTTENHELM, R. T. E. & WIGGERS, A. J. (eds) *The Quaternary History of the North Sea*. University of Uppsala, 115–142.
KIRBY, R. & OELE, R. 1975. The geological history of the Sandettie–Fairy Bank area, southern North Sea. *Philosophical Transactions of the Royal Society of London A*, **279**, 257–267.
LAMBECK, K. 1991. A model for Devensian and Flandrian glacial rebound and sea-level change in Scotland. *In*: SABADINI, R., LAMBECK, K. & BOSCHI, E. (eds) *Glacial Isostasy, Sea Level and Mantle Rheology*. Kluwer Academic Publishers, Dordrecht, 33–62.
——1993*a*. Glacial rebound of the British Isles. I. Preliminary model results. *Geophysical Journal International*, **115**, 941–959.
——1993*b*. Glacial rebound of the British Isles. II. A high-resolution, high-precision model. *Geophysical Journal International*, **115**, 960–990.
——1995. Late Devensian and Holocene shorelines of the British Isles and North Sea from models of glacio-hydro-isostatic rebound. *Journal of the Geological Society of London*, **152**, 437–448.
——, SMITHER, C. & JOHNSTON, P. 1998. Sea-level change, glacial rebound and mantle viscosity for northern Europe. *Geophysical Journal International*, **134**, 102–144.
LUDWIG, G., MULLER, H. & STREIF, H. 1981. New dates on Holocene sea-level changes in the German Bight. *Special Publications of International Association of Sedimentologists*, **5**, 211–219.
METCALFE, S. E., ELLIS, S., HORTON, B. P., INNES, J. B., MCARTHUR, J. J., MITLEHNER, A., PARKES, A., PETHICK, J. S., REES, J. G., RIDGWAY, J., RUTHERFORD, M. M., SHENNAN, I. & TOOLEY, M. J. 2000. The Holocene evolution of the Humber Estuary: reconstructing change in a dynamic environment. *This volume*.
PEACOCK, J. D. 1995. Late Devensian to early Holocene palaeoenvironmental changes in the Viking Bank area, northern North Sea. *Quaternary Science Reviews*, **14**, 1029–1042.
—— & HARKNESS, D. D. 1990. Radiocarbon ages and the full-glacial to Holocene transition in seas adjacent to Scotland and southern Scandinavia: a review. *Transactions of the Royal Society of Edinburgh, Earth Sciences*, **81**, 385–396.
PELTIER, W. R. 1996. Mantle viscosity and ice age ice sheet topography. *Science*, **273**, 1359–1364.
——1998. Postglacial variations in the level of the sea: implications for climate dynamics and solid-earth geophysics. *Reviews of Geophysics*, **36**, 603–689.
PLATER, A. J., RIDGWAY, J., RAYNER, B., SHENNAN, I., HORTON, B. P., HAWORTH, E. Y., WRIGHT, M. R. & WINTLE, A. G. 2000. Sediment provenance and flux in the Tees estuary: the record from the Late Devensian to the present. *This volume*.
PUGH, D. T. 1987. *Tides, Surges and Mean Sea Level*. John Wiley, Chichester.
REES, J. G., RIDGWAY, J., ELLIS, S., KNOX, R. W. O'B., NEWSHAM, R. & PARKES, A. 1999. Holocene sediment storage in the Humber Estuary. *This volume*.
RIDGWAY, J., ANDREWS, J. E., ELLIS, S., HORTON, B. P., INNES, J. B., KNOX, R. W. O'B., MCARTHUR, J. J., MAHER, B. A., METCALFE, S. E., MITLEHNER, A., PARKES, A., REES, J. G., SAMWAYS, G. M. & SHENNAN., I. 2000. Analysis and interpretation of Holocene sedimentary sequences in the Humber Estuary. *This volume*.
SCOURSE, J. D. & AUSTIN, R. M. 1995. Palaeotidal modelling of continental shelves: marine implications of a land bridge in the Straits of Dover during the Holocene and Middle Pleistocene. *In*: PREECE, R. C. (ed.) *Island Britain: a Quaternary Perspective*. Geological Society, Special Publications, **96**, 75–89.
SHENNAN, I. 1986. Flandrian sea-level changes in the Fenland. II: Tendencies of sea-level movement, altitudinal changes and local and regional factors. *Journal of Quaternary Science*, **1**, 155–179.
——1989. Holocene crustal movements and sea-level changes in Great Britain. *Journal of Quaternary Science*, **4**, 77–89.
——1994. Clastic sedimentary environments. In: Waller, M. (ed.) *The Fenland Project, Number 9: Flandrian environmental change in the Fenland*. Archaeology Report Number 70, Fenland Project Committee, Cambridgeshire County Council, Cambridge, 36–38.
—— & ANDREWS, J. 2000. An introduction to Holocene land–ocean interaction and environmental change around the western North Sea. *This volume*.
——, LAMBECK, K., HORTON, B. P., INNES, J. B., LLOYD, J. M., MCARTHUR, J. J. & RUTHERFORD,

M. M. 2000. Holocene isostasy and relative sea-level changes on the east coast of England. *This volume*.

——, ——, ——, ——, ——, ——, PURCELL, A. & RUTHERFORD, M. M. in press. Late Devensian and Holocene records of relative sea-level changes in northwest Scotland and their implications for glacio–hydro-isostatic modelling. *Quaternary Science Reviews*.

TROELS-SMITH, J. 1955. Characterization of unconsolidated sediments. *Danmarks Geologiske Undersogelse, Series IV*, **3**, 38–73.

WALLER, M. 1994. *The Fenland Project, Number 9: Flandrian Environmental Change in Fenland, East Anglian*. Archaeology Report Number 70, Fenland Project Committee, Cambridgeshire County Council, Cambridge.

WINGFIELD, R. T. R., EVANS, C. D. R., DEEGAN, S. E. & FLOYD, R. 1978. *Geological and Geophysical Survey of the Wash*. Report of the Institute of Geological Sciences, 78/18.

Index

Page numbers in italic, e.g. *122*, signify references to figures. Page numbers in bold, e.g. **264**, denote references to tables.